Darwin Day in America

Darwin Day in America

*How Our Politics and Culture Have Been
Dehumanized in the Name of Science*

John G. West

ISI
BOOKS

Wilmington, Delaware

West, John G.

   Darwin day in America : how our politics and culture have been
   dehumanized in the name of science / John G. West. — 1st ed. —
   Wilmington, Del. : ISI Books, 2007.

      p. ; cm.
      ISBN: 978-1-61017-135-9
      Includes bibliographical references and index.

         1. Science — Social aspects — United States. 2. United States —
   Politics and government. 3. Darwin, Charles, 1809–1882 —
   Influence. 4. Social values — United States. 5. Religion and politics —
   United States. I. Title.

Q175.52.U6 W47 2007   2007933849
303.48/3073 — dc22   0710

ISI Books
Intercollegiate Studies Institute
3901 Centerville Road
Wilmington, Delaware 19807
www.isibooks.org

to Sonja Elise West, the love of my life

Contents

Preface

At the dawn of the last century, leading scientists and politicians giddily predicted that modern science—especially Darwinian biology—would supply solutions to all the intractable problems of American society, from crime to poverty to sexual maladjustment.

Instead, politics and culture were dehumanized as a new generation of "scientific" experts began treating human beings as little more than animals or machines:

+ In criminal justice, these experts denied the existence of free will and proposed replacing punishment with invasive "cures" such as the lobotomy.

+ In welfare, they proposed eliminating the poor by sterilizing those deemed biologically unfit.

+ In business, they urged the selection of workers based on racist theories of human evolution and the development of advertising methods to more effectively manipulate consumer behavior.

+ In sex education, they advocated creating a new sexual morality based on "normal mammalian behavior," without regard to long-standing ethical or religious imperatives.

This book explores the far-reaching consequences for society when scientists and politicians deny the fundamental differences between human beings and the rest of the natural world. It also exposes the disastrous results that ensue when experts claiming to speak for science turn out to be wrong. Finally, the book presents a plea for democratic accountability in an age of experts.

This book is not anti-science, although some dogmatic champions of current scientific theories may try to claim otherwise. Modern science has brought tremendous benefits to human life, ranging from wonder-working medicines to the personal computer on which I am writing this preface. But science is not God. It has not perfected human nature nor has it curbed the human abuse of power. Like every human enterprise, science is partial, corruptible, and prone to error, and it needs to be subject to the checks and balances of a free society and the moral law.

My interest in exploring the impact of modern science on public policy and culture was sparked long ago by my reading of C. S. Lewis's *That Hideous Strength* and *The Abolition of Man*. In the late 1990s, I embarked on a long-term research project to document in America the impacts of reductionist science on public policy and culture so perceptively identified by Lewis. This book is the result.

It has been a long journey. Along the way, I discovered that many of the things I thought I knew about the history of science and public policy were wrong, or at least misleading. When I started, for example, I assumed that Darwin's theory of evolution and "Social Darwinism" were substantially distinct, and that "Social Darwinism" was a twisting of Darwin's theory in a way that Darwin would not have approved. Then I read Charles Darwin's *Descent of Man* and realized that my simplistic dichotomy between Darwin and "Social Darwinism" could not be maintained. The results of my investigation into Darwin's own application of his theory to human beings and society can be found in chapter 2.

Similarly, I started out believing that Darwinism was championed by ruthless businessmen during the Gilded Age to justify the worst excesses of nineteenth-century capitalism. The nineteenth-century capitalist Darwinist served as a nice counterpoint to the socialists and progressives who also championed Darwin. However, there was a major problem with this widely held view. If you want to know what it is, please read chapter 6.

In the area of crime and punishment, I began with the background assumption that the use of science to diminish criminal responsibility was primarily used to justify the "liberal" view that criminals should not be punished for their crimes, because crime is tantamount to a disease. Little did I realize that the same scientific debunking of personal responsibility that justified liberals' leniency toward criminals also layed the groundwork for a more radical "conservative" approach to crime that sought to ensure that criminals would never re-offend. I explain the paradoxical impact of reductionist science on the criminal justice system in chapter 5.

I approach this book as a political scientist and a scholar, but I am also a participant in current debates over Darwinian evolution and intelligent design. As a result, chapters 10 and 11 as well as the new afterword draw on firsthand knowledge as well as academic research in exploring the controversies over the teaching of evolution in schools.

When I began research for this book in earnest in 1998, I did not realize it would be a marathon. Because the scope of the book encompasses so many disciplines and more than a century of public policy, researching each section was like starting a new book. I am indebted to Bruce Chapman, president of Discovery Institute, who encouraged me to begin the project; to Stephen Meyer, director of Discovery Institute's Center for Science and Culture, who made research funding from Discovery Institute a priority; and to Seattle Pacific University for allowing me to take a leave of absence during the 2003–4 academic year so I could complete an early draft of the manuscript. I am especially grateful to Jeremy Beer of ISI Books for his enthusiastic interest in publishing my manuscript.

I further wish to express my appreciation for a number of fine research assistants over the years. These include Matt and Lisa Cooper, who gathered stacks of primary sources for me during the summer of 1998. Others who provided significant research help include Greg Piper, Anna Sperling, Joshua Littlefield, Logan Gage, Marshall Sana, Marsha Michaelis, Cindy Lin, Benjamin Miller, and Joe Miceli. Additional thanks should go to Mike Perry, publisher of Inkling Books, who drew my attention to the controversy in Bottineau, North Dakota (discussed in chapter 7), through a letter he had located in the Margaret Sanger papers; the interlibrary loan office at Seattle Pacific University; the American Philosophical Society library in Philadelphia, which holds the papers of Charles Davenport, the Eugenics Record Office, and the American Eugenics Society; and the California Institute of Technology archives in Pasadena, which houses the records of the Human Betterment Foundation.

I also profited from the careful comments and suggestions I received from a number of reviewers of all or part of the manuscript, including Richard Weikart, professor of history at California State University, Stanislaus; Robert Clinton, professor of political science, Southern Illinois University at Carbondale; Edward Larson, Herman E. Talmadge Chair of Law and Richard B. Russell Professor of American History at the University of Georgia; Larry Arnhart, professor of political science at Northern Illinois University; Bruce Chapman and Jonathan Wells of Discovery Institute; Francis Beckwith, associate professor of philosophy at Baylor University;

Reed Davis, professor of political science at Seattle Pacific University; Linda Jeffrey; Fred Foote; and my wife, Sonja West. Of course, the standard disclaimer applies: Any errors of fact or interpretation are my own fault.

Finally, I would like to thank those who helped during the preparation of the final manuscript: Anika Smith, whose expert care ensured the proper formatting of hundreds of endnotes; and Grant Hackett, who created the index.

Introduction

✦ A popular college psychology textbook informs students that all of their decisions are predetermined by their genes and their environment. According to the author, defenders of free will may claim "that people sometimes make decisions not controlled by their genetics, their past experiences, or their environment. But what is left besides genetics and environment?"[1]

✦ "There is obviously some evolutionary basis to rape just like there is some evolutionary basis to all aspects of living things," claims anthropologist Craig Palmer, coauthor of *A Natural History of Rape*. "Rape might be an adaptation. There might have been selection favoring males who raped under some circumstances in the past. And therefore there might be some aspects of male brains designed specifically to rape under some conditions."[2]

✦ Scientists at Rice University and the Baylor College of Medicine have tried to identify the origins of altruism by studying "a slime-mold crowd of single-celled amoebas." Funded by $5 million from the National Science Foundation, "The Evolution of Biological Social Systems" project used "genetic technology . . . to identify the genes and molecular pathways that underlie . . . social behavior among amoebas."[3]

"An age of science is necessarily an age of materialism," wrote Hugh Elliot early in the last century. "Ours is a scientific age, and it may be said with truth that we are all materialists now."[4] One does not have to look far to discover the continued accuracy of Elliot's assessment. Scientific materialism—the claim that everything in the universe can be fully explained as the products of unintelligent matter and energy—has become the operating assumption for much of American politics and culture. We are repeatedly told today that our behaviors, our emotions, even our moral and religious longings are reducible to some combination of physical processes interacting with our environment.

Scientific materialism is akin to what others have called "scientism"[5] and includes not only strict philosophical materialism but also "positivism."[6] While philosophical materialists flatly deny the existence of anything except matter and energy (including God), many positivists merely claim that human reason is unable to obtain genuine knowledge about anything other than the material world. Whatever the cogency of this distinction in metaphysics, in politics and culture it has been a distinction without much of a difference. Just like philosophical materialists, writes Neal Gillespie, positivists have "typically used mechanistic or materialistic models of causality, rejected supernatural, teleological, or other factors . . . and—whatever the desires or beliefs of individual practitioners, many of whom were theists or even good Christians—embraced and promoted those far-reaching cultural tendencies conventionally known as secularism and naturalism."[7]

Granting in theory that something beyond the material world might exist, positivists have maintained in practice that anything beyond matter cannot be rationally investigated, making it irrelevant to the conduct of human life. As a result, positivists no less than philosophical materialists have tried to explain everything in human life in terms of material causation. Given the essential agreement between philosophical materialists and positivists on how to explain the natural and social worlds, "scientific materialism" is used here to describe any effort to explain human behavior or beliefs wholly in terms of material causes—regardless of the personal metaphysical beliefs of those who offer such explanations. A lawyer who argues that crime is due to bad heredity rather than free choice may or may not be a philosophical materialist, but he is offering a fundamentally materialistic explanation for human behavior. The same is true of a biologist who claims that morality is simply an evolutionary adaptation. What matters is the attempt to treat material causes as the only relevant explanations when trying to understand human life or formulate public policy.

Probably the most powerful justification for scientific materialism today is supplied by the neo-Darwinian theory of evolution, especially as it is advanced by proponents of what is often called "sociobiology" or "evolutionary psychology." At least in theory, Darwinian biologists such as E. O. Wilson find in modern genetics and neuroscience complete explanations for all domains of human behavior, from literature and art to religion and politics.[8]

Support for the application of Darwinism to human affairs spans the political spectrum. Writing for the conservative *National Review,* law professor John McGinnis has urged conservatives to jump on the bandwagon of sociobiology and its Darwinian framework. According to McGinnis, discoveries during the last two decades have led to a convincing revival of Darwinism, and "any political movement that hopes to be successful must come to terms with the second rise of Darwinism."[9] McGinnis is not alone among intellectuals in championing a rapprochement between conservatism and modern Darwinism. More nuanced versions of this view can be found in the writings of James Q. Wilson and in political theorist Larry Arnhart's provocative book, *Darwinian Conservatism.* Arnhart declares that "conservatives need Charles Darwin. They need him because a Darwinian science of human nature supports conservatives in their realist view of human imperfectibility and their commitment to ordered liberty." Arnhart goes so far as to suggest that conservatism is doomed unless it embraces Darwinian theory. "The intellectual vitality of conservatism in the twenty-first century will depend on the success of conservatives in appealing to advances in the biology of human nature as confirming conservative thought."[10] Notably, conservative theorists like Arnhart who defend Darwinian theory try to avoid the virulent reductionism championed by many in the scientific community.

On the other side of the political divide, Princeton professor Peter Singer has declared that "it is time to develop a Darwinian left." Of course, there always has been a Darwinian left (as will be discussed in this book's section titled "Wealth and Poverty"), but Singer has called for the Left to integrate the findings of modern Darwinism into its agenda. "It is time for the left to take seriously the fact that we have evolved from other animals; we bear the evidence of this inheritance, not only in our anatomy and our DNA, but in what we want and how we are likely to try to get it."[11]

Ideas have consequences, of course, and this book will argue that scientific materialism has been one of the most powerfully consequential ideas in American politics and culture during the past century. That does not mean it has been the only powerful idea, or that other ideas may not have exerted equally profound influences on the public square. Still less does it mean that the public policies discussed in this book are the results of scientific materialism alone, or even that they are the inevitable result of scientific materialism. The causes of human behavior are rich, varied, and complex. Nevertheless, particular ideas can have particular consequences, and this book tries to untangle some of the consequences of scientific materialism in such fields as criminal justice, economics, welfare, architecture, education, and medicine.

The effort to apply scientific materialism to American public policy began in earnest more than a century ago with high hopes. Around the turn of the twentieth century, defenders of scientific materialism issued increasingly lofty claims about how science's understanding of the material world could be enlisted to solve the problems of human society. The same scientific advances that produced inventions like the steam engine and medical breakthroughs like the germ theory of disease were also supposed to supply the basis for eliminating a host of social ills ranging from poverty to crime to unproductive workers.

Writing in the journal *Science* in 1903, J. McKeen Cattell, president of the American Society of Naturalists, argued that science's previous achievements in helping man to subjugate the natural world were just a foretaste of the future power science would bestow on man to control human nature: "The nineteenth century witnessed an extraordinary increase in our knowledge of the material world and in our power to make it subservient to our ends; the twentieth century will probably witness a corresponding increase in our knowledge of human nature and in our power to use it for our welfare."[12]

Charles Eliot, president of the American Association for the Advancement of Science (AAAS), similarly predicted in 1915 that "biological science" would open the door to the "prevention as well as cure" of the "bodily defects" that caused such antisocial behaviors as murder, robbery, forgery, and prostitution. "These are all biological problems; and the progress of biological inquiry during the past fifty years is sufficient to afford the means of solving on a large scale these fundamental social problems."[13]

Now that the twentieth century is over, it is appropriate to take stock of such grand predictions and the mind-set that helped shape them. What have been the real contributions of scientific materialism to public policy over the past century? Has scientific materialism promoted or impeded the solution of social problems? Has it advanced or detracted from human dignity? Has it encouraged social progress or social dissolution? These are some of the questions that this volume will address. While the book delves into a variety of current issues, its focus is less on current events than on telling the story leading up to those events. The goal is to impart a deeper understanding of the intellectual and historical roots of current controversies.

The book begins in section one ("Origins of Scientific Materialism") by describing how materialism came to be enshrined at the heart of contemporary culture. Tracing the development of materialism as an idea from ancient Greece to the rise of Darwin's theory in the nineteenth century, chapters 1 and 2 argue that the work of Charles Darwin ultimately supplied the empirical basis for a robust materialism finally to take hold. Darwin's theory made materialism credible for many intellectuals by explaining how man and his moral beliefs could have developed through an unplanned material process of natural selection. In the words of nineteenth-century German physiologist Emil Du Bois–Reymond, "the evolution theory in connection with the doctrine of natural selection forces upon [man] . . . the idea that the soul has arisen as the gradual result of certain material combinations."[14]

The following sections of the book explore how scientific materialism has had a transformative impact on specific areas of public policy and culture, including criminal justice (chapters 3–5); economics, business, and welfare (chapters 6–9); education (chapters 10–13); and medicine (chapters 14–15).

While Darwin's theory is featured prominently in several chapters (and the book's title), the scope of this study is broader than just Darwin. The overall aim is to examine the impact of materialistic reductionism on public policy and culture, and Darwinism is only one part of that larger story.

Moreover, the intent of each chapter is not to provide a comprehensive analysis of the issues discussed, but only to explore the ways in which scientific materialism influenced those issues. For example, personal autonomy is an important theme in debates over the right to die, but it does not figure prominently in this book's discussion of euthanasia because the link to scientific materialism is limited. Similarly, while the roots of racism are many and diverse, this book only examines the relationship between racism and scientific materialism. The goal of such a limited focus is not to deny the importance of other ideas on public policy, but to make as clear as possible the particular impact made by scientific materialism.

This book is intended to offer a sober warning about what can happen when policymakers—and the general public—uncritically accept a materialistic understanding of human nature advanced in the name of science. In many ways, the book might be considered an effort to trace in the history of American public policy the consequences of scientific materialism so presciently delineated by C. S. Lewis in *The Abolition of Man*. Lewis was prophetic to warn that "if man chooses to treat himself as raw material, raw material he will be: not raw material to be manipulated, as he fondly imagined, by himself, but by mere appetite . . . in the person of his dehumanized Conditioners."[15]

Origins of Scientific Materialism

1

Nothing Buttery
from Atomism to the Enlightenment

"In the last ten years we have come to realize humans are more like worms than we ever imagined," declares biologist Bruce Alberts, former president of the National Academy of Sciences.[1] Geneticist Glen Evans at the University of Texas Southwestern Medical Center agrees. Genetically speaking, he reports, "the worm represents a very simple human."[2]

Other biologists prefer to think that humans are actually oversize *Drosophila melanogaster*—fruit flies. "In essence, we are nothing but a big fly," insists Charles Zuker, genetic researcher and professor of biology at the University of California at San Diego.[3]

Or perhaps humans beings are merely overgrown mice. After all, "we share 99 percent of our genes with mice, and we even have the genes that could make a tail," claims British scientist Jane Rogers, who worked with other researchers to publish the mouse genome in 2002. According to one journalist, "the genetic blueprint of the mouse" produced by Rogers and other scientists demonstrates that "there isn't much difference between mice and men."[4]

Apparently there is even less difference between men and chimps. "We humans appear as only slightly remodeled chimpanzee-like apes,"[5] insists Morris Goodman of Wayne State University, coauthor of a 2003 study which concluded that "chimpanzees are more closely related to people than to gorillas . . . and probably should be included in the human branch of the family tree."[6] Goodman complains about the traditional view of human beings that "emphasizes how very different humans are from all other forms of life," deriding it as contaminated by "anthropocentric bias." In reality, according to Goodman, "the molecular genetic view . . . places all the living apes (gibbons, orangutans, gorillas and chimpanzees) with humans in the same family and within that family barely separates chimpanzees from humans."[7]

Few people today would deny that human beings can learn important things about themselves by studying other animals, but many scientists want to go considerably further. They seem to think that studying worms or flies or chimps will reveal in time all the mysteries of the human soul. They are inspired in this belief by a powerful assumption enshrined at the heart of modern science: the assumption that the material universe is the sum of reality. Scientists "have a prior commitment . . . to materialism," asserts Harvard biologist Richard Lewontin, adding that scientists are compelled by their "*a priori* adherence to material causes to create an apparatus of investigation and a set of concepts that produce material explanations, no matter how counter-intuitive, no matter how mystifying to the uninitiated. Moreover, that materialism is absolute."[8] This commitment to what some have called "scientific materialism" has led ineluctably to reductionism in the natural and social sciences during the past century, as those speaking in the name of science have attempted to show how everything that exists—including human action and thought—can be fully explained as the products of unintelligent matter and motion.

To borrow the words of political philosopher Leo Strauss, scientific materialism has tried "to understand the higher in terms of the lower: the human in terms of the subhuman, the rational in terms of the subrational."[9] According to scientific materialism, the parts are more important than the whole—to the degree that sometimes the whole seems to be a lot less than the sum of its parts. The end result is what the late British neurophysicist Donald MacKay liked to call the "nothing buttery" syndrome,[10] the attempt to explain every fact of human existence as "nothing but" some nonrational material process—as in, "Our enjoyment of symphonies is nothing but the conversion of mechanical energy into electrical signals by the cells in our inner ear,"[11] or "Morality . . . is merely [i.e., 'nothing but'] an adaptation put in place to further our reproductive ends,"[12] or "your joys and your sorrows . . . your sense of personality and free will, are in fact no more than [i.e., 'nothing but'] the behavior of a vast assembly of nerve cells and their associated molecules."[13] Even the realms of mind and spirit cannot escape from the reductionists' onslaught, because in their view "matter is the ground of all existence; mind, spirit, and God as well, are just [i.e., 'nothing but'] words that express the wondrous results of neuronal complexity."[14]

While the triumph of materialistic reductionism over nearly every field of human endeavor is of recent vintage, both materialism and reductionism have deep roots in Western culture. The story of their rise to ascendancy is a tangled and multifaceted one, and this book cannot tell it exhaustively. But it can present some of the highlights. It is usually perilous to try to pinpoint the originator of an idea. The very attempt presumes a universal knowledge of human history that finite human beings probably can never attain. But stories must have a beginning, and so if we are to tell the story of materialism's development, we must start sometime and somewhere. The sometime is more than two millennia ago. The somewhere is Greece.

Atoms and Void: Materialism in Greece and Rome

It was a hot midsummer's day, and three elderly pilgrims walked the dusty road from the city of Cnossus on the island of Crete to the grotto of Zeus. Situated on top of Mount Ida, the grotto was the legendary birth-place of the deity who was chief in the pantheon of the Greek gods. One of the pilgrims, Cleinias, was a native Cretan. The second pilgrim, Megillus, was a Spartan. The third man came from Sparta's historic enemy, Athens.[15]

The walk was a long one, and so the men passed the time in conversation. Eventually their discussion turned toward the gods, and the Athenian spoke of philosophers who claimed that planets and stars, animals, and even human beings arose by "nature and chance" from the material elements of fire, water, earth, and air.[16]

"In this manner the whole heaven has been created, and all that is in the heaven, as well as animals and all plants . . . not by the action of mind, as they say, or of any God, or from art, but as I was saying, by nature and chance only."[17] For such philosophers, the Athenian continued, the fundamental reality is material, and everything that is real is a by-product of that material reality, even the human soul. As for religion and morality, the naturalistic philosophers claimed that they have no basis in nature and are transient and changeable according to the dictates of the time.

"These, my friends, are the sayings of wise men, poets and prose writers, which find a way into the minds of youth. They are told by them that the highest right is might, and in this way, the young fall into impieties, under the idea that the Gods are not such as the law bids them imagine."[18]

"What a dreadful picture . . . ," replied the Cretan Cleinias in shock, "and how great is the injury which is thus inflicted on young men to the ruin both of states and families."[19]

The Athenian went on to indict the naturalistic philosophers for reversing that which is first and last. They wrongly think that justice, beauty, and the human soul are by-products of the physical body, when in fact "the soul is prior to the body." Because the soul is prior to the body, "characters and manners, and wishes and reasonings, and true opinions, and reflections, and recollections"—in short, all the qualities that typify mind—are also prior to matter, and it is an error to try to reduce mind to matter rather than treat it as something with independent dignity.[20]

This imaginary conversation between Cleinias, Megillus, and the unnamed Athenian appears in Book X of Plato's *Laws,* probably one of the last dialogues written by the great Athenian philosopher. Plato (427?–347 B.C.) and his pupil Aristotle (384–322 B.C.) were the most compelling critics of materialism and reductionism in the ancient world.[21] Some of the "pre-Socratic" philosophers whom Plato and Aristotle critiqued were possibly not materialists in the strict sense of the term.[22] But they supplied fertile soil out of which a full-blooded materialism grew, and by the fifth century B.C. Leucippus and his student Democritus (460?–357 B.C.) were spreading a wholly materialistic philosophy that became known as atomism.

While trustworthy knowledge about Leucippus and Democritus is scant, they apparently believed that everything—plants, animals, the operations of the senses, and mind itself—could be reduced to atoms in motion.[23] In their view, sense perception and thought were purely mechanical processes. According to Aëtius, Leucippus and Democritus believed that "sensations and thoughts are alterations of the body" that "take place by the impact of images from outside. Neither occurs to anyone without the impact of an image."[24] Thus, in the atomists' view, a person can see a tree because the tree gives off atoms that ultimately enter his eyes.[25] The problems with this mechanistic view of mind were apparent even to the ancients. Ridiculing the atomists' conception, Cicero a few centuries later asked "whether . . . whenever he thought about Britain, an 'image' had detached itself from that island to come and hit him in the chest."[26]

Determinism was another key feature of the philosophy of the early atomists. In the words of Leucippus, "nothing occurs at random, but everything for a reason and by necessity."[27] Democritus likewise believed that "everything happens according to necessity."[28] It is easy to see how their determinism followed from their atomism. According to Leucippus and Democritus, the atoms in the void were a sort of perpetual motion machine, the motion of each atom being produced by a prior collision with another atom. Almost by definition there was no room for freedom in their system.

In his magisterial history of Greek philosophy, W. K. C. Guthrie suggests that Leucippus and Democritus "encouraged the faith of all who in later ages have been attracted by the notion of man as a machine." The two atomists further sparked debate over whether the universe displayed the hallmarks of design. "Has matter formed itself unaided into organisms of an almost incredible complexity, delicacy and adaptability to purpose, or has this order and efficacy been imposed from outside by a rational agent working to a plan?" According to Guthrie, "it was Democritus who first compelled philosophers to take sides by his detailed exposition of a system in which intelligence, direction and purpose were epiphenomena emerging at a late stage from nothing but the undesigned clash and recoil of individually inanimate particles."[29]

After the deaths of Leucippus and Democritus, atomism was revived by Epicurus (342–270 B.C.), who likewise maintained that "the universe is bodies and space," with the primary bodies being "indivisible and solid" atoms.[30] For Epicurus, everything that exists was ultimately reducible to atoms in motion, including the human soul. Those who insisted that "the soul is incorporeal" were "talking idly."[31] The soul is just as corruptible as the rest of the human body, and so "when death comes . . . we do not exist."[32]

Epicurus did make one major break with Leucippus and Democritus: He defended free will. The better man, according to Epicurus, is one who "laughs at destiny" and "thinks that with us lies the chief power in determining events," because though some events "happen by necessity and some by chance," there are still some remaining events that "are within our control."[33] Epicurus seemed to provide a theoretical foundation for free will within his atomic system by suggesting that atoms occasionally swerved unpredictably, so that there were at least some atomic movements that were not dictated by prior collisions with other atoms.[34] Whether Epicurus's concession to indeterminacy actually allows for human freedom and responsibility (or even makes sense, given his overall system of physical causes) is another matter. As Frederick Lange pointed out in his *History of Materialism*, "the unconscious arbitrariness" of the atomic behavior in question seems to refute "any intimate connection between the actions of a person and his character."[35]

Epicurus attempted to connect his ethical teachings to his atomic theory, grounding his ethics on the sensations of pleasure and pain. To pursue pleasure and avoid pain was the primary maxim of his ethical philosophy. But he denied that he was a defender of unbridled hedonism.[36]

The materialistic philosophy of Epicurus was later propagated in Rome by the Roman poet Lucretius (96?–55 BC), who made the materialistic conception of the universe the heart of his poem *De Rerum Natura* ("On the Nature of Things"). Running more than seven thousand hexameter lines, *De Rerum Natura* was a remarkable achievement under any circumstances, but especially so considering that Lucretius (if one can trust the account of Jerome) underwent fits of insanity while writing it and committed suicide before bringing the poem to completion.[37]

Like the Greek atomists before him, Lucretius saw human beings and the rest of nature as products of the mindless collisions of atoms. The first and essential things were atoms in motion, and everything else came out of them:

Neither by design did the primal germs
'Stablish themselves, as by keen act of mind,
Each in its proper place; nor did they make,
Forsooth, a compact how each germ should move;
But since, being many and changed in many modes
Along the All, they're driven abroad and vexed
By blow on blow, even from all time of old,
They thus at last, after attempting all
The kinds of motion and conjoining, come
Into those great arrangements out of which
This sum of things established is create . . .[38]

De Rerum Natura has been called "the greatest monument in Latin poetry,"[39] and its language is echoed repeatedly by the celebrated Virgil.[40] Despite Lucretius's literary influence on Virgil, however, the philosophy he championed failed to thrive. It is not difficult to understand why. The tremendous gulf between living and nonliving matter, the equally cavernous gap between rational and irrational living beings, all seemed to make incredible the notion that intelligence and the rational ordering of the natural world could be produced by blind forces operating on irrational matter. The idea that mind was preeminent over matter seemed confirmed by common sense, and Plato and Aristotle and their followers pressed this point against the early materialists with vigor. Indeed, their alternative theories of mind and matter were so persuasive that they supplied the foundations for intellectual life during the next two millennia.

Plato and Aristotle differed in some of their metaphysical principles, but they were united in opposing the materialist attempt to reduce mind to matter. In Aristotle, the defense of mental independence comes out clearly in his famous theory of causation. Aristotle taught that there are four primary kinds of causes: material, efficient, formal, and final.[41] The *material* cause of something is the matter from which it is made. The material cause of a bronze statue is the metal from which it was constructed, just as the material cause of a human being is the chemical and cellular materials that make him up.[42] The *efficient* cause of something is "that which brings about a change" in it.[43] The efficient cause of a baby, for example, is his or her parents. Modern science generally proceeds as if material and efficient causes together provide an exhaustive explanation of the world around us. But Aristotle would argue that two of the most important kinds of causes—the formal and the final—cannot be understood simply as matter or matter acting on matter.

The formal cause of something is the pattern or idea (i.e., the "form") that makes the thing what it is.[44] This pattern or idea is not itself material, and it is what allows us to differentiate between a stack of cedar logs and a handmade cedar chest. Both the logs and the chest are fashioned from the same material. What differentiates them, then, is not a difference in matter but a difference in form. As Harry Jaffa observes, "the fact that the form of a thing can be separated from its matter is the very heart of human understanding, and of human intelligence. Without this possibility, modern science itself would not be possible, because all science presupposes the detachment of the mind from its object as a condition of human speech about the object."[45]

Aristotle's final cause (or *telos*) also transcends the merely material. The final cause of something is the reason for which it exists—in other words, its purpose. The final cause of someone exercising at a health club might be health.[46] In many instances, the formal cause of something is also its final cause. The final cause of a human fetus, for example, is to grow into the mature form of a human being.

While materialism tends to generate reductionistic explanations, formal and final causation tend to encourage teleological explanations. Teleology means "study of purpose," and it is the effort to understand things in terms of their overall ends. While reductionism tries to understand the whole in light of its parts, teleology tries to comprehend the parts in light of the whole. Consider the oft-cited example of an oak tree. According to the Aristotelean approach, one can understand more about what an oak tree is by looking at the attributes of a fully grown oak tree than by trying to describe it as nothing more than an overgrown acorn. Similarly, the best way to understand human beings is to look at who they are as an integrated whole, rather than breaking them down into their material or psychological building blocks.

For much of modern science, which thrives on reductionism, teleology is a bad word. Despite the modern dismissal of teleology, however, there remains something appealing about the approach, as a simple example may demonstrate. Imagine a table with two baking pans on it. One pan is filled with raw eggs, flour, baker's chocolate, vegetable oil, and sugar. The other pan is filled with a German chocolate cake that has just been baked. In a reductionistic framework, the contents of both pans are essentially equivalent, for they are reducible to the same raw ingredients. Yet what person would really think it adequate to describe the cake as "nothing but" raw eggs, flour, baker's chocolate, vegetable oil, and sugar? Most people would find such a description woefully incomplete in explaining what a chocolate cake really is. As most people understand almost intuitively, the cake is more than the sum of its parts. That is why if offered a choice between a serving of the unmixed mess in pan one or a piece of the finished cake in pan two, they will invariably prefer the latter. If one can apprehend why the cake in the example is qualitatively better than its raw unmixed ingredients, he will have grasped why reductionistic explanations are far from a complete description of reality—and why the teleological theories of Plato and Aristotle exerted an almost insurmountable influence on the Western mind for so long.

After Lucretius, the rise of Christianity posed a further obstacle to the widespread acceptance of materialism. Christian beliefs about personal immortality, the Holy Spirit dwelling inside believers, and God's eternal nature were hard to square with materialist accounts of the universe, although a few thinkers such as Tertullian attempted a partial reconciliation.[47] Materialism became so discredited that it is difficult to find a prominent thoroughgoing materialist for well over a thousand years after the birth of Christ.[48] Not until the 1500s did materialism once again become a hotly contested public issue.

Nature the Machine: Materialism in the Sixteenth and Seventeenth Centuries

By anyone's estimation, the boy was brilliant. Forbidden by his father to study mathematics during his childhood, he taught himself the rudiments of geometry without the benefits of either books or tutors, working out the first thirty-two propositions of Euclid by age twelve. A few years later he joined meetings of a group of distinguished mathematicians in Paris. At sixteen he turned inventor, beginning work on a machine to compute sums. Within three years he had produced the world's first digital calculator.[49]

The boy's sister would later recall that the calculator "was looked upon as a thing perfectly new in Nature" for it "brought . . . within the small Compass of a Machine" complex logical operations that previously had been reserved for the human mind.[50] The device was called the Pascaline—after its creator, Blaise Pascal (1623–62), later famous for the religious and philosophical reflections collected in his *Pensées*.

One of those who learned about the young Pascal's miraculous device was an English expatriate living in Paris. Reflecting upon the revolutionary implications of the calculator, the Englishman reportedly declared: "Brass and iron have been invested with the functions of brain and instructed to perform some of the most difficult operations of mind."[51]

The declaration may be apocryphal, but the Englishman was real, and it is certain that he had become intrigued by the machine-like features of the human brain. Whether he gained inspiration for his views from Pascal's calculator we may never know for sure, but when he eventually published his own views on the human mind, the Englishman shocked many of his contemporaries by his boldness. This was despite his self-proclaimed timidity.[52]

The Englishman was Thomas Hobbes (1588–1679). Perhaps best remembered today for his observation that the life of man in a state of nature is likely to be "solitary, poore, nasty, brutish, and short,"[53] Hobbes also was probably the most thoroughgoing materialist of his day. Believing that "life is but a motion of Limbs," he openly wondered why devices such as watches that "move themselves by springs and wheeles" could not be said to have "an artificial life." After all, wrote Hobbes, "what is the *Heart*, but a *Spring*; and the *Nerves*, but so many *Strings*; and the *Joynts*, but so many *Wheeles*, giving motion to the whole Body . . . ?"[54] Hobbes's theory of mind came straight out of Democritus and Epicurus. In his view, human cognition was nothing more than the residue of the impact of external objects on our bodies.[55]

Given Hobbes's mechanical conception of man, it should be no surprise that he embraced a deterministic psychology. Hobbes did not claim that a man could not "act according to his *will*," but he insisted that the "will" itself was already determined. To claim that a man could determine his own will was "absurd."[56] "Every act of mans [sic] will, and every desire, and inclination proceedeth from some cause, and that from another cause, which causes in a continual chaine . . . proceed from necessity."[57] Thus, *"free from necessitation . . .* no man can be."[58]

Although regarded by many as an atheist during his own lifetime, Hobbes did not openly reject Christianity. Nevertheless, his theology was far from orthodox. He argued that it was incoherent to believe in an incorporeal being, and so concluded that God himself must be corporeal.[59] He thought that human beings ceased existence upon the death of their bodies and God recreated them later at a general resurrection.[60] He redefined Satan as "any Earthly enemy of the Church,"[61] and he radically reinterpreted the Trinity, identifying the Holy Spirit with the apostles who were appointed to represent God after Jesus's resurrection.[62]

Hobbes was the forerunner not only of much of modern materialism, but of modern theories of biblical criticism. In his own day, however, he "was a rather lonely figure," as Stanley Jaki points out.[63] Other thinkers of the era approached the materialism of Hobbes, but they stopped short of asserting that man was merely a machine. For example, Francis Bacon (1561–1626), who preceded Hobbes, paved the way for a mechanistic view of nature, but he was cautious about applying his view to human beings.[64]

The seventeenth century is commonly referred to as the age of the Scientific Revolution. Given the robust natural science that existed in the Middle Ages, this appellation is probably a misnomer.[65] Nevertheless, it is surely the case that by the 1600s increasing interest was being expressed by many thinkers in understanding the material basis of the natural world. An interest in matter, however, does not necessarily make one a materialist. To believe that the natural world has a material basis is not the same thing as claiming that everything must be reduced to materiality. Accordingly, some thinkers who argued most convincingly for viewing the natural world as a machine insisted with just as much vigor that the human mind could never be reduced simply to a material mechanism. The most famous philosopher of the period who adopted this view was undoubtedly René Descartes (1596–1650), who bequeathed a tangled legacy that is still energetically disputed.[66]

On the one hand, later materialists took inspiration from Descartes' mechanical account of the natural world, which claimed that "there is nothing in the whole of nature . . . which is incapable of being deductively explained on the basis of" material principles relating to atoms in motion.[67] Descartes expanded his materialism to encompass the human body as well. In his *Treatise on Man,* he implied that the human body was "nothing but a statue or machine made of earth."[68] On the other hand, defenders of the independent reality of mind could take comfort from Descartes' insistence that "the soul is entirely distinct from the body"[69] and that "our knowledge of thought is prior to, and more certain than, our knowledge of any corporeal thing."[70] The coherence of Descartes' view may be debated, but it is difficult to regard him as a materialist on his own terms.

The same is true for John Locke (1632–1704) in England, who argued in his *Essay Concerning Human Understanding* that it was not logically impossible for matter to be endowed with the property of thought. Some concluded from this statement that Locke actually believed in the existence of thinking matter.[71] But that is not what he wrote. Locke claimed that it was not a logical contradiction to believe that God could endow matter with the property of thought.[72] But he nowhere asserted that this was in fact the case. Indeed, Locke's main point was that it was impossible for human beings to gain certain knowledge by reason about whether human rationality inhered in an immaterial soul or a thinking body. According to Locke, both alternatives are "clogged with equal difficulties," and we should be content to remain in a state of ignorance about the ultimate nature of the mind.[73] We should accept the fact that we are thinking beings and proceed from there.

If Locke had said nothing else, it still would not have been accurate to regard him as an avowed materialist. But in other comments he made clear how profound the difference was between his views and a materialist understanding of mind. While Locke did not regard it as a contradiction for God to endow matter with thought, he did regard it as a contradiction to believe that matter in and of itself produced the ability to think.[74] Repudiating the views of the ancient atomists, Locke wrote that it is impossible to believe that intelligence can somehow arise from the chance organization of matter. Indeed, it is

> as impossible that things wholly void of knowledge, and operating blindly, and without any perception, should produce a knowing being, as it is impossible that a triangle should make itself three angles bigger than two right ones. For it is as repugnant to the idea of senseless matter, that it should put into itself sense, perception, and knowledge, as it is repugnant to the idea of a triangle, that it should put into itself greater angles than two right ones.[75]

According to Locke, intelligence must proceed from another mind, not matter. In his view, that mind must be God.[76] Locke also indicated that while we cannot know with certainty whether human beings have an immaterial soul or a thinking body, we can draw a conclusion based on probability, and here the evidence clearly pointed to the existence of something immaterial. In another section of the *Essay*, Locke defended at length the idea of immaterial beings, and in a response to one of his critics he said that "it is to the highest degree probable" that the thinking substance inside human beings is "immaterial."[77]

Like Descartes and Locke, Sir Isaac Newton (1642–1727) is often identified with the effort to reduce the universe to matter in motion. George Gilder, for example, says that Newton "had a vision of the world that was profoundly materialistic and determinist. He believed that at the basis of the universe were solid, immutable, impenetrable, mindless bits of matter. Further, he maintained that the universe was built up from these bits of matter in a great determinist machine, like the workings of a clock."[78] Newton did think he could explain the movements of the planets by mechanical laws, and in his famous treatise *The Principia* (1686) he expressed his "wish" that "we could derive the rest of the phenomena of Nature by the same kind of reasoning from mechanical principles."[79] But whatever the materialistic implications of his system, Newton himself was far from a thoroughgoing materialist. He clearly believed that blind, mechanical causes were not sufficient to explain the design of the universe and accepted God as an incorporeal being who was omnipresent in the universe.[80] Moreover, like Locke and Descartes, Newton never had the presumption to try to reduce human beings to mere matter in motion.

This reluctance to reduce man to a wholly material being was widely shared during the seventeenth century. Some of the reluctance may have been due to fear of upsetting popular prejudices, but another factor was likely the sheer mystery of sentient life, which made the very idea of mind being produced by mindless matter seem incredible. As the following century would show, however, the reluctance in applying materialism to the human soul was not to last.

Man the Machine: Materialism during the Enlightenment

Situated on the banks of the Danube, the Bavarian university town of Ingolstadt was already centuries-old by the eighteenth century, when it became the setting for scientific experiments by a young college student that would shock the civilized world. For two years, the student had worked with an obsessiveness that verged on madness. He spent his days and nights with cadavers and body parts, trying to unlock the mysteries of life and discover how to infuse life into an inanimate body. When he finally found what he wanted, he put his knowledge to the test by creating a living being.

"It was on a dreary night of November that I beheld the accomplishment of my toils," the student wrote later.

> With an anxiety that almost amounted to agony, I collected the instruments of life around me, that I might infuse a spark of being into the lifeless thing that lay at my feet. It was already one in the morning; the rain pattered dismally against the panes, and my candle was nearly burnt out, when, by the glimmer of the half-extinguished light, I saw the dull yellow eye of the creature open; it breathed hard, and a convulsive motion agitated its limbs . . . For this I had deprived myself of rest and health . . . but now that I had finished, the beauty of the dream vanished, and breathless horror and disgust filled my heart.[81]

The student was Victor Frankenstein, the fictional mad scientist in Mary Shelley's famous novel.

Written early in the nineteenth century, but set in the 1700s, *Frankenstein* depicted a nightmare vision of taking human power over nature too far. Though fanciful, the story intentionally drew on the scientific experiments of the time, especially the ongoing attempts to inject life into lifeless matter. These efforts to unlock the mystery of life dated back at least to 1768, when the wife of Italian physiologist Luigi Galvani (1737–98) informed her husband of the convulsive movements of a dead frog in his laboratory. The frog had accidentally come into contact with an electrified scalpel. From this fact Galvani eventually theorized that a type of electricity was an animal's animating force.[82] The quest to reanimate dead bodies took off from there. By the turn of the century, Galvani's nephew Giovanni Aldini was performing macabre experiments on decapitated oxes, horses, lambs, and humans.[83] "The unenlightened part of mankind are apt to entertain a prejudice against those . . . who attempt to perform experiments on dead subjects," Aldini later acknowledged, but he maintained that such experiments were justified because the object was to improve human welfare. "It is . . . an incontrovertible fact, that such researches in modern times have proved a source of the most valuable information."[84]

Despite his assured tone, even Aldini admitted to having to overcome a sense of "repugnance" when he first turned to experiments on human beings.[85] Determined to understand the workings of what he called "the human animal machine,"[86] Aldini knew that he needed to procure bodies while they were still fresh and "retained . . . the vital powers in the highest degree of preservation." His solution? "I was obliged, if I may be allowed the expression, to place myself under the scaffold, near the axe of justice, to receive the yet bleeding bodies of unfortunate criminals, the only subjects proper for my experiments."[87]

Aldini's first human experiments were conducted on the bodies of two young brigands who had been decapitated in Bologna in January 1802. Aldini started by applying electricity to various parts of their decapitated heads. This produced "the most horrid grimaces. The action of the eyelids was exceedingly striking."[88] Through further applications of electricity, the mouth of one of the heads generated a small amount of saliva, and the tongue moved back into the mouth after having been pulled out. Aldini also dissected the brains and applied electricity to their various components, producing further facial convulsions. Finally, Aldini applied electricity to the trunks of the bodies, which resulted in one of the corpses raising its forearm—"to the great astonishment of those who were present."[89]

By 1803, Aldini was in London experimenting on the body of a British criminal who had been hanged. Aldini praised the "enlightened" British legislators who permitted the bodies of criminals executed in England to be handed over for medical experimentation.[90] His British experiment "surpassed our most sanguine expectations," he later wrote enthusiastically, adding that it had been so successful "vitality might, perhaps, have been restored, if many circumstances had not rendered it impossible."[91] Aldini's experiments sparked imitators, as scientists across Europe rushed to discover the secret of animation.[92]

In many respects, these experiments in the early 1800s were a fitting culmination of the natural philosophy of the previous century. For it was during the 1700s that scientific thinkers had pushed the boundaries of material explanations even into the human soul itself. By the mid-eighteenth century, the ideas that Hobbes had proclaimed like a voice in the wilderness in the 1600s had gained numerous disciples in both England and on the continent. Given the reduction of man to a completely material mechanism, it seemed eminently reasonable to try to discover the material process that would allow one to reanimate dead bodies.

Nowhere was the rise of an all-encompassing materialism more apparent than in France, where natural philosophers during the eighteenth century relentlessly pushed materialism to its logical conclusions. Two Frenchmen, in particular, laid the foundations for a vigorous and consistent materialism: Julien Offray de la Mettrie (1709–51) and Paul Henri Thiery, Baron d'Holbach (1723–89).

A physician, La Mettrie's major contribution to materialist thought was his tract *Man a Machine* (*L'Homme Machine*), published in 1748. According to La Mettrie, "the human body is a machine which winds its own springs,"[93] and the "diverse states" of the human mind "are always correlative with those of the body."[94] In other words, human beings are mechanisms whose rational life is completely dependent on physical causes. Those causes include everything from heredity to raw meat. In what has to be one of the more interesting passages in culinary analysis, La Mettrie opined: "Raw meat makes animals fierce, and it would have the same effect on man. This is so true that the English who eat meat red and bloody, and not as well done as ours, seem to share more or less in the savagery due to this kind of good."[95]

Climate, age, blood circulation, inheritance, and gender were other physical factors dictating a human being's rational nature.[96] According to La Mettrie, the physical weakness of women leads to their natural tenderness, emotionalism, and superstitions. "Man, on the other hand, whose brain and nerves partake of the firmness of all solids, has not only stronger features but also a more vigorous mind."[97] In sum, "everything depends on the way our machine is running."[98]

Applying his model of physical causes to crime, La Mettrie argued that "there are a thousand hereditary vices and virtues which are transmitted from parents to children as those of the foster mother pass to the children she nurses."[99] Unlike some future materialists, La Mettrie believed that even the hereditarily vicious were not completely outside the bounds of morality. He had "no doubt that the wildest and most savage have some moments of repentance."[100] Nevertheless, he did believe that a person might be forced to commit an "involuntary crime" that he only regretted after the fact. In order to sort out involuntary crimes from the rest, La Mettrie proposed that "physicians might be the only judges. They alone could tell the innocent criminal from the guilty. If reason is the slave of a depraved or mad desire, how can it control the desire?"[101] In advocating this turnover of the courts to the medical profession, La Mettrie said that he did not "mean to say that all criminals are unjustly punished; I only maintain that those whose will is depraved, and whose conscience is extinguished, are punished enough by their remorse when they come to themselves."[102]

A substantial part of La Mettrie's treatise was devoted to attacking the belief that an unbridgeable gulf separated human beings from animals. "The transition from animals to man is not violent, as true philosophers will admit," contended La Mettrie.[103] "Man is not moulded from a costlier clay; nature has used but one dough, and has merely varied the leaven."[104] According to La Mettrie, the traits traditionally identified as the basis for man's uniqueness—his intellectual and moral capacities—are not unique to man and are displayed by animals as well.

The intellectual capacities of animals are demonstrated by their "varying capacity . . . to learn." Some animals "learn to speak and sing; they remember tunes, and strike the notes as exactly as a musician. Others, for instance the ape, show more intelligence."[105] Yet here La Mettrie admitted a problem. Apes are definitely more intelligent than many other members of the animal kingdom, but he acknowledged that they cannot learn music. What could be the reason for this strange inability in an otherwise intelligent creature? La Mettrie thought it was "some defect in the organs of speech."[106] But he also thought that the defect could be remedied. If human deaf mutes can be taught to talk and converse, surely apes can as well.[107] According to La Mettrie, there was nothing mysterious in how to raise apes to men, because there was nothing mysterious in how men acquired their own rational faculties. "Man has been trained in the same way as animals. He has become an author, as they became beasts of burden. A geometrician has learned to perform the most difficult demonstrations and calculations, as a monkey has learned to take his little hat off and on, and to mount his tame dog."[108]

Animals also demonstrate the capacity for morality. A dog who bit his master "seemed to repent a minute afterwards; it looked sad, ashamed, afraid to show itself, and seemed to confess its guilt by a crouching and downcast air." Similarly, there was "a lion which would not devour a man abandoned to its fury, because it recognized him as its benefactor."[109] According to La Mettrie, animals show evidence of a natural moral law just as much—or perhaps more than—human beings.

Lest some people think that he was insulting human beings by implying that they are on the same level as animals, La Mettrie assured his readers that the opposite was in fact the case. He was actually doing man "honor to place him in the same class" as animals.[110] That is because men are in so many ways lower than animals in their native abilities. What many animals know by instinct men must learn by education through trial and error.[111] Only education, concluded La Mettrie, "raises us from the level of the animals and lifts us above them." But not all human beings are in fact higher than the animals. In a chilling passage, he asks:

> Shall we grant this same distinction [of being above the animals] to the deaf and to the blind, to imbeciles, madmen, or savages, or to those who have been brought up in the woods with animals; to those who have lost their imagination through melancholia, or in short to all those animals in human form who give evidence of only the rudest instinct? No, all these, men of body but not of mind, do not deserve to be classed by themselves.[112]

Underlying every part of *Man a Machine* is La Mettrie's steadfast belief that there is no mystery in attributing mind to matter. Rejecting Locke's ambivalence about how blind matter could produce mind, La Mettrie asserted that "given the least principle of motion, animated bodies will have all that is necessary for moving, feeling, thinking, repenting."[113] Even so, La Mettrie tempered this assertion by refusing to claim that the universe as a whole had to be the product of blind chance. He admitted the logical possibility of a creator who infused the universe with purpose, although he chose to remain agnostic on the subject. He thought the arguments both for and against God were inconclusive.[114] Similarly, he did not deny outright the possibility that material bodies might have the attribute of immortality, but he again thought the relevant arguments were inconclusive.[115]

Baron d'Holbach, author of the ponderous tract *The System of Nature* (1770), was considerably less ambiguous on the subject of God. In part two of a treatise that runs for more than five hundred pages, d'Holbach makes clear that all religions can be debunked as "superstition . . . founded upon manifest contradictions," because they mistakenly attribute natural events to "an intelligent cause, distinguished from nature." According to d'Holbach, man creates gods after his own image, "clothing his gods with his own imbecile qualities."[116] Even simple theists who subscribe to a minimalist belief in a God who rules the universe by his intelligence are fools according to d'Holbach.

Like La Mettrie, d'Holbach argued that man is nothing more than a "machine"[117] and his "soul is nothing more than the body."[118] In sum, "man is a being purely physical,"[119] bound by "laws of Nature" from which "he cannot deliver himself" and from which "he cannot step beyond . . . even in thought."[120]

Those laws of nature, wrote d'Holbach, make man but the "feeble plaything in the hands of necessity."[121] Man "is never the master of the determination of his own peculiar will," and hence "he never acts as a free agent."[122] D'Holbach did not deny that human beings think that they have free will, but he concluded that this belief is an illusion. "Man . . . resembles a swimmer who is obliged to follow the current that carries him along; he believes himself a free agent, because he sometimes consents, sometimes does not consent, to glide with the stream; which, notwithstanding, always hurries him forward."[123] It follows from this that people who act viciously do so from necessity. In d'Holbach's view, "the wicked are never more than men who are either drunk or mad," and "it is not until tranquility is re-established in their machine" that they are able to "see the consequences of their actions" and experience remorse.[124]

D'Holbach denied that his view created difficulties for the punishment of criminals. He argued that whatever the cause of a man's crime, "society possesses the right to crush the effects, as much as the man whose land would be ruined by a river, has to restrain its waters by a bank."[125] D'Holbach thus redefined justice as the deterrence of crime rather than the need to hold someone morally accountable for his or her actions. The criminal justice system is utilitarian; any policy that reduces crimes presumably serves its purpose.

However, after claiming that his fatalistic view did not raise an obstacle to punishing criminals, d'Holbach proceeded to contradict himself by stating that "the law has not acquired the right to punish [a criminal] . . . if it has failed to present to him the motives necessary to have an influence over his will."[126] D'Holbach went on to criticize society for punishing "those propensities of which it is itself the author, or which its negligence has suffered to spring up in the mind of man."[127] According to d'Holbach, society itself is the main fount of crime because of its failure to supply education and economic opportunities for everyone.

One suspects that the point of d'Holbach's criticisms was to encourage the leaders of society to restructure social institutions in order to eradicate crime. In other words, d'Holbach wanted people to use their newfound knowledge about the material causes of human behavior to try to change that behavior by manipulating material causes. Yet d'Holbach never quite said this, probably because he recognized that such advice would imply the existence of a free choice. After all, if human beings were completely determined by their physical causes, how could they be urged to step outside those causes in order to reform them?

Despite d'Holbach's reluctance to encourage people to reform the material causes that drive social ills, he was hopeful that such a reformation would in fact occur. Nature had bestowed tender hearts on many people, and they would be led to share their humanitarian doctrine with others—under the conviction that "it will, by degrees, become a certain remedy for their sufferings, that it will produce those necessary effects which it is of its essence to operate."[128] One can already see in these statements the seeds of the doctrine of inevitable progress. It was a doctrine that would reappear with a vengeance in future decades.

As articulated by La Mettrie, d'Holbach, and others, the materialistic philosophy of the eighteenth century promised to expand man's knowledge by unlocking the secrets of life itself. Like the mechanic who understands the innermost operations of a machine, the materialist scientist looked forward to understanding the most intricate inner workings of man. Armed with this knowledge, he could hope to obtain the kind of knowledge that had previously been reserved for God.

In a lecture on probability delivered at the end of the eighteenth century, French mathematician Pierre-Simon Laplace (1749–1827) proclaimed that an intelligence that could grasp "at a given instant . . . all the forces by which nature is animated and the respective situation of the beings that make it up" could derive a mathematical formula that would explain everything that would ever happen.[129] "For such an intelligence nothing would be uncertain, and the future, like the past, would be open to its eyes." Laplace regarded science as the pursuit of omniscience. While he acknowledged that man would "always remain infinitely distant from this intelligence," he held forth God-like knowledge as the lofty goal to which scientists should aspire. As much as possible, scientists should seek to reduce the behavior of everything in the material universe to mechanical laws that could be expressed in terms of mathematics. With the promise of such knowledge came the possibility of incredible new powers, even power over life itself, which was the Holy Grail pursued by scientists like Giovanni Aldini who were trying to reanimate dead bodies.

It would be wrong to suggest that everyone in the eighteenth century was smitten by the return of materialism, for they weren't. La Mettrie, d'Holbach, and their compatriots received more than their share of notoriety and opposition. Most philosophers and scientists continued to insist on man's uniqueness and ostracized wholesale materialists.[130] La Mettrie was even attacked by his own publisher, who wrote a treatise in response titled *Man, More than a Machine*.[131] But unlike in the 1600s, when almost no one took materialism to its logical conclusions, the materialist vision in the eighteenth century was clearly making inroads among the elites. By the end of the century it had even gained a muse. In England, a latter-day Lucretius was enshrining the materialist account of the creation of life in his poetry. Describing the generation of the first life on earth, he wrote:

First HEAT from chemic dissolution springs,
And gives to matter its eccentric wings . . .
ATTRACTION next, as earth or air subsides,
The ponderous atoms from the light divides . . .
Last, as fine goads the gluten-threads excite,
Cords grapple cords, and webs with webs unite . . .
Hence without parent by spontaneous birth
Rise the first specks of animated earth.[132]

The poet who penned this ode to spontaneous generation was Erasmus Darwin (1731–1802). Actually, he was a part-time poet. A physician by profession, he dabbled in all things scientific, even becoming an early proponent of biological evolution.[133] During the century about to unfold, his grandson Charles (1809–82) would help spread materialism to the masses by seeming to place it on a solid scientific footing.

2
Darwin's Revolution

The first half of 1869 had proved frustrating for Charles Darwin. An article in the April issue of the *Quarterly Review* supplied the initial irritant. While the essay's author praised Darwin's theory of evolution by natural selection, he blasted the notion that it could explain the development of mind. In his view, "neither natural selection nor the more general theory of evolution can give any account whatever of the origin of sensational and conscious life."[1] Moreover, "the moral and higher intellectual nature of man is as unique a phenomenon as was conscious life on its first appearance in the world, and the one is almost as difficult to conceive as originating by any law of evolution as the other."[2]

The *Quarterly Review* author observed that the brain size of prehistoric man was nearly the same as modern man, even though the mental needs of prehistoric man would have been "very little above those of many animals." This fact created a quandary for the theory of natural selection, which selected for traits and abilities that allowed the organism to survive in its existing environment. "The higher moral faculties and those of pure intellect and refined emotion are useless" to the lowest savages, he pointed out, and they

> have no relation to their wants, desires, or well-being. How, then, was an organ developed so far beyond the needs of its possessor? Natural selection could only have endowed the savage with a brain a little superior to that of

an ape, whereas he actually possesses one but very little inferior to that of the average members of our learned societies.[3]

The only reasonable conclusion, according to the essayist, was to "admit the possibility that in the development of the human race, a Higher Intelligence has guided the same laws for nobler ends."[4]

Had the author been anyone else, Darwin might not have been so disturbed. But the essayist arguing for the limits of Darwinism was one of Darwin's closest supporters, naturalist Alfred Russel Wallace (1823–1913), who shared credit with Darwin for developing the theory of evolution by natural selection.[5] Darwin quickly wrote Wallace about his dismay. While he liked the rest of Wallace's essay, the remarks about man were so disheartening that Darwin would not have believed that Wallace could have written them. "If you had not told me I should have thought that they had been added by some one else . . . I differ grievously from you, and I am very sorry for it. I can see no necessity for calling in an additional and proximate cause in regard to man."[6]

More bad news was to come. When Darwin wrote to famed geologist Charles Lyell (1797–1875) that he had been "dreadfully disappointed" with Wallace's disclaimer about man, Lyell wrote back that he agreed with Wallace. "As I feel that . . . evolution cannot be entirely explained by natural selection, I rather hail Wallace's suggestion that there may be a Supreme Will and Power which . . . may guide the forces of law and nature."[7]

The same year former Darwin enthusiast St. George Mivart published anonymous articles critiquing Darwin's theory. A gifted zoologist, Mivart would eventually publish a volume titled *The Genesis of Species*, an influential book that raised serious questions about the limits of natural selection, especially in its application to man. Far from rejecting Darwin wholesale, Mivart continued to embrace evolution and believe that the physical capacities of human beings had developed from the lower animals. But he continued to insist—like Wallace—that man was radically unique from the rest of creation and had a soul. Egged on by Thomas Huxley, Darwin became increasingly bitter over his former disciple's criticisms, despite Mivart's attempts to be personable in private correspondence and his public praise of the "invaluable labours and active brains of Charles Darwin and Alfred Wallace."[8] Huxley eventually tried to blackball Mivart from the scientific community, and Darwin vowed to cut off communication after Mivart criticized an article by Darwin's son George that advocated eugenics.[9]

Darwin's unease with those who refused to adopt a completely materialistic theory of evolution is difficult to square with how he is sometimes portrayed today. "Darwin was not a materialist," insists Dr. Audrey Chapman of the American Association for the Advancement of Science. "I think that's a major misunderstanding."[10] Kenneth Miller and Joseph Levine go further in the teacher's edition of their popular biology textbook. According to Miller and Levine, not only was Darwin not a materialist,

> He was a very religious man. He had a degree in theology from Cambridge and even thought about spending his life as a country preacher. He remained a devout Christian all of his life. He saw no conflict—and no less awe or majesty—in a God that ruled through natural laws rather than through supernatural means.[11]

There is some support for this view of Darwin. As a young adult, Darwin did consider becoming a clergyman at the behest of his freethinking (but pragmatic) father, and he ultimately did earn a degree from Christ's College, Cambridge, in preparation for a clerical career.[12] Later in life, when asked in a letter by a German student about the relationship between evolution and God, Darwin had a member of his family respond for him that in his view "the theory of Evolution is quite compatible with the belief in a God." The family member then added rather ambiguously: "but . . . you must remember that different persons have different definitions of what they mean by God."[13] In *The Origin of Species* (1859), Darwin refrained from discussing how much his theory explained about the origin and development of man, though he predicted that in the future "much light will be thrown on the origin of man and his history."[14] Darwin also went out of his way in the book to talk several times about the "Creator" and "creation"[15]; he explicitly noted that he saw "no good reason why the views given in this volume should shock the religious feelings of anyone."[16] Indeed, the final sentence of *The Origin of Species* clearly seemed to imply that evolution was merely God's method of creation: "There is grandeur in this view of life, with its several powers, having been originally breathed by the Creator into a few forms or into one."[17]

However, those who accept such comments without digging deeper are likely to miss just how momentous—and radical—Darwin's intellectual project really was. Part of the problem is that many who read Darwin today look only at *The Origin of Species*. While it is true that Darwin called this book "the chief work of my life,"[18] it is probably not the best text from which to glean Darwin's mature views about his theory. It is certainly not the best text from which to derive the implications of Darwin's theory for human society. For information on that topic one needs to turn to *The Descent of Man* (1871), which Darwin wrote even as Wallace and others were trying to articulate limits to Darwinian explanations. It was in this book that Darwin most publicly applied his theory to human beings and human culture. In chapters 1 and 4, Darwin made the case that human beings shared a common ancestor with other animals. In chapters 2, 3, and 5, he turned to far more controversial matters such as the development of mind and morality and the application of natural selection to human societies.[19] It is in these chapters that the extraordinary reach of Darwin's theory finally becomes plain.[20]

Mind from Matter

According to Darwin, his purpose in chapter 2 of *The Descent of Man* was "to shew that there is no fundamental difference between man and the higher mammals in their mental faculties."[21] The chapter was devoted to presenting a blizzard of examples to prove that every mental attribute of human beings can be discovered in other animals.

Darwin believed that men and animals share not only lower instincts such as self-preservation and "sexual love," but the whole range of emotions and intellectual faculties. Like men, the lower animals express terror, suspicion, courage, timidity, and revenge.[22] Like men, they exhibit love: "The love of a dog for his master is notorious."[23] Animals also display complicated emotions such as jealousy: "Every one has seen how jealous a dog is of his master's affection, if lavished on any other creature; and I have observed the same fact with monkeys."[24] Darwin additionally thought one could find pride and magnanimity in animals. "They love approbation or praise; and a dog carrying a basket for his master exhibits in a high degree self-complacency or pride . . . A great dog scorns the snarling of a little dog, and this may be called magnanimity."[25] In a later edition of *Descent*, Darwin argued that animals even display a sense of humor:

> Dogs shew what may be fairly called a sense of humour, as distinct from mere play; if a bit of stick or other such object be thrown to one, he will often carry it away for a short distance; and then squatting down with it on the ground close before him, will wait until his master comes quite close to take it away. The dog will then seize it and rush aways in triumph, repeating the same manoeuvre, and evidently enjoying the practical joke.[26]

Turning to man's intellectual faculties, Darwin argued that one can find imitation, imagination, and reason among animals. Darwin pointed out that "birds imitate the songs of their parents, and sometimes of other birds; and parrots are notorious imitators of any sound which they often hear."[27] Equating imagination with dreaming (quoting the German writer Jean Paul Richter, who said that "the dream is an involuntary art of poetry"), Darwin maintained that many animals have acquired the faculty of imagination because they dream. "As dogs, cats, horses, and probably all the higher animals, even birds have vivid dreams . . . and this is shewn by their movements and voice, we must admit that they possess some power of imagination."[28] In a revised edition of *Descent*, Darwin invoked yet another canine example to prove his point, claiming that "there must be something special, which causes dogs to howl in the night, and especially during moonlight, in that remarkable and melancholy manner called baying."[29] QED, dogs show imagination. Even reason itself can be found in the animal kingdom, a fact which Darwin said "only a few persons now dispute . . . Animals may constantly be seen to pause, deliberate, and resolve. It is a significant fact, that the more the habits of any particular animal are studied by a naturalist, the more he attributes to reason and the less to unlearnt instincts."[30]

Darwin acknowledged that despite all of these similarities between men and animals, "many authors have insisted that man is separated through his mental faculties by an impassable barrier from all the lower animals."[31] These authors usually supplied a list of traits that they thought proved man's uniqueness from the rest of creation. Darwin spent the rest of the chapter discussing these supposedly unique traits of man and trying to show that they could in fact be found in animals.

According to Darwin, some argued that "man alone" was "capable of progressive improvement."[32] Darwin denied this, pointing out that animals are able to learn to avoid traps. They also "both acquire and lose caution in relation to man or other enemies."[33] Another trait claimed as demonstrating man's uniqueness was his use of tools, but Darwin described how "a chimpanzee in a state of nature cracks a native fruit, somewhat like a walnut, with a stone." Other animals employ sticks and stones as weapons. The orangutan will "cover itself at night with the leaves of the Pandanus," and a baboon will "protect itself from the heat of the sun by throwing a straw-mat over its head." To Darwin such acts signified the "first steps towards some of the simpler arts, namely rude architecture and dress."[34]

Language too could be accounted for through a purely materialistic process. Despite the admitted gulf in this area between the abilities of men and animals, Darwin wrote that he could not "doubt that language owes its origin to the imitation and modification, aided by signs and gestures, of various natural sounds, the voices of other animals, and man's own instinctive cries."[35] The process could have originated with an "unusually wise ape-like animal" who thought to imitate "the growl of a beast of prey, so as to indicate to his fellow monkeys the nature of the expected danger . . . [T]his would have been a first step in the formation of a language."[36]

Darwin argued that the vocal organs of such ape-like creatures could have developed over time "through the inherited effects of use."[37] Darwin repeatedly appealed to the concept of the inherited effects of use in *The Descent of Man*.[38]

Darwin next sought to debunk the idea that self-consciousness and abstract thought were somehow uniquely human. "Can we feel sure," he asked, "that an old dog with an excellent memory and some power of imagination, as shewn by his dreams, never reflects on his past pleasures in the chase?"[39] Darwin then seemed to suggest that human beings did not universally possess these traits anyway. Citing outspoken German materialist Ludwig Büchner as his authority, Darwin asked "how little can the hard-worked wife of a degraded Australian savage, who uses hardly any abstract words and cannot count above fours, exert her self-consciousness, or reflect on the nature of her own existence"[?][40]

At the end of chapter 2, Darwin finally turned to the claim that humans were distinguished from the lower animals by a universal belief in an all-powerful creator. All human beings do not share such a belief, countered Darwin. "There is no evidence that man was aboriginally endowed with the ennobling belief of an Omnipotent God. On the contrary there is ample evidence" to show that "numerous races have existed and still exist, who have no idea of one or more gods."[41] Darwin was willing to admit that there was an "almost universal" belief in some sort of spiritual world among primitive peoples, but he placed such beliefs on the level of a dog who barks and growls upon seeing an open parasol moved by the wind. Just as the dog mistakenly believes that some intelligent agent moved the parasol, primitive peoples surmised that unseen beings were behind all manner of natural phenomena.[42]

Darwin inserted in his discussion of religion a statement that the lack of a universal belief in God did not mean that God does not exist, and "the highest intellects that have ever lived" have affirmed his existence.[43] However, this disclaimer was followed three pages later by the claim that the "same high mental faculties" which led man to believe in monotheism also produced his beliefs in "unseen spiritual agencies," "fetishism," "polytheism," and "various strange superstitions and customs" such as "the sacrifice of human beings to a blood-loving god; the trial of innocent persons by the ordeal of poison or fire"; and "witchcraft."[44] If man's intellectual faculties were capable of embracing such fallacious and horrible beliefs in the past, why should one believe that they reached the truth in embracing monotheism? Darwin did not raise this question explicitly, but it lurks behind his words as one of the implications of his naturalistic account of the rise of religion. Darwin's disclaimer to the contrary, his discussion in *The Descent of Man* tended to depict religion as a mere by-product of biology and cultural development. He nearly said as much at the conclusion of the book: "The idea of a universal and beneficent Creator of the universe does not seem to arise in the mind of man, until he has been elevated by long-continued culture."[45]

Biology and Morality

Having argued that man's mind could have arisen naturally from the capacities of animals, Darwin in chapter 3 proceeded to claim the same thing about the development of man's moral faculties. Like a long list of moral philosophers before him, Darwin believed that human beings had a "moral sense" that instructed them about right and wrong. Some have tried to argue from this that Darwin was a traditional moral theorist seeking to defend traditional morality. Larry Arnhart sees Darwin as trying to recapture the natural basis of morality à la Aristotle and Aquinas.[46] There is something to be said for this argument. Darwin not only claimed that man's social instincts arise out of his biology, but that these social instincts "naturally lead to the golden rule, 'As ye would that men should do to you, do ye to them likewise.'" According to Darwin, this maxim "lies at the foundation of morality."[47] Rather than promoting the law of the jungle, evolution promotes the morality of Jesus.

However, before one rushes to install Darwin in the pantheon of defenders of a natural moral law, one needs to scrutinize the account of morality Darwin gave prior to his conclusion supporting conventional morality. While according to Darwin nature has led to the golden rule, it did not do so because the golden rule is somehow intrinsically right. It did so because the golden rule ultimately is connected to self-preservation. At the base of the golden rule, in Darwin's view, are the social instincts, and these developed primarily because they promote survival: "Those communities, which included the greatest number of the most sympathetic members, would flourish best and rear the greatest number of offspring."[48] In the conclusion to *The Descent of Man*, Darwin made this point even more clearly, stating that "the . . . origin of the moral sense lies in the social instincts, including sympathy; and these instincts no doubt were primarily gained, as in the case of the lower animals, through natural selection."[49]

But what happens in cases where traditional morality does *not* happen to promote survival? If human morality is ultimately grounded in the struggle to survive, it seems optimistic in the extreme to think that the by-product will always be something akin to traditional Judeo-Christian morality. Darwin himself provided exhaustive evidence on this point. While he tried to show that traditional virtues such as courage and love were products of nature, he also demonstrated that a great many vices are no less a product of the struggle for existence. Maternal instinct is natural, but so is infanticide.[50] Care toward family members is natural, but so is euthanasia of the feeble, even if they happen to be one's parents:

> That animals sometimes are far from feeling any sympathy is too certain; for they will expel a wounded animal from the herd, or gore or worry it to death. This is almost the blackest fact in natural history, unless indeed the explanation which has been suggested is true, that their instinct or reason leads them to expel an injured companion, lest beasts of prey, including man, should be tempted to follow the troop. In this case their conduct is not much worse than that of the North American Indians who leave their feeble comrades to perish on the plains, or the Feegeans, who, when their parents get old or fall ill, bury them alive.[51]

Throughout his discussion of morality, Darwin repeatedly referred to "higher" and "lower" moral impulses as if there were some transcendent standard of morality to which he compared human and animal behavior. Darwin wrote as if conventional virtues such as kindness and courage were objectively preferable to conventional vices such as cruelty and lust. But it is difficult to make sense of such comments in terms of Darwin's own system, which clearly portrayed morality as ultimately reducible to that which promotes biological survival. In the current set of circumstances Darwin could believe that his view meant the extension of benevolence "to the men of all races, to the imbecile, the maimed, and other useless members of society, and finally to the lower animals."[52] He could even hope that "looking to future generations, there is no cause to fear that the social instincts will grow weaker, and we may expect that virtuous habits will grow stronger, becoming perhaps fixed by inheritance . . . and virtue will be triumphant."[53] But even Darwin would have to acknowledge, if pressed, that given a different set of circumstances, a radically different conception of morality might be dictated. At one point, he said as much: "If, for instance . . . men were reared under precisely the same conditions as hive-bees, there can hardly be a doubt that our unmarried females would, like the worker-bees, think it a sacred duty to kill their brothers, and mothers would strive to kill their fertile daughters; and no one would think of interfering."[54]

Temperate and considerate in his own life, Darwin may have sincerely believed that his biology supported traditional morality.[55] Nevertheless, the internal logic of his theory did not allow any permanent foundation for ethics other than the struggle to survive, and for that reason his attempt to square a biological understanding of ethics with traditional morality is ultimately unpersuasive. Far from supplying the basis for a traditional conception of morality, Darwin's account of the moral sense ultimately implies the abolition of transcendent moral standards.

Darwin's account of morality also raised serious questions about personal responsibility. His discussion of human behavior largely explained behavior as the function of various predetermined—and largely antisocial—instincts. For all of Darwin's praise of man's social instincts, he wrote that "it cannot be maintained that the social instincts are ordinarily stronger in man, or have become stronger through long-continued habit, than the instincts . . . of self-preservation, hunger, lust, vengeance, &c."[56] What did this mean in practice? "At the moment of action, man will no doubt be apt to follow the stronger impulse; and though this may occasionally prompt him to the noblest deeds, it will far more commonly lead him to gratify his own desires at the expense of other men."[57] Darwin tried to soften the implications of his view by going on to claim that men will learn to regret their impulsive actions and eventually this regret will create in them a conscience. However, Darwin did not convincingly explain why conscience would trump the instincts he earlier depicted as so overwhelming. Even if conscience is able to counteract the antisocial instincts in some men, presumably those who act antisocially are only following their own strongest instincts. If this be the case, how responsible are those who act against society? In *The Descent of Man* Darwin does not address the consequences of his account for free will and criminal responsibility. Followers such as Ludwig Büchner in Germany were not so reticent; nor was Darwin in unpublished notebooks written earlier in his career. There he wrote that "the general delusion about free will [is] obvious," and that one ought to punish criminals "solely to *deter* others"—not because they did something blameworthy. "This view should teach one profound humility," wrote Darwin. "One deserves no credit for anything . . . nor ought one to blame others." Darwin denied that such a fatalistic view would harm society, because he thought that ordinary people would never be "*fully* convinced of its truth," and the enlightened few who did embrace it could be trusted.[58]

If Darwin's account of morality in chapter 3 was at least implicitly subversive of traditional ethics, his treatment in chapter 5 of how natural selection worked in human societies was explicitly so. Drawing on the writings of W. R. Greg, Alfred Wallace, and Francis Galton, Darwin layed out a rationale for what would later become known as eugenics, the attempt to improve the human race through breeding.[59] The opening of Darwin's discussion of this topic is so remarkable that it deserves quotation in full:

With savages, the weak in body or mind are soon eliminated; and those that survive commonly exhibit a vigorous state of health. We civilised men, on the other hand, do our utmost to check the process of elimination; we build asylums for the imbecile, the maimed, and the sick; we institute poor-laws; and our medical men exert their utmost skill to save the life of every one to the last moment. There is reason to believe that vaccination has preserved thousands, who from a weak constitution would formerly have succumbed to small-pox. Thus the weak members of civilised societies propagate their kind. No one who has attended to the breeding of domestic animals will doubt that this must be highly injurious to the race of man. It is surprising how soon a want of care, or care wrongly directed, leads to the degeneration of a domestic race; but excepting in the case of man himself, hardly any one is so ignorant as to allow his worst animals to breed.[60]

Darwin spent much of the rest of the chapter backtracking from the radical implications of this passage. Reverting again to the proper Victorian, Darwin assured his readers that whatever our reason might tell us about the harmful consequences of counteracting natural selection, our sympathy compels us to be compassionate. Indeed, we could not "check our sympathy, if so urged by hard reason, without deterioration in the noblest part of our nature . . . [I]f we were intentionally to neglect the weak and helpless, it could only be for a contingent benefit, with a certain and great present evil."[61]

Darwin then seemed to imply that modern society can afford to be compassionate because natural selection continued with its work despite our compassion. For example, "the weaker and inferior members of society" do not marry "so frequently as the sound," presumably leading to fewer children from the unfit.[62] Better conditions and more plentiful food promote the "better development of the body."[63] And society and nature continue to work together to dispose of the morally unfit: "Malefactors are executed, or imprisoned for long periods, so that they cannot freely transmit their bad qualities. Melancholic and insane persons are confined, or commit suicide. Violent and quarrelsome men often come to a bloody end . . . Profligate women bear few children, and profligate men rarely marry; both suffer from disease."[64]

In this contention, Darwin was perhaps at his least convincing. Later in the chapter, he himself indicated otherwise. "The reckless, degraded, and often vicious members of society, tend to increase at a quicker rate than the provident and generally virtuous members."[65] Darwin hoped that the higher mortality rate of such classes might prevent them from overwhelming "the better class of men." But he was not so bold as to promise such a result. Instead, he stated that if natural checks did not prevent the inordinate growth of degraded classes, "the nation will retrograde, as has occurred too often in the history of the world. We must remember that progress is no invariable rule . . . Natural selection acts only in a tentative manner."[66]

Perhaps fearing that this chapter was too grim for Victorian sensibilities, in his revised edition of *Descent* Darwin tacked on a happy conclusion to this discussion, asserting that "the more intelligent members within the same community will succeed better in the long run than the inferior, and leave a more numerous progeny, and this is a form of natural selection."[67] That statement verged on the disingenuous. Not only did it cut against much of the evidence Darwin marshalled in the chapter, but it contradicted Darwin's own private views during the latter part of his life. According to Alfred Wallace, during one of his last conversations with his friend, Darwin

> expressed himself very gloomily on the future of humanity, on the ground that in our modern civilisation natural selection had no play, and the fittest did not survive. Those who succeed in the race for wealth are by no means the best or the most intelligent, and it is notorious that our population is more largely renewed in each generation from the lower than from the middle and upper classes.[68]

It is possible that Darwin became more negative on this subject after the revised edition of *The Descent of Man* was published, and this increased negativism was reflected in Wallace's account; but it seems more likely that Darwin harbored the same doubts while he wrote *Descent*. He simply repressed them when writing for the public.

Chapter 5 is one of the most unsettling chapters of *The Descent of Man*. Despite Darwin's lip service to the need for compassion toward the weak, it is difficult to square such comments with the clear import of the rest of the chapter. If Darwin really believed that society's efforts to help the impoverished and sickly "must be highly injurious to the race of man" (note the word "must"), then the price of compassion, in his view, appeared to be the survival of the human race. Darwin the Victorian might have been willing to pay such a high price; but one must wonder how many others would have agreed with him once his ideas gained currency. If most people thought that their efforts for the poor and sickly would destroy the human race, how many of them would be willing to press the case for compassion?

In chapter 5 and again in the book's conclusion, Darwin cautioned against efforts to curtail the struggle for survival, for it was this very struggle that had raised man above the lower animals. Rather than seek to check rapid population growth, modern society should accept the necessity of a growing population for the improvement of the race. Darwin recognized that such population growth generated a multitude of evils, such as infanticide in "barbarous tribes" and "in civilised nations . . . abject poverty, celibacy, and . . . the late marriages of the prudent." But this was the price of progress, and "as man suffers from the same physical evils with the lower animals, he has no right to expect an immunity from the evils consequent on the struggle for existence. Had he not been subjected to natural selection, assuredly he would never have attained to the rank of manhood."[69] It was only when the many were forced to compete for limited resources that natural selection could produce optimal results:

> Man, like every other animal, has no doubt advanced to his present high condition through a struggle for existence consequent on his rapid multiplication; and if he is to advance still higher he must remain subject to a severe struggle. Otherwise he would soon sink into indolence, and the more highly gifted men would not be more successful in the battle of life than the less gifted. Hence our natural rate of increase, though leading to many and obvious evils, must not be greatly diminished by any means. There should be open competition for all men; and the most able should not be prevented by laws or customs from succeeding best and rearing the largest number of offspring.[70]

Nevertheless, although Darwin generally opposed attempts to curtail population growth (including contraception),[71] he endorsed efforts to reduce the propagation of the unfit through voluntary restrictions on marriage. According to Darwin, "both sexes ought to refrain from marriage if in any marked degree inferior in body or mind." Darwin recognized that this goal was "Utopian," but he implied that it might be "partially realised" when "the laws of inheritance are thoroughly known." In Darwin's view, "all do good service who aid towards this end."[72]

Marriage was only touched on in chapter 5, but it became a central focus later in the book, as Darwin turned to examine what he called "sexual selection," or how mating and reproduction promote evolution.[73] Here Darwin offered a materialistic account of marriage and family life, just as he had offered a materialistic account of morality. Rejecting the Judeo-Christian conception of marriage as an institution ordained by God from the inception of humanity, Darwin depicted marriage as yet another evolutionary product of the struggle to survive. "It seems certain that the habit of marriage has been gradually developed, and that almost promiscuous intercourse was once extremely common throughout the world."[74] Darwin discussed at length evidence that humans originally held their mates in common, at one point stating that "all those who have most closely studied the subject, and whose judgment is worth much more than mine, believe that communal marriage was the original and universal form throughout the world, including the intermarriage of brothers and sisters."[75]

After staking out this radical view of the matter, Darwin characteristically pulled back and concluded, based on the lifestyles of apes, that the first humans probably practiced some form of polygamy or monogamy.[76] By monogamy, though, he meant serial monogamy, where "the pairing may not last for life, but only for each birth."[77] Only when human beings had developed further—relying less on their instincts and more on their intellectual faculties—did practices develop such as promiscuity, infanticide, and the treatment of women as slaves.[78] While Darwin's conclusion that the earliest men practiced polygamy or serial monogamy rather than communal sex made his theory more palatable to Victorian sensibilities, there was no getting around the fact that in his system there could be not be one naturally superior form of sexual relations between human beings. According to Darwin's explanation, there was nothing sacred or absolute about any of the forms of family life. They were adaptations to the environment humans faced, and presumably when the environment changed, so would sexual patterns.

The Descent of Man was finely calculated to persuade, and it remains a masterpiece of scientific rhetoric. The book demonstrated Darwin's almost encyclopedic knowledge of the natural world, and it presented a nearly overwhelming array of examples, observations, and statistics to prove his points. Darwin went out of his way to sift the evidence, to present evidence for positions he did not ultimately favor, and to deal with possible objections. He also smoothed over the harsh edges of his theory, recognizing the limits of Victorian propriety. His tone, moreover, remained civil and non-threatening.

Yet for all of Darwin's fair-mindedness, comprehensiveness, and civility, the logic of *The Descent of Man* remains unsatisfying. Many of the book's conclusions do not seem to follow from Darwin's premises. This is especially noticeable in Darwin's chapter on the mental capacities of men and animals. When it came to the higher mental faculties such as imagination, reasoning, and self-consciousness, Darwin seemed to succeed in making his case only by redefining various faculties in the most reductive terms. If imagination really is no more than dreaming, then perhaps Darwin was right and animals possess the same kind of imaginative faculty as human beings. If magnanimity really is revealed when one dog ignores another, then perhaps a dog is magnanimous. But it is not at all clear that such faculties and attributes can be reduced to Darwin's definitions.

Then there is Darwin's anthropomorphism, which sometimes seems to rival that of a Walt Disney film. Darwin's discussion of the canine sense of humor is illustrative: Dogs engage in behavior that seems humorous when interpreted according to human conventions; therefore, they may be said to have a sense of humor. But Darwin supplied no evidence to show that dogs understand what they are doing in human terms. He merely assumed that they do—assuming the exact point he was trying to prove. The outward antics that Darwin described do not necessarily entail the inward mental states he was attempting to attribute to dogs. Even Darwin had to acknowledge in his revision of *The Descent of Man* "the impossibility of judging what passes through the mind of an animal."[79]

One must also wonder whether Darwin fully appreciated how a vast difference in degree may in fact be a difference in kind. Even if one were to accept in toto all of the examples supplied by Darwin to prove the mental faculties of the lower animals, one would still be left with considerably less than a man. With some justice, G. K. Chesterton parodied the Darwinian attempt to reduce the gulf between men and animals by producing his own comparison between birds and men. Pointing out that birds, like men, build homes for themselves, Chesterton argued that the process of nest-building, far from showing a bird's kinship with man, demonstrates just how huge the rift is between them. A bird is perfectly content to reproduce the same type of nest that had been constructed by his parents. But suppose, wrote Chesterton, someone saw a bird

> build as men build. Suppose in an incredibly short space of time there were seven styles of architecture for one style of nest . . . Suppose the bird made little clay statues of birds celebrated in letters of politics and stuck them up in front of the nest. Suppose that one bird out of a thousand birds began to do one of the thousand things that man had already done even in the morning of the world . . . [then] we can be quite certain that the onlooker would not regard such a bird as a mere evolutionary variety of the other birds; he would regard it as a very fearful wild-fowl indeed.[80]

Chesterton was pointing out that one could accept the natural data presented by Darwin without embracing his conclusion that man could be accounted for by a purely natural material process. Looking at the evidence marshalled by Darwin, Chesterton could contend that it actually reinforced the dramatic divergence of man from every other living being on earth. "Man is not merely an evolution but . . . a revolution."[81] Several of Darwin's defenders—including Alfred Wallace, Charles Lyell, and Asa Grey—would have agreed with such an assessment, much to Darwin's dismay.

The Descent of Man was an attempt to portray human rationality, morality, and society as ultimately the products of purely materialistic processes. Indeed, much of *Descent* reads like an updated and expanded version of La Mettrie's *Man a Machine*. Like La Mettrie, Darwin contended that there was no cavernous gulf between men and the rest of the animal world. Like La Mettrie, he tried to supply a naturalistic explanation for the development of language. Like La Mettrie, he argued that animals exercised moral faculties.[82] But Darwin was considerably more convincing than La Mettrie because of his masterly compilation of examples.

Darwin, the Reluctant Materialist?

Given the underlying materialism of *The Descent of Man*, what is one to make of Darwin's accommodations to religion in his public writings? Some of his comments were undoubtedly designed to disarm popular prejudices. For example, we know that his concluding sentence in *The Origin of Species*, where he spoke of "life . . . having been originally breathed by the Creator into a few forms or into one,"[83] was intended to make his book more palatable to the public. After a hostile reviewer tried to employ this passage from Darwin for his own purposes, Darwin wrote to a friend: "I have long regretted that I truckled to public opinion, and used the Pentateuchal term of creation, by which I really meant 'appeared' by some wholly unknown process."[84] As for Darwin's statement in *The Descent of Man* that the existence of God had been affirmed "by the highest intellects that have ever lived," not only did Darwin's account of the development of religion undercut the seriousness of his claim; the statement may also reflect the judicious editing of Darwin's daughter Henrietta, who according to some scholars was charged by her father with toning down the manuscript of *The Descent of Man*.[85]

In his private writings, Darwin was more direct about his religious skepticism. In his *Autobiography*, a document originally intended for use by his family, Darwin told of how he "gradually came to disbelieve in Christianity as a divine revelation" and had "never since doubted even for a single second that my conclusion was correct." Indeed, by the end of his life, Darwin could "hardly see how anyone ought to wish Christianity to be true; for if so the plain language of the text seems to show that the men who do not believe, and this would include my Father, Brother and almost all my best friends, will be everlastingly punished. And this is a damnable doctrine."[86]

Darwin's disbelief eventually spread beyond Christianity to include any sort of belief in God. While writing the first edition of *The Origin of*

Species, he claimed that he had probably been a theist because he saw the "impossibility of conceiving this immense and wonderful universe, including man . . . as the result of blind chance or necessity."[87] But that belief too had gradually eroded. "The old argument of design in nature . . . which formerly seemed to me so conclusive, fails, now that the law of natural selection has been discovered."[88] Moreover, it seemed inconceivable to Darwin that an omnipotent God could sanction the cruelties inherent in nature's struggle for existence—"the sufferings of millions of the lower animals throughout almost endless time."[89] Having demystified the world through natural selection, Darwin was no longer filled with "higher feelings of wonder, admiration, and devotion" when looking at nature. "I well remember my conviction that there is more in man than the mere breath of his body. But now the grandest scenes would not cause any such convictions and feelings to rise in my mind."[90]

Despite the persuasiveness with which he articulated the case for atheism in his *Autobiography*, Darwin was loath to take the final step and explicitly embrace the doctrine. He waffled: "The mystery of the beginning of all things is insoluble by us; and I for one must be content to remain an Agnostic."[91] Darwin's reluctance to unreservedly affirm either atheism or materialism was more pronounced in his interactions outside his inner circle. Throughout his life he continued to tell correspondents that it seemed incredible that the universe was simply the product of blind chance. To Asa Gray, his Christian defender in America, Darwin wrote in 1860 that his views were "not at all necessarily atheistical" and "I cannot . . . be contented to view this wonderful universe, and especially the nature of man, and to conclude that everything is the result of brute force. I am inclined to look at everything as resulting from designed laws, with the details . . . left to . . . chance." Darwin continued that he saw "no reason why a man, or other animal, may not have been aboriginally produced by other laws . . . that . . . may have been expressly designed by an omniscient Creator, who foresaw every future event and consequence."[92]

Such statements were typically written to friendly theists, and there would have been ample reason for Darwin to tailor his answers to what they wanted to hear. Moreover, Darwin's proto-theistic statements were often embedded in a highly ambivalent context. For example, while telling Asa Gray that his views were consistent with the idea that God created the laws according to which men and animals developed, Darwin also declared that he could not see

as plainly as others do . . . evidence of design and beneficence on all sides. There seems to me too much misery in the world. I cannot persuade myself that a benevolent and omnipresent God would have designedly created the Ichneumonidæ with the express intention of their feeding within the living bodies of caterpillars, or that a cat should play with mice.

Darwin ended his letter to Gray with the disclaimer that "the more I think the more bewildered I become; as indeed I have probably shown by this letter."[93] In a similar letter to W. Graham in 1881, Darwin praised Graham for articulating "my inward conviction . . . that the Universe is not the result of chance." But in the same letter, Darwin denied that "the existence of so-called natural laws implies purpose" and qualified his rejection of chance with a statement of materialistic skepticism:

> But then with me the horrid doubt always arises whether the convictions of man's mind which has been developed from the mind of the lower animals, are of any value or at all trustworthy. Would any one trust in the convictions of a monkey's mind, if there are any convictions in such a mind?[94]

To Julia Wedgewood he likewise wrote in 1881 that while "the mind refuses to look at this universe, being what it is, without having been designed . . . yet" when it came to "the structure of a sentient being, the more I think on the subject, the less I can see proof of design."[95] In such letters, Darwin supplied the proof for a complete materialism, but refrained from taking the final step.

By the latter part of Darwin's life, the weight of the evidence suggests that the main obstacle that kept Darwin from outright atheism was not intellectual. It was the desire to spare the feelings of theistic family members and deflect public criticism. Writing to Edward Aveling, one of his disciples who was an outspoken atheist, Darwin made clear the practical roots of his religious reticence:

> direct arguments against christianity & theism produce hardly any effect on the public; & . . . freedom of thought is best promoted by the gradual illumination of men's minds, which follow[s] from the advance of science. It has, therefore, been always my object to avoid writing on religion, & I have confined myself to science. I may, however, have been unduly biassed [sic] by the pain which it would give some members of my family, if I aided in any way direct attacks on religion.[96]

He made a similar point to Aveling when the disciple sent Darwin a copy of a book called *The Student's Darwin*, which drew out the atheistic implications of Darwinism. Darwin responded that he could not stop writers from pressing his views "to a greater length than seems to me safe."[97] Carefully note Darwin's wording here: He protests that atheists were advancing his views further "than seems to me *safe*," not that they were taking his views further "than seems to me *true*." He did not object to the veracity of their pronouncements, only to their prudence. Despite his concerns about staying within the bounds of Victorian propriety, at least once Darwin hazarded to align himself publicly with religious skepticism, endorsing a tract by American freethinker Francis Abbot that predicted "the extinction of faith in the Christian Confession." Darwin wrote Abbot that he "admire[d]" the contents of Abbot's tract "from my inmost heart, & I agree to almost every word." Darwin then allowed Abbot to use his endorsement in Abbot's weekly magazine, the *Index*. Several years later, Darwin withdrew permission for public use of his endorsement, fearing it might be used against him in England.[98]

Further evidence that Darwin's reticence in this area derived more from prudence than genuine ambivalence can be found in private notebooks written by him during the period of 1836–44. Although produced well before Darwin's *The Origin of Species*, these notebooks contain the seeds of much of his later work. They also show how Darwin's private speculations far outpaced both his public statements and comments to his own family. Thought is depicted in these notebooks purely as the product of bodily organization. "Brain makes thought,"[99] wrote Darwin, and "thought . . . seems as much function of organ, as bile of liver."[100] Thought in animals is produced "as soon as [the] brain [is] developed . . . no soul superadded."[101] The implication of this statement was clear: If animals did not need a soul to think, why should man? Darwin also attacked free will in his notebooks, arguing that human choice is the product of bodily organization interacting with circumstances and education. He thought heredity was a major factor: "Verily the faults of the fathers, corporeal & bodily are visited upon the children."[102]

In his notebooks, Darwin was ambivalent about whether his materialism tended toward atheism. At one point he suggested that it did not, but at another point he indicated that it did. The belief that human choice was dictated by heredity and environment, wrote Darwin, "would make a man a predestinarian of a new kind, because he would tend to be an atheist."[103] In light of Darwin's later writings, perhaps the most intriguing comment in his notebooks concerned how far he should admit to being a materialist: "To avoid stating how far, I believe, in Materialism, say only that emotions, instinct degrees of talent, which are heredetary [sic] are so because brain of child resemble, parent stock."[104] That comment should give pause to anyone trying to interpret the real intent of Darwin's later writings. How much did he choose to conceal because he wished to avoid persecution?

Whatever the reasons for Darwin's reluctance to explicitly affirm materialism later in his career, the theory he expounded in *The Descent of Man* purported to explain human beings and their social relations in wholly materialistic terms. Stephen Jay Gould was right to argue that "Darwin applied a consistent philosophy of materialism to his interpretation of nature," and that according to his theory matter was "the ground of all existence; mind, spirit, and God as well, are just words that express the wondrous results of neuronal complexity."[105] According to Darwin's account, natural selection and the laws of heredity acting on the material world produced mind, morality, and civilization. Whether or not Darwin wished to call himself a materialist, his theory had the consequence of making a materialistic understanding of man and society finally credible. By describing in detail natural mechanisms that could produce the complexity of life as we know it, Darwin helped transform materialism from a fantastic tale told by a few thinkers on the fringe of society to a hallowed scientific principle enshrined at the heart of modern science.

This is not to claim that the cultural triumph of Darwinian materialism was immediate. By Darwin's death in 1882, the theory of organic evolution had become firmly entrenched in both science and culture, but Darwin's claim that evolution could be explained wholly through unguided material causation remained shaky. Unlike some of Darwin's aggressive popularizers in England and Germany, many scientists and intellectuals did not push Darwinism to its logical conclusion. This was especially true in America, where some defenders of Darwin went out of their way to disentangle the theory from its seeming materialism. Defending Darwin from an attack by Princeton theologian Charles Hodge, Harvard botanist Asa Gray referred to Darwin as a "confessed theist" who did not "deny 'purpose, intention, or the cooperation of God' in nature."[106] Accordingly, Gray thought that "evolution may be as profoundly and as particularly theistic as it is increasingly probable."[107] Naturalists in the National Academy of Sciences during the latter part of the nineteenth century generally squared their evolutionary beliefs with their religious convictions.[108]

By the early part of the twentieth century, that situation had changed. According to a survey published in 1916, only 16.9 percent of prominent American biologists believed in God.[109] Materialistic explanations of man and society were on the rise, fueled in part by vigorous popularizers of Darwinian biology. By 1919, Hugh Elliot could baldly assert that science was necessarily materialistic and "it may be said with truth that we are all materialists now."[110]

Elliot exaggerated, but he captured the spirit of his age. Darwin helped spark an intellectual revolution that sought to apply materialism to nearly every area of human endeavor. This new, thoroughly "scientific" materialism affected the entire span of culture, from economics and politics to education and the arts. As will be seen in the next section, nowhere was the triumph of scientific materialism more consequential than in the field of crime and punishment.

Crime and Punishment

3

Criminal Science

Born into families of wealth and privilege, Nathan "Babe" Leopold and Richard "Dickie" Loeb were graduate students from Chicago who decided to commit the perfect crime.[1] One spring day in 1924 they lured fourteen-year-old Bobby Franks into a car and murdered him by stabbing his head with a chisel and suffocating him with a gag. The young collegians subsequently hid the boy's body in a culvert, pouring hydrochloric acid over the nude corpse to make it harder to identify. Leopold and Loeb then sent a ransom note to Franks's family demanding money for the return of their son alive. The "perfect" crime quickly turned out to be not so perfect after all when Leopold's eyeglasses were discovered near where the body had been left. Subjected to vigorous interrogation by the police, both young men eventually confessed and implicated each other.

In an effort to save their sons, the families of Leopold and Loeb enlisted famed attorney Clarence Darrow to defend them. The prospects for the defense were bleak. First, prosecutors had accumulated a large amount of evidence against Leopold and Loeb in addition to their confessions. Second, standards governing the use of the insanity defense at the time were narrow and likely insurmountable (prosecutors prudently had their own psychiatrists examine Leopold and Loeb right away in order to circumvent such a defense). Finally, Chicago was being whipped into a frenzy over the killing by the newspapers, and it would be difficult to find a sympathetic jury.

Realizing that he probably could not achieve an outright acquittal, Darrow lowered his sights and worked to spare his clients from the gallows. In order to avoid having the case tried before a hostile jury, Darrow had Leopold and Loeb plead guilty to the crimes of murder and kidnapping. The guilty plea prevented a trial and moved the case to the sentencing hearing presided over by Judge John R. Caverly. Darrow thought he would have a better chance if he could make his arguments to the judge alone.

Darrow's closing speech for the sentencing hearing has been called "this century's most eloquent courtroom argument."[2] While that is probably an exaggeration, Darrow's final statement did apparently bring the judge to tears.[3] More importantly, Darrow's speech brilliantly encapsulated the scientific ideas about crime that were coming to be widely accepted among America's intellectual elite.

According to Darrow, the real question before the court was whether it would embrace "the old theory" that "a man does something . . . because he willfully, purposely, maliciously and with a malignant heart sees fit to do it" or the new theory of modern science that "every human being is the product of the endless heredity back of him and the infinite environment around him."[4] For Darrow this was a battle between superstition and reason, the dark ages and the age of enlightened progress. Deriding the "old theory" that man was responsible for his crimes as originating with a superstitious belief in demonic possession, Darrow appealed for the court to embrace the truth that human beings are machines determined wholly by their heredity and environment.

In Darrow's eyes, it was Leopold and Loeb who were the "victims" of this tragedy.[5] "They killed [Bobby Franks] . . . because they were made that way. Because somewhere in the infinite processes that go to the making up of the boy or the man something slipped."[6] Darrow relentlessly drove home his message of unvarnished scientific materialism, especially when discussing Richard Loeb. "Is he to blame that his machine is imperfect?" Darrow asked the court.[7]

Darrow described how in British courts horses, dogs, and pigs used to be convicted of committing crimes. "I have in my library a story of a judge and jury and lawyers trying and convicting an old sow for lying down on her ten pigs and killing them." In Darrow's view, Richard Loeb had no more responsibility than that pig. "Do you mean to tell me that Dickie Loeb had any more to do with his making than any other product of heredity that is born upon the earth?"[8]

For all of his invocations of heredity, Darrow did not claim to actually know what part of Loeb's "machine" was broken. He merely knew that it had to be broken. Why else would he have committed the murder? "I know that somewhere in the past that entered into him something missed.

It may be defective nerves. It may be a defective heart or liver. It may be defective endocrine glands. I know it is something. I know that nothing happens in this world without a cause."[9] Darrow would not even entertain the notion that human beings could cause their own actions by freely choosing to do something.

In defending Loeb, Darrow based most of his argument on heredity, although he suggested that Loeb's reading of detective stories also contributed to his crime. When defending Leopold, Darrow reversed himself and focused almost exclusively on his environment. While clearly stating that he thought Leopold suffered from a "diseased mind," Darrow spent much of his time arguing that Leopold had been driven to murder by the philosophy of Nietzsche, which he had imbibed while a student at the University of Chicago. According to Darrow, Leopold was deluded into thinking that he was Nietzsche's "superman" and therefore that the ordinary rules of morality did not apply to him. "Your Honor," Darrow told the judge, "it is hardly fair to hang a nineteen-year-old boy for the philosophy that was taught him at the university. The university, the scholars and publishers of the world would be more to blame than he is." Darrow quickly added that he did "not believe that the universities are to blame" either. "I do not think they should be held responsible."[10]

In fact, Darrow came very close to stating that no one could ever be held responsible for anything: "I know that every influence, conscious and unconscious, acts and reacts on every living organism, and that no one can fix the blame."[11] Although Darrow made this particular argument in order to spare his clients from the death penalty rather than to exempt them from punishment entirely, the views he propounded to the judge were susceptible of a much wider application, and he knew it. Indeed, it was the wider application of his scientific principles of criminology that partly motivated Darrow's lengthy presentation. As he told the judge:

> I am not pleading so much for these boys, as I am for the infinite number
> of others to follow, those who perhaps cannot be as well defended as these
> have been, those who may go down in the tempest, without aid. It is of them
> I am thinking, and for them I am begging of this court not to turn backward
> toward the barbarous and cruel past.[12]

If Darrow was right and criminals were programmed for crime by material forces over which they had no control, why should one stop with abolition of the death penalty? If it was really true that no one was responsible for their actions, the very idea of punishment seemed unjust.

Darrow had thrown down his gauntlet to the traditional criminal justice system. Robert Crowe, the state's chief prosecutor in the case, clearly grasped what Darrow was trying to do. Seeing that Darrow had turned the Leopold-Loeb case into a platform from which to advance his ideas about crime and responsibility, the prosecutor was not about to allow Darrow's claims to go unchallenged. In Crowe's view, what was really on trial during the hearing was "Darrow's dangerous philosophy of life," and he read extensive excerpts from another Darrow speech to prove his point.[13] The speech had been delivered to prisoners at the county jail more than two decades earlier, and it showed just how far Darrow was prepared to press the implications of his scientific doctrines. Embarrassed by having his own words quoted back at him, Darrow protested to the judge that his old speech didn't have "anything to do with this case."[14] As a matter of legal evidence Darrow was undoubtedly correct. But as a matter of legal philosophy, his earlier comments certainly exposed the full implications of his view for the traditional criminal justice system.

Darrow had told the prisoners that there was no difference whatever in moral condition between themselves and those who were free. "I do not believe people are in jail because they deserve to be," he declared. "They are in jail simply because they cannot avoid it, on account of circumstances which are entirely beyond their control, and for which they are in no way responsible."[15] When speaking to Judge Caverly, Darrow focused much of his criticism on the operation of the death penalty. But speaking to the prisoners, Darrow did not feel so limited. "There ought to be no jails," he informed the prisoners, "and if it were not for the fact that the people on the outside are so grasping and heartless in their dealing with the people on the inside, there would be no such institutions as jails."[16] He added that he knew why "every one" of the prisoners committed his crime, even if he did not know the reason himself: "You did these things because you were bound to do them."[17] Those prisoners who thought they made a choice to commit a crime were simply deluded. "It looked to you at the time as if you had a chance to do them or not, as you saw fit; but still, after all, you had no choice." As if such statements were not inflammatory enough, Darrow even suggested that the police were the real criminals, and he concluded by claiming that pleasure was the ultimate basis for morality: "I believe that progress is purely a question of the pleasurable units that we get out of life. The pleasure-pain theory is the only correct theory of morality, and the only way of judging life."[18]

Crowe responded that the philosophy Darrow propounded to the prisoners was indistinguishable from the doctrines of his client Leopold, and he charged that society would self-destruct if the justice system ever adopted such ideas. Crowe lamented that it would have been better for Bobby Franks's murder to go unsolved than for the judge to place his official sanction upon the doctrines of anarchy preached by Clarence Darrow . . . Society can endure, the law can endure, if criminals escape; but if a court such as this court should say that he believes in the doctrines of Darrow, that you ought not to hang when the law says you should, a greater blow has been struck to our institutions than by a hundred, aye, a thousand murders.[19]

Despite Crowe's prediction of doom, Judge Caverly rejected the prosecution's call for the death penalty and gave the defendants life imprisonment instead. Nevertheless, the judge carefully refrained from embracing Darrow's scientific speculations. While noting that the scientific analysis presented by the lawyers and various expert psychiatrists at the hearing raised "broad questions of human responsibility and legal punishment," Caverly observed that these questions were more appropriate for legislative rather than judicial consideration. In the end, the judge said he based his decision largely on the young age of the defendants.[20]

From a narrow legal standpoint, the effect of the Leopold-Loeb case was negligible. Conducted as a sentencing hearing, the case set no precedent, and Judge Caverly carefully refused to break new ground in his decision. Yet from a broader cultural standpoint the case was a watershed. It publicized the far-reaching nature of the challenge posed by science to the traditional justice system. While Darrow's ideas likely sounded far-fetched to the great majority of Americans at the time, the ideas themselves were far from uncommon among the growing numbers of college-educated doctors and social scientists who were committed to the scientific study of crime and its causes. In fact, the Leopold-Loeb case merely brought out into the open the scientific ideas that were already forging a revolution in the traditional understanding of both crime and punishment. That revolution had deep roots in nineteenth-century biology.

Rise of the Scientific Study of Crime

A criminal "must first be found to be deserving of punishment before any consideration is given to the utility of this punishment for himself or for his fellow citizens," wrote Immanuel Kant at the close of the eighteenth century.[21] Kant's statement articulated the traditional underpinning of punishment in Western jurisprudence, according to which criminals were punished not simply to protect society, nor to deter future crimes, but because their actions were morally blameworthy.[22] Implicit in this view of punishment is the idea that most criminals have the freedom to choose whether or not to engage in criminal behavior. It is this moral freedom that makes it just for society to hold criminals morally accountable for their actions.

By Kant's time, reformers such as Cesare Beccaria and Jeremy Bentham were already deemphasizing the concept of punishment as the deserved penalty for committing a crime, preferring to defend punishment on utilitarian grounds of safety and deterrence.[23] But even the utilitarian reformers continued to assume that most criminals had the ability to choose whether or not to engage in crime, and that they were somehow morally blameworthy.

Nineteenth-century science challenged those assumptions head-on. In his notebooks Charles Darwin struggled with the serious consequences that scientific materialism posed for human free will and responsibility, but for the most part he chose to keep his misgivings to himself. Some of Darwin's followers, such as Ludwig Büchner, were more daring. A German medical doctor who became president of the Congress of the International Federation of Freethinkers, Büchner was an outspoken atheist and authored one of the nineteenth century's best-selling materialist tracts, *Force and Matter.* By the early 1890s, the book had gone through fifteen editions in German and four in English.[24]

In a chapter titled "Free Will," Büchner tried to show how humans had considerably less freedom than they thought. More nuanced than some of the eighteenth-century materialists, Büchner was willing to grant that "the more highly a man is intellectually developed and trained, the stronger is his will and the greater his responsibility."[25] Yet he still resorted to simplistic material explanations that sound like quackery to modern readers, such as suggesting that the "restless and almost feverish excitability" of Americans could be traced to "the excessive dryness of the air" in the United States.[26]

Turning his attention to crime, Büchner stated that "the vast majority of those who offend against the laws of the State and of Society ought to be looked upon rather as unfortunates who deserve pity than as objects of execration."[27] In his view, most criminals were compelled to break the law because of physically defective brains, lack of education and mental training, or poverty. "What is the good of free will to him who, from necessity or acting under the impulse of the irresistible tendency of self-preservation, lies, steals, robs, and murders? To what extent can a man be held accountable for his actions whose destructiveness and whose leaning towards cruelty are great, while his powers of reason are small?"[28] Büchner added that the brain abnormalities in many criminals showed that they were throwbacks to "the brains of pre-historic men."[29] He concluded by citing another writer who advised that "we should do best in neither judging nor condemning anyone." According to Büchner, future generations "will look back on the criminal trials of the present time with feelings akin to those with which we regard the trials for witchcraft and the inquisition of the

Middle Ages."[30] These themes would be repeated in subsequent decades by legions of scientifically inclined doctors, social scientists, and jurists.

Born Criminals

By the end of the nineteenth century, American scholars were talking with excitement about the "new school of criminal anthropology," which sought to use modern science to identify the causes of crime. Leading the way was Italian criminologist Cesare Lombroso (1835–1909), whose book *Criminal Man* (1876) remains a landmark work in the field of criminology. Lombroso and his disciples contended that criminal behavior could be explained largely as a throwback to earlier stages of Darwinian evolution. According to Lombroso, infanticide, parricide, theft, cannibalism, kidnapping, and antisocial actions can all be found throughout the animal kingdom, as well as among human savages.[31] In earlier stages of development such behaviors aided survival and were therefore bred into animals by natural selection. As William Noyes, one of Lombroso's American disciples, explained, "[I]n the process of evolution, crime has been one of the necessary accompaniments of the struggle for existence."[32] While crime no longer served a necessary survival function in civilized societies, many modern criminals could be considered atavists—humans in whom characteristics of earlier stages of evolutionary development reappeared. According to Lombroso, such atavists were "born criminals," exhibiting from birth the physical as well as behavioral characteristics of savages. Physical markers of such individuals included "abundant hair," "sparse beard[s]," "enormous frontal sinuses and jaws," "broad cheekbones," a "retreating forehead," and "voluminous ears."[33]

Lombroso was criticized for focusing on biology to the exclusion of environmental factors, and in subsequent works he made clear that he did not think that every criminal was a "born criminal." He also explored a variety of environmental determinants of crime, including climate, food, education, poverty, religion, and the use of alcohol. Nevertheless, Lombroso maintained that organic factors accounted for 35 to 40 percent of criminal activity, observing that the "so-called [environmental] causes of crime" were "often only the last determinants" of a crime, while "the great strength of congenital impulsiveness [was] the principal cause."[34]

Lombroso and his followers repudiated the traditional idea that "crime involved . . . moral guilt." Invoking modern science in general and the work of Charles Darwin in particular, Italian jurist Enrico Ferri (1856–1929), one of Lombroso's most celebrated disciples, argued that it was no longer reasonable to believe that human beings could make choices outside the normal chain of material cause and effect. Portraying the controversy over Darwin's ideas as nothing less than a battle between the forces of enlightenment and "the lovers of darkness," Ferri applauded Darwin for showing "that man is not the king of creation, but merely the last link of the zoological chain, that nature is endowed with eternal energies by which animal and plant life . . . are transformed from the invisible microbe to the highest form, man." Ferri looked forward to the day when punishment and vengeance would be abandoned and crime would be treated as a "disease."[35]

Although Lombroso and his followers wished to sever the link between crime and moral guilt, they did not think that criminals should be left alone to pursue their antisocial activities. Agreeing with the utilitarians that society must be able to defend itself against criminals, they merely insisted that measures taken against criminals be curative rather than punitive. The goal of criminal justice should not be to punish but to cure—at least in those cases where cures were possible. Lombroso and his supporters described their approach as much more humane than a system based on retribution, and pointed out the horrific practices (such as torture and capital punishment for even minor offenses) that the traditional theory of punishment had sometimes allowed. Yet the humanitarian theory of justice espoused by Lombroso had a darker side of its own.

Lombroso's emphasis on rehabilitation rather than punishment seemed to give doctors and psychiatrists almost absolute power over offenders and created the prospect that many criminals would be warehoused in perpetuity until their doctors certified them as "cured." And what happened to the criminals who could never be cured? Lombroso believed that rehabilitative efforts were completely useless when applied to the major proportion of offenders he categorized as "born criminals."[36] In his view, such persons might have to be locked up permanently.[37] But given Lombroso's view that born criminals were throwbacks to earlier stages of evolution, why should they even be allowed to live? "You have shown us fierce and lubricious orangutans with human faces," wrote one admirer to the Italian criminologist. "It is evident that as such they cannot act otherwise. If they ravish, steal, and kill, it is by virtue of their own nature and their past, but there is all the more reason for destroying them when it has been proved that they will always remain orangutans."[38] Lombroso quoted the words of his correspondent in one of his books without criticism.

Clearly, the superior humanitarianism of the new criminology was dubious. In the traditional view, adult criminals generally were regarded as responsible individuals who had the ability to make choices. In the new view, adult criminals were reduced to the level of "children" and "moral invalids" who needed to be treated by experts.[39] In the traditional view, sentences for non-capital offenses were supposed to be specific and limited. In the new medical model of criminal justice, sentences should be indefinite, because one could not tell beforehand how long it would take to effect a cure. In the traditional view, with its sharp distinction between guilt and innocence, procedural safeguards were regarded as essential lest innocent people be wrongly convicted. But in Lombroso's scientific justice system, these safeguards were regarded as an inefficient interference with the scientific enterprise, and thus he attacked such basic components of the traditional justice system as juries, public trials, the right of appeal, and the adversarial process that allows for a separate prosecution and defense.[40]

Defenders of the traditional view found the new approach anything but humanitarian. Writing after the medical model for criminal justice gained ascendancy in the twentieth century, C. S. Lewis lamented that the new view, "merciful though it appears, really means that each one of us, from the moment he breaks the law, is deprived of the rights of a human being."[41]

Whatever the deficiencies of his theories, Lombroso was a seminal figure in the founding of the scientific study of crime. Perhaps his most important role was helping to inaugurate criminology's quest for the Holy Grail—the search for the material basis of crime. Although many of Lombroso's particular findings were quickly superceded, the professional literature of the last hundred years is littered with studies purporting to identify the biological, chemical, psychological, and environmental causes of crime. That literature makes for interesting reading, because it shows the lengths to which social scientists were willing to go in applying the tenets of scientific materialism, even on the thinnest of evidence.

American researchers inspired by Lombroso were soon making the case for the physical and hereditary foundations of crime. In 1894, American physician James Weir published a study in the *Medical Record* seeking to corroborate Lombroso's approach of identifying criminals by bodily characteristics. Weir paid special attention to ear size. Noting that an Italian researcher had found "that fully eighty per cent of criminals have abnormal ears, and that large ears indicate a murderous tendency, while small ears indicate a thievish nature," Weir stated that his own observations confirmed these findings. "I have never seen a large ear in an habitual thief, nor have I ever seen a small ear in a murderer."[42] Weir's study contained numerous other nuggets of wisdom, including a claim that "with an abnormal brain we may always expect to find an abnormal ear. Wardens and keepers of penitentiaries have often told me that they could tell a thief from a murderer by the size and shape of the ears."[43] Weir further reported that criminals had a "great lack of sensibility to pain," and according to prison doctors, even prefer to be operated on without anesthetic (at least in "minor operations")! Moreover, criminals are "like the lower animals" in the "rapidity with which [they] recover from wounds."[44]

Embracing Lombroso's theory of atavism, Weir asserted that the "instinctive criminal . . . is antisocial simply because he is morally and mentally an atavism . . . Atavism has hurled him back thousands and thousands of years, and has placed him beside his pithecoid ancestor."[45] Weir ended by contrasting the consideration given the "intellectually insane," who are housed in "asylums where they are presumably treated with kindness," with the treatment of the criminally insane, who are either executed or incarcerated in "penitentiaries, prisons, and workhouses." "Poor unfortunates!" he exclaimed. "Are they to be held responsible for Nature's vagaries? Enlightened humanity has a difficult problem to solve, when it is asked the question, 'What will you do with the Criminal?' "[46]

In 1913, British prison doctor William Goring discredited many of Lombroso's specific findings about the physical characteristics of criminals with a study of his own comparing criminals to noncriminals. However, Goring's study was far from a rejection of the biological approach to crime. Despite its criticisms of Lombroso, the study reaffirmed the existence of certain physical differences between criminals and noncriminals; and Goring estimated that 60 percent of criminal behavior could be assigned to heredity as opposed to environment.[47]

While biological theories of crime remained controversial during the early 1900s, the general idea that there was a hereditary component to crime became widespread in America. The result was a vigorous movement to curtail childbearing by criminals and others considered socially undesirable, lest the undesirables bequeath their antisocial tendencies to their progeny.[48] By the 1920s and '30s, the search for physical factors in crime gradually turned from heredity to body chemistry, with particular attention being paid to endocrine glands, which are involved in regulating the body's metabolism. Psychiatrists testifying for the defense in the Leopold-Loeb case argued that endocrine abnormalities played a role in the mental condition of the defendants.[49] In a paper presented to the Society for Medical Jurisprudence in 1931, Dr. Louis Berman proposed that given the link between endocrine glands and crime, "it should be possible to modify, palliate, or even completely abolish criminal behavior by means of correction of the discovered glandular malfunctioning."[50] Berman urged that every child be subjected to periodic inspections of his or her endocrine glands in order "to prevent the appearance of criminal tendencies and criminals." In his view, "early diagnosis and treatment in childhood and adolescence will eliminate the potential criminal."[51]

In the 1950s, psychiatrist Edward Podolsky updated the findings relating to body chemistry in a journal article titled "The Chemical Brew of Criminal Behavior." Stating that "not all human behavior is chemically determined," Podolsky added somewhat contradictorily that "life, in the final analysis, is a series of chemical reactions." He went on to catalog various chemical triggers of crime, including low blood sugar ("a father almost strangled his daughter during one of these sugarless episodes"), a lack of calcium ("Quite often violent tempers have their roots in calcium deprivation"), and low levels of lime ("the person whose lime level is below normal has a tendency to shout and scream and strike").[52] Acknowledging that "the biochemical evaluation of the criminal personality and of criminal behavior" was "still in its infancy," Podolsky held great hopes for its future efficacy in the treatment and eradication of crime.[53]

Psychosocial Explanations of Crime

Despite continued interest in the biology of criminal behavior, as the twentieth century progressed American criminologists increasingly turned to psychological and sociological approaches to explain crime. This shift away from biology was probably accelerated by the backlash against the horrific eugenics program implemented by Nazi Germany; moreover, as James Q. Wilson and Richard Herrnstein point out, sociological causes of crime were regarded as easier to fix than biological causes.[54] Yet from the perspective of the traditional theory of punishment, the new approaches were just as problematic. Modern psychology and sociology were largely founded on the tenets of scientific materialism, and neither discipline left much room for free will or moral responsibility.

Freudian psychology, which became prominent in the field of criminology, was a case in point. Freud's theory of psychic determinism was just as materialistic as explicitly biological explanations of human behavior. Indeed, Freud took Darwinian biology as his foundation. Praising Darwinian biologists for demonstrating man's "ineradicable animal nature," he made clear that psychoanalysis was designed to expose the unconscious roots of human behavior in man's biological instincts, especially his drives for sex and self-preservation.[55] Freud drew inspiration from the popular biology of his time for a number of his specific ideas. For example, he adapted for his purposes the now-discredited "law of ontogenesis," maintaining that "each individual repeats in some abbreviated fashion during childhood the whole course of the development of the human race."[56]

Freud even posited the existence of "primal phantasies," in which a person recalls "true prehistoric experiences" that have been embedded in his heredity through the evolutionary process.[57] The popular fascination with this latter idea in America can be seen in Jack London's fantasy novel *Before Adam* (1907), in which the narrator relives the life of one of his prehistoric ancestors through his nightmares. "It was not till I was . . . at college," writes the fictional protagonist,

> that I got any clew [sic] to the significance of my dreams, and to the cause of them . . . [A]t college I discovered evolution and psychology, and learned the explanation . . . For instance, there was the falling-through-space dream . . . This, my professor told me, was a racial memory. It dated back to our remote ancestors who lived in trees. With them, being tree-dwellers, the liability of falling was an ever-present menace. Many lost their lives that way . . . Now a terrible fall [that was] averted . . . was productive of shock. Such shock was productive of molecular changes in the cerebral cells. These molecular changes were transmitted to the cerebral cells of progeny, [and] became, in short, racial memories.[58]

Freudians who applied psychoanalysis to crime were explicit about the material nature of the drives that determined human behavior and ruled out free will. "The source of the instincts is our physical condition, the chemical-physical state of our body," wrote Freudian physician David Abrahamsen in *Who Are the Guilty?*, published in 1952.[59] Abrahamsen demonstrated how the psychoanalytical approach to crime could be just as reductionist as a strictly biological approach. Offering up story after story to show that criminals are largely the victims of their unconscious drives, Abrahamsen argued that "the criminal very rarely knows completely the reason for his conduct. Only when he has learned about himself (specifically, through psychoanalysis) will he become aware of his motivations."[60]

Discussing the case of a teenage boy who repeatedly stole cars ostensibly "to get even with people," Abrahamsen shared how psychoanalysis revealed the real cause of the boy's crime. Under the professional guidance of Abrahamsen, the boy came to understand that he stole cars so that he could search for the mother he never knew. At an even deeper level, Abrahamsen came to believe that "a car in itself had a meaning" for the boy because "in a symbolic way it represented his mother. Since he could not get her, he had to have a substitute, and a car was that substitute. Hence his thrill every time he drove a car; hence his continuous stealing." While Abrahamsen acknowledged that "to the average reader such an explanation may seem farfetched," he assured his readers that "clinical experience shows that a car often stands for a woman." His evidence for this claim? "In daily language we often call a car 'she' " and "when we have the tank filled with gasoline we say 'Fill her up!' "[61]

If Freudian psychology undermined the case for moral accountability, behavioral psychology, championed in the twentieth century by such figures as John Watson and B. F. Skinner, was even less open to what Skinner dismissively called the "autonomous man."[62] Behavioral psychology made no apologies for describing human beings as completely determined by their heredity in conjunction with environmental conditioning.

While there have been recent changes in the discipline of psychology, its underlying materialism—with the concomitant suspicion of personal responsibility—remains largely unchallenged.[63] "Nearly all psychologists believe that behavior is completely determined by heredity and environment," noted John Staddon, professor of psychology at Duke University during the 1990s. "A substantial majority agree with Skinner that determinism rules out the concept of personal responsibility. This opposition between determinism and responsibility is now widely accepted, not just by behaviorists but by every category of mental-health professional."[64]

The sociological study of crime was further removed from biology than the psychological approach, but it too was reductionistic and rested largely on the foundations of scientific materialism. In the place of biological determinants, sociologists studying crime stressed poverty, alcohol, age, urbanization, family, culture, education, and other allegedly causal factors.

The class-conflict model was typical of sociological explanations of crime, and it was just as materialistic as the typical biological and psychological models. According to the proponents of this view, the fierce struggle for material goods accounted for much, if not most, of criminal behavior. A sociology text published in 1949 told readers about research indicating that "a materialistic culture in which people admire the winners of a bitter, competitive struggle and which allows some of those who are defeated in the strife to collect in slums, tends to produce just about the criminals that are to be found in a nation at any given time."[65] The author added that when wealthy businessmen "make millions out of speculation or gambling, when their high finance causes thousands to lose hard-earned savings, and when some . . . run away from their country in order to escape trial and punishment . . . the average man is likely to feel that such examples justify him in almost any antisocial conduct."[66]

Despite the ascendancy of psychological and sociological approaches to crime, interest in more directly biological explanations revived during the 1960s and '70s. During this period, research into the biomedical aspects of behavior gained adherents even among those who preferred other approaches, such as psychoanalysis.

The Return of Biology

Writing in the *California Law Review* in 1962, prominent psychoanalyst Bernard Diamond acknowledged that brain studies had now convinced him that many "habitual criminals" and "sociopaths" were "suffering from an organic, neuro-physiological disease of the brain, which completely dominates their behavior. Their appearance of normalcy, their apparent ability to exercise free will, choice, and decision . . . is purely a façade, an artifact that conceals the extent to which they are victims of their own brain pathology."[67]

Diamond predicted that the criminal justice system was about to undergo a revolution; "within ten years biochemical and physiological tests will be developed that will demonstrate beyond reasonable doubt that a substantial number of our worst and most vicious criminal offenders are actually the sickest of all." The demonstration of this fact would force both legal professionals and the general public, "whether they like it or not," to "accept the fact that large numbers of individuals who now receive the full, untempered blow of social indignation, ostracism, vengeance, and ritualized judicial murder are sick and helpless victims of psychological and physical disease of the mind and brain."[68]

Interest in the connections between brain pathology and crime continued to grow, and in 1970 psychiatrist Frank Ervin and neurosurgeon Vernon Mark published their provocative book *Violence and the Brain*, which claimed that brain dysfunction might be at the root of some inner-city violence.[69] There also was a revival of attempts to find a genetic basis for crime, including controversial studies suggesting that men who had an extra "Y" chromosome were more likely to be criminals.[70] By the 1980s and '90s, the biological model of criminal behavior was back with a vengeance, bolstered by the burgeoning field of sociobiology and government-funded research in genetics and neurobiology.[71]

Science and the Denial of Responsibility

While the past hundred years have seen a variety of approaches to the scientific study of crime, perhaps the most striking fact about criminology during this period is not its variation but its uniformity. Regardless of the particular models adopted, criminologists almost universally agreed that free moral agency had nothing to do with crime, and that it was therefore nonsensical to talk as if criminals were somehow morally blameworthy for their actions. These social scientists self-consciously contrasted their "scientific" study of crime with earlier "speculations" by theologians and

philosophers. Sociologist J. P. Shalloo complained in 1942 that such speculations about the supposed sinfulness of criminals "probably set back our
understanding of human conduct at least 500 years." In Shalloo's view, it
was the "world-shaking impact of Darwinian biology, with its emphasis
upon the long history of man and the importance of heredity for a clear
understanding of man's biological constitution," that finally opened the
door to a truer understanding of crime.[72]

Social-science textbooks from the 1930s onward show just how widespread the rejection of traditional notions of moral accountability was.
Sometimes the debunking of free will and accountability was explicit, especially among those enamored of biological and physiological approaches to human behavior. In *Crime: Criminals and Criminal Justice* (1932), University of Buffalo criminologist Nathaniel Cantor ridiculed "the grotesque
notion of a private entity, spirit, soul, will, conscience or consciousness
interfering with the orderly processes of body mechanisms." According to
Cantor, man is just as much a "part of nature" as rocks or birds. "Certain
protoplasm feeds on milk, grows feathers and cackles. Other protoplasmic
organizations feed on milk, grow skin and talk. Physiological, biological
and neurological law underlies one as well as the other. The mechanisms
of human behavior, though perhaps more complex, are subject to the same
laws of cause and effect as the sun, moon, and other stars."[73] Because of
this, "man is no more 'responsible' for becoming wilful and committing a
crime than the flower for becoming red and fragrant. In both cases the end
products are predetermined by the nature of protoplasm and the chance
of circumstances."[74]

Discussions of crime in sociology textbooks were less likely to contain
direct denials of free will and personal responsibility, but they often left
readers with little basis on which they might continue to believe in such
concepts. If the idea of personal responsibility was broached at all, it was
likely to be treated as inconsequential or qualified so severely as to be
meaningless. Consider the discussion in *Sociology: A Synopsis of Principles*
(1947) by John Cuber. Under a subsection titled "Personal Responsibility,"
Cuber explained:

> Many persons fail to measure up to the minimum standards required by
> the society . . . In common parlance, some of the inadequacies are presumed
> to be the person's "faults," like crime, while others are presumed not to be
> the person's fault, like illness. Objective examination of the facts, however,
> readily reveals that in numerous instances these assumptions completely
> break down. Some criminals may commit criminal acts as a result of a
> chain of circumstances over which they have had no real opportunities to
> exert control . . . The better acquainted one becomes with the environmental

forces which operate through culture and unique experience the less inclined he is to speak glibly about a person's "responsibility" for either his
inadequacies or his distinctions.

After this debunking of personal responsibility, Cuber added a tangled
disclaimer that "[i]t is not meant to deny that humans are not in some
measure responsible for their acts."[75] Framed as a triple negative, the
disclaimer was little more than window-dressing. Nevertheless, even this
mild disclaimer was apparently too robust for Cuber, for in future editions
of his text it was watered down. Instead of declining "to deny that humans
are not in some measure responsible for their acts," now Cuber was only
willing to state that he declined "to deny that humans are not in some
measure apparently responsible for their acts."[76]

For much of the early part of the twentieth century, the abandonment
of traditional concepts of moral responsibility by social scientists studying
crime had a limited effect on the operation of the criminal-justice system.
Although several reforms proposed by criminologists were adopted (such
as an increased emphasis on rehabilitation, and the implementation of parole and probation), the criminal-justice system in America remained substantially punitive in nature, and judges and legislators resisted opening
the door too widely to psychiatrists and other social scientists. In 1911, a
distinguished committee of legal scholars under the auspices of the American Institute of Criminal Law and Criminology arranged for the publication of Cesare Lombroso's *Crime: Its Causes and Remedies* in English as
part of a general series of international works in the field of criminology.
In their introduction to the series, committee members complained that
"the public in general and the legal profession in particular have remained
either ignorant of . . . or indifferent to the entire scientific movement" to
study crime.[77]

By midcentury, however, the new scientific understanding of crime began to exert more influence, as courts increasingly grappled with issues of
insanity, drug addiction, and alcoholism.

4

Crime as Mental Illness

"I am a man of destiny," declared Charles Guiteau to a packed courtroom in the District of Columbia in 1881. "I am a man of destiny as much as the Savior, or Paul, or Martin Luther . . . [T]he Deity never employs a fool to do His work."[1]

By almost any definition, Charles Guiteau was batty. Convinced he was an instrument in the hands of God, Guiteau believed that the Almighty had given him a special commission to kill President James Garfield. He purchased a handgun, and shortly thereafter he walked up to President Garfield at Grand Union Station and shot him. Guiteau's lifelong erratic behavior, as well as his repeated strange outbursts throughout the trial, bore witness that, mentally, something was not quite right. The defense called medical experts who were up to date on the latest scientific theories to declare that Guiteau was in fact insane. Yet the prosecution was able to overwhelm the defense with its own medical and scientific testimony proving Guiteau's sanity under the legal standards of the time. It took the jury less than an hour to decide: Guiteau should be hanged.

The Guiteau trial displayed how the new scientific understanding of criminal responsibility was still struggling for legal and social respectability in the latter half of the nineteenth century. Even by the 1920s, when Leopold and Loeb were prosecuted, the scientific understanding of criminal causation had not had much effect on the determination of legal culpability. Clarence Darrow was able to present a much more impressive array of experts articulating the new science's case for diminished responsibility. But he only succeeded in sparing his clients from the death penalty, and Judge Caverly indicated that it was the defendants' age, not the scientific testimony, that convinced him to sentence Leopold and Loeb to life imprisonment.

To the frustration of many social scientists, the American legal system was slow, skeptical, and cautious about rewriting the legal standards of insanity and responsibility to make them consistent with the findings of modern social science. The "M'Naghten rules" constituted the primary bulwark against change. These rules were formulated in an 1843 English case involving a delusional defendant named Daniel M'Naghten, who had attempted to assassinate the British prime minister because he thought the PM was conspiring against him. M'Naghten mistakenly killed the prime minister's secretary instead. The M'Naghten rules were quickly adopted by American courts and placed sharp limits on who could assert the insanity defense. The most basic M'Naghten rule (also known as the "right-wrong" test) was that a defendant invoking the insanity defense must show that at the time of the crime he "was labouring under such a defect of reason, from disease of the mind, as not to know the nature and quality of the act he was doing; or, if he did know it, that he did not know he was doing what was wrong."[2] One can see how narrow the M'Naghten standard was by applying it to the Leopold-Loeb case. Leopold and Loeb knew that they were committing murder and that murder was considered wrong. Those facts could be enough to convict them under M'Naghten. Proof that their murder of Bobby Franks was the product of a compulsion arising from mental disease would be legally irrelevant.[3]

Almost from the start, defense attorneys and medical experts condemned the M'Naghten test as too limited and urged judges to allow testimony about a much wider range of mental defects. American courts slowly accommodated themselves to the new findings of science. While most state courts continued to apply the M'Naghten test, some jurisdictions began to experiment with looser definitions of insanity. In the early 1870s, New Hampshire courts concluded that "no man shall be held accountable, criminally, for an act which was the offspring and product of mental disease."[4] New Hampshire's rule boldly junked M'Naghten's "right-wrong" test and gave juries maximum discretion to make their own determinations about what constituted insanity without interference from judges.

Other courts pursued more incremental reforms. Some jurisdictions supplemented M'Naghten with versions of the "irresistible impulse" defense, allowing defendants to claim that a mental condition deprived them of their wills and compelled them to perpetrate a crime.[5] Many judges found even this change too radical, and the concept of "irresistible impulse" was rejected in many states.[6] A more subtle method for expanding protections for the criminally insane was to allow evidence of mental illness to prove a lack of criminal intent. For example, a defendant in a murder case might claim that a mental defect deprived him of the requisite criminal intent needed to show murder in the first degree.[7]

Yet another method of loosening the M'Naghten standards was to shift the burden of proof for the legal determination of sanity. Under M'Naghten, defendants were typically presumed sane. If they wished to employ the insanity defense, the burden was on them to affirmatively show that they were insane during the commission of the crime.[8] Under the new approach, the defendant was automatically presumed insane once any evidence was presented to suggest insanity. It then became the state's obligation to prove the defendant sane beyond a reasonable doubt. New Hampshire adopted this rule in 1861, and the Supreme Court embraced it as a rule for federal courts in *Davis v. United States* (1895).[9]

By the 1950s, thirty states still clung to the M'Naghten test, but more changes were on the horizon.[10] In *Durham v. United States* (1954), the United States Court of Appeals for the District of Columbia Circuit adopted a standard similar to the New Hampshire test, ruling that "an accused is not criminally responsible if his unlawful act was the product of mental disease or mental defect."[11] This new standard seemingly opened the door to an almost limitless range of claims under the insanity defense, and the state had the burden to prove not only that the defendant perpetrated the crime but that the defendant was sane at the time of the crime. The court declared that as long as there is " 'some evidence' that the accused suffered from a diseased or defective mental condition," jurors must find the defendant not guilty unless the state showed beyond a reasonable doubt either that the defendant "was not suffering from a diseased or defective mental condition, or that the act was not the product of such abnormality."[12]

The *Durham* Legacy

Acquittals in insanity cases in the District of Columbia skyrocketed after *Durham* was announced. In 1954, a mere 0.4 percent of cases tried in the District resulted in verdicts of "not guilty by reason of insanity." By 1961, an astonishing 14.4 percent of cases yielded similar acquittals.[13]

Yet the D.C. court of appeals continued to press the limits of the insanity defense in the 1962 case *Campbell v. United States*.[14] Eugene Campbell had been convicted of two counts of robbery. At his trial, defense psychiatrists from St. Elizabeth's hospital testified that Campbell suffered from an "emotionally unstable personality," a disorder that had been recognized by the hospital only since 1957. The basis of the diagnosis included findings that Campbell was "extremely emotional," "tense and anxious," and "perspiring profusely." One psychiatrist added that Campbell "may demonstrate poor judgment" and that he "has difficulty in his relationships with the people around him."[15] Presumably one does not need to be a trained psychiatrist to ascertain that a robber shows "poor judgment" and has "difficulty in his relationships."

To its credit, the court of appeals was unimpressed by the quality of the flimsy expert testimony. Nevertheless, the court did not reject "emotionally unstable personality" as a legitimate mental disease covered by the *Durham* test. Nor did it affirm Campbell's conviction, overturning the jury's verdict because of an allegedly faulty jury charge by the trial judge. Here is where things get "curiouser and curiouser," to use the words of Lewis Carroll's Alice. According to the appellate court, the trial judge was wrong to instruct the jury that if the defendant "could have controlled" his impulse to rob "and refrained from doing the act" then he was to be considered legally "responsible for his act."[16] The appellate court was upset that "the jury was . . . in substance told that the defendant should be held criminally responsible if he could have controlled an impulse to rob." The court branded this idea "patently erroneous."[17]

"What can be 'patently erroneous' about telling a jury that if the defendant 'could have controlled an impulse to rob' he 'should be held criminally responsible'?" demanded Appellate Judge Warren Burger in exasperation.[18] In a blistering dissent, the future chief justice of the U.S. Supreme Court excoriated his colleagues on the appellate bench for repudiating "everything this court has uttered on the subject of free will and criminal responsibility."[19] "The majority view appears to be that an act can be the product of disease even though the defendant could have controlled it by the exercise of will. This simply is not correct."[20] Burger complained that the majority's holding was "a long step in the direction of substituting a determinist philosophy for the belief . . . that all law in Western civilization is 'guided by a robust common sense which assumes the freedom of the will as a working hypothesis.'"[21]

Burger further chastised the majority for "giving judicial approval to the extraordinary process by which a small segment of the medical profession alters the scope of the law of criminal responsibility by the simple device of 'administratively' expanding the definition of mental disease from time to time."[22] Burger pointed out that the only basis for accepting "emotionally unstable personality" as a mental disease was an administrative decision of the director of St. Elizabeth's hospital to reclassify it as such. No medical evidence had been presented justifying the reclassification or demonstrating that it represented a consensus of the medical profession. Burger was clearly worried that such arbitrary reclassifications were becoming a habit for the staff at St. Elizabeth's. In 1957, the hospital had reclassified one disorder right in the middle of a trial. According to one account,

> in what came to be know[n] as the "weekend flip-flop case," a St. Elizabeth's psychiatrist on Friday afternoon testified that the defendant, diagnosed as a psychopath, was not suffering from a mental disease. The following Monday morning the St. Elizabeth's staff decided to classify psychopathic personality as a mental disease. Returning to the stand Monday afternoon, the psychiatrist changed his testimony and declared that the defendant was suffering from a mental disease.[23]

Given that these reclassifications did not result in any change in treatment for affected patients, Burger suggested that psychiatrists at the institution had "opened themselves to the dual charges of abandoning medical concepts and usurping judicial functions in order to bring about acquittals which they or their superiors consider appropriate."[24] Thus did Burger skillfully expose the credulity with which his colleagues accepted claims made in the name of science, even on the thinnest evidence.

Few jurisdictions embraced the expansive insanity test pushed by the reformers on the D.C. circuit court, but another proposal for a liberalized insanity test proved to be much more influential. The same year *Campbell* was decided, the American Law Institute (ALI) proposed the final draft of a sweeping new insanity standard of its own in its Model Penal Code. According to the ALI test, a defendant should not be held responsible for conduct resulting from "mental disease or defect" if "he lacks substantial capacity either to appreciate the criminality [wrongfulness] of his conduct or to conform his conduct to the requirements of the law."[25] The "substantial capacity" standard opened the insanity defense to a much greater variety of claims than had previous rules, although drafters of the ALI test tried to prevent abuse by adding that "the terms 'mental disease or defect' do not include an abnormality manifested only by repeated criminal or otherwise antisocial conduct."[26]

But perhaps the most ambitious initiative to overturn the traditional approach to legal responsibility in the 1960s began with a case that on the surface had nothing to do with insanity. Instead, the issue was illegal drugs, and the controversy would go all the way to the United States Supreme Court.

Innocent by Addiction

On the night of February 4, 1960, police officers arrested Lawrence Robinson in Los Angeles on suspicion of being a user of narcotics.[27] Robinson was subsequently charged with violating California's health and safety code, which forbade the nonmedical use of narcotics, being "under the influence of" narcotics, and being "addicted to the use of narcotics."[28]

Prosecutors sought to show at trial that Robinson had used narcotics and that he was addicted to narcotics. To this end, the arresting officer testified that Robinson had admitted his repeated narcotics use to him, and a member of the Narcotics Division of the Los Angeles Police Department who examined Robinson the morning after his arrest offered expert testimony about the freshness of needle marks and scabs on the defendant's arms. The narcotics officer also testified that Robinson admitted using narcotics to him. At trial, Robinson denied using narcotics and insisted that the discolorations on his skin were caused by an allergic condition.

The judge instructed jury members that in order to convict Robinson they must either conclude that he had used narcotics within the jurisdiction of Los Angeles County or that "he was addicted to the use of narcotics." Jurors were told that for a conviction they needed to reach unanimous agreement on one of these two offenses, but they did not need to reveal which offense they had agreed upon in their verdict. Robinson was subsequently convicted of a misdemeanor and sentenced to two years of probation, the first ninety days of which were to be served in jail, and he was required to submit to periodic drug testing to ensure that he stayed drug-free during his probation.

After exhausting all of the options for appeal within the state of California, Robinson's attorney, Samuel Carter McMorris, filed an appeal with the United States Supreme Court. McMorris argued that criminal penalties for drug addiction violated several provisions of the U.S. Constitution, including the ban on cruel and unusual punishment.[29] In McMorris's view, the fundamental question posed by the case was simple: "May the legislature punish as a crime a mental and physical illness? That such is the proper definition of addiction the court may take judicial notice from unanimous medical and scientific treatment of the subject."[30] McMorris argued in his brief to the Supreme Court that California was unjustly punishing people for their "status" as drug addicts rather than for their conduct.[31] Pressing his argument further, McMorris drew a connection between addiction and insanity, going so far as to claim that "addiction is in fact insanity," and "to penalize the condition of addiction as such is about as advanced as the burning at the stake of 'witches' (insane persons) during colonial days."[32] In a rather disconcerting section of his brief, McMorris heaped praise on the laws of Nazi Germany for having adopted the legal standard that he was arguing for, pointing out that the German civil code in the 1940s defined insanity to include "the degeneration of character of the morphine addict."[33] (Presumably McMorris didn't realize that being classified as mentally ill made one a target for forced sterilization and even euthanasia under Germany's eugenics program.[34])

McMorris's arguments raised many more questions than they answered. To begin with, his grounds for distinguishing "status" and "conduct" were murky at best. California courts interpreted "addiction" to mean habitual use, and at trial the prosecution had presented evidence of Robinson's repeated use of narcotics. Thus, the so-called "status" of addiction in California was explicitly defined by conduct, which seemed to make hash of McMorris's attempt to sharply distinguish between the two concepts. Moreover, if status and conduct were so intertwined, it was hard to see how the application of McMorris's underlying rationale could be confined only to crimes of "status." If it was inherently unjust (and unconstitutional) for the state to criminalize the "status" of addiction because it was a mental illness, then how could the state justifiably criminalize the conduct resulting from that addiction—or from any other mental or physical illness, for that matter?[35]

In his oral argument before the Supreme Court, McMorris agreed that his underlying rationale would apply to behaviors generated by a variety of mental maladies. But that was not a drawback as far as he was concerned. Making clear that his agenda was far broader than simply ending the punishment of "status" crimes, McMorris let his purported distinction between status and conduct unravel as he argued that the government in fact had no right to punish thefts arising from kleptomania or fires traceable to pyromania. "If a person does in fact commit the act of theft, or an act of arson, or any act because of a provably mental condition, such persons, I think, should be treated and not punished," he insisted to the justices. According to McMorris, this exemption from punishment should apply especially in cases of alcohol and narcotics addiction, "because these are clearly and totally mental and physical illnesses which cry aloud for help, for treatment, for hospitalization, rather than for the type of punishment which we here applied."[36] Not a single justice challenged McMorris's expansive claims.

McMorris also told the court that he favored civil-commitment laws to compel addicts to get treatment for their addictions. How serious he was in this proposal can be questioned. Presumably the goal of treatment is to effect a cure, yet McMorris acknowledged that under existing medical knowledge narcotics addiction might be incurable, that it might be a "lifetime condition."[37] Did that mean that addicts could be forcibly committed for the rest of their lives? It is hard to believe that McMorris would have espoused such a view. It is more likely that he would have used the futility of treatment to argue against civil commitment, notwithstanding his argument before the court.[38]

On the surface, the result sought by McMorris in *Robinson v. California* was relatively narrow: the invalidation of drug addiction as a crime. Given that most states preferred to punish possession and use rather than addiction, the initial effect of invalidating drug-addiction statutes might have been negligible.[39] Yet the stakes were considerably higher than the specific facts of the case suggested. The arguments employed by McMorris could be applied to a whole range of cases dealing with mental illness and criminal responsibility. More importantly, by striking down addiction laws as unconstitutional the court would transform the issue of criminal responsibility from a state and local matter into a constitutional issue to be litigated in federal courts. States that refused to keep up with the findings of modern psychiatry and expand the applicability of the insanity defense would now be subject to second-guessing by federal judges. Defendants employing the mental-illness defense would have another ground for appeal to the federal courts after exhausting state remedies.

Robinson v. California had the potential to be a nightmare for prosecutors, but the Supreme Court under Chief Justice Earl Warren was entering one of the most judicially creative phases in its history, and it was in no mood to be cautious. From the 1960s to the early 1970s, the court issued an array of landmark decisions giving federal courts dramatically expanded authority over many areas formerly handled almost wholly at the state and local levels. These decisions included *Mapp v. Ohio* (1961), excluding illegally seized evidence from state courts on the basis of the Fourth Amendment; *Abington v. Schempp* (1963), banning state-sponsored Bible-reading as a violation of the Establishment Clause of the First Amendment; *Miranda v. Arizona* (1966), requiring the police to advise arrestees of their Fifth Amendment right against self-incrimination; and *Roe v. Wade* (1973), striking down state laws forbidding abortion as unconstitutional on the basis of an unenumerated "right of privacy." In fact, on the very day the court issued its ruling in *Robinson*, it also handed down *Engel v. Vitale*, striking down teacher-led prayer in public schools.[40]

The court's ruling in *Robinson* was no less dramatic. Writing for the majority, Justice Potter Stewart announced that California's law against addiction violated the Constitution's ban on cruel and unusual punishment because it sought to criminalize the "illness" of narcotics addiction. According to Stewart, laws against narcotics addiction were equivalent to making "it a criminal offense for a person to be mentally ill, or a leper, or to be afflicted with a venereal disease."[41] The court further embraced Mc-

Morris's strained distinction between "status" and behavior and claimed that the California law was impermissible because it tried to criminalize a "status," despite the fact that this "status" was explicitly defined by the trial court in terms of behavior.[42]

Seeking to assuage the concerns of state governments, the court did list a variety of other ways that they could continue to address the drug problem. "A State might impose criminal sanctions . . . against the unauthorized manufacture, prescription, sale, purchase, or possession of narcotics within its borders." It might also "establish a program of compulsory treatment for those addicted to narcotics. Such a program of treatment might require periods of involuntary confinement. And penal sanctions might be imposed for failure to comply with established compulsory treatment procedures."[43]

Only Justices Tom Clark and Byron White filed dissents. Justice White challenged the court for adopting the "status" argument in this case, noting that "as defined by the trial court, addiction is the regular use of narcotics and can be proved only by evidence of such use."[44] In White's view, the claim that California had criminalized a "status" distinct from a behavior was simply untenable on the facts. In addition, White questioned the reach of the court's ruling, saying that "the Court's opinion bristles with indications of further consequences. If it is 'cruel and unusual punishment' to convict appellant for addiction, it is difficult to understand why it would be any less offensive . . . to convict him for use on the same evidence of use which proved he was an addict." But perhaps the court intended to prevent states from punishing the use of narcotics as well, suggested White. "It is significant that, in purporting to reaffirm the power of the States to deal with the narcotics traffic, the Court does not include among the obvious powers of the State the power to punish for the use of narcotics. I cannot think that the omission was inadvertent."[45] White finally chastised the court's majority for "imposing its own philosophical predilections upon state legislatures" and Congress. "I deem this application of 'cruel and unusual punishment' so novel that I suspect the Court was hard put to find a way to ascribe to the Framers of the Constitution the result reached today."[46]

The decision came too late to matter to Lawrence Robinson. The man who had steadfastly insisted at trial that he did not use narcotics died in a Los Angeles alley of an apparent drug overdose on August 5, 1961. That was nearly a year before the Supreme Court decided his case, and nearly nine months before oral arguments were presented.[47] Members of the court were unaware that Robinson was no longer alive when they heard his appeal, and Robinson's attorney, Samuel McMorris, did nothing to inform them of the fact. Indeed, during oral arguments McMorris was asked by a justice about Robinson's current status, and McMorris replied, "He's still on probation at this time."[48] McMorris later claimed, incredibly, that he had been unaware of his client's demise.[49] A more likely scenario was that McMorris wasn't about to squander his chance at overturning the addiction statute by admitting that he no longer had a live case. Whatever the real story, like many classic Supreme Court cases, *Robinson v. California* was less about justice for the individual defendant than about establishing a beachhead in the law for a certain legal principle. Eventually informed of Robinson's death, the Supreme Court refused to vacate its decision, despite the objections of three of its members—including Justice Stewart, who had written the majority opinion.[50]

As soon as the court issued its decision in *Robinson v. California*, law reviews began to percolate with interest about the ways in which the case could be applied to new situations. Several thought that *Robinson* made prosecutions of the use or possession of narcotics dubious,[51] and a number referenced the applicability of *Robinson* to laws involving alcoholism.[52] Other issues to which *Robinson* was thought to apply included homosexuality,[53] the actions of kleptomaniacs and pyromaniacs,[54] and mental-illness defenses in general.[55] While some law-review authors were cautious in predicting the impact of *Robinson*, others thought the decision should spark a fundamental reevaluation of the criminal-justice system. "Robinson makes it incumbent that the objectives and methods of our whole criminal law system be re-examined," urged an article in the *Nebraska Law Review*. "Indeed, to the extent that nearly all criminal activity is the result of emotional or social maladjustment, perhaps some future generation will see the termination of penal sanctions as we know them today."[56]

Because of *Robinson*, courts in at least five states as well as the District of Columbia soon had to deal with constitutional challenges to laws prohibiting the possession and use of narcotics.[57] A California defendant even claimed immunity from prosecution for operating a motor vehicle while addicted to narcotics.[58] However, the courts in these cases refrained from extending *Robinson*, and so lawyers seeking to expand the decision's reach had to look elsewhere. Attention turned to laws criminalizing public drunkenness, and in *Driver v. Hinnant* (1966), the United States Court of Appeals for the Fourth Circuit overturned as "cruel and unusual punishment" the conviction of a chronic alcoholic in North Carolina for public intoxication.[59] Claiming that *Robinson* "sustain[ed] if not command[ed]" their decision, the Court of Appeals drew a strict parallel between the two cases: "The California Statute criminally punished a 'status'—drug addiction—involuntarily assumed; the North Carolina Act criminally punishes an involuntary symptom of a status—public intoxication."[60] Shortly thereafter, the Court of Appeals for the District Columbia invalidated prosecutions for public intoxication in its jurisdiction.[61] These decisions were widely praised in the legal community.[62]

Following the lead of the federal judiciary, state courts in Georgia, Pennsylvania, and Maryland subsequently ruled that public drunkenness was an illness and could not be punished as a crime, although courts in California, Washington, and Michigan disagreed.[63] Additional decisions pressed the logic of *Robinson* even further. Both a federal and state court ruled that it was unreasonable to require an alcoholic to stop drinking as a condition of probation.[64] In a criminal action for failure to pay child support, a Wisconsin court opened the door to using alcoholism as a defense for failure to pay support, although the defendant was not successful in proving this defense in the case before the court.[65] The Supreme Court declined to accept a public-intoxication case for review in 1966, but in 1967 it accepted a similar case, *Powell v. Texas*, brought on behalf of sixty-six-year-old Leroy Powell.[66] Most legal observers thought that the Supreme Court would extend the application of Robinson and overturn Powell's conviction.[67] The case inspired one law professor to commit poetry:

> In a Texas town Leroy Powell was found,
> Intoxicated, helpless, and lying on the ground.
> And so began an unusual tale
> Revolving poor Leroy in and out of jail.
> From a humble tavern, where shoes he would shine,
> The proceeds of which he invested in wine,
> To the U.S. Supreme Court he finally went,
> In an effort to overthrow precedent.[68]

There is little question that Leroy Powell had a serious problem with alcohol. He apparently drank every day, and according to police records, over the course of his life he had been arrested "72 times for public drunkenness and once for disturbing the peace."[69] By his own account, he had been arrested more than one hundred times. "Whole episodes of his life have been lost in an alcoholic fog," wrote a lawyer for the American Civil Liberties Union in its amicus brief, and he was probably right.[70]

Legal arguments presented during the appeal of *Powell v. Texas* showed just how far scientific materialism had penetrated the thinking of the American legal community. While it was to be expected that *Powell's* defenders would depict his actions as the involuntary symptoms of a disease, even the prosecutors charged with defending the Texas law were hesitant to argue outright for the reality of free will and moral responsibility. They did point out that according to a defense medical expert, Powell had the power to abstain from drinking for certain periods, and they noted that he had shown restraint on the morning before his trial, only having one drink, "since he knew he had to appear in court that afternoon."[71] But for the most part, the state either tried to square the Texas law with a deterministic outlook or claim that the debate over free will and determinism could be ignored.

In its brief to the Supreme Court, the state actually maintained that "the traditionally acceptable purposes of the criminal law" were perfectly compatible with scientific determinism. According to the state, those purposes were "treatment, isolation, and deterrence."[72] Noticeably absent from the state's list was retribution or atonement, which lay at the foundation of the traditional theory of punishment. The state was apparently unwilling to defend the outdated notion that criminals should be punished because they were morally blameworthy.

For those who might still see a tension between science and justice, the state claimed that the debate over free will and determinism was irrelevant to the case. Even if Powell's actions were involuntary, he had to be held responsible for the safety of society. "For the purposes of the criminal law it should not matter whether from an Olympian point of view, appellant's decision may be attributed to unconscious impulses originating in experiences in his early life, or in socio-cultural factors. Otherwise there would be no criminal law and society would have no protection from behavioral deviants."[73] The state pointed out that according to the "presently dominant psychiatric view of man," every choice of every human being is "determined by prior factors."[74] If one accepts the argument that such determinism in Powell's case should immunize him against punishment, then it is difficult to see how any criminal's choices could be punished: "the principle urged by appellant here, if accepted with full force, would exculpate all crimes if one accepted a deterministic view of man."[75]

This was arguing the point purely on utilitarian grounds: Society cannot restrict punishment to the morally blameworthy, because under the modern scientific view, virtually no one is blameworthy and society would be overrun by the criminals. During the late 1960s, rates of violent crime in America were skyrocketing, and such an argument was no doubt convincing to the general populace. But it ceded the moral high ground to Powell's attorneys, who condemned the injustice of trying to punish someone who was the victim of an involuntary condition.

The debate over *Powell* displayed with peculiar clarity the tension between the traditional legal system and the modern philosophy of scientific materialism. The traditional legal system was based on the assumption that most people had some ability to choose whether or not to commit crimes; thus, those who commited crimes were regarded as morally blameworthy and deserving of punishment. Scientific materialism, by contrast, presumed that all behaviors could be reduced to material causes rather than the free choices of the individual; according to this view, it was unclear whether anyone could ever be considered "morally blameworthy" in the classical sense. The scientific view threatened to undo the Western conception of criminal justice. On the one hand, if the justice system clung to the notion that only the blameworthy should be punished, then it increasingly had to abandon punishment, for science showed that no one was really blameworthy (this was the view embraced by Powell's defenders). On the other hand, if the justice system clung to the idea that punishment must be maintained, then it increasingly had to sever the connection between punishment and blameworthiness (this essentially was the position espoused by the state attorneys in *Powell*). Either way, the traditional justice system would be abolished.

In June 1968, the Supreme Court shocked the legal community and upheld Powell's conviction. But it was a fractured decision, the court splitting three ways. Announcing the judgment of the Court (but speaking for only four of its members), recently appointed Justice Thurgood Marshall declared that "traditional common law concepts of personal accountability and essential considerations of federalism lead us to disagree with the appellant."[76] Marshall and those who joined his opinion clearly feared that *Robinson v. California* had opened Pandora's box, and they wanted to limit the previous decision's reach. In Marshall's view, the Court was being asked to invent nothing less than a "constitutional doctrine of criminal responsibility" that would substitute the court's judgment for "the centuries-long evolution of the collection of interlocking and overlapping concepts which the common law has utilized to assess the moral accountability of an individual for his antisocial deeds."[77] According to Marshall, it was the province of the states, not the Supreme Court, to adjust the doctrines of criminal responsibility according to the "changing religious, moral, philosophical, and medical views of the nature of man."[78] In a concurring opinion, Justice Hugo Black was even more doubtful about intruding on state authority to define criminal responsibility. "This Court," he complained, ". . . is asked to set itself up as a board of Platonic Guardians" to decide matters for local communities in an area totally beyond the court's expertise.[79] Indeed, should the court accept the arguments of Powell's attorneys, charged the plainspoken justice, it "would be venturing far beyond the realm of problems for which we are in a position to know what we are talking about."[80]

Justice Marshall added that the Supreme Court was being asked to intervene in an area where even the professionals could not agree. "There is no agreement among doctors and social workers with respect to the causes of alcoholism, there is no consensus as to why particular treatments have been effective in particular cases, and there is no generally agreed-upon approach to the problem of treatment on a large scale."[81] How was the court supposed to construct a new theory of criminal responsibility out of such confusion?

To those who proposed civil commitment as a substitute for criminal penalties, Marshall questioned whether commitment would be more humane to alcoholics. Pointing out that jail terms for public drunkenness were usually quite short in duration, Marshall noted that " 'therapeutic civil commitment' lacks this feature; one is typically committed until one is 'cured,' " creating the potential that alcoholics could be committed for "an indefinite period . . . with . . . no prospect of periodic 'freedom.' "[82]

But perhaps the most important section of Marshall's opinion was its radical reinterpretation of Justice Stewart's muddled reasoning in *Robinson*. Marshall drew a bright line between status and behavior and argued that *Robinson* was inapplicable to public-intoxication statutes because such laws sought to curtail a "public behavior" rather than a "mere status."[83] Sharply circumscribing the reach of Robinson, Marshall implied that laws penalizing behavior by definition were not laws against status. Black, in his concurrence, was even bolder, insisting that the court had "explicitly limited" its holding in *Robinson* "to the situation where no conduct of any kind is involved."[84]

Justice Abe Fortas, writing for the four dissenters, rejected Marshall and Black's narrow reading of *Robinson* and proposed that the previous decision be extended to any person who finds himself in a condition he is "powerless to change."[85] In Fortas's view, "a person may not be punished if the condition essential to constitute the defined crime is part of the pattern of his disease and is occasioned by a compulsion symptomatic of the disease."[86] Fortas maintained that his extension of Robinson could be limited to conduct that is "a characteristic and involuntary part of the pattern of the disease," and thought that it was "not foreseeable" that this doctrine would be applied "in the case of offenses such as driving a car while intoxicated, assault, theft, or robbery."[87]

"That is limitation by fiat," shot back Justice Marshall in his opinion, writing that the logic of Fortas's opinion could not be confined "within the arbitrary bounds which the dissent seems to envision."[88]

The justice holding the balance of power in *Powell v. Texas* turned out to be Byron White, who had dissented in *Robinson*. White voted to uphold the conviction of Powell, but refused to join Marshall's opinion. In his own concurring opinion, he adopted a reading of *Robinson* that was similar to the dissenters' view. "If it cannot be a crime to have an irresistible compulsion to use narcotics," wrote White,

I do not see how it can constitutionally be a crime to yield to such a compulsion . . . Unless *Robinson* is to be abandoned, the use of narcotics by an addict must be beyond the reach of the criminal law. Similarly, the chronic alcoholic with an irresistible urge to consume alcohol should not be punishable for drinking or for being drunk.[89]

White's concurrence in *Powell* makes for interesting reading, if only as a contrast to his dissent in *Robinson*. In the previous case, White had attacked the court for imposing its "novel" view of cruel and unusual punishment on the states, had rejected its "status" analysis as inapplicable, and had worried about the impact on laws against the use of narcotics, although he predicted that the court would never actually forbid criminal penalties against the use of narcotics. In *Powell*, White seemed like a different justice. Not only did he think now that the use of narcotics was constitutionally protected, he thought that applying *Robinson*'s rationale to cases of public drunkenness "would hardly have radical consequences."[90] As a practical matter, however, White did demand a higher standard of evidence than the dissenters to establish a constitutional wrong. He insisted that Powell show not only that he was a compulsive alcoholic but that he was compelled to appear in a public place while drunk. That second fact had not been proven at trial, therefore White was content to let Powell's conviction stand.[91]

Law reviews seized upon White's concurring opinion to argue that the majority of the court had embraced the rationale advocated by the dissent.[92] Looking to the future, commentators speculated that White's reasoning could be used to invalidate the application of public-intoxication laws to skid-row alcoholics.[93] Unlike Leroy Powell, many such alcoholics might have no homes, and so they could be regarded as being "compelled" to appear in public places while intoxicated. White himself had implied that alcoholics who faced such a situation would meet his evidentiary standard.[94]

Reformers' hopes of extending *Robinson* in a future case proved to be illusory. *Powell v. Texas* turned out to be the high-water mark in the effort to get the court to impose the scientific-materialist view of criminal responsibility as a matter of constitutional law. That did not stop experimentation at state and local levels, where defense attorneys continued to push for expanded versions of the insanity defense, and a number of high-profile cases involving claims of mental illness continued to capture media attention.

Backlash against the Insanity Defense

In 1978, former San Francisco supervisor Dan White murdered Mayor George Moscone and fellow supervisor Harvey Milk. Claiming "diminished capacity" because of manic-depressive illness, White was found guilty of a reduced charge of manslaughter. In 1981, John Hinckley attempted to assassinate President Ronald Reagan, and in 1982 he was declared not guilty by reason of insanity (he was subsequently committed to a mental institution).[95] In the 1990s, the first trial of Erik and Lyle Menendez ended with deadlocked juries after defense lawyers claimed that the defendants were driven to kill their parents because of psychosexual abuse (the brothers were later convicted in a new trial).[96]

Defenders of the insanity defense commonly insisted that it was invoked in less than 1 percent of criminal trials, and that only in a fraction of those trials did it actually prevail.[97] But these claims were misleading. The usage of insanity pleas during the 1970s and '80s varied widely according to the jurisdiction. A study of nearly a million felony indictments between 1976 and 1985 in eight states calculated insanity plea rates (insanity pleas as a proportion of total felony indictments in each state) ranging from 0.29 percent in New York to 5.74 percent in Montana.[98] Given the variation in plea rates, and the limited number of states included in the study, it is difficult to extrapolate these findings to other states. Focusing on insanity pleas as a proportion of all felony indictments (for both violent and nonviolent felonies) is also problematic, because the vast majority of insanity pleas are for violent offenses.[99] Thus, these statistics likely understate the impact of the insanity defense in the prosecution of violent crimes. An even more serious problem is that the focus on insanity pleas ignores the more common use of psychological evidence to reduce charges against a defendant or to show that the defendant did not have the requisite intent to commit the crime for which he had been charged. Even so, by the late 1970s and early '80s a hundred or more defendants a year were being acquitted in some states by invoking the insanity defense, and there is evidence to suggest that the number of insanity acquittals as a proportion of the population rose significantly from 1970 to the early 1980s.[100] Given these facts, it is difficult to claim that popular concern over the growth of insanity-type defenses was entirely misplaced.

Nor was the public wrong to think that, when they were invoked, insanity pleas were often successful. In one study claiming to demythologize "inaccurate perceptions of the insanity defense," researchers faulted the public for believing that nearly 44 percent of insanity pleas result in acquittals, when in fact only 26 percent of them did, according to 1976–85 data.[101] But the 26 percent figure the researchers cited was an average; the actual acquittal rates in the underlying study ranged from 7.31 percent to 87.36 percent. Half of the states studied, in fact, had acquittal rates of nearly 40 percent or more.[102] Hence, public perceptions of the success of insanity pleas during the 1970s and '80s were much closer to the truth than some scholars wanted to admit.

Popular fears of the insanity defense's affront to free will and personal responsibility eventually provoked a legislative backlash. "Between 1978 and 1990, there were 124 different attempts to reform the insanity defense in thirty-four jurisdictions."[103] Unlike the initiatives of the 1960s, most of these new proposals were designed to curtail the use of the insanity defense, not expand it. Several states tried to abolish the insanity defense; voters in California reenacted the "right-wrong" test in 1982; and Congress reinstituted a version of the M'Naghten rules in federal courts in 1984.[104] By the 1990s, those who hoped to redefine criminal responsibility according to the findings of modern science lamented the "legal system's consistent rejection of psychodynamic principles."[105] The American legal system's grand experiment with the scientific approach to criminal responsibility seemed to be over.

Or was it? Despite the tightening of legal standards of responsibility, insanity defenses linger. During the 1980s and '90s, defense attorneys and legal theorists continued to proffer creative legal defenses derived from psychology, including posttraumatic stress disorder, battered women's syndrome, urban psychosis, urban survival syndrome, television intoxication, black rage, neonaticide syndrome, and postpartum psychosis and depression.[106] Postpartum defenses became especially popular in infanticide cases. Sheryl Massip ran over her baby with a car. Though she was convicted by a jury, the judge overturned the verdict because of postpartum psychosis. Ann Green, a nurse, killed two of her newborns and attempted to kill a third; she was declared not guilty by reason of insanity (postpartum psychosis again). Green did not even have to spend time confined to a mental institution. She was assigned treatment as an outpatient.[107]

By the end of the century, appeals courts at the state level continued to grapple with mental-illness defenses in their jurisdictions. In Massachusetts, a woman tried to overturn her conviction for killing her son in part by claiming that she suffered from "battered women's syndrome," allegedly caused by her boyfriend's emotional abuse.[108] In Hawaii, a man who hammered another man to death outside a Burger King argued on appeal that he should not have been found criminally responsible because he suffered from a mental illness induced by illegal drug use.[109] In Texas, a man claimed that "temporary insanity" induced by intoxication led him to shoot his common-law wife.[110] In Washington, a previously convicted sex offender argued that he kidnapped and sexually assaulted his psychotherapist because of multiple personality disorder.[111] In Virginia, a fourteen-year-old student claimed that he stabbed a schoolmate due to a form of schizophrenia.[112] Most such appeals fail, but in some states there are still nearly a hundred or more insanity acquittals a year.[113] Moreover, defense attorneys continue to regard the insanity plea as a common defense likely to be helpful to the defendant.[114]

The repeated articulation of these kinds of excuses in the courtroom and in the news media saps the cultural foundations of the traditional theory of punishment. This is true even if insanity defenses are primarily reserved for a handful of high-profile cases. "The rarity of high-profile cases does not lessen their philosophical and tutelary significance," points out social scientist James Q. Wilson. "It is precisely when all eyes are focused on a case that the law itself, and not simply the defendants, is on trial."[115] Courtroom depictions of human beings as the mere products of their biology or environment help reshape how we view ourselves. Even if such portrayals ultimately fail to secure acquittals, they contribute to a new cultural understanding of humans as victims of their biology and environment, rather than as morally accountable agents who are responsible for their choices.

Such mental-illness defenses also affect how the justice system deals with criminals after they are convicted. While claims of non-responsibility due to biology or environment may rarely produce verdicts of "not guilty" in today's legal system, their impact on the sentencing and punishment of offenders can still be significant—as the next chapter will show.

5

Turning Punishment into Treatment

Mary Kay Letourneau was tearful as she addressed the judge at her sentencing hearing. "I did something I had no right to do," the former grade-school teacher told the judge in November 1997. "Morally or legally, it was wrong. I give you my word that it [won't] . . . happen again." Letourneau was convicted of twice raping a sixth-grade boy who had been her student. She later gave birth to the boy's baby. The prosecution sought a heavy sentence due to the gravity of the offense. The judge imposed a seven-and-a-half-year prison term, then suspended all but six months of the sentence and ordered Letourneau to receive outpatient treatment in a sex-offender program.[1]

The judge was convinced by expert psychiatric testimony indicating that Letourneau was not fully responsible for her actions. Psychiatrist Julie Moore, who evaluated Letourneau for sentencing, explained that unconscious drives drove Letourneau to substitute the boy as "the man in her life" after her father was diagnosed with cancer and her marriage began to falter. But the ultimate root of Letourneau's actions, according to Moore, was bipolar disorder, which leads people to engage in risky behavior regardless of the consequences. "It's like she has a happy button and a love button and a hypersexual button in her brain and . . . when it's pressed, there is little room for self-reflection."[2]

Letourneau's sentencing to outpatient treatment rather than imprisonment exemplified what has been called the "rehabilitative ideal."[3]

According to this view, the paramount purpose of the criminal-justice system is not to punish criminals, but to reform them. Rehabilitation has always been an important theoretical purpose of the American criminal justice system.[4] But early formulations of the rehabilitative ideal in America, heavily influenced by Christianity, emphasized the moral guilt of the offender and the need for repentance and atonement.[5] Retribution and atonement were not regarded as separate from rehabilitation. Nor was rehabilitation regarded as somehow contrary to personal responsibility. In order to be truly rehabilitated in early America, a criminal was expected to own up to his actions and repent. It was no accident that prisons in early America, became known as "penitentiaries." That is precisely what they were designed to be: places of penitence.[6] At the same time, there was a healthy skepticism among penal officials in early America about the prospects of rehabilitation.[7] They knew that criminals would not abandon a life of crime unless they themselves wanted to change.

The rehabilitative ideal was transformed by the rise of the scientific study of crime and its utopian claims. Criminologists believed that science had uncovered the material foundations of human behavior, and they began to believe that it was only a matter of time before techniques would be developed to scientifically rehabilitate even the most hardened criminals through manipulation of their bodies and environments. According to the modern rehabilitationists, once one replaced punishment with treatment the crime problem would be solved. "The sharp improvement which takes place in the attitude and behavior of the offender when he is handled as an object for treatment instead of an object for punishment is always definite," proclaimed one doctor at the National Conference of Social Work in 1926.[8]

The implicit materialism of the new scientific rehabilitation movement can be seen in a report on the penal system drafted by a federal advisory committee in the early 1930s. The advisory committee waxed enthusiastic about the prospects for truly scientific rehabilitation based on modern "medicine, biology, psychiatry, psychology, and sociology."[9] According to the committee, prisons for all practical purposes should be turned into treatment facilities where there would be "diagnosis of the causes of antisocial acts or series of acts, and treatment designed to remove those causes."[10] The committee acknowledged that its ideas were newfangled, so the attitudes of many judges, lawyers, and law professors "must be changed" to be in accord with the "new philosophy." Most of all, "there must . . . be radical revision in fundamental ways of thinking about treating criminals held by the main body of the public."[11]

That "radical revision" occurred, and by the 1960s the "new philosophy" of rehabilitation had superseded retribution as the guiding ideal in

American criminal justice. Yet from early on the scientific rehabilitative ideal attracted serious criticism from across the political spectrum.[12] Conservatives attacked rehabilitation efforts for being soft on criminals by deemphasizing punishment and personal accountability. Liberals argued that rehabilitation programs, far from being soft on criminals, were overly coercive because they turned prisoners into guinea pigs and subjected them to dehumanizing experiments. Both sets of critics had a point.

Treating Criminals like Victims

The "soft" side of rehabilitation could be seen in the expanded use of parole and probation, reforms long championed by advocates of the scientific approach to crime. While in theory neither parole nor probation necessarily undercut the concept of criminal responsibility, in practice professionals in these fields tended to view criminals as victims of heredity and environment rather than as responsible agents.

Writing in 1948, G. I. Giardini, superintendent for the Pennsylvania Board of Parole, asserted that the behavior of criminals like Al Capone was largely determined by the "environmental circumstances under which they were reared . . . Under a different set of circumstances, they might have turned out to be honest businessmen." In Giardini's view, human beings were so much the victims of their environment that "one glass of beer . . . at a given moment . . . may turn a law-abiding individual into a murderer."[13]

A 1964 textbook for parole and probation officers contained an entire chapter devoted to "Fundamentals of Human Behavior," in which criminology professor Reed Clegg tried to inculcate in the criminal-justice professional a scientific understanding of human behavior. The author acknowledged that "our penal codes envision a free deliberating sort of [criminal] . . . and attempts to set punishments accordingly," but he also declared that "our knowledge of human beings reveals that no such person exists."[14] The rest of the chapter acquaints parole and probation officers-in-training with a host of psychological terms, including psychosis, insanity, schizophrenia, neurosis, sublimation, repression, fixation, compulsions, and obsessions. According to Clegg, the root causes of delinquency were man's natural desires for security, love, approval, and new experiences. "When these desires are thwarted, the individual is unhappy, unadjusted, and seeks unusual means of obtaining them."[15]

In keeping with the philosophy that criminals were victims, Clegg cautioned probation and parole officers against the sin of being perceived as condemning or judgmental. If the offender abused alcohol, the corrections officer could try to get him to recognize that "he is an alcoholic and needs help." But the officer should avoid "lecturing and discussion of the problem," because the offender might feel "he is being condemned and that no one understands him." Similarly, the probation or parole officer should avoid "lecturing . . . and condemning" the sexually deviant offender, although the officer could at least have a "common sense discussion" with the offender about "his problem and what society expects of him."[16] Another book warned corrections officers against placing many behavioral restrictions on the offender, because such demands "are wholly out of line with current thinking and practice. Their arbitrary character reveals their punitive motivation."[17] The modern corrections officer should view the offender more as a patient to be treated than as a wrongdoer who needed to repent and be held accountable for his actions. Rehabilitation increasingly was disconnected from the need to acknowledge one's own culpability, to make atonement both to society and to one's victims, and to attempt to reform one's behavior.

Of course, probation and parole officers were still part of the criminal-justice system, and so they had to temper their scientific paradigm with political realities. Take the case of the "sexual deviate," which Clegg defined as anyone who violated society's sexual norms, including child molesters, rapists, promiscuous heterosexuals, and homosexuals. "The advice of many psychiatrists is [for the deviants] to accept the fact of abnormality," wrote one professional. "This relieves tension and serious conflict" and helps avoid harmful "repression" of an offender's sexual conflict, which could produce "psychotic reactions." Nevertheless, the author warned his readers that it was "difficult to reconcile this advice with what the public and the courts expect of a probationer or parolee." In other words, the corrections officer who tells the deviant to accept his urges might be out of a job. The best a probation or parole officer could do is to advise the offender "that he cannot help being what he is, but he can exercise control as normal individuals control their heterosexual desires."[18]

Given the growing propensity of criminal-justice professionals to view criminals as the victims of material causes, it is not surprising that criminals began to see themselves in the same way. As early as 1942, one criminologist wrote about his encounter with some prison inmates who were familiar with the latest theories. When asked why they or others committed crimes, they would "discuss crime causation in the jargon of academic criminologists, with whose writings they have some acquaintance and whose views on the interrelationship of multiple causative factors they expound with conviction."[19] By the 1960s, this mentality was parodied in the musical *West Side Story*, which featured gang members ridiculing the notion that their antisocial activities are the products of mental disease and environment. At one point in the song "Gee, Officer Krupke," a gang member imitates a judge who berates the police for their outdated belief that juvenile criminals ought to be punished: "Officer Krupke, you're really a square / This boy don't need a judge, he needs an analyst's care! / It's just his neurosis that oughta be curbed / He's psychologic'ly disturbed!" Later another gang member exclaims, "I'm depraved on account I'm deprived."[20]

The point of view parodied by *West Side Story* remains prevalent. Defendants are given incentives to deny responsibility because they know that diminished-capacity claims may reduce their punishment. Consider again the case of Mary Kay Letourneau. Although she was convicted of her crime, she received a dramatically reduced sentence. Legal experts asked to comment at the time of her sentence said that such alternate sentencing was not unusual.[21] They were right. Probation rather than jail is common for those who assert mental-capacity defenses, no matter how horrific their crimes.

Latrena Pixley, who smothered her baby, was sentenced by a District of Columbia judge to a mere 156 weekends in jail; a year later, even that sentence was suspended, and she was placed on probation. The reason for the leniency? Her defense attorney explained that she suffered from postpartum depression. Itsumi Koga drowned her newborn son in a Michigan pond. She spent seven months in a mental facility and received five years' probation.[22] Mental illness can also shield a criminal from having to pay restitution to his victim. In Arizona, an appeals court overturned the imposition of restitution on a defendant found "guilty except insane" of attempted arson.[23]

Such evidence is anecdotal, but quantitative studies corroborate it. In New Jersey and Wisconsin, nearly 40 percent of those found guilty after asserting an insanity defense are released rather than imprisoned after conviction. In Georgia, Ohio, Washington, and California, the release rate is 29 to 34 percent.[24] These figures are for cases where insanity pleas are actually filed during the trial. But regardless of how a defendant pleads, he can raise psychological claims to mitigate his punishment once he is convicted. This is especially the case in states like Montana, where legislators have attempted to abolish the insanity defense outright. In those jurisdictions, psychiatric evidence is simply redirected to other phases of the criminal-justice process.[25]

Ascertaining the pervasiveness of diminished-capacity claims during the sentencing phase is difficult, but evidence suggests that they are not uncommon, particularly in cases where prosecutors have not offered some sort of plea agreement to the defendant. In jurisdictions with sentencing guidelines, the most common reasons cited by judges for imposing sentences below the standard sentencing range are connected to plea bargains, guilty (or no contest) pleas, and cooperation with the prosecution. These are prosecution-driven reasons for sentence reduction, and they tell us little about how important diminished-capacity claims are for defense attorneys in cases where the prosecutors do not want sentence mitigation. In Pennsylvania, once one excludes prosecution-connected factors, the fourth- and fifth-most-cited reasons by judges for imposing sentences below the standard range are "mental/emotional/psychological problems" and drug dependency.[26]

A similar pattern emerges at the federal level. Once one excludes prosecution-connected sentence reductions (as well as sentence reductions because a prisoner is being deported), "diminished capacity" consistently ranks as the fifth- or sixth-most-frequently cited reason for a federal judge to issue a below-standard sentence. "Mental and emotional conditions" is usually the tenth-most cited reason.[27] The assertion of diminished-capacity claims in federal sentencing hearings was made easier by a 1998 revision in the sentencing guidelines that expanded the types of cases where such claims can be heard.[28]

Following the revision, Alan Ellis, former president of the National Association of Criminal Defense Lawyers, urged his colleagues to be more creative in using "diminished capacity" to get sentence reductions for their clients in federal white-collar crime cases. To those who might wonder "how . . . a securities fraud offender who bilked investors out of millions of dollars . . . [could] have been mentally ill," Ellis explained that mental disease does not discriminate against lawyers and other professionals. After all, "Abraham Lincoln, it is reported, suffered from a bipolar disorder, manic depression. Fortunately, he did not need a downward departure [from federal sentencing guidelines]. In short, bright, competent people can suffer from significant mental disorders."[29]

Yet if the scientific model of crime can encourage too much leniency by diminishing criminal responsibility, it can also do the opposite, as liberal critics of the rehabilitative ideal have claimed. The utopian belief that criminal tendencies can be "cured" can lead to forms of treatment that are nothing short of horrific. Procedures that would be regarded as cruel and barbaric as punishments gain new justification when designed to "treat" an offender rather than "punish" him.

Treatment or Torture?

"Castrate the criminal, cut off both ears close to his head and turn him loose to go where he will," advised Dr. Jesse Ewell in 1907.[30] Though his prescription for curing criminals may seem shocking today, Dr. Ewell was not a crank, and the views he propounded in "A Plea for Castration to Prevent Criminal Assault" were not unusual. From the 1890s into the early years of the twentieth century, a growing number of American doctors advocated castration for habitual criminals, rapists, and murderers.[31] Proponents of castration like Frank Lydston derided the failed rehabilitation efforts of the "sentimentalist and his natural ally, the preacher."[32] Lydston argued that "asexualization" surgery would produce results by preventing criminals from passing down their criminal tendencies to their children, by striking fear into noncastrated criminals, and by changing the personality of the castrated criminal: "The murderer is likely to lose much of his savageness; the violator loses not only the desire, but the capacity for a repetition of his crime, if the operation be supplemented by penile mutilation according to the Oriental method."[33] Lydston's views were grounded forthrightly in scientific materialism. "The attempt to reduce criminology to a rational and materialistic basis has constituted a great step in advance—one which marks a distinct epoch in scientific sociology," he proclaimed in 1896.[34]

Some doctors went beyond talk and actually unsheathed the scalpel. Dr. F. Hoyt Pilcher operated on forty-four boys and fourteen girls of the Kansas State Home for the Feeble-Minded during the 1890s. Though he had to curtail his castrations due to public outcry, the board of trustees of his institution was unrepentant, insisting that "those who are now criticizing Doctor Pilcher will, in a few years, be talking of erecting a monument to his memory."[35]

Martin Barr, chief physician of the Pennsylvania Training School for Feeble-Minded Children, advocated "asexualization . . . not as [a] penalty for crime but [as] a remedial measure preventing crime and tending to the future comfort and happiness of the defective." Barr desired the operation to "become common for young children immediately upon being adjudged defective by competent authorities properly appointed."[36] In 1901, Barr and two other doctors urged the Pennsylvania state legislature to allow both castration and sterilization for "idiots and imbecile children," and a bill was actually enacted by the legislature, though it was "returned by the governor for the correction of some trifling technicality" and did not become law.[37] According to Barr, fears about compulsory castration for the "feeble-minded" were unfounded. Those who thought that "the operation for unsexing" would make patients "less vigorous or manly" should remember that "a gelding or ox loses nothing but becomes in every respect more docile, more useful, and better fitted for service."[38]

Castration never became widespread as a method of crime control, but the enthusiasm with which some in the medical community promoted the procedure was indicative of how powerful the materialistic ideal was becoming. According to scientific materialism there was a physical basis for everything, and so why shouldn't there be a physical basis for crime? And if there was a physical basis for crime, why shouldn't scientists be able to find and correct it? Scientific materialism made the castration hypothesis seem eminently plausible. The tentative acceptance of castration within the medical community also foreshadowed the next surgical panacea for social ills: sterilization.

During the 1890s, medical researchers developed the vasectomy and the salpingectomy, operations that sterilized men and women without affecting their gender. Reformers were soon pressing for compulsory sterilization of those deemed unfit. The call for surgical sterilization was part of the broader eugenics movement, which was directly inspired by nineteenth-century Darwinian biology. Darwinists emphasized the heritability of a wide array of biological traits, and although it was not until the early twentieth century that they had a better grasp of genetics, the basic idea was clear enough. It was second nature for Darwinists to think that bad behaviors as well as bad physical traits (such as defective eyesight or mental abilities) could be passed down from parents to children.

"Eugenics" was supposed to be the science that would allow man to take control of his own evolution by breeding a better race. In the words of Horatio Hackett Newman, a University of Chicago zoology professor, through eugenics "man might control his own evolution and save himself from racial degeneration."[40] "Positive eugenics" focused on encouraging those deemed the most fit to reproduce more, while "negative eugenics" focused on curtailing reproduction by those deemed unfit, including mental defectives and criminals.

Although eugenics is sometimes regarded as a perversion of Darwinian biology, Charles Darwin himself praised the idea of voluntary eugenic restrictions on marriage in *The Descent of Man* (as we saw in chapter 2), and his sons George and Leonard actively promoted the eugenics agenda, with Leonard becoming the president of the Eugenics Education Society, the main eugenics group in Great Britain.[41]

But it was Darwin's cousin Francis Galton who is justly considered the founder of the modern eugenics crusade. Inspired by *The Origin of Species*, Galton set about to apply his cousin Charles's theory to the rise of human genius. After researching the family connections of members of the British elite, Galton announced in articles and then in books that intellectual and artistic talent was largely hereditary.[42] Thus, if society wanted to guarantee its future improvement, it needed more children from the "fit" and fewer from the "unfit." By the 1880s, Galton had coined the term *eugenics* (adapted from a Greek root meaning "good in birth")[43] and was urging efforts to improve the race through better breeding.

While Galton stressed the need for positive eugenics (in order to cultivate the geniuses needed for society to thrive and progress), he also favored negative eugenic measures and thought that those deemed unfit could be segregated in institutions where they would not be allowed to reproduce.[44] (A fuller account of the ideology of eugenics is presented in chapter 7, where eugenics as a method of abolishing poverty is discussed at length.)

Eugenicists were hampered initially by scientists' misunderstanding of heredity. At the time Darwin wrote, many scientists—including Darwin—believed that behaviors and characteristics learned during a person's lifetime could be inherited by his children. As Mark Haller has pointed out, "As long as people believed in the inheritance of acquired characters, there was little need for eugenics; instead all efforts could be devoted to betterment of present health and education, with the assurance that the improvement would appear in the hereditary endowment of succeeding generations."[45] Galton rejected the idea that acquired characteristics were heritable, but it was not until the early years of the twentieth century that this view became popular in the scientific community.[46]

By the early 1900s, eugenists could claim with confidence that "nearly all forms of mental deficiency are incurable," and that according to "most biologists . . . acquired characters are not inherited."[47] Once this new view of heredity was embraced, the rationale for eugenics became much more powerful, for the only way to eliminate biological defectives was to make sure they did not breed at all. Given this sort of thinking, proposals for forced sterilization seemed perfectly reasonable, and it is no surprise that the new century opened with a flurry of efforts to enact eugenics legislation across the United States.

Indiana enacted the first compulsory sterilization law in 1907. The bill had been promoted by a doctor and superintendent at the Indiana Reformatory, and it mandated sterilization for "confirmed criminals, idiots, imbeciles, and rapists in state institutions when recommended by a board of experts."[48] After passage of the bill, Dr. Harry Sharp of the Indiana Reformatory declared that he was proud "that Indiana is the first State to enact such a law, and that the Indiana Reformatory is the pioneer in this work."[49] By the early 1930s, thirty states had enacted sterilization laws, "and in twenty-seven states the laws were still on the books, if not always enforced."[50] These statutes commonly included "criminals" and/or "habitual criminals" as targets for operations.[51] As a general rule, "criminals and degenerates and weaklings are born . . . not . . . made by their environment," asserted Dr. H. E. Jordan of the University of Virginia in 1911. Consequently, "to be permanently rid of deficients, they must be prevented from reproducing their defective kind. Our only hope for racial improvement lies in prohibiting the continual replenishment in frightful numbers of our degenerate and criminal classes."[51] Several decades later, when many people had abandoned eugenics altogether, one could still find a few die-hard activists proposing sterilization as a solution to crime. In a letter to the Human Betterment Federation of Des Moines in 1951, Margaret Sanger of Planned Parenthood urged that the unfit be sterilized, because their "children form our army of delinquents and become social burdens, ending their lives in institutions, such as reform schools and penitenturies [sic]."[52]

By 1931, more than twelve thousand Americans had been sterilized under various state laws, a figure that rose to more than sixty thousand by 1958.[53] Most of those sterilized were classified as insane or feeble-minded rather than as criminals, and so one might conclude that sterilization never became a popular method of crime prevention. But insanity and feeble-mindedness were commonly linked to crime by eugenists, and so sterilization was regarded as serving the cause of crime prevention whether or not the sterilized person was a convicted criminal. A study of a dysfunctional extended family published in 1916 reported that members of the family who were criminals were also "feebleminded." Thus, "the eradication of crime in defective stocks depends upon the elimination of mental deficiency."[54] A year earlier, the editor of the *Journal of Heredity* described research showing that "the so-called 'criminal type' . . . is merely a type of feeblemindedness . . . driven into criminality for which he is well fitted by nature."[55]

Eventually, the popularity of eugenics waned in America. Although still widespread, the movement attracted increased opposition during the 1930s, when the application of eugenics to criminals became especially controversial.[56] In *Skinner v. the State of Oklahoma* (1942), the Supreme Court struck down a state law authorizing forced sterilization for certain kinds of habitual criminals, although it did so on narrow grounds and declined to overrule a previous decision approving forced sterilization for noncriminals.[57]

But the decline of the eugenics crusade did not end the quest to prevent antisocial behavior by attacking its material foundations. Reformers simply latched onto the next cure offered up by scientific materialists: psychosurgery, the surgical destruction of healthy brain tissue in order to pacify disturbed individuals.

Carving Up the Brain

Psychosurgery on human beings had taken place as early as the 1880s in Switzerland,[58] but the technique did not become popular until the 1930s, when Portuguese neurologist Antonio Egas Moniz began promoting it. Moniz gained inspiration from an international neurological meeting in London in 1935. Attended by such luminaries as legendary Russian physiologist Ivan Pavlov, the conference included a presentation about the removal of prefrontal lobes from the brains of two chimpanzees. According to the American researchers who conducted the experiment, the chimpanzees became happy and free from anxiety after the surgery, although their mental faculties deteriorated. Moniz was intrigued. If the procedure worked in monkeys, he asked the researchers, why not humans?[59] Two months later, Moniz and a colleague were conducting experiments on human subjects, cutting into their brains and destroying brain tissue. Moniz claimed that his procedure—the "leucotomy"—was "always safe" and produced dramatic results in mentally ill patients, reducing anxiety, agitation, and delusions.[60]

Neurologist Walter Freeman and neurosurgeon James Watts renamed Moniz's procedure the lobotomy and quickly adapted it for American patients. In the 1940s, Freeman transformed the lobotomy into an in-office procedure utilizing a modified ice pick. (The ice pick would be driven into the brain just over the eyelid, then moved around in order to destroy tissue in the prefrontal lobes.) A tireless showman, Freeman promoted the new miracle operation in mental hospitals across the country, often with dramatically staged demonstrations.[61] Despite its detractors, the lobotomy gained widespread acceptance. In 1949 Antonio Egas Moniz was awarded a Nobel Prize for his role in developing the prefrontal leucotomy, and by the 1950s an estimated forty to fifty thousand lobotomies had been performed in the United States alone.[62]

Unfortunately, in their eagerness to find a surgical panacea for mental illness, promoters of the lobotomy glossed over the operation's serious consequences for the patient's personality. The prefrontal lobotomy "often rendered the patient not only calm, but also apathetic, irresponsible and asocial. In addition, intellects were blunted, judgment was impaired and creativity was stifled."[63] Late in his career, even Walter Freeman had to acknowledge that "psychosurgery reduces [a patient's] creativity, sometimes to the vanishing point."[64] (Sometimes it did far more than that. Freeman's operation on John F. Kennedy's sister Rosemary left her so severely incapacitated that she had to be institutionalized for the rest of her life.)[65]

Although not originally designed for criminals, the lobotomy became so popular that it was perhaps inevitable that it would be used on them. In 1944, a man with a history of violence was lobotomized in Hawaii and then released as cured.[66] In 1946 in Pennsylvania, habitual burglar Millard Wright sought to avoid a long prison term by agreeing to a lobotomy as part of a plea bargain.[67] Wright received his lobotomy, but a new judge appointed in the case decided that he shouldn't simply be let off. The judge worried that "if we were to accede to the request that Wright be absolved . . . we would soon be overwhelmed by similar requests from many other habitual criminals who are now incarcerated . . . and who would be willing to submit to similar operations."[68] Nevertheless, the judge did impose a significantly reduced sentence. Neither the lobotomy nor the lighter sentence helped Wright, who committed suicide while in prison.[69]

Experts disagreed about how helpful psychosurgery would be in preventing crime. But Walter Freeman, who performed lobotomies on children as young as four, saw no reason why the operation couldn't be extended to criminals as well, and he lobotomized prisoners from at least the 1940s to the early 1960s.[70] Describing one of his success stories in 1949, Freeman told about a man with "at least twenty crimes of violence charged against him" who after his lobotomy was able to "work outside the institution" and no longer gave anyone "trouble." According to Freeman, "extensive statistics" demonstrated that "people after lobotomy have an extraordinary adherence to the moral code of society. Lacking abstract morals . . . they carry out what is expected of them in a rather remarkable, socially competent way. They do not offend."[71]

With the development of psychoactive drugs by the 1950s, psychosurgery fell out of favor,[72] and as a method of crime prevention it never became widespread. But during the late 1960s second-generation psychosurgery techniques (which utilized electrodes to destroy brain tissue) began to be promoted explicitly as a solution to violent crime. In 1967, neurosurgeons Vernon Mark and William Sweet, along with psychiatrist Frank Ervin, published a letter in the *Journal of the American Medical Association* suggesting that urban riots might be linked to "brain dysfunction in the rioters." The authors proposed "intensive research and clinical studies of the *individuals* committing the violence. The goal of such studies would be to pinpoint, diagnose, and treat those people with low violence thresholds before they contribute to further tragedies."[73]

Mark and Ervin elaborated on this proposal in their book *Violence and the Brain* (1970). In a chapter titled "Violence Prevention," they proposed assigning violent criminals (as well as violent noncriminals) to special research facilities that would subject them to detailed psychological, physiological, neurological, and genetic testing in the hope of discovering the root causes of their violent behavior.[74] Those with "both temporal lobe epilepsy and uncontrollable violent behavior" would undergo additional testing with "implanted brain electrodes." After diagnostic testing, the inmates of these facilities would be subjected to experimental treatments that might include behavior modification ("à la Pavlov"), drugs, and, of course, psychosurgery.[75] The goal would be to perfect "an 'early warning test' of limbic brain function to detect those humans who have a low threshold for impulsive violence," and then to develop "more effective methods of treating them once we have found out who they are."[76]

While such a proposal may seem the stuff of science fiction, California was already experimenting on some of its prisoners. In 1968, doctors at the Vacaville penitentiary performed psychosurgery on three prisoners in their twenties, and in 1971 the director of corrections for the state outlined a project to perform "surgical and diagnostic procedures" on violent inmates in order "to locate centers in the brain which may have been previously damaged and which could serve as the focus for episodes of violent behavior." If the diagnostic procedures located brain areas that "were indeed the source of aggressive behavior, neurosurgery would be performed."[77]

Psychosurgery was only one aspect of the biologically oriented antiviolence initiatives of the early 1970s. Some in the criminal-justice community advocated setting up diagnostic centers around the country that would both identify persons biologically predisposed to violence and develop treatments to suppress their violent urges. The prototype for this effort was to be the Center for the Study and Reduction of Violence in California, announced in 1972. To be run by the Neuropsychiatric Institute of the University of California at Los Angeles, the center was to be funded by the federal Law Enforcement and Assistance Administration as well as the state of California.[78] According to the original proposal for the center, the goal was to explore "the largely unexplained interface between biological and psychological aspects of violent behavior."[79] State penal facilities would supply the center with prisoners, who would be screened for genetic, biochemical, and neurophysiological factors which might be connected to violence. The center would also seek to develop antiviolence drugs and a test to enable "large-scale screening that might permit detection of violence predisposing brain disorders prior to the occurrence of a violent episode."[80]

The proposed Los Angeles center sparked angry opposition and floundered, but many prisoners around the country were already being subjected to invasive behavior-modification programs in the name of rehabilitation. These coercive programs employed such methods as electric-shock therapy and drugs to induce vomiting and paralysis.[81] "There have been programs that have used drugs such as apomorphine which, when injected into an individual, causes profound nausea and vomiting," testified Dr. Bernard Diamond from the University of California before the U.S. Senate in 1973. "There [also] have been uses of electro-shock therapy for the purposes of behavioral control."[82] At the Vacaville state prison in California, "violence-prone inmates were given a drug which temporarily paralyzed them to the extent that they couldn't breathe. During the terrifying moments when the men were suffocating, they were told to associate their agony with their misbehavior."[83] In a Maryland prison known as the Patuxent Institution, indeterminate sentencing exacerbated the horrors of coercive rehabilitation. A prisoner sentenced to Patuxent could be kept there until a psychiatrist certified him as no longer dangerous to the community. "According to the ACLU . . . many individuals have been picked up for such infractions as joyriding, given a two-year sentence, and held at Patuxent for as long as eighteen years."[84]

Indeterminate sentencing, of course, was a proposal dating back to Cesare Lombroso, who thought that criminals should be imprisoned as long as it took to cure them. In the 1930s, the Wickersham Commission advocated the eventual implementation of indeterminate sentences in the United States as part of a criminal-justice system run on truly scientific principles. After all, stated the commission, "it is not possible to require the prison to rehabilitate the offender if its hands are tied by an obligation to release him at a time when it feels that such release is contrary to the public interest."[85]

Backlash against Rehabilitation

By the early 1970s, psychiatrist Peter Breggin began to raise objections to the revival of psychosurgery. Breggin was concerned that the psychiatric community was about to repeat the debacle that had occurred with its largely uncritical acceptance of lobotomies in the 1940s and '50s.[86] Breggin's case was considerably strengthened when he met Thomas R., a patient who had been experimented upon by neurosurgeon Vernon Mark and psychiatrist Frank Ervin.

Thomas R. was touted by Mark and Ervin as one of the great success stories of the new electrode-based psychosurgery. According to *Violence and the Brain*, Thomas had been a "brill[i]ant 34-year-old engineer with several important patents to his credit . . . but his behavior at times was unpredictable and even frankly psychotic." His "chief problem was his violent rage" directed "mostly . . . toward his wife and children."[87] After drug therapy failed to work, Mark and Ervin surgically implanted electrodes within Thomas R.'s brain and used the electrodes to stimulate various areas. Then they suggested using electrodes to destroy some of the brain tissue. In *Violence and the Brain*, Mark and Ervin claimed that "four years have passed since the operation, during which time Thomas has not had a single episode of rage. He continues, however, to have an occasional epileptic seizure with periods of confusion and disordered thinking."[88] After

Breggin began to speak out about the renewed use of psychosurgery, he was contacted by Thomas R.'s mother. Mark and Ervin, it turned out, had been less than forthcoming about the results of their patient's treatment.

According to Breggin, Thomas R.'s mother "wept as she explained how her son had lost everything as a result of the surgery: his employment, his wife, his children, his freedom, and his mind."[89] Far from solving the troubled man's problems, the surgery apparently had made him more violent as well as delusional. After the horrific experiments with electrodes planted within his brain, he continued to think that the electrodes were still there, controlling him.[90] After examining Thomas R.'s medical records, Breggin charged that Mark and Ervin had seriously exaggerated their patient's problems before his surgery while glossing over the demented state produced by his treatment. "It was as if the doctors had reversed their story, mixing up the before and the after," he wrote later.[91]

The revelation that Mark and Ervin may have misrepresented the results of their experiments, as well as their controversial proposals for screening and treating individuals thought to have violent tendencies, sparked fierce controversy, as did reports that the federal government was funding such efforts. Soon Congress was investigating federal funding of psychosurgery and other behavior-modification programs, and a National Commission for the Protection of Human Subjects of Biomedical and Behavioral Research was established to study and make recommendations about psychosurgery and other forms of human experimentation.[92]

The growing public debate exposed how wedded psychosurgery proponents were to scientific materialism. In *Violence and the Brain*, Mark and Ervin grounded their argument for the control of violence in an explicitly materialistic account of the human mind, declaring that "the more brain scientists have discovered about the brain, the clearer it has become that brain responses and the parallel behavior are not separate events—that is, there is no aspect of human behavior that is supernatural." In their view, dualistic accounts that sought to preserve an immaterial element of the human mind were no longer tenable. "Once we are aware that thinking is a physical process, Aristotle's 'soul' and Descartes' 'mind' are no longer useful concepts."[93]

Yale neurophysiologist José Delgado, who developed some of the electrodes used to treat Thomas R.,[94] went further. In Delgado's view, "human behavior, happiness, good and evil" are all "products of cerebral physiology," and man should seek to control the human brain in order to produce a better society.[95] There was nothing sacred or inviolable about the brain; it was simply another bodily organ to be manipulated for the common good. According to Delgado, altering the brain finally offered man the hope of taking charge of his own evolution. "We are only the blind product of a blind evolution, without human purpose," Delgado told one interviewer. "What I'm proposing is to give a human purpose to man himself."[96]

Rejecting the reductionism of the psychosurgeons, critics hearkened back to an older view of human beings as more than merely the sum of their physical parts. The brain was the seat of the soul, and it should be approached with a sense of sacred awe. Surgical intervention in the brain therefore should be approached with great caution. Peter Breggin articulated this view in his testimony before Congress: "What goes on in your mind is a very different thing from what goes on in your kidney or your heart . . . I believe your eternal spirit passes through your brain and lives there for a while and then goes on. Tampering here in the brain is considerably different than tampering with the heart or the lungs. It must impair the expression of your spiritual self."[97] Because psychosurgery by definition destroyed healthy tissue (as opposed to removing an abnormal growth such as a tumor), psychosurgeons were claiming for themselves a godlike power to redesign the human brain.

When the national commission on biomedical and behavioral research finally issued its report on psychosurgery in 1977, it recommended that psychosurgery be allowed, but only under a number of restrictions.[98] The commission's report was largely moot, however, because public opposition (fueled in part by literary and cinematic depictions of psychiatric abuses such as Ken Kesey's novel *One Flew Over the Cuckoo's Nest*, released as an acclaimed film in 1975) had made the procedure so controversial that few doctors were willing to promote it any longer, and the Law Enforcement Assistance Administration had already curtailed federal funding of psychosurgery research and strictly limited the type of medical research it would support in the future.[99] Moreover, a court in Michigan struck down proposed "voluntary" psychosurgery experiments on inmates of psychiatric hospitals.[100] Although the decision applied only in Michigan, it cast doubts on the legality of the practice elsewhere. Psychosurgery and invasive behavioral-modfication programs as a method of criminal rehabilitation had clearly become suspect.

At the same time that civil libertarians and critics of psychiatry were questioning the ethics of rehabilitation, others were questioning its efficacy. In an influential essay for the *Public Interest* in 1974, social scientist Robert Martinson presented the findings of a massive study of the effectiveness of rehabilitation programs. His conclusions were devastating: There was little empirical evidence that any rehabilitative treatment methods worked consistently.[101] Martinson's article (and a book he coauthored a year later) jolted criminologists and sparked a backlash in academic circles about the failure of rehabilitation.[102] Parallel with this backlash among academics was a backlash among ordinary citizens fearful of violent crime, which seemed to explode even as rehabilitation efforts were expanded. Both developments inspired legislators and criminal-justice professionals to reassert traditional ideas about retribution and moral blame as the foundations of the penal system.[103] In 1985, the *American Bar Association Journal* proclaimed "The Return of Retribution" in a feature story.[104]

But the rehabilitative ideal was too attractive to be scrapped completely. By the 1990s, revisionist criminologists had reexamined the case for rehabilitation and found that many programs worked after all.[105] Or so they claimed. The average difference in recidivism rates between treated offenders and nontreated offenders was only 10 percent—not exactly overwhelming.[106] Still, these marginally positive findings helped spawn a second-generation rehabilitation movement.

The Return of Rehabilitation

The most prominent initiative of the new rehabilitationists has been the "drug court," which gives drug offenders the option to avoid prison by undergoing intensive monitoring and treatment for a prescribed period.[107] According to supporters, drug courts can dramatically reduce re-offense rates, although there continues to be debate over the long-term effectiveness of such programs.[108]

Many of the new rehabilitation initiatives distance themselves from the accept-no-blame scientific reductionism of the past. Rather than treating criminals as the mere products of their biology and environment, they treat them as if they are personally responsible for their actions. This emphasis on personal accountability can be seen in how these programs are publicly described. In Georgia, for example, drug courts are known as "Accountability Courts," and their purpose is described as "promot[ing] personal responsibility and family values by holding the participant accountable for his/her actions and behavior. It is not a get-out-of-jail-free card."[109] Similarly, Marc Levin of the Texas Public Policy Foundation emphasizes that drug courts are available only "for those offenders willing to take responsibility for their actions," adding, "Drug courts are not soft on crime. Instead of isolating an offender in prison, they force participants to confront their addiction and repair the damage they have done to themselves, their families, and their communities."[110]

Rather than focusing only on the material factors associated with crime, the new rehabilitation programs take seriously the notion that human beings are rational creatures who can make real choices based on their underlying beliefs. Recent research does suggest that rehabilitation can work, but the programs that work best identify the "antisocial beliefs of the offender " and "in a firm yet fair and respectful manner . . . [point] out to the offender that the beliefs in question are not acceptable."[111] Offenders who show change are rewarded; those who persist in their antisocial beliefs face penalties.

Even more consistent with the older view of free will and personal responsibility is the "restorative justice" movement, which focuses on having offenders make atonement and restitution for their crimes. "Implicit in restorative justice initiatives," writes a former director of the National Institute of Justice, is the idea "that an important purpose of the criminal sanction is reintegrating the offender into the community following his acceptance of personal responsibility for the harm done to victim and community and his 'payment' of appropriate penance."[112] In other words, the operating assumption of restorative justice is that accepting moral responsibility is a necessary part of one's rehabilitation. Similar assumptions have led to proposals by some legal scholars for punishments designed to shame defendants or educate them as to the moral consequences of their actions.[113] A related trend in prison management is the restructuring of prisons so that prisoners have to make active choices about how they organize their lives. Prisoners are then held accountable for the choices they make.[114] In sum, the current wave of rehabilitation is less a continuation of scientific materialism than a repudiation of it.

Of course, the old thinking also persists, as evidenced by the lax punishment meted out to Mary Kay Letourneau and other offenders claiming diminished capacity. (Letourneau herself was eventually sent to prison for violating her probation. Even the judge lost patience after discovering that Letourneau had continued a sexual relationship with her former student.[115]) Perhaps of greater concern are growing efforts to identify and treat the supposed biomedical foundations of violence and antisocial behavior.

In 1992, a branch of the National Academy of Sciences issued a report advocating "systematic searches for neurobiologic markers for persons with elevated potentials for violent behavior."[116] The same year, controversy broke out over plans by the federal government to sponsor a "national violence initiative" that would focus in part on identifying biological markers for violence. According to draft plans of the initiative, antiviolence centers were to be set up around the country to screen and identify children who were predisposed to violence and antisocial behavior. The identified children would then be steered into various kinds of preventive treatment, including drug therapy. Peter Breggin, the psychiatrist who had led the campaign against psychosurgery in the 1970s, again raised an outcry, and the national violence initiative was shelved.[117]

Yet many in the medical, education, and law-enforcement communities have continued to view antisocial behavior in largely biological terms, and professionals in the criminal-justice system of the future will have to decide just how far they want to go in utilizing the biological model of behavior to control human conduct. Already some in the justice community are exploring this issue in earnest. A report issued in May 2000 by the National Institute of Justice discussed both the "promises" and the "uncertain threats" of the emerging "technocorrections" movement, which seeks to apply the newest technologies to crime control and prevention. Author Tony Fabelo described how the "genetic roots of human behavior" might eventually be identified, leading to "genetic or neurobiologic tests" to assess the likelihood a given individual will become a drug addict, sex offender, or other kind of violent criminal.[118] Persons identified as being at higher risk for violent or addictive behaviors could then be held in "preventive incarceration" or treated with drugs to curb their criminal propensities. Potential criminals might also be subjected to twenty-four-hour surveillance through satellite-based tracking systems, miniature cameras, and computer chips implanted in their bodies.[119] Because of the possibilities for abuse, Fabelo urged policymakers to begin to think through the ethical challenges posed by the emerging technocorrections movement while there is still time.[120]

Fabelo focused most of his attention on how such technologies might be applied to offenders already convicted of some crime in order to prevent their recidivism. But conceivably the same techniques could be used for those never convicted of any crime at all.

An Ounce of Prevention

If "Bobby Smith" has a gene that predisposes him to violence, why should society have to wait to intervene until after he has assaulted or murdered someone? How much better it would be to identify Bobby's inherent criminal tendencies while he was still a child and then medically intervene to prevent his criminality from being realized.

Those who consider such a scenario far-fetched should examine the movement already underway to treat the biological roots of antisocial behavior among children. One of the more extreme proponents of this movement is child psychiatrist Harold Koplewicz, a featured speaker at the White House Conference on Mental Health convened by President Bill Clinton in 1999. At that conference, Dr. Koplewicz rejected the idea that "inadequate parenting and bad childhood trauma" could cause psychological problems in children and attacked those who retained the "antiquated . . . notion that somehow absent fathers, working mothers, overpermissive parents are the cause of psychiatric illness in children." In

Koplewicz's view, mental illness is no less "physiological" than diabetes or asthma.[121] While the empirical evidence for this contention is slim,[122] the biological model of mental illness has been widely accepted in both the medical and educational communities, and it has serious repercussions for how we deal with antisocial behavior in children. Once it is assumed that the causes of all mental disorders must be physiological, it naturally follows that the prescribed cures should be physiological as well. Accordingly, Dr. Koplewicz champions drug therapy for children with behavior problems, seeking to fix their supposed brain abnormalities through psychiatric medications. He dismisses worries that doctors might overmedicate children. "I actually think we're not medicating kids enough," he told one interviewer.[123]

Peter Breggin views Koplewicz's comments at the White House conference as a prime example of what is wrong with the biological model of psychiatry. Insisting on reducing every thought and behavior to its material basis, the biological model goes against both common sense and empirical research. Referring to Koplewicz's claims that absentee parents and childhood trauma (presumably including physical and sexual abuse) play no role in causing mental disorders, Breggin says he was "shocked that anyone in the mental health field would dare to deny the mountain of clinical and research evidence that confirms the devastating effect of broken relationships and traumatic events on the lives of children."[124]

Yet Koplewicz is not a fringe figure, as his prominent role at the White House conference attests. Nor is he alone in thinking that Americans do not medicate their antisocial children enough. During the past decade, the number of children receiving psychiatric medications in the United States has skyrocketed.[125] Chief among the drugs being prescribed is the stimulant Ritalin (methylphenidate), used to treat Attention-Deficit/Hyperactivity Disorder (ADHD). ADHD is an ill-defined condition manifested by such symptoms as a dislike for homework, fidgeting, and talking and running "excessively."[126] Because the diagnostic criteria are so broad, the ADHD label has been faulted for being subjective and arbitrary. It is certainly true that diagnosis rates differ markedly across communities and doctors.[127] A physical cause of ADHD has yet to be demonstrated conclusively,[128] and critics complain that ADHD has become a catchall diagnosis for otherwise normal (even bright) children with behavior problems arising from a variety of social factors, such as abusive parents and schools with low expectations. Even some professionals who regard ADHD as a legitimate diagnosis have begun to raise serious questions about the rampant use of psychiatric drugs as a quick fix to treat the disorder.[129]

Despite public assurances by some educators that Ritalin is employed only as a last resort, the facts suggest otherwise. During the 1990s, the sale of methylphenidate in the United States jumped almost 500 percent.[130] By the end of the decade, an estimated 4 million America children were on the drug and doctors were issuing 11 million prescriptions for it each year.[131] In some schools, the proportion of students on Ritalin and similar stimulants has reached 30 or 40 percent.[132] America consumes nearly 90 percent of the world's production of Ritalin, and the drug is now routinely prescribed for children below the age of three, despite the fact that is only approved for use on children age six or older.[133] Pharmacologically related to cocaine, Ritalin is a potentially dangerous drug that can have serious side effects, including psychotic reactions.[134]

Medical doctors as well as educators are apparently finding it hard to resist the lure of the materialistic quick fix offered by Ritalin and other psychiatric drugs. Treating a behavior problem as solely a physical disorder of the brain is much simpler than trying to investigate a child's dysfunctional relationships. It also avoids the messy situation of having to get people to accept responsibility for their actions. In the biological model, the child need not accept responsibility for his misbehavior because he is ultimately controlled by his disordered brain chemistry. Similarly, educators need not take responsibility for creating a stimulating learning environment, because student boredom must be caused by organic brain disease rather than, say, the curriculum. Above all, parents need never feel guilty about failing in their obligations as mothers and fathers, because in the biological model parents' actions are unrelated to their children's distress.

"When we diagnose and drug our children, we disempower ourselves as adults," Peter Breggin testified at congressional hearings in 2000. "While we may gain momentary relief from guilt by imagining that the fault lies in the brains of our children, ultimately we undermine our ability to make the necessary adult interventions that our children need."[135]

Apart from the issue of drug safety, there is the question of how well psychostimulants actually work, and whether the same results might not be achieved with nondrug interventions. The federal government funded a large study during the 1990s that purported to show that drug therapy was significantly more effective than behavioral therapy in treating children with ADHD.[136] Incredibly, however, the study included no control group of untreated children, making it impossible to determine to what extent the study's observed effects were due to the treatments employed. The study also failed to include a placebo treatment group.[137] It is well known that the simple act of taking a pill—even if the pill has no medicinal value—can produce improvements in a patient's physical health. If a patient thinks that the pill has medicinal value, that belief is enough to stimulate physical improvements. This is especially the case with psychoactive drugs.[138] Even more problematic is the way the study measured the improvement of ADHD symptoms in children. Most of the behavioral evaluations were not "blind." Instead, they were filled out by teachers and parents who knew beforehand which children were receiving treatment with drugs. As a result, parents and teachers who believed that drugs were likely to be effective may have been predisposed to report positive effects of drug treatment even when they did not exist.

Interestingly, the study did include some blind third-party observations of the behavior of children enrolled in the study. Unlike the teacher and parent ratings, these blind observations showed no significant behavioral differences between children in any of the treatment groups, which casts doubt on the supposed differences found in the nonblind ratings.[139] The comparison between drug therapy and behavioral therapy was further skewed because the behavioral therapy decreased in intensity over time, while the drug therapy did not.[140] Further analysis of some of the data from the federal study that tried to correct for this imbalance indicated that the relative benefits of drug therapy over behavioral therapy may be considerably less than previously claimed.[141]

A recent federal study of preschool students with ADHD has raised additional red flags. Although the study corroborated the view that methylphenidate can reduce the symptoms of ADHD, it did not demonstrate the drug's efficacy in actually bringing ADHD into remission. Indeed, the study found that there was no statistically significant difference in remission rates between preschoolers who were on the drug and preschoolers who were on a placebo.[142] The study's investigators also discovered that the "tolerability" of methylphenidate "was lower than expected, with 11 [percent] discontinuing it because of a higher rate of medication-related" adverse events.[143]

Unfortunately, the push to put increasing numbers of children on psychoactive drugs has continued unabated, raising concerns about coercive treatment. In 2003, the President's New Freedom Commission on Mental Health recommended comprehensive screening and testing of children for mental-health problems and argued that schools should play a key role in such assessments.[144] In many school districts teachers and school staff—often with little formal training—are encouraging parents to put their children with behavioral problems on Ritalin or other medications.[145] What happens when parents refuse these recommendations? In some jurisdictions they face legal threats.

In New York state, Michael and Jill Carroll wanted to take their seven-year-old son off of Ritalin for a two-week trial period because of the drug's serious side effects. They were reported to the state for child abuse by school authorities, and they then had to fight to clear themselves in family court.[146] According to New York City attorney David Lansner, the Carrolls' experience is not unique.[147] A member of Colorado's state board of education has described similar stories from her state in testimony before Congress.[148] Given public fears about violent crimes perpetrated by disturbed students, one can easily imagine how parental rights in this area may increasingly give way to the perceived needs of public safety. In an effort to safeguard the rights of parents and children to make their own medical decisions, Congress in 2003 held a hearing on a bill that would have prevented educators from requiring students to take medication as a condition of attending school, but the bill was actively opposed by mental-health advocates.[149]

None of this means that psychoactive drugs should never be used to address behavioral problems in children and adults. Clearly such drugs can have a positive impact. But the tendency to reduce all behavioral problems to brain disorders that should be solved primarily through drugs is indicative of just how uncritically our society has embraced the philosophy of scientific materialism. Scientific materialism adds plausibility to the idea that all behavioral problems, including criminal activity, can be fixed through biological and chemical interventions, even if research has yet to sustain such a claim.

The consequences of this overselling of science as the solution to crime and misbehavior have been tragic. Instead of improving the human condition, it has led to dehumanization, with criminal defendants and delinquents treated as guinea pigs for experiments in the service of scientistic ideology.

But the dehumanizing effects of scientific materialism reach far beyond the criminal-justice system. Reductionist thinking has been applied to the fields of business, economics, and welfare—with equally grim results. In the next section, we will look at the pervasive impact of Darwinism and scientific materialism on conflicts over wealth and poverty in America.

Wealth and Poverty

6
Law of the Jungle

In Oliver Stone's *Wall Street* (1987), Gordon Gekko is the consummate corporate raider of the 1980s. Cunning, ruthless, and a master of manipulation, he swallows up companies and people without regard to law or ethics. At a memorable juncture in the movie, Gekko explains his survival-of-the-fittest philosophy to a stockholders' meeting of the troubled Teldar Paper corporation, which he is trying to take over. He rails at the flabbiness of company management and yearns for the good old days of business leadership exercised by "the Carnegies, the Mellons, the men that built this great industrial empire."

"The new law of evolution in corporate America seems to be survival of the unfittest," Gekko complains to the stockholders in disgust. "Well, in my book you either do it right or you get eliminated!"

Gekko then launches into a remarkable ode to greed that remains one of the most striking orations in cinematic history. "Greed is right," he tells the stockholders. "Greed works. Greed clarifies, cuts through, and captures the essence of the evolutionary spirit. Greed, in all of its forms— greed for life, for money, for love, knowledge—has marked the upward surge of mankind!"

The stockholders applaud in enthusiastic agreement, and Gekko takes over the company.

Gordon Gekko is the quintessential champion of what is commonly called "Social Darwinism," the application of Darwin's theory of natural selection to business and economics. Of all the impacts of scientific materialism on American politics and culture, Social Darwinism is likely the most widely known. Even Americans with only a smattering of history have heard about how Darwin's biology was used to justify cutthroat corporate competition starting in America's "Gilded Age" in the late nineteenth century. During an era of robber barons, sweatshops, and corporate greed, Darwinism is supposed to have supplied the perfect rationale for laissez-faire capitalism. Propagated by countless novels and popular accounts of the era, this view of American history has become an integral part of popular folklore, not to mention textbooks.

"According to social Darwinism, the success of huge industries and great industrialists was a sign of their superiority and therefore inevitable," explains one recent history text.[1] "Social Darwinism accounted for brutal business practices and underhand methods, justifying them as the natural 'law of the jungle,'" claims another.[2] "Social Darwinists . . . long claimed that since society operated like a jungle . . . efforts to improve social conditions were misguided attempts to interfere with the 'natural' order," asserts a third textbook.[3] "The new doctrine thus opposed poor relief, housing regulations, and public education and justified poverty and slums," explains another volume.[4]

These same textbooks depict American businessmen as some of the most enthusiastic boosters of Social Darwinism. "Even as they schemed to control market forces, the captains of industry . . . endorsed the principles of social Darwinism, or survival of the fittest . . . that purportedly explained, and justified, why some Americans grew rich while others remained poor."[5] "Defenders of business . . . eagerly embraced the doctrine of Social Darwinism, which seemed to justify aggression in human society."[6] "Social Darwinism appealed to businessmen because it seemed to legitimize their success and confirm their virtues."[7]

No scholar did more to cultivate this view of the fusion between Darwinism and capitalism in the post–Civil War economy than historian Richard Hofstadter. In his classic work *Social Darwinism in American Thought*, originally published in 1944, Hofstadter contended that "it was those who wished to defend the political status quo, above all the laissez-faire conservatives, who were first to pick up the instruments of social argument that were forged out of the Darwinian concepts."[8] Preeminent among the Social Darwinists identified by Hofstadter were the British social theorist Herbert Spencer and the American sociologist William Graham Sumner.

An evolutionist before Darwin, Spencer was the one who actually coined the phrase "survival of the fittest," which Darwin later appropriated as a description of natural selection. (Indeed, Darwin eventually described "survival of the fittest" as "more accurate" than his own term of "natural selection."[9]) Applying this doctrine to human society, Spencer argued that government aid to the poor was counterproductive because it interfered with the natural process of the elimination of the unfit. "The whole effort of nature is to get rid of such, to clear the world of them, and make room for better."[10]

Yale professor Sumner, who Hofstadter called "the most vigorous and influential [S]ocial Darwinist in America," propounded similar themes.[11] He argued that "millionaires are a product of natural selection," and said that "if we do not like the survival of the fittest, we have only one possible alternative, and that is the survival of the unfittest."

According to Hofstadter, America's business leaders enthusiastically embraced the ideology espoused by Spencer and Sumner. In support of this thesis, Hofstadter supplied a collection of seemingly irrefutable quotes from leading men of business who championed Darwinian capitalism. Railroad magnate James J. Hill declared that "the fortunes of railroad companies are determined by the law of the survival of the fittest."[12] Oil baron John D. Rockefeller told a Sunday school group that "the growth of a large business is merely a survival of the fittest," comparing the salutary effects of economic competition to a wise gardener who prunes his roses. "The American Beauty rose can be produced in the splendor and fragrance which bring cheer to its beholder only by sacrificing the early buds which grow up around it," said Rockefeller. "This is not an evil tendency in business. It is merely the working-out of a law of nature and a law of God."[13] Steel magnate Andrew Carnegie proclaimed that Darwin's law of natural selection was essential to human society, no matter the hardships it may pose for specific individuals: "While the law may sometimes be hard for the individual, it is best for the race, because it insures the survival of the fittest in every department."[14]

Hofstadter's claim that Social Darwinism supplied the business community's justification for cutthroat competition in the economy eventually became the popularly received view of late-nineteenth-century capitalism. According to this view, Darwinism's most pernicious social effect happened to be the powerful quasi-scientific rationalization it handed businessmen for old-fashioned greed and injustice.

There is just one problem with Hofstadter's thesis. It is largely untrue.

The Non-Darwinian Foundations of American Capitalism

Fifteen years after Hofstadter published his book, University of Wisconsin historian Irvin Wyllie urged his fellow historians to reevaluate Hofstadter's case for Social Darwinism in the business community. In an address offered at a meeting marking the centennial of *The Origin of Species*, Wyllie argued that the evidence for Darwinism's influence on the American business community during the Gilded Age was spotty at best.

To be sure, there were Darwinists in America who tried to apply Darwin's theory to society, but they "were for the most part scientists, social scientists, philosophers, clergymen, editors, and other educationally advantaged persons."[15] Because most businessmen at the time were not well educated, it would have been strange to find many who were even knowledgeable about the ideas of Darwin and Spencer, let alone advocates of those ideas.

Wyllie acknowledged that the lives of a few business leaders such as Carnegie did reflect the influence of Spencer and Darwin, but he argued that the most salient point was that there were so few examples among the business class.[16] Moreover, some of the examples that had been adduced by Hofstadter were either ambiguous or false. Pointing out an embarrassing misattribution by Hofstadter, Wyllie noted that it was not John D. Rockefeller who defended business as "a survival of the fittest," but John D. Rockefeller Jr., the industrialist's son. "This sentiment, uttered by John D. Rockefeller Jr. in 1902 at an address to the YMCA at his alma mater . . . may prove that the university-trained son knew how to use Darwinian phraseology, but it does not prove that his Bible-reading father was a Spencerian in the Gilded Age."[17]

Nineteenth-century American businessmen and economists did praise free enterprise, vigorous competition, and laissez faire. But these ideas did not originate with Darwin. Instead, they sprang from such classical economists as Adam Smith and Frederic Bastiat.[18] Darwin was not only unnecessary to defend free enterprise and competition, his model conflicted with prevalent ideas about the morality of commerce. Most businessmen of the period would have found it thoroughly unpalatable to justify economic success as the product of an amoral struggle for survival in which only the strongest and most ruthless competitors can survive. Far from believing that commerce was amoral or harsh, nineteenth-century American businessmen argued that business success was squarely based on moral virtue. According to Wyllie,

> anyone who examines the voluminous nineteenth-century literature of business success cannot fail to be impressed that businessmen who talked about success and failure took their texts from Christian moralists, not from Darwin and Spencer. In the race for wealth they attributed little influence to native intelligence, physical strength, or any other endowment of nature, and paramount influence to industry, frugality, and sobriety—simple moral

virtues that any man could cultivate. They urged young men to seek the business way of life in the Bible, not in *The Descent of Man* or *The Principles of Sociology*. The problem of success was not that of grinding down one's competitors, but of elevating one's self—and the two were not equivalent.[19]

Since business success was supposed to be based on "good character," one businessman's success did not have to be built on top of another businessman's failure. Instead, "opportunities for success, like opportunities for salvation, were limitless; heaven could receive as many as were worthy."[20]

At its core, the American resistance to invoking Darwinism as a rationale for laissez faire reflected a rejection of the pessimism at the heart of Darwin's theory, which had been inspired by the Rev. Thomas Malthus's dour *Essay on the Principle of Population* (1798).[21] By Darwin's own account, it was his reading of Malthus that stimulated him to develop his theory of natural selection, and in *The Origin of Species*, Darwin claimed that the "struggle for existence" of which he spoke was merely "the doctrine of Malthus applied with manifold force to the whole animal and vegetable kingdoms."[22] That statement was not quite right, because Malthus had indicated that his theory applied to the entire biological world, not just human beings.[23] But Darwin was correct to suggest that he emphasized the broader implications of Malthus's theory of population more than had Malthus himself.

Malthus argued that men, animals, and plants all tend to reproduce more offspring than nature can support. The inevitable result of this over-population is widespread death until the population is reduced to a level that nature can support. Hence, plants and animals die off due to lack "of room and nourishment, which is common to animals and plants, and among animals by becoming the prey of others."[24] Human beings are similarly killed off by disease, famine, and vice. There is little that can be done to counteract this struggle for survival because it is grounded in the laws of nature. "The perpetual tendency in the race of man to increase beyond the means of subsistence is one of the general laws of animated nature which we can have no reason to expect will change."[25]

Darwin adopted this struggle for existence articulated by Malthus as the foundation for his theory of evolution by natural selection. Darwin wrote that while reading Malthus, "it at once struck me that under these circumstances [of the struggle for existence] favourable variations would tend to be preserved and unfavourable ones to be destroyed. The result of this would be the formation of new species."[26]

Applied to the world of commerce, Malthusian theory presented economics as a zero-sum game. Additional people almost inevitably meant privation for others. The more people there are, the less food there will be to go around. The more laborers there are, the lower the standard wage will be.[27] While Malthus noted some exceptions to this rule, he suggested they were temporary. In America, for example, "the reward of labour is at present . . . liberal," but "it may be expected that in the progress of the population of America, the labourers will in time be much less liberally rewarded."[28] In the Malthusian view, economic progress for the few could only be purchased at the price of misery for the many.

American defenders of capitalism during the latter 1800s explicitly repudiated the Malthusian view of economics, which meant that they also had little desire to invoke Social Darwinism as a defense of free enterprise. In 1879, for example, Harvard political economist Francis Bowen inveighed against "Malthusianism, Darwinism, and Pessimism" in the *North American Review*. Bowen generally supported laissez faire, but he was anything but a Malthusian or Social Darwinist. Contra Malthus, Bowen argued that "the bounties of nature are practically inexhaustible."[29] Therefore starvation and misery among human beings were not inevitable consequences of overpopulation but the products of human ignorance, indolence, and self-indulgence. "It is not the excess of population which causes the misery, but the misery which causes the excess of population," he insisted.[30] Bowen noted that "since 1850 . . . English writers upon political economy have generally ceased to advocate Malthusianism and its subsidiary doctrines," and he observed how ironic it was that "in 1860, at the very time when this gloomy doctrine of 'a battle for life' had nearly died out in political economy . . . it was revived in biology, and made the basis in that science of a theory still more comprehensive and appalling than that which had been founded upon it by Malthus."[31]

American businessman and economist Edward Atkinson likewise denounced any attempt to wed economics to Malthus's theory of population or to David Ricardo's similarly pessimistic theory of rent. Regarding both theories as premised on a false "law of diminishing returns," he also objected that they promoted the spurious view "that the struggle for life must inevitably become more difficult and more violent, and must inevitably fail."[32] In fact, asserted Atkinson, capitalism allowed mankind to alleviate the struggle for existence by creating greater abundance for everyone. Instead of being a destructive struggle for static resources, commerce in a free society could generate sufficient new resources to raise the standard of living for ever greater numbers of people:

> Through competition among capitalists, capital itself is every year more effective in production, and tends ever to increasing abundance. Under its working the commodities that have been the luxuries of one generation be-

come the comforts of the next and the necessities of the third . . . The plane of what constitutes a comfortable subsistence is constantly rising, and as the years go by greater and greater numbers attain this plane.[33]

Atkinson argued that capitalism's abundance was brought about chiefly by harnessing "the mental faculties of man" through new technologies.[34] Malthus's theory "finds as yet no warrant in the experience of men" in large part because "invention and discovery have yielded greater and greater abundance for each given portion of time devoted to the work of procuring subsistence."[35]

So far from promoting the destruction of the many on behalf of the success of the few, commerce could lead to progress for all. "In a free state, governed by just laws, the more the few increase in wealth the more the many gain in welfare."[36] And this increased wealth was generated not by ruthless and amoral competition but by mutual service, which in Atkinson's view lay at the heart of economics. "Commerce is an occupation in which men serve each other; it is an exchange in which both parties in the transaction gain something which they desire more than the thing they part with."[37] Atkinson's optimistic defense of free enterprise was the standard view adopted by businessmen during the Gilded Age. According to Irvin Wyllie, "because American businessmen operated in a land blessed with an abundance of resources they rejected the Malthusian idea that chances were so limited that one man's rise meant the fall of many others. Theirs was a more optimistic view, that every triumph opened the way for more."[38]

If few businessmen or economists during the Gilded Age invoked Darwinism to justify capitalism, how did the idea of the Social Darwinist defender of business arise? Part of the answer undoubtedly lies in the influence of notable intellectuals outside the business community, such as Herbert Spencer and William Graham Sumner. Most businessmen may not have relied on "survival of the fittest" as a justification for their economic views, but social theorists like Spencer and Sumner certainly did.[39] So did a number of prominent biologists, including Charles Darwin himself.[40] Although it is commonly claimed that Darwin had nothing to say about the economic applications of his theory, historian Richard Weikart has ably documented how Darwin drew on natural selection to criticize the work of trade unions and cooperatives, and how he even endorsed a book by German biologist Ernst Haeckel that employed Darwinism to refute socialism.[41] Darwin's colleague Thomas Huxley likewise invoked Darwinian principles to defend laissez-faire.[42]

Still, neither the laissez-faire views of some biologists nor the economic Darwinism of thinkers like Spencer and Sumner seem sufficient to explain the close connection between Darwinism and capitalism in the public mind. Although Charles Darwin was sympathetic to laissez-faire capitalism and occasionally connected it with his biological views, economic policy was hardly a central theme in his public writings.[43] In the late 1860s, when a German economist sent him an article defending economic liberty on Darwinian grounds, Darwin expressed interest but stated that "it did not occur to me formerly that my views could be extended to such widely different, and most important, subjects."[44] Furthermore, while Spencer and Sumner were influential figures in the nineteenth century, it is not clear why their Darwinist view of capitalism should triumph so decisively in the public mind over the non-Darwinian view of many other defenders of capitalism. Thus, the question remains: Why has the link between Darwinism and capitalism persisted for so long in the popular imagination? The answer lies in the persuasive power of the critics of free enterprise.

For the most part, it was not capitalism's defenders but its detractors who identified capitalism with Darwin's theory. Outside of academics such as Spencer and Sumner, those who most often equated capitalism with "the survival of the fittest" were opponents of laissez-faire economics. According to historian Robert Bannister, those on the left believed that the "survival of the fittest" metaphor was an accurate description of the evil realities of unbridled capitalism: "New Liberals and socialists asserted in almost a single voice that opponents of state activity wedded Darwinism to classical economics and thus traded illicitly on the prestige of the new biology."[45]

Arguably the era's most searing depiction of capitalism as Social Darwinism was supplied by socialist Upton Sinclair in his novel *The Jungle* (1906), a lurid expose of the meat-packing industry in Chicago.[46] Although Darwin's name did not appear in the novel, the Darwinian struggle for survival was an underlying theme of the entire book, which told the sad tale of Lithuanian immigrant Jurgis Rudkis and his attempt to make his fortune in Chicago's slaughterhouses. The area surrounding the packing houses was depicted as a vision straight from hell. Stench and flies permeated the air, and the landscape was pockmarked with fetid pools of stagnant water, garbage dumps, and ramshackle houses. Black smoke billowed from the smokestacks of the packing plants, where every day thousands of cattle, hogs, and sheep were scientifically butchered, processed, and packaged.

Death clung to the place. But Jurgis Rudkis was strong and healthy, and he was certain he could earn his own way in the world. When older hands tried to warn him that the meat-packing plants butchered more than hogs and cattle, that the workers were broken and crushed just as surely as the animals they killed, Jurgis would not listen. "That is well enough for men like you . . . puny fellows—but my back is broad," he replied confidently.[47]

That was before he saw his father, his wife, and his son die; before he lost his own job and faced famine, prison, and worse. Only then did Jurgis Rudkis finally begin to understand the grim reality of life under capitalism. "It was a war of each against all, and the devil take the hindmost."[48] Men were pitted in a brutal struggle for survival in which the law of the jungle was the only code. "It was true . . . here in this huge city, with its stores of heaped-up wealth, human creatures might be hunted down and destroyed by the wild-beast powers of nature, just as truly as ever they were in the days of the cave men!"[49]

When Jurgis had arrived at the slaughterhouses, "he had stood and watched the hog-killing, and thought how cruel and savage it was, and come away congratulating himself that he was not a hog." But now he saw clearly "that a hog was just what he had been—one of the packers' hogs. What they wanted from a hog was all the profits that could be got out of him; and that was what they wanted from the workingman . . . What the hog thought of it, and what he suffered, were not considered." It was then that Jurgis realized that the Beef Trust, which controlled the packing plants, "was the incarnation of blind and insensate Greed. It was a monster devouring with a thousand mouths, trampling with a thousand hoofs; it was the Great Butcher—it was the spirit of Capitalism made flesh."[50]

Other writers and polemicists pressed the same themes. A "Declaration of Principles" published by the Nationalist Club in 1889 charged that "the principle of competition is simply the application of the brutal law of the survival of the strongest and most cunning."[51] Henry Demarest Lloyd wrote that "the 'survival of the fittest' theory . . . is practically professed" by American businessmen.[52] Socialist writer Jack London observed that "the key-note to laissez faire" was "everybody for himself and devil take the hindmost . . . It is the let-alone policy, the struggle for existence, which strengthens the strong, destroys the weak, and makes a finer and more capable breed of men."[53]

Nor were popular polemicists the only ones promoting the new view of American capitalism. "Academic economists and sociologists . . . added their authority to the developing stereotype," writes Bannister.[54] Thus, the primary use of Darwinism was not to justify capitalism, but to delegitimize it. As Wyllie points out, "all the leading muckrakers sensed that there was no better way to discredit a businessman than to portray him as a renegade of the jungle."[55]

The Rise of the Darwinian Metaphor in Economics

After critics in the nineteenth century effectively stigmatized capitalism by affixing to it the Social Darwinist label, more businessmen and economists began dancing to the enemy's tune by using the Darwinist metaphor as a defense for free enterprise. By the 1920s, articles in the business press regularly appealed to the Darwinian process as a justification for competition or as a reason to reject government intervention.[56]

"It is the nature of the evolutionary process to discard that which is inefficient or impractical and retain only that which experiment has proved successful," wrote one business journalist. "Most of the changes in the industry and commerce therefore have been for the better."[57] In the same vein, the president of Youngstown Pressed Steel Company argued in another business publication that "survival of the fittest" was one of the "fundamental economic laws," so that "if there are four producers of the same product operating in a field which can only absorb the output of three, then it naturally follows that the profits of all will be reduced until the particular company least able to bear the pressure of extreme competition will gradually become inactive, or . . . pass into the control of its more able competitor."[58] The vice president of Great Northern Railway likewise claimed that "in the process of evolution" some farmers "dropped out and gave way to more competent farmers. In other words the rule of survival of the fittest applied."[59]

At the same time, there was continued resistance to any wholesale appropriation of the Darwinian metaphor among capitalism's most articulate defenders. In his influential work *Socialism* (first published in English in 1932), economist Ludwig von Mises chided the attempt to apply Darwinism and the struggle for existence to economic relations. Although men do engage in a struggle against the "natural environment" in order to survive, the purpose of society is to replace that struggle with social cooperation. "Society . . . in its very conception . . . abolishes the struggle between human beings and substitutes the mutual aid which provides the essential motive of all members united in an organism. Within the limits of society there is no struggle, only peace."[60] While it is true that the "task" of "economic competition" is "selection of the best," Mises argued that the Darwinian metaphor was peculiarly inapt as a description of this process, because competition properly understood "is an element of social collaboration," not social warfare.[61] Accordingly, Mises believed it was utterly inappropriate to equate the destruction of uncompetitive businesses with a Darwinian war for survival:

People say that in the competitive struggle, economic lives are destroyed. This, however, merely means that those who succumb are forced to seek in the structure of the social division of labour a position other than they one they would like to occupy. It does not by any means signify that they are to starve. In the capitalist society there is a place and bread for all. Its ability to expand provides sustenance for every worker. Permanent unemployment is not a feature of free capitalism.[62]

F. A. Hayek expressed similar skepticism about the importation of Darwinian doctrines into the social sciences, complaining that "such conceptions as 'natural selection,' 'struggle for existence,' and 'survival of the fittest' . . . are not really appropriate" in these disciplines. "The whole episode of 'social Darwinism' has, in this field, merely tended to discredit an indispensable intellectual tool."[63]

Darwinian theory did continue to attract the support of some economic thinkers, but the attraction was based on more than a crude "Social Darwinism." Starting in the 1950s Milton Friedman and other economists tried to use the concept of "natural selection" to support their hypothesis that individual business firms "behave as if they were seeking rationally to maximize their expected returns."[64] Even this limited utilization of the Darwinian metaphor was controversial. Writing in the *American Economic Review*, political economist Edith Penrose argued that such "biological analogies contribute little either to the theory of price or to the theory of growth and development of firms and in general tend to confuse the nature of the important issues."[65] More recently, Alexander Rosenberg has claimed that "Darwinian theory is a remarkably inappropriate model, metaphor, inspiration or theoretical framework for economic theory. The theory of natural selection shares few of its strengths and most of its weaknesses with neoclassical theory, and provides no help in any attempt to frame more powerful alternatives to that theory."[66]

A more profound link between Darwinian theory and economics has been made by those who emphasize the ability of economic systems to generate complex order without an overarching designer. Here Darwin's emphasis on the unguided nature of evolution was regarded as the most relevant application for economics. Hayek, who wrote dismissively of "Social Darwinism," championed "the emergence of order as the result of adaptive evolution." This was the belief that "complex and orderly and, in a very definite sense, purposive structures might grow up which owed little or nothing to design, which were not invented by a contriving mind but arose from the separate actions of many men who did not know what they were doing."

Although this view is surely consonant with Darwinism, Hayek emphasized that it originated not with Darwin but rather with the "Scottish

moral philosophers led by David Hume, Adam Smith, and Adam Ferguson." Thus, it was incorrect to say that social sciences such as economics
were Darwinian; it would be more accurate to claim that Darwinism was
an outgrowth of ideas espoused by early social scientists about how societies evolve over time without centralized control. "There can be little doubt
that it was from the theories of social evolution that Darwin and his contemporaries derived the suggestion for their theories," wrote Hayek.[67]

Political theorist Larry Arnhart adopts Hayek's idea of "spontaneous
order" as a key plank in the platform of what he calls "Darwinian conservatism." In Arnhart's view, "spontaneous order" is not only grounded
in the truths of Darwinian biology, but it also flatly contradicts the major
assumption of "intelligent design": "the fundamental premise of the 'intelligent design' argument is that complex order in the living world must be
the deliberately contrived work of an intelligent designer, which denies
Hayek's notion of spontaneous order."[68]

More recent popularizers of a "Darwinian" view of business also stress
the centrality of unguided evolution in business, highlighting in particular the claims of "complexity theory" that complex systems in nature have
"self-organizing" properties that can naturally produce even greater levels
of complexity.[69] Some libertarians see in complexity theory at least a partial vindication of traditional laissez faire. "On the surface, the computer-
assisted discovery of spontaneous order would appear to be a triumphant
vindication of libertarian social theory in general and the Austrian School
of economics in particular," wrote William Tucker in the libertarian magazine *Reason*.[70] Tucker added that "at the heart of complexity theory . . . lies
the notion of freely evolving systems, including social and economic systems."[71] But Tucker also noted that economists who embrace complexity
theory have used it to support government intervention in the economy
rather than laissez faire. He attributed this to their ideological beliefs,
which blinded them to the logical implications of their research.[72]

However, there is a deeper problem with trying to analogize from Darwinism to free markets. The Darwinian process in nature is supposed to
be blind to intelligence and to the future. Random (i.e., non-guided) variations are in the driver's seat. Variations in the business world, however, are
driven by human beings exercising their intelligence and foresight. This
intelligence and foresight may well be limited, but it is neither purposeless
nor completely blind to the future. Moreover, cultural variations are not
transmitted in a manner equivalent to biological inheritance.

Cultural (as opposed to biological) inheritance depends on learning,
teaching, and choice, not on a mechanical process of genetic transmission.
If anything, so-called Darwinian analogies applied to business could be
better described as Lamarckian analogies, because they involve the trans-

mission of characteristics acquired through an organism's conscious efforts to adapt to its environment.

Although it is true that social cooperation may not be guided by a single designer, that is not because the process is driven by random variations but because it results from the intelligent choices of innumerable designers interacting with each other. It is therefore somewhat misleading to say that human order arises "spontaneously." That term suggests a lack of conscious thought, yet human social order arises out of the intentional actions of individuals and groups to associate with each other, to exchange goods, and to improve their environments. Just because these actions are decentralized does not mean they are not designed. This sort of design-driven cooperation is alien to the Darwinian mechanism. It is not alien, however, to the teachings of the classical school of economics that predated Darwin.

Although Larry Arnhart asserts that "a spontaneous order is an unintended order" and that "spontaneous order is design without a designer,"[73] he implicitly acknowledges that this is not literally true. At one point he writes about "allowing social order to arise spontaneously through the mutual adjustment of individuals and groups seeking their particular ends."[74] But order that arises through the "mutual adjustment of individuals and groups" who are pursuing "their particular ends" does not come about through a purposeless interaction of chance and necessity. It comes about through the rational actions of many intelligent designers.[75]

Myths aside, Darwinism has offered little genuine support for laissez-faire capitalism. Few apologists for laissez-faire actually used "Social Darwinism" to justify their views, and those who have cited Darwinian theory often turn out to be applying to their fields of inquiry something other than genuine Darwinism. However, the limited role played by "Social Darwinism" in defending capitalism has not stopped capitalism's critics from continuing to use the metaphor for their own rhetorical purposes.

Survival of the Fittest Rhetoric

During the 1980s, opponents of Ronald Reagan frequently denounced his administration's economic policies as little more than Social Darwinism. Columnist Molly Ivins sniped that Reagan's "mind is mired somewhere in the dawn of social Darwinism," while political analyst Kevin Phillips condemned Reagan for promoting a system of "economic Darwinism" in which "corporate raiders like T. Boone Pickens and Carl C. Icahn perform a social and economic service by weeding out the weak—like wolves in the forest."[76] *Los Angeles Times* editorial writer Ernest Conine likewise asserted

that Reagan "brought with him a crew of ideologues who preached a doctrine of social and economic Darwinism—a dog-eat-dog, survival-of-the-
fittest dogma that included little sympathy for victims of the revolutionary
changes occurring in the U.S. economy."[77]

During the presidential election campaign of 1984, New York governor
Mario Cuomo proclaimed that "President Reagan told us from the beginning that he believed in a kind of social Darwinism, survival of the fittest,"
and Democratic presidential candidate Walter Mondale assured audiences
that unlike his Republican opponent, he believed "in social decency, not
Social Darwinism."[78]

During the 1990s, similar charges were levelled at Newt Gingrich and
the Republican Congress. Columnist Robert Scheer attacked Gingrich's
program of tax cuts and welfare reform as representing "a social Darwinism that holds that only the strong deserve to live."[79] Occidental College political science professor Roger Boesche equated the Republicans' ideas on
welfare reform with the "social Darwinism" of William Graham Sumner,
who believed (in Boesche's words) that "survival of the fittest and poverty
for the least fit are simply laws of nature."[80] And in what was supposed to
be an objective news article, the *Los Angeles Times* described the California Libertarian Party's platform as "a synthesis of social Darwinism, individualism and laissez-faire economics."[81] (A party official retorted that
in his view the Libertarian Party "clearly takes a position in opposition to
contemporary social Darwinism.")[82]

The Social Darwinist label remains a powerful rhetorical tool for anyone wishing to criticize capitalism and limited government. Writing in
2001, former Clinton administration official Robert Reich dismissed the
past policies of Britain's Margaret Thatcher as "Lady Thatcher's small-
government Social Darwinism."[83] In the midst of the financial accounting
scandals of 2002, Kevin Phillips resurrected his Reagan-era critique: "The
1980's and 1990's have imitated the Gilded Age in intellectual excesses of
market worship, laissez-faire and social Darwinism."[84]

These ritual invocations of Social Darwinism have strayed increasingly far from reality. Whatever the actual demerits of the Reagan administration's economic policies, the economic Bible of the Reagan administration
was not supplied by Herbert Spencer but George Gilder, whose best-seller
Wealth and Poverty (1981) articulated a view of economics that was anything
but Darwinian.[85] In Gilder's view, the driving force behind capitalism is
not greed, but something close to altruism. "Capitalism begins with giving," wrote Gilder, and the products generated by capitalists "will succeed
only to the extent that they are altruistic and spring from an understanding of the needs of others."[86] According to Gilder, capitalism not only leads
to greater material plenty for everyone, but properly understood cultivates

the nonmaterial virtues of faith, hope, and love.[87] While Gilder's optimistic assessment of capitalism can be debated, it is not Social Darwinism.

It is just as incongruous to label Newt Gingrich a Social Darwinist. Gingrich was abrasive as a politician, but his vision of a "conservative opportunity society" was not premised on a Darwinian struggle for existence in which only a handful of rich people prosper.[88] Instead, Gingrich preached economic growth as a method of empowering the poor, and welfare reform as a way of helping the poor gain dignity and self-sufficiency. The thinker most associated with this agenda was not William Graham Sumner but Marvin Olasky, whose book *The Tragedy of American Compassion* (1992) argued that welfare should be rethought because it had failed to help the poor. Olasky explicitly condemned "Social Darwinism" toward the poor.[89] Olasky later became the architect behind George W. Bush's "compassionate conservatism." "We believe in social mobility, not social Darwinism," President Bush declared at the University of Notre Dame in 2001.[90]

In the realm of economics, then, Darwinism has had its most pronounced impact in shaping the rhetoric used to criticize, not defend, capitalism. It wasn't merely that Social Darwinism supplied critics with a convenient label with which to denounce capitalism. It was that Darwinian theory supplied critics with a powerful positive rationale for expanding the government's role in the economy.

Socialist Darwinism

"The cardinal tenet of Socialism is that forbidding doctrine, the materialistic conception of history," wrote novelist Jack London in the early 1900s. "Men are not the masters of their souls. They are the puppets of great, blind forces."[91] Many on the far left were attracted to Darwinian theory first of all because it seemed to confirm their materialistic understanding of the natural world. According to Frederick Engels, "Darwin first developed in connected form" the "proof . . . that the stock of organic products of nature environing us today, including man, is the result of a long process of evolution from a few originally unicellular germs, and that these again have arisen from protoplasm or albumen, which came into existence by chemical means."[92] Darwin's materialistic theory was also praised by socialists for banishing purpose from nature. According to Marx, "despite all shortcomings, it is here [in Darwin's work] that, for the first time, 'teleology' in natural science is not only dealt a mortal blow but its rational meaning is empirically explained."[93] Engels agreed. Before Darwin, he wrote, "there was one aspect of teleology that had yet to be demolished, and that has now been done."[94]

In the former Soviet Union, Darwinism ultimately became fused with Marxism as a way of buttressing the state's reigning ideology of materialism. By "the 1930s, Darwinism had become blended with Marxism" in Soviet ideology, writes Russian historian of science Nikolai Krementsov. "The Marxist classics considered Darwin's theory the materialist explanation of biological evolution and praised it to the skies . . . Major pronouncements on Darwinism had been given by . . . leading party figures . . . Most importantly, evolutionary doctrine was taught as a part of the official state ideology, dialectical materialism."[95] The rhetorical prestige of Darwin was so high in the Soviet Union that when Soviet scientist T. D. Lysenko attacked Mendelian genetics and tried to revive a form of Lamarckian evolution, his attack was framed as an effort not to undermine Darwinism but to defend what he called its "rational core."[96] Lysenko inveighed against Mendelian genetics explicitly on the grounds that it "disagree[d] . . . with the Darwinian theory of the development of plant and animal forms."[97] Lysenko's supporters no less than his critics "appealed to the sacral authority of Darwinism," and "both camps repeatedly accused each other of being 'anti-Darwinist.' "[98]

In addition to providing a convincing materialistic account of the natural world, some on the left thought that Darwin's theory of natural selection provided a biological foundation for the class struggle in human society. "Although it is developed in the crude English fashion," *The Origin of Species* "is the book which, in the field of natural history, provides the basis for our views," Karl Marx wrote Frederick Engels in 1860.[99] "Darwin's work is most important and suits my purpose in that it provides a basis in natural science for the historical class struggle," he added to another correspondent in 1861.[100] While Marx ultimately expressed ambivalence about the relevance of Darwin's theory to human affairs, he was not above invoking natural selection to help explain specific economic struggles in *Das Kapital* (1867).[101] Other socialists drew similar connections.[102] "The second great tenet of Socialism . . . is the class struggle," wrote Jack London, who described the class struggle as part of "the social struggle for existence," which itself was based on the Darwinian struggle to survive.[103]

In the socialist version of "the survival of the fittest," it was the members of the proletariat who would prove themselves fittest by joining together to topple their capitalist overlords. In the words of Engels, "the struggle for existence can then consist only in this: that the producing class takes over the management of production and distribution from the class that was hitherto entrusted with it but has now become incompetent to handle it, and there you have the socialist revolution."[104] Upton Sinclair expounded this theme at the end of *The Jungle*. Sinclair recounted a speech by a Socialist newspaper editor who provided a Darwinian argument for why the triumph of the working class was inevitable:

It was a process of economic evolution, he said . . . Life was a struggle for existence, and the strong overcame the weak, and in turn were overcome by the strongest. Those who lost in the struggle were generally exterminated; but now and then they had been known to save themselves by combination—which was a new and higher kind of strength. It was so that the gregarious animals had overcome the predaceous; it was so, in human history, that the people had mastered the kings. The workers were simply the citizens of industry, and the Socialist movement was the expression of their will to survive. The inevitability of the revolution depended upon this fact, that they had no choice but to unite or be exterminated; this fact, grim and inexorable, depended upon no human will, it was the law of the economic process.[105]

Reinforcing the class struggle with Darwinism was not universal among those on the left,[106] and even those who made the connection were not arguing for permanent social conflict of the kind found in the natural world. As Gertrude Himmelfarb points out, "the struggle for existence that Darwin took to be a permanent condition of animal life, Marx saw as a condition only of particular epochs in human history." Marx further assumed that this struggle "would be suspended or transcended to permit the emergence of the classless society."[107] So while Darwin's theory might help explain the ultimate roots of class conflict, it did not provide a prescription for social policy. The point of socialism and Marxism was to change the world, not merely to accept the status quo.[108] For socialists and Marxists, the materialistic understanding of history did not lead to a bleak endorsement of current conditions, but to scientific knowledge about how to reshape the material world to fulfill mankind's boldest dreams.

For mainstream liberals, the major lesson derived from Darwin was somewhat different. What they found most useful about Darwin's theory was the case it made for the necessity—indeed, the inevitability—of change. According to these liberals, nations and their economic systems are subject to evolution just as much as are plants and animals; and like plants and animals, nations must adapt to new conditions or die. Governments must evolve in order to deal with new challenges. The roots of the liberal idea of social and political evolution were supplied not by Darwin but by Hegel and the political science of the German administrative state.[109] But Darwin was honored for showing that the truths preached by the political philosophers had been substantiated by biology.

This faith in the necessity of social evolution underlay the Progressive movement in America. While laissez-faire may have been required by the exigencies of a previous era, new social conditions called for new economic policies, according to the Progressives, and therefore American government must evolve and grow in order to meet the demands of the new circumstances. In adopting their evolutionary view of government, the Progressive reformers had to discard the political theory of the American Founders just as much as they did the teachings of laissez-faire capitalism. The Founders had claimed that government existed to secure certain unchanging natural rights. Because these rights did not change, neither did the purposes of government. The Founders' conception of government was thus more stationary than evolutionary. But such a conception was tantamount to heresy in the new scientific view of the world.[110]

One of the most articulate spokesmen for the new view was a political scientist from New Jersey, who argued that "in our own day, whenever we discuss the structure or development of a thing . . . we consciously or unconsciously follow Mr. Darwin."[111] The political scientist was Woodrow Wilson, then president of Princeton and soon to be governor of New Jersey and eventually president of the United States. During the presidential election campaign of 1912, Wilson explicitly invoked Darwin to justify an evolutionary understanding of the U.S. Constitution that would allow the federal government to dramatically expand its powers over the economy.

According to Wilson, the Constitution betrayed the Founders' "Newtonian" view that government was built on unchanging laws like "the law of gravitation."[112] In truth, however, government "falls, not under the theory of the universe, but under the theory of organic life. It is accountable to Darwin, not to Newton. It is modified by its environment, necessitated by its tasks, shaped to its functions by the sheer pressure of life." Hence, "living political constitutions must be Darwinian in structure and in practice. Society is a living organism and must obey the laws of Life . . . [I]t must develop." Wilson averred that "all that progressives ask or desire is permission—in an era when 'development,' 'evolution,' is the scientific word—to interpret the Constitution according to the Darwinian principle."

The doctrine of the evolving Constitution articulated by Wilson and other Progressives opened the door to much greater regulation of business and the economy, eventually paving the way for the New Deal. But once the door was opened to regulate the rich, there was no reason increased government powers should not also be used to regulate the poor. The stage was thus set for a different kind of Social Darwinism, one based not on laissez faire but on the idea that government should scientifically plan and regulate even the most intimate questions of family life.

7

Breeding Our Way out of Poverty

According to her teachers, Deborah Kallikak could "run an electric sewing machine, cook, and do practically everything about the house."[1] Although it might take her half an hour to memorize four lines, once she learned something, she retained it. She was "cheerful," "affectionate," and learned "a new occupation quickly." She also exhibited an independent spirit. "Active and restless," she was "inclined to be quarrelsome." At the same time, she was "fairly good-tempered."

"The description . . . is one that millions of parents might give of their own teenage daughters," notes a recent writer.[2] But in the eyes of psychologist Henry Goddard there was nothing normal about Deborah. In fact, there was something horribly wrong.

Goddard was convinced that this free-spirited young girl who was kind to animals, loved music, and "was bold towards strangers," was nothing less than a menace to the future of American civilization.

Goddard, who holds the dubious honor of introducing the term "moron" into the English language,[3] was obsessed with how "feebleminded" Americans were degrading their country's racial stock. Deborah was his case in point, and in 1912 he presented his indictment of the dangers she and her family posed to America's survival in his provocative book *The Kallikak Family: A Study in Feeble-Mindedness*. According to J. David Smith, "Goddard's book . . . was received with acclaim by the public and by much of the scientific community," and "it went through several editions."[4] Interest was even expressed in turning it into a Broadway play.

Goddard invoked the Kallikak family to show that the underclass was produced more by bad heredity than bad environment. Born to an unmarried woman on welfare, Deborah Kallikak came from what Goddard believed was a long line of biological defectives. In an effort to prove that her feeble-mindedness was hereditary, Goddard and fellow researchers at the Training School for Backward and Feeble-minded Children in New Jersey zealously tracked down Deborah's relatives and researched her ancestors in search of other defectives.

According to Goddard, a field investigation of the area surrounding the "ancestral home" of Deborah's family "showed that the family had always been notorious for the number of defectives and delinquents it had produced." Indeed, the more Kallikak family members the investigators located, the more deficient the family's bloodline appeared to be.

"The surprise and horror of it all was that no matter where we traced them, whether in the prosperous rural district, in the city slums . . . or in the more remote mountain regions, or whether it was a question of the second or the sixth generation, an appalling amount of defectiveness was everywhere found."[5] Goddard eventually traced the family line all the way back to one Martin Kallikak Sr., a Revolutionary War soldier whose affair with a tavern girl produced an illegitimate son. Of the 480 descendants to come from this son, Goddard claimed to have "conclusive proof" that 143 "were or are feeble-minded, while only forty-six have been found normal. The rest are unknown or doubtful."[6]

Goddard believed that members of the Kallikak family were especially dangerous to America's racial stock because on the surface many of them did not appear to be particularly deficient. "A large proportion of those who are considered feeble-minded in this study are persons who would not be recognized as such by the untrained observer," acknowledged Goddard, whose observations were nothing if not trained.[7]

Deborah was in this category. Goddard complained about "the unwillingness of . . . [Deborah's] teachers to admit even to themselves that she is really feeble-minded,"[8] but he noted that this refusal to face reality was common with teachers. Faced with a "high-grade" feeble-minded girl like Deborah who is "rather good-looking, bright in appearance, with many attractive ways, the teacher clings to the hope, indeed insists, that such a girl will come out all right. Our work with Deborah convinces us that such hopes are delusions."[9]

Published during the same year presidential candidate Woodrow Wilson was campaigning for an evolutionary understanding of the Constitution, Goddard's book urged the nation to apply biological science to its social-welfare policies as well. Calling the family history of the Kallikaks a "ghastly story," Goddard went on to declare that "there are Kallikak families all about us. They are multiplying at twice the rate of the general population, and not until we recognize this fact . . . will we begin to solve these social problems."[10]

In Goddard's view, heredity rather than charity was the key to eliminating the underclass and its associated social ills. By 1912, his message was striking a chord with American policymakers, social scientists, and cultural leaders. New books advocating eugenics were being published, a Broadway play on the subject was in preparation, and professional societies were taking up the topic in earnest.

In Washington, D.C., Dr. Woods Hutchinson of the New York Polyclinic preached eugenics at the annual meeting of the American Public Health Association. Hutchinson proposed that all American schoolchildren be given a eugenics inspection by their third year in school. "As soon as the 2 to 3 per cent of all children who are hereditarily defective are determined they should be given such a training as will fit them for the part they are likely to play in life. Then they should either be segregated in open-air farm colonies or sterilized."[11] A few days later, Dr. L. F. Barker of Johns Hopkins University lectured the International Hygiene Congress about the importance of "providing for the birth of children endowed with good brains" and "denying, as far as possible, the privilege of parenthood to the manifestly unfit."[12]

Eugenics also made inroads in the churches, with the Episcopal hierarchy in Chicago announcing in 1912 that henceforth "no persons will be married at the [city's] cathedral unless they present a certificate of health from a reputable physician to the effect that they are normal physically and mentally and have neither an incurable nor a communicable disease."[13] As Christine Rosen has documented ably in her book *Preaching Eugenics*, many liberal Protestants, Catholics, and Jews became enthusiastic boosters of eugenics.[14]

The eugenics agenda was promoted by a growing number of national organizations, including the American Breeders Association (established 1903), the Eugenics Record Office (established 1910), the Race Betterment Foundation of Battle Creek, Michigan (established 1911), and the American Eugenics Society (established 1923).[15] The American Breeders Association (later renamed the American Genetic Association) was organized at the instigation of President Theodore Roosevelt's Secretary of Agriculture James Wilson and Assistant Secretary of Agriculture W. Hays.[16] Besides publishing an influential periodical eventually titled the *Journal of Heredity*, the association helped create the Eugenics Record Office (ERO) in Cold Spring Harbor, New York. The goal of the ERO was to collect comprehensive eugenics information on "a large portion of the families of America," records which would be stored permanently in the group's fireproof vaults and could be consulted by those who wanted to ensure that their prospective mates were eugenically fit.[17] Secretary Wilson praised those "assembling the genetic data of thousands of families" for "making records of the very souls of our people, of the very life essence of our racial blood."[18]

The American eugenics movement was so well established by 1912 that it was drawing favorable notices in Europe. In fact, in July of that year American eugenists played a starring role in the first International Eugenics Congress held in London.[19] At that event, Professor G. Ruggeri from Italy publicly recognized the American contribution to eugenics, declaring that "thanks to recent researches in the United States, it was now certain that the races of man acted in exactly the same way as the races of animals."[20]

Social-welfare agencies in America struggled to come to grips with the new movement. Traditionally, American charities had focused on ending poverty both by instilling in their clients good moral character and by counteracting the influences of a bad environment.[21] Premised on the idea that people in poverty had the capacity for self-improvement, these efforts operated on the assumption that no one was beyond the possibility of help. However, the eugenists' creed raised a substantial challenge to such an optimistic position, and at the 1912 meeting of the National Conference of Charities and Corrections (NCCC), the nation's social-welfare establishment grappled with what the new biological view would mean for social programs. A trio of prominent eugenists called on the social-service establishment to fundamentally revise the nation's approach to helping the underclass.

Bleecker Van Wagenen of the American Breeders Association opened the discussion by outlining "The Eugenic Problem," describing the increasing burden on American society caused by hundreds of thousands of defectives, including the blind, the deaf, the feeble-minded, the insane, paupers, criminals, and juvenile delinquents. "It is impossible to measure the industrial and social handicap caused by these individuals," said Van Wagenen, "but just as the great leaders of successful human endeavor exert an influence altogether incommensurate with their number, so, doubtless, these classes constitute a correspondingly heavy drag upon society."[22] Professor Robert Yerkes of Harvard University next described the "Scientific Basis and . . . Program" of eugenics, advocating, among other things, the establishment of "a federal department of public welfare" within which would be located a bureau of eugenics.[23] Finally, Charles Davenport of the Eugenics Record Office discussed the relationship between "Eugenics and Charity."

Previously a professor of zoology at the University of Chicago, Davenport was one of the most well-known American propagandists for eugenics. Historian Edward Larson calls him "the universally acknowledged spokesperson for the American eugenics movement."[24] Speaking to the NCCC, Davenport attacked traditional charitable efforts as useless in solving the problem of poverty, and promoted eugenics as the replacement.

"People are unequal less because of unequal external conditions and opportunities than because of unequal innate equipment," he asserted. "In fact it is futile to hope to supply innate deficiency by means of improved environment. Even better schools, more churches, better living conditions, better food, sunlight, air and hours of work will not make strong those without the elements for mental and physical development. The only way to secure innate capacity is by breeding it."[25]

It was a hard sell, and not everyone in the audience was convinced. A blind music teacher from Cleveland was incensed by how the speakers lumped together the blind with criminals and the mentally deficient. "As an intelligent woman but handicapped by blindness I do not in the least object to the classification which has associated us with criminals and [the] feeble-minded," she announced sarcastically.[26] She added that "in my fearful struggle for human life against such a terrible handicap, and in my experience as a teacher I have more than once wished that we had that fearlessness of conscience which would permit us by a painless anesthetic to send every little blind baby back to eternity." She then rebuked the speakers for thinking that handicapped people were necessarily a drag on society. "I would remind you that 60 per cent of the blind people sent out from the schools are self-supporting." She also said that the eugenists should revisit the question of who in society really was most unfit: "When I observe the idle, selfish, shallow sons and daughters of the rich spending their days in worthless pursuits, making no contribution of life and service to society, no answer to the great cry of humanity, I ask myself the questions—who, in the sight of God, are the unfit?"

Some religious social-welfare workers raised even more fundamental objections to the materialistic determinism preached by the eugenists. The claim that the poor were captives of their heredity was something many religious charity workers were loath to admit. They preached empowerment and the ability to overcome one's circumstances, not enslavement either to biology or to the social conditions in which one grew up.[27] Social-gospel proponent Jacob Riis, author of *How the Other Half Lives* (1890), had little patience for eugenist propaganda. Attending a conference on "race betterment," Riis was disgusted with the eugenists' obsession with heredity. "The word has rung in my ears until I am sick of it. Heredity, heredity! There is just one heredity in all the world that is ours—we are children of God, and there is nothing in the whole big world that we cannot do in his service with it."[28] Washington Gladden, another leader of the social-gospel movement, declared in one of his sermons: "Heredity is no excuse . . . Your heredity is from God. He is your Father. Deeper than all other strains of ancestral tendency is this fact that your nature comes from God . . . Environment is no excuse for you . . . God is the great first fact in all our environment, no matter where you may be. There is no place of temptation in which he is not nearer to you than any human influence can be."[29]

Despite its detractors, the American eugenics movement continued to gain both strength and influence among America's elites, inspired by its boosters' almost boundless faith in science and by their almost overpowering fear that without eugenics society was on the road to racial ruin. The mixture of blind optimism and unrelenting fear supplied the recipe for a potent ideological blend.

Sinning against Natural Selection

The eugenics movement drew direct inspiration from Darwinian biology. Yet today the Darwinian roots of eugenics tend to be downplayed both by the popular media and by some scholars. When Darwin's theory is mentioned at all, a sharp distinction is often drawn between Darwin's own views and the "Social Darwinism" of the eugenists, who supposedly extended Darwin's theory into realms unanticipated by him. In the recent book *War Against the Weak* (2003), for example, Edwin Black argued that "Darwin was writing about a 'natural world' distinct from man"; others were to blame for "distilling the ideas of Malthus, Spencer and Darwin into a new concept, bearing a name never used by Darwin himself: social Darwinism."[30] Black seemed unaware that Darwin wrote extensively about the application of natural selection to human beings in *The Descent of Man*. But at least Black acknowledged the influence of Darwinism. Sometimes the connection between Darwinian biology and eugenics is evaded altogether.

On the "Understanding Evolution" website funded by the National Science Foundation, users will find a cartoon showing Charles Darwin yelling "Get out of my house!" to a proponent of eugenics.[31] The intended point is clear: Darwin opposed eugenics. A similar claim was made by a much-heralded museum exhibit on Charles Darwin sponsored by the American Museum of Natural History in New York in 2005.[32] Incredibly, one educator writing recently about eugenics not only failed to mention Darwinian biology, he traced the eugenists' beliefs instead back to the Bible. In his view, eugenics embodied "the biblical concept that 'like breeds like,' to which eugenics researchers provided a scientific gloss."[33]

Yet it was society's violation of the law of natural selection, not biblical doctrine, that provided the operating premise for the eugenists' ideology.

The eugenists' underlying fear was the same as the one Charles Darwin had articulated so clearly in *The Descent of Man*: By saving the weak through medicine and charity, and by allowing defective classes to reproduce, civilized societies were counteracting the law of natural selection to the detriment of the human race.[34]

Time and again American eugenists lamented their country's sins against natural selection. According to former governor of Illinois Frank Lowden, "in a state of nature" defective individuals "would long ago have disappeared from the face of the earth. Starvation, disease, and exposure, if they had been left to their own resources, would have eliminated them long ago. Man's interference with natural laws alone save them from perishing."[35] Harvard biologist Edward East agreed:

> Nature eliminates the unfit and preserves the fit . . . Her fool-killing devices were highly efficient in the olden days before civilisation came to thwart her. It is man, not Nature, who has caused all the trouble. He has put his whole soul to saving the unfit, and has timidly failed to do the other half of his duty by preventing them from perpetuating their traits.[36]

Edwin Conklin, professor of biology at Princeton University, added that while nature may still kill off the worst defectives, "nevertheless a good many defectives survive in modern society and are capable of reproduction who would have perished in more primitive society before reaching maturity."[37] Such defectives survive "in the most highly civilized States" because they "are preserved by charity, and . . . are allowed to reproduce . . . [T]hus natural selection, the great law of evolution and progress, is set at naught." For this reason some eugenists criticized efforts to reduce infant mortality by improving sanitation, hygiene, and prenatal care. According to these critics, such efforts merely postponed the deaths of many defective babies, and those defective babies who did survive into adulthood would drag the race down by perpetuating "another strain of weak heredity, which natural selection would have cut off ruthlessly in the interests of race betterment."[38] Hence, "from a strict biological viewpoint" efforts to reduce infant mortality by improving environmental influences were "often detrimental to the future of the race." Professor H. E. Jordan of the University of Virginia made the same point more generally: "What sanitary science and hygiene seek to accomplish by attention to external conditions alone largely defeats its own ends by counteracting the working of the principle of selection."[39]

Harvard's Edward East argued that "eugenic tenets are strict corollaries" of "the theory of organic evolution," which helps explain why leading eugenists were among the most prominent public defenders of the theory of evolution.[40] In the anti-evolution controversies of the 1920s, for example, the American Association for the Advancement of Science (AAAS) appointed a special committee to publicly defend evolutionary theory. Its membership consisted of three scholars who were also leaders of the eugenics movement: Charles Davenport, Henry Fairfield Osborn, and Edwin Conklin.[41]

Despite their law-of-the-jungle rhetoric, American eugenists did not advocate going back to the days when "war . . . poverty, disease, and capital punishment did a fairly thorough if not a very beautiful piece of work before we began to civilize them away."[42] Instead, they argued that "some substitute has to be found for natural selection." That substitute was the directed selection of eugenics. "Natural selection's death rate of the jungle helped to purify the primitive race by destroying the weak and permitting only the strong to live and reproduce. Eugenicists hope to arrive at the same result by the selective birth rate."[43] Man had to take control of his own evolution by encouraging the "best" to breed more and the "worst" to breed less. According to the eugenists, human beings were essentially no different from horses, dogs, or blackberries, and so the techniques perfected to breed animals and plants could easily be applied to men and women with just as much success. "Man is an organism—an animal," declared Charles Davenport, "and the laws of improvement of corn and of race horses hold true for him also."[44] "All life is conditioned by the same fundamental laws of nature," agreed H. E. Jordan. "It would seem, then, that the same methods that man now employs in producing a high quality breed of dogs, or birds, or cattle, or horses, he must apply to himself."[45] "If the human race is to be permanently improved in its inherited characteristics," wrote Princeton biologist Edwin Conklin, "there is no doubt that it must be accomplished in the same way in which man has made improvements in the various races of domesticated animals and cultivated plants."[46] And since breeders of animals and plants are experts in heredity, the public should let them determine how humans should breed. According to inventor (and eugenist) Alexander Graham Bell, "All recognize the fact that the laws of heredity which apply to animals also apply to man; and that therefore the breeder of animals is fitted to guide public opinion on questions relating to human heredity." Bell said that this represented "an opportunity for the members of the American Genetic Association . . . Most of the disputed questions of human heredity can be settled by them, and their verdict will be acquiesced in by the general public."[47]

The eugenists were thoroughgoing biological reductionists. In their view, social problems like poverty and unemployment were rooted in man's biology rather than his environment or free choices. One eugenist described going into a prosperous town in Iowa and visiting families whose houses were "truly the dirtiest, most ill-smelling places I have ever seen."[48] "Now honestly, my uplifting environmental friend," asked the eugenist, "what can you do for such people? They had plenty of money and ample opportunity. They went to picture shows, and their children attended, or rather were forced to attend, school . . . But their poverty was pure biological poverty, inborn, ineradicable. Their real poverty was poor heredity." If such biological defectives moved into the cities they would "fall naturally into the slums."[49] Similarly, another eugenist proclaimed that "we know that some by no means small proportion of the unemployed were really destined to be unemployable from the first, as for instance by reason of hereditary disease. It were better for them and for us that they had never been born."[50] Those who objected to this view of man as a biologically determined machine were told to stop standing in the way of scientific progress. "Science seeks to explain phenomena in terms of mechanism, and no other interpretation now brings entire satisfaction," argued Charles Davenport. "If human behavior can be brought under a mechanical law instead of being conceived of as controlled by demons or by a 'free' will . . . why should we regret it?"[51]

Many eugenists acknowledged that environment played some role in social problems, but they insisted that heredity was more decisive. "An understanding of the facts of biology leads us to expect that heredity should be nearly all-powerful and the force of environment slight," proclaimed one essayist in the *Journal of Heredity*.[52] "The number of social problems whose solution lies with genetics rather than with ordinary sociology is far greater than anyone except the eugenist realizes," claimed another article.[53]

Because of the primacy of heredity, some eugenists even questioned the utility of universal education. Many students may be biologically unfit for education, they claimed. "The expensive 'special classes' of the public schools are filled with children a large part of whom are morons," reported the *Journal of Heredity*, which complained that "an attempt is made to educate" such students "when an examination of their ancestry would show that it is humanly impossible to educate them, in the way that their playmates are educated."[54]

Not all eugenists were quite so strident, and some endorsed the importance of "euthenics"—trying to improve human beings by improving social conditions. But even the more moderate eugenists maintained that eugenics was required to make such social efforts fruitful. According to Paul Popenoe and Roswell Johnson, "Eugenics is, in fact, a prerequisite of euthenics, for it is only the capable and altruistic man who can contribute to social progress; and such a man can only be produced through eugenics."[55] Accordingly, eugenists like Charles Davenport encouraged philanthropists to shift money from traditional charities to eugenics programs. "Vastly more effective than ten million dollars to 'charity' would be ten million dollars to eugenics," declared Davenport. "He who by such a gift, should redeem mankind from vice, imbecility and suffering would be the world's wisest philanthropist."[56]

If the fear of being swamped by biological defectives was a powerful motivator for eugenists, the hope of achieving biological perfection was equally inspiring. The eugenists' naïve faith in modern science spawned a virulent utopianism. Dressed up in quasi-religious terminology, the eugenics faith promised to create heaven on earth through the magic of human breeding. The utopian vision had been a key part of the eugenics crusade from its inception. Francis Galton had promoted the goal of "gradually raising the present miserably low standard of the human race to one in which the Utopias in the dreamland of philanthropists may become practical possibilities."[57]

American eugenists were no less optimistic about what could be accomplished. "The Garden of Eden is not in the past, it is in the future," promised Albert Wiggam.[58] A "rigidly applied eugenics" eventually will produce an "ideal state of human society!" seconded H. E. Jordan, adding that "thoroly [sic] healthy bodies could develop the highest ranges of mental capacity. There would be little suffering, weakness, sickness, crime, or vice."[59] These benefits of eugenics "may seem utopian . . . But by all the signs of the times, this day is coming . . . And it behooves us as intelligent, moral men and women to do our share . . . to hasten the time of this life more abundant in this kingdom of heaven on the earth."[60] Maynard Metcalf similarly expressed "entire confidence that we shall in time almost banish physical, mental and moral invalidism, which today are most prominent characteristics of the human species."[61] Indeed, eugenics could rid human beings of original sin, allowing society to reengineer human nature and "build a race that is physically sound, intellectually keen and strong and whose natural impulses are wholesome! Not a race of men who are decent because they are restrained from following their natural bent, but a race whose natural quality is wholesome, who need not so much to restrain as to develop themselves."[62] Metcalf urged people to make eugenics their religion. "The people who make eugenics part of their religion and are loyal to its truth will have found . . . the fountain of youth," he declared. Eugenists seemed certain that once man took control of his own evolution, he could do an even better job than nature. "It has taken Mother Nature long, long ages to turn fierce greedy hairy ape-like beasts into such people as we are," wrote feminist eugenist Charlotte Perkins Gilman. "It will take us but two or three close-linked generations to make human beings far more superior to us than we are to the apes."[63]

Some eugenists qualified their utopian rhetoric, but such reservations often seemed halfhearted at best. Wiggam conceded that "the Eden of eugenics can never be attained," but he also urged people to pursue it as their goal, so that "the passion for it, the going toward it, the belief in it, the training and education of men for it, [will] constitute that 'new religion' of a better humanity which Galton said would 'sweep the world.' "[64] At the same time Conklin doubted man's ability to create "a race of supermen," he insisted that "there is no doubt that something may be gained by eliminating the worst human kinds from the possibility of reproduction, even though no great improvement in the human race can be expected as a result of such a feeble measure."[65]

Herbert Walter acknowledged that giving society the power to decide who can bear children might be abused—in theory. "One needs only to recall the days of the Spanish Inquisition or of the Salem witchcraft persecution to realize what fearful blunders human judgment is capable of."[66] Nevertheless, Walter was sanguine that in an age of modern science nothing similar would recur. "It is unlikely that the world will ever see another great religious inquisition, or that in applying to man the newly found laws of heredity there will ever be undertaken an equally deplorable eugenic inquisition."

The attitude of James Wilson, U.S. secretary of agriculture under Theodore Roosevelt, was probably typical of many eugenists. Although he admitted that the promise of eugenics "at first seems like an Utopian vision," he went on to assure people that its goals might be attainable after all. "Like world peace . . . it may come, and may we not all ask . . . Why should it not come? Must science stop in its beneficence with the plant and the animal? Is not man, after all, the architect of his own racial destiny?"[67]

Confident that modern biology had revealed to them how to breed a better race, eugenists set about putting their scientific ideas into action. A few months before the first International Eugenics Congress in 1912, a British eugenist aptly summarized the practical outlook of many eugenists around the world, including those in America. She observed that research had generated "fairly authoritative opinions about certain defects and the method of their transmission. The present necessity . . . is to convert these opinions into social action and legislation."[68] American supporters of eugenics were already well on their way to achieving that goal.

Restrictions on Marriage and Immigration

Marriage laws represented the first wave of eugenics legislation in America. While states had long regulated who could marry, eugenists advocated strengthening legal standards to prevent the "feeble-minded" and others with hereditary defects from marrying, lest they spread their defective germ plasm to the next generation. Connecticut enacted the first eugenic marriage law in 1896.[69] Several other states adopted similar laws soon after the turn of the century, "so that by 1914 more than half of the states had imposed new restrictions on the marriage of persons afflicted with mental defects," writes Edward Larson.[70] Some of these new laws were difficult to enforce, but others enlisted medical professionals as gatekeepers. In Wisconsin, for example, couples could only marry if they obtained a health certificate from a doctor verifying that they were free from physical and mental defects and communicable diseases.[71]

Liberal clergy enthusiastically embraced proposals for health certificates, not only supporting legislation but sometimes imposing their own health requirements on couples seeking a church wedding.[72] Because the new marriage restrictions typically sought to prevent the spread of venereal disease as well as hereditary defects, some eugenists frowned on them, believing that eugenics should not be confused with the effort to prevent communicable diseases.[73] But even if the marriage laws were not purely eugenic, eugenics was unquestionably one of their primary objectives.

Immigration policies were also targeted by some eugenists who believed that biological defectives from foreign countries contributed disproportionately to America's social-welfare problems. Eugenists were by no means the only advocates of immigration restrictions, of course, but their invocation of science provided a powerful new rationale for the restrictions. Writing in 1913, eugenist Herbert Walter urged Americans to select new immigrants in the same way that they might select a new horse. Just as the "wise breeder" looks into the "pedigree of his prospective stock" when "selecting horses for a stock-farm," wrote Walter, "it is to be hoped that the time will come when we, as a nation," will demand "knowledge of the germplasm" of "the foreign applicants who knock at our portals."[74] Walter proposed sending "trained inspectors" to the home countries of prospective immigrants so that they could "look up the ancestry of prospective applicants and . . . stamp desirable ones with approval." After all, "the United States Department of Agriculture already has field agents scouring every land for desirable animals and plants to introduce into this country, as well as stringent laws to prevent the importation of dangerous weeds, parasites, and organisms of various kinds. Is the inspection and supervision of human blood less important?"

The eugenists' anti-immigration arguments attracted the attention of members of Congress, and in 1920 the U.S. House of Representatives held hearings on the "Biological Aspects of Immigration" featuring testimony by Harry H. Laughlin of the Eugenics Record Office. "Our failure to sort immigrants on the basis of natural worth is a very serious national menace," Laughlin testified at the hearings before the House Committee on Immigration and Naturalization.[75] "By setting up an eugenical standard for admission demanding a high natural excellence of all immigrants regardless of nationality and past opportunities, we can enhance and improve the national stamina and ability of future Americans." Laughlin was subsequently appointed "expert eugenics agent" of the House Committee on Immigration, and in that capacity he carried out research and advised Congress as it developed the new immigration law adopted in 1924.[76] That law, which sharply curtailed the number of immigrants allowed from southern and eastern Europe, was hailed by some eugenists and criticized by others.

Those eugenists who supported the law saw it as an important "step forward" in applying—albeit crudely—the principles of selection to immigration, while those who opposed the law pointed out that the act effectively excluded specific racial groups rather than selecting the most eugenically fit immigrants from among all groups.[77] The *New York Times* sided with the critics, arguing that "in every race the great mass is, eugenically speaking, so much deadweight or worse."[78] Thus, the United States should implement an immigration law that would select only the top 10 percent "of all applicants, quite independent of geography." In short, according to the *Times*, the new immigration law was not nearly eugenic enough.

Marriage laws and immigration restrictions, however, were only part of the eugenists' agenda to eradicate chronic poverty and associated social ills. Even more far-reaching was the effort to identify biological defectives throughout America so that they could be incarcerated and sterilized.

The Eugenic Solution to Welfare

This is the law of Mendel.
And often he makes it plain,
Defectives will breed defectives
And the insane breed insane.
Oh, why do we allow these people,
To breed back to the monkey's nest,
To increase our country's burdens
When we should only breed the best?
     —Joseph DeJarnette, Virginia physician[79]

Carrie Buck seemed destined for a life of heartache.[80] Born to parents who were regarded as unfit, she was placed in a foster home at age four. By the time she was ten, her parents had divorced and her mother was labelled mentally defective and incarcerated in the Virginia Colony for Epileptics and the Feebleminded. One can only imagine how Carrie felt about the social stigma of her family background. However, she made the best of her circumstances. She performed well in school—at least for the five grades she had the opportunity to attend—and she attended church and sang in two church choirs.

Then came the terrible summer of 1923, which would change the rest of her life. At age seventeen, she was raped by the nephew of her foster parents. A pregnancy resulted. Instead of holding their nephew accountable, Carrie's foster parents blamed her. Apparently wishing to cover up the scandal, they had Carrie committed to the Virginia Colony for Epileptics and Feebleminded. Shortly before being institutionalized, Carrie gave birth to her daughter Vivian, who was put in a foster home. In September 1924, the board of the Virginia Colony decided that Carrie Buck should be sterilized under Virginia's newly enacted sterilization law, and Carrie suddenly found herself entangled in a court case with national implications.

By the early part of the twentieth century, forced sterilization had become the preeminent policy objective of the eugenics movement. Marriage laws were widely regarded as ineffective, and immigration restrictions did nothing to stop procreation by defectives already in the United States. Permanent segregation of defectives from the general population was theoretically possible, but it was also prohibitively expensive. Sterilization, by contrast, was seen as cheap, safe, and permanent. Some eugenists even argued that after sterilization many feeble-minded persons could be released from their institutions and live productive lives. Albert Priddy, superintendent of the state institution in which Carrie Buck was confined, stated that after the operation Buck "could go out, get a good home under supervision, earn good wages, and probably marry some man of her own level and do as many whom I have sterilized for disease have done—be good wives—be producers, and lead happy and useful lives in their spheres."[81] Priddy was apparently blind to the irony of his comments: On the one hand, he demonized feeble-minded persons like Carrie Buck as menaces to society, while on the other he admitted that they had the ability to "earn good wages," "be good wives," and "lead happy and useful lives." Given such contradictory claims, discerning persons might have wondered whether the feeble-minded really were such a threat to the nation after all.

Lawmakers, however, eagerly embraced the new cure offered in the name of science, and by 1917, sixteen states had enacted sterilization statutes.[82] But then the movement stalled. In some states sterilization laws were invalidated by the courts on procedural grounds, while in others the "laws were in such dispute as to have been de facto suspended in their operation," notes historian Daniel Kevles.[83] Despite the extensive propaganda efforts undertaken by eugenists, opposition to forced sterilization remained potent in many areas of the country. New eugenics legislation was vetoed by the governor of Idaho in 1919, the governor of Pennsylvania in 1921, and the governor of Maine in 1923.[84]

Something had to be done to circumvent hostile court decisions and regain momentum, so in the early 1920s Harry Laughlin of the Eugenics Record Office drafted a new model statute, one he thought could survive a court challenge.[85] The Virginia law adopted in 1924 closely followed Laughlin's model. Eugenists wanted the courts to pass judgment on the law's constitutionality before actually implementing it. They thought they had found the perfect defendant in Carrie Buck.

Under the new law, Virginia's welfare authorities were not allowed to order Buck's sterilization on their own. On paper, at least, the statute provided for due process, and so a lawyer, Irving Whitehead, was duly appointed to defend Buck. Whether Whitehead actually defended Carrie's interests is doubtful. He had served as a board member of the Virginia Colony and had even helped hire Priddy as superintendent. He was also a longtime friend of Aubrey Strode, the former state legislator who drafted Virginia's sterilization law. Historian Paul Lombardo suggests that Whitehead acted in collusion with the advocates of the sterilization law and that he intended to lose the case.[86] Whether or not that is true, Whitehead certainly put on an incompetent defense in rebutting the evidence presented by the state.

Leaving nothing to chance, Priddy and his fellow eugenists did their best to construct an airtight case against Buck. She was depicted as promiscuous, even though she had been raped. She was portrayed as feebleminded, even though she had a earned a good record in school. A number of scientific and medical authorities provided expert testimony on behalf of the state. Without even meeting Buck, Harry Laughlin sent the court a deposition damning her as hereditarily defective. For his deposition, Laughlin drew on "facts" supplied by Superintendent Priddy, including a claim that Carrie and her ancestors "belong to the shiftless, ignorant, and worthless class of antisocial whites of the South."[87] Based on Priddy's report, Laughlin stated that Carrie fit the "typical picture of a low-grade moron," and concluded that "the chances of Carrie Buck being a feeble-minded person through environmental and non-hereditary causes, are exceptionally remote."[88] Laughlin also made clear that sterilization was the appropriate remedy for such hereditary defectiveness. "Modern eugenical sterilization is a force for the mitigation of race degeneracy which, if properly used, is safe and effective. I have come to this conclusion after a thorough study of the legal, biological and eugenical aspects [of the problem]."[89]

Other medical experts reinforced Laughlin's claims. Arthur Estabrook of the Eugenics Record Office performed an infant IQ test on Carrie Buck's daughter Vivian and determined that she was below normal in intelligence.[90] Estabrook also testified to the court that germ plasm, not individuals, was the important unit of analysis in the new age of eugenical science. "We look upon individuals now as merely offshoots of the stock—the germ plasm is what goes through."[91] Superintendent Priddy, meanwhile, praised sterilization as "a blessing" not only for "society" but for "the individuals on whom the operation is performed."[92] When asked whether his patients objected to the operation, Priddy insisted, "They clamor for it."

Buck's lawyer Whitehead made little effort to challenge any of these claims during the trial, leaving the testimony of the state's scientific experts was left uncontradicted. In the view of the state's experts, Carrie Buck was a link in three generations of hereditary defectives (her mother Emma also allegedly had been feeble-minded). Eugenical science therefore dictated that Carrie Buck be sterilized, removing her defective germ plasm from the population.

By the time *Buck v. Bell* reached the U.S. Supreme Court, the evidentiary record was so skewed against Carrie Buck that it might have been difficult for the justices to rule in her favor even had they been sympathetic to her plight. But most of them probably were not sympathetic. Justice Oliver Wendell Holmes, who wrote the opinion in the case, certainly was not. A religious skeptic and a thoroughgoing Darwinist, Holmes believed that society's only hope for achieving "wholesale social regeneration" lay in "taking in hand life and trying to build a race," by which he meant "restricting propagation by the undesirables and putting to death infants that didn't pass the examination."[93]

In *Buck v. Bell*, Holmes not only upheld the sterilization of Carrie Buck, he lauded the wisdom of Virginia's compulsory-sterilization statute. In the next to last paragraph of the decision, he opined that "it is better for all the world if, instead of waiting to execute degenerate offspring from crime or to let them starve for their imbecility, society can prevent those who are manifestly unfit from continuing their kind."[94] He ended the paragraph with a declaration that remains one of the most chilling statements ever penned by a Supreme Court justice: "Three generations of imbeciles are enough." Holmes spoke for a nearly unanimous court. Only Justice Pierce Butler, a politically conservative Roman Catholic, dissented.

Writing a friend a few weeks after completing his opinion in *Buck v. Bell*, Holmes observed with satisfaction that after writing the decision, "I felt that I was getting near to the first principle of real reform."[95] If reviving the sterilization movement was Holmes's objective, he certainly achieved it. Historian Edward Larson recounts that after *Buck v. Bell* "the flow of new [sterilization] legislation turned into a flood."[96] According to Daniel Kevles, from the 1920s to the end of the 1930s, the national sterilization rate jumped fivefold, from 2–4 per hundred thousand to 20 per hundred thousand.[97]

Eugenists in the 1920s marketed sterilization as the cure to what they depicted as a looming welfare crisis. In a 1926 speech at Vassar College promoting sterilization, Margaret Sanger spoke in near-apocalyptic terms about the ruinous costs to taxpayers of welfare spending to care for defectives. "In 1923 over nine billions of dollars were spent on state and federal charities for the care and maintenance and perpetuation of these undesirables," she complained. "Year by year their numbers are mounting. Year by year their cost is increasing. Huge sums—yes, vast fortunes—are expended on these, while the normal parents and their children are compelled to shift for themselves and compete with each other." She added that "the American public is taxed, heavily taxed, to maintain an increasing race of morons, which threatens the very foundations of our civilization."[98] In her bestselling book *The Pivot of Civilization* (1922), Sanger likewise tried to alert Americans to alarming expenditures on social-welfare programs for the mentally defective, urging readers that "our eyes should be opened to the terrific cost to the community of this dead weight of human waste."[99]

Eugenists' tendency to depict the underclass almost exclusively as a threat represented a sharp break with the humanitarian principles espoused by traditional philanthropy. Heavily influenced by Judeo-Christian idealism, traditional welfare workers viewed those at the bottom of the social ladder as fellow human beings worthy of sympathy, mercy, care, and exhortation. Eugenists, by contrast, branded them as enemies of civilization that needed to be eradicated. Despite occasional claims that sterilization would be good for presumed defectives as well as for society, the eugenists' rhetoric clearly dehumanized the poor and tended to depict them as subhuman.

According to Charles Davenport, such persons represented "animalistic strains" from earlier stages of evolution and carried along with them "a torrent of defective and degenerate protoplasm."[100] Harvard biologist Edward East dubbed them "the parasitic fraction of the population," saying they were "like a cancerous growth . . . on the healthy issues of society."[101] Margaret Sanger said they were a "menace . . . to the race" and compared them to "weeds."[102] Plant breeder Luther Burbank reportedly made the same comparison. Speaking about the inmates of "insane asylums and similar institutions where we nourish the unfit and criminal instead of exterminating them," he declared: "Nature eliminates the weeds, but we turn them into parasites and allow them to reproduce."[103] Tart-tongued doctor Lena Sadler from Illinois piled on the lurid metaphors, vilifying defectives as a "viper of degeneracy," a "monster [that] will grow to such hideous proportions that it will strike us down," and "an army of the unfit [that] will increase to such numbers that they will overwhelm the posterity of superior humans and eventually wipe out . . . civilization."[104]

Eugenists also criticized traditional welfare programs for ignoring biological reality and relying instead on sentimental ideals of human equality. Margaret Sanger warned of the "dangers inherent in the very idea of humanitarianism and altruism, dangers which have today produced their full harvest of human waste, of inequality and inefficiency."[105] Lena Sadler prophesied that "civilization is doomed if we continue to drift down the stream of a few more generations on the defenseless raft of mistaken brotherly love and blinded sentimentalism."[106] Edward East attacked as unscientific the idea that "man is created in the image of God,"[107] and further suggested that the claim that all human beings have equal worth is ludicrous. "One of our prominent social workers is quoted as saying that every child is worth $5,000 to society," wrote East. "Stuff and nonsense! Some of them are not worth 5,000 Soviet roubles—they are liabilities, not assets; others are worth golden millions. If prosperity is to be promoted, the assets should be increased and the liabilities reduced."[108] Sadler also questioned the moral validity of treating all children as if they are worth helping. In her view, society could only afford such largesse by demanding from defectives something in return: "If my profession continues to try to save every weak child that is born into the world; if we continue to serve the unfit baby in our welfare stations, dispensaries and clinics, and if this coddled, protected weakling grows to adolescence and shows" itself "manifestly defective" and "likely to produce only unfit individuals," then society must tell the child, "we will do our full duty by you, but there must be no more like you."[109]

The extreme rhetoric of sterilization proponents worked. Many states began to employ sterilization as an important tool to eradicate poverty and reduce welfare spending. In Virginia, state authorities raided welfare families in rural mountain communities and took the women to be sterilized at a state facility. A former county official later recalled that "everybody who was drawing welfare then was scared they were going to have it done on them . . . They were hiding all through these mountains, and the sheriff and his men had to go up after them."[110] In Delaware, the state legislature enacted a sterilization law at the urging of the State Board of Charities, which later declared that the law was "producing remarkable results" and was "one of the most important laws on our statute books."[111] In Vermont, there were regular eugenic surveys to identify defectives among the poor and trace their bad heredity. The surveys eventually led to the adoption of a sterilization statute in the state.[112] The Abenaki Indians were especially ravaged by the Vermont eugenics program. "Many members of Abenaki families who were investigated by the Eugenics Survey were also incarcerated in institutions and subsequently sterilized,"[113] reports historian Nancy Gallagher. Similar efforts to identify and sterilize defectives were undertaken in Indiana by its Committee on Mental Defectives.[114]

During the Great Depression, some eugenists even championed sterilization as a solution to the unemployment problem, which they blamed in part on unlimited procreation by defectives. In 1932, a doctor from the "Essex County Mental Hygiene Clinic" in New Jersey told delegates at the Third International Congress of Eugenics that the "present picture of millions of unemployed" provided evidence for the idea that "our population has already attained a greater number than is necessary for efficient functioning of the race as a whole." He further suggested that "a major portion of this vast army of unemployed are social inadequates, and in many cases mental defectives, who might have been spared the misery they are now facing if they had never been born." Indeed, "it would certainly be understandable" if such people "prefer[red] not to have been born, if they could have known what was in store for them on this earth where the struggle for existence and the urge toward the survival of the fittest makes it necessary for all those who would survive to possess a native endowment of at least average intelligence."[115]

By 1940, almost thirty-six thousand men and women had been sterilized in public institutions across the United States.[116] Nearly half of the operations occurred in California, which performed more than 14,500 sterilizations. Next in line was Virginia, which sterilized nearly four thousand people. Seven other states (Indiana, Kansas, Michigan, Minnesota, North Carolina, Oregon, and Wisconsin) performed more than one thousand sterilizations each. All told, government-sponsored sterilizations took place in thirty states, and 46 percent of the operations were performed on those classified as "feebleminded."

How many of these allegedly "feeble-minded" persons really were mentally handicapped is hard to know. The eugenists were certain they all were. At the beginning of the 1930s, Harry Laughlin claimed that "no one has yet suggested . . . a single instance" in which a state had "made an eugenical error; that is, that it . . . [had] prevented reproduction by an individual whose offspring would, by any token of biology or statistics, probably have been a credit to the state."[117] Before the decade was over, Laughlin could no longer maintain such a fiction. In one particularly notorious case in the mid-1930s, the mother of heiress Ann Cooper Hewitt allegedly had her daughter sterilized in an attempt to steal her inheritance, which would revert to the mother if the daughter had no children.[118] According to a subsequent lawsuit, Mrs. Hewitt "subjected her daughter to a battery of intelligence tests when she was ill with appendicitis."[119] Despite the questionable circumstances surrounding the tests, "a [p]sychiatrist in the California State Department of Public Health declared that Ann had a mental age of eleven, making her a high-grade moron. With this determination in hand . . . Mrs. Hewitt had Ann sterilized."[120]

In the decades following the heyday of eugenics, scholars and journalists exposed just how shaky and subjective the diagnosis of "feeblemindedness" could be. Consider the plight of the "Kallikak" family made infamous by psychologist Henry Goddard. Goddard hid the real identities of the Kallikaks, making it impossible for other scholars to try to verify his account. But through meticulous scholarly detective work, J. David Smith was finally able to identify the family in the 1980s. He conclusively showed that Goddard's assessment was more a product of prejudice than unbiased scientific investigation. The Kallikaks were not hereditarily unfit at all. They had their share of social misfits, but they also had their "strengths and successes. The tragedy of the disfavored Kallikaks is that their story was distorted so as to fit an expectation. They were perceived in a way that allowed only their weaknesses and failures to emerge."[121] What was true of the Kallikaks in general was true of Deborah Kallikak in particular. She may have had certain learning disabilities in the area of language, but she was nonetheless highly capable in other areas and able to successfully and responsibly perform a variety of complex tasks.[122] "Visitors and new employees often expressed disbelief when told that she was mentally retarded."[123] One person even mistook her for the teacher of the kindergarten class at the institution. Despite her evident abilities, Deborah was institutionalized until her death in the late 1970s.[124]

Ironically, the person who had done the most to stigmatize Deborah as a menace to the nation may have come to regret his role in the affair. By the end of the 1920s, Henry Goddard had moderated some of his earlier views. He had come to believe that education could help the "feebleminded" and that they "do not generally need to be segregated in institutions."[125] It was a stunning reversal, but it provided scant comfort to the victims of his earlier work.

Carrie Buck was another example of the slipshod way in which people were labelled "feeble-minded" and selected for sterilization. By the time of her death in the early 1980s, she was no longer considered mentally unfit. She was said to be "an avid reader," and she wrote perfectly coherent letters.[126] She married, joined the Methodist Church, and returned to singing in the church choir. Her first marriage lasted nearly a quarter of a century, ending with her husband's death in 1956. Her second marriage lasted until her own death in 1983. In an interview with reporters in 1980, she revealed that she was never informed by state authorities about the purpose of her operation. "All they told me was that I had to get an operation . . . I never knew what it was for. Later on, a couple of the other girls told me what it was. They said they had it done on them."[127] She regretted not being able to have children, but she wasn't bitter. "I tried helping everybody all my life, and I tried to be good to everybody. It just don't do no good to hold grudges."[128]

"She spent most of her adult life helping others," wrote J. David Smith.[129] "She was a trusted caregiver to elderly people and one of her employers told me that Carrie could not have been mentally retarded. Her competence was obvious, she said, in the quality of care she gave to those who depended on her. 'There was nothing wrong with that woman's mind,' said the employer."

Carrie Buck's sister Doris was also sterilized by the state, although she did not discover that fact until she was in her late sixties. She had been told her operation was an appendectomy.[130] She and her husband had spent years trying to have children without success, and she was heartbroken when she finally learned the true reason she could not bear children. "I broke down and cried. My husband and me wanted children desperately. We were crazy about them. I never knew what they'd done to me."[131] Like her sister Carrie, Doris Buck was no longer considered mentally defective in the later years of her life.

There were other Carrie and Doris Bucks around the country. After the death of their father from pneumonia, Fred Aslin and six of his siblings were taken from their mother and confined to the Lapeer State School, an institution for mental defectives. Aslin was branded a "feeble-minded moron," and eventually he, four brothers, and a sister were sterilized.[132] "They termed us feeble-minded idiots, and wrote that our children would be like us or even worse," he recalls today.[133] "My bandleader came to me and said, 'They want you to sign papers to get sterilized. You might as well go along with it. They say if you don't sign, they'll get your mother to sign.' I said, 'No, I don't want it! I don't want anyone cutting on me!'"[134] But he was sterilized anyway at age eighteen. This alleged moron was a model student at the school who earned praise from his teachers. "Fred is the best reader in the group," wrote one teacher.[135] "Fred is decidedly the leader of his group," wrote another.[136] The state continued to confine Aslin after his operation, and he had to hire a lawyer in order to get released.[137] He later won a Purple Heart for military service in Korea and went on to marry a widow and adopt and raise her two children as his own.[138] Needless to say, he is not regarded as a "moron" today. Neither is his brother Ted, another victim of sterilization. Ted also married and adopted a son. Michigan later licensed him "to be a foster parent to roughly 100 children over the course of a decade."[139]

Carrie and Doris Buck were poor whites, but the Aslins were Native Americans, which has led some to suggest that race played a role in their being targeted by state welfare authorities.[140] There is probably more to that charge than idle speculation. Although feeble-mindedness was supposedly diagnosed without regard to color, the campaign clearly had racist overtones. So did the eugenics movement as a whole, which regularly drew on the teachings of Darwinian biology to stir up fears about "race suicide."

Eugenics, Darwinism, and Race Purity

Eugenists believed that natural selection had produced races of human beings with unequal capacities. "The more we study this process of selection, the more we realize why one race differs from another in temperament and mentality as well as in physique," wrote Yale geographer Ellsworth Huntington in his book *The Character of Races* (1924).[141] Biologist Charles Davenport claimed that racial differences arose as evolutionary adaptations: "Each race of man that has long persisted in a distinct environment has gained, by preservation of useful mutations, certain adaptations to that environment. The useful phaenotypical adaptations have enabled their possessors to survive and the genotype that produced them continues the characters of the race."[142] For example, "the high intelligence and the ambition of the European races" could be regarded "as an adaptation to the competition and crowding arising in a life largely devoted to barter and commerce." Similarly, "the fear of darkness, in the negro race," could be explained as an adaptation to "a country where lions and other predaceous animals prowl at night." As the latter example suggests, not all evolutionary adaptations remained beneficial in civilized society, according to eugenists. As a result, some races were better equipped by evolution to deal with the challenges of modern life than others. "We have abundant evidence today of an innate difference in capacity of learning, of forming judgments, of profiting by experience in different strains of humans," wrote Davenport. "In fact it seems probable that in the same country we have, living side by side, persons of advanced mentality, persons who have inherited the mentality of their ancestors of the early Stone Age, and persons of intermediate evolutionary stages."[143]

Bluntly put, the evolutionary process had led to the development of superior and inferior races. Americans were able to cite Charles Darwin himself in support of this idea—and did. Although Darwin opposed slavery and according to some scholars "personally opposed programs premised on the permanent inferiority of nonwhites,"[144] he did maintain in *The Descent of Man* that human intellectual development was the product of natural selection and that there were significant differences in the mental faculties of "men of distinct races."[145] In the same book, Darwin disparaged blacks and observed that the break in evolutionary history between apes and humans fell "between the negro or Australian and the gorilla," indicating that he considered blacks the most ape-like humans.[146] Darwin also predicted that "at some future period, not very distant as measured by centuries, the civilised races of man will almost certainly exterminate and replace throughout the world the savage races."[147]

The racist cast of Darwin's thought is difficult to deny, but that has not stopped some scholars from doing their best to downplay it. The evidence for Darwin's racism "rests largely on guilt by association and scattered quotations," concludes historian Robert Bannister.[148] Darwin's "ideas may have been 'pervasive,' but neither his doctrine of natural selection nor his personal philosophy was inherently racist," echoes another scholar.[149]

Of course, I am not claiming that Darwin's theory created racism, nor even that racism was its inevitable result. Defenders of Darwin are correct to emphasize that racism preceded Darwin's work, and that certain critics of Darwinism have been just as racially bigoted as Darwinists. There are even scattered examples of people trying to use Darwin's theory to undercut racism.[150] These ambiguities are important to point out, but they do nothing to erase Darwin's own racist statements. Nor do they refute the overwhelming evidence that American eugenists regularly drew on Darwin's theory as a powerful scientific justification for racism and as a rationale for racist public policies.

The Darwinian justification for racial inequality was so culturally pervasive a century ago that it was embraced even by scholars who were skeptical of the period's more dogmatic assertions of white supremacy. University of Oklahoma sociologist Jerome Dowd professed to be sympathetic to the plight of American blacks and criticized claims of Nordic supremacy.[151] But in *The Negro in American Life* (1926) he nevertheless concluded that racial equality could not be accepted except in a very "limited sense" because such an idea ran counter to the clear findings of evolutionary biology. Even if "we have reason to believe that all races of men have the same mental faculties, and that in general ability to learn they differ in no important degree," the fact remains that "due to many centuries of natural selection, the races of men have not now equal capacity to adapt themselves to the same environmental conditions, nor to attain to the same accomplishments."[152] Thus, to argue for "racial equality" as a general principle would be tantamount to rejecting modern biology: "Racial equality means that, whereas differences in hereditary value exist among all varieties of plants and animals, the races of men form an exception to the rule . . . It means that the biological principle of natural selection does not apply to human beings . . . It means that sexual selection is inoperative among men . . . It means that the science of eugenics is 'bunk.'"

Dowd concluded that to embrace such an incredible view would be "a complacent philosophy, which no man of the first order of ability has ever believed in."[153]

Unquestionably the most blatant example of racism in the eugenics movement was its strident opposition to racial interbreeding. Because eugenists maintained that "fundamentally . . . racial differences are gene differences,"[154] they were especially concerned about the consequences of allowing interbreeding between "superior" and "inferior" races. In *The Passing of the Great Race* (1921), Madison Grant denounced the American ideal of the "melting pot" and insisted that the inevitable result of race-crossing was the degeneration of the superior race. "The result of the mixture of two races, in the long run, gives us a race reverting to the more ancient, generalized and lower type. The cross between a white man and an Indian is an Indian; the cross between a white man and a Negro is a Negro; the cross between a white man and a Hindu is a Hindu."[155]

Grant ended his book with a dire warning that "the altruistic ideals which have controlled our social development during the past century and the maudlin sentimentalism that has made America 'an asylum for the oppressed,' are sweeping the nation toward a racial abyss."[156] In his view, America's only hope was to study the history of evolution and then apply what was learned to guide the future development of the human race. "We may be certain that the progress of evolution is in full operation to-day under those laws of nature which control it and that the only sure guide to the future lies in the study of the operation of these laws in the past."[157]

Eugenists applied their concerns about "race-crossing" with special virulence to blacks, who they thought represented a more primitive stage of human evolution, or at the very least, the product of evolution gone awry. Claiming that "wherever the negro has been placed he has . . . failed miserably and utterly by the white man's standards," biologist Edward East said that such a record lent credence to the conclusion of British eugenist Karl Pearson that "the negro lies nearer to the common stem" of man's evolutionary tree "than the European."[158] Charles Davenport explained that the reason "a smaller proportion" of blacks than whites exhibited "self control," a "special regard for property rights," and an "appreciation of cause and effect" was that "the Negro from Africa . . . had not evolved in the direction of these traits."[159] Davenport further implied that blacks brought to America on slave ships had been fitted by nature for slavery. "Scores of thousands of black men from the interior of Africa . . . had been kidnapped by the more enterprising natives that lived along the coast. These negroes represented some of the mentally feeblest races of the globe, with an inborn docility and fidelity which made them good slaves."[160]

To substantiate their claims of Negro mental inferiority, Davenport, East, and others cited the results of Army intelligence tests of recruits during World War I.[161] After those tests were discredited, Davenport trumpeted new research in Jamaica purporting to show that "in tests involving some organization, foresight and planning . . . the negroes seem to be inferior to the whites."[162]

This supposed biological inferiority of blacks supplied the scientific rationale for preventing intermarriage between blacks and whites. Unlike Madison Grant, who seemed to condemn any type of race-mixing out of hand, biologists Davenport and East conceded that race-crossing was not necessarily harmful.[163] After all, the whole idea of eugenics was to breed better strains of human beings, and the experience of plant and animal breeders showed that hybridization could produce superior strains. But Davenport and East argued that trying to hybridize widely separated stocks could produce seriously defective offspring, and they implied that this was what would happen if whites and blacks interbred. According to East, it was foolhardy to allow "racial crossing even between widely separate races of equivalent capacity simply because the operation of the heredity mechanism holds out only a negligible prospect of good results against a high probability of bad results."[164] How much more so, then, should interbreeding be discouraged between races of markedly unequal capacities such as white and blacks.[165] "The negro race as a whole is possessed of undesirable transmissible qualities both physical and mental, which seem to justify not only a line but a wide gulf to be fixed permanently between it and the white race."[166]

East believed that the greatest danger confronting white society from race-crossing came from "the mulatto," not "the pure black."[167] Drawing again on research from Jamaica, Davenport contended that mulattoes were spoiled by their white blood:

> While the full blooded negro is, for the most part, easily satisfied with his lot and is loyal and devoted, the mulatto is dissatisfied and often rebellious. This difference is probably due to a disharmony introduced by the cross. The mulatto shows an ambition and push, combined often with an

intellectual inadequacy, which makes him dissatisfied with his lot and a nuisance to others.[168]

In Davenport's view, white germ plasm made blacks just intelligent enough to be dissatisfied, but not intelligent enough to be able to better themselves. Hence, society should err on the side of caution when race-mixing was concerned:

> In general, we have enough evidence of disharmony in human hybrids to urge that it is on the whole bad when wide crosses are involved. Valuable new combinations might possibly arise through hybridization; but society has not yet worked out a plan by which such better combinations may be encouraged to reproduce, while the worse combinations should remain sterile. Until it does race crossing is not to be encouraged.[169]

Davenport and East helped supply a scientific justification for more restrictive anti-miscegenation laws that targeted mixed-race persons. In 1924, Virginia enacted "An Act to Preserve Racial Integrity" that redefined white persons as those who have "no trace whatsoever of any blood other than Caucasian," excepting persons "who have one-sixteenth or less of the blood of the American Indian and have no other non-Caucasic blood."[170] The law made it unlawful for a white person to marry anyone not meeting the new, stricter definition of "pure white." A prime backer of the law was W. A. Plecker, Virginia's registrar of vital statistics and a fervent eugenist. Plecker was not above using intimidation to stop marriages between whites and anyone with even a trace of Negro blood, or to ensure that no children with any African ancestry (no matter what their actual color) attended schools reserved for whites.[171]

A member of the American Eugenics Society, Plecker corresponded with Charles Davenport, and he delivered a paper about Virginia's efforts to keep the white race pure at the Third International Congress of Eugenics in New York.[172] He began his presentation with the presumption "that no one in this audience will dispute the wisdom and desirability of preserving the different races of man in their purity . . . [T]he preservation of racial purity is one of the fundamental objects of eugenic endeavor."[173]

Although Davenport favored anti-miscegenation laws, he saw them as only a small part of the solution to the problem posed by American blacks.[174] Such restrictions, after all, only curtailed the damage blacks inflicted on the germ plasm of the white race. But what about the drag on society caused by defective blacks who bred among themselves? In notes for a talk titled "A Biologist's View of the Negro Problem," Davenport argued that the most effective solution to the South's racial problems would be to focus on the feeble-minded rather than directly on race.[175] According to Davenport, blacks supplied a disproportionate number of the South's feeble-minded. So if Southern states segregated and/or sterilized all their feeble-minded, they would eliminate the most burdensome blacks in the process.

Blacks were not the only biologically inferior race according to eugenists. Asians, Native Americans, and whites from southern and eastern Europe were also denigrated by eugenist reformers. Underlying their contempt for other races was their adulation of the "Nordic races." Madison Grant wrote that the Nordics were "above all" a race "of rulers, organizers and aristocrats."[176] Although scientists like Edward East distanced themselves from the rhetoric of Grant, Grant was far from a pariah in the American scientific community. He served as chairman of the New York Zoological Society, as a board member of the prestigious American Museum of Natural History, and as councilor of the American Geographical Society.[177] His book *The Passing of the Great Race* went through multiple editions, each with a congratulatory preface by zoologist Henry Fairfield Osborn of Columbia University. Nor were appeals to white supremacy limited to Grant. Several years before the publication of Grant's book, the *American Breeders Magazine* informed readers that the "aryo-germans which all through history have proved to be carriers of culture and civilization can assure themselves of the continuance of their dominance in world's [sic] affairs, and of the permanence and even brilliant expansion of the splendid civilization they have created, by scientifically directing their evolution."[178] During debates over forced sterilization in Louisiana, meanwhile, one of the main proponents of a proposed sterilization law declared: "If something of this sort is not done soon, our nordic civilization is gone."[179]

Given the views of African inferiority and Nordic supremacy held by some American eugenists, it is not difficult to see why they were blind to the horrors that unfolded in Germany during the 1930s. Indeed, when Germany adopted sterilization legislation in the 1930s, a number of American eugenists praised the measure, while German eugenists acknowledged their debt to previous American sterilization laws.[180] Hitler told one of his colleagues that he had "studied with great interest the laws of several American states concerning prevention of reproduction by people whose progeny would, in all probability, be of no value or be injurious to the racial stock."[181] And in 1936, the German consul in Los Angeles, Dr. G. Gyssling, issued a letter expressing his government's thanks "to all those American organizations and men who have worked in the line of Human Betterment" (meaning eugenics). Observing that when Germany "passed its National Hygiene Legislation, it was well aware of the work which had been done already in this field in the United States," Gyssling highlighted the research and writing of California eugenists E. S. Gosney and Paul Popenoe, reporting that their work "proved to be a valuable contribution to the considerations which led to the legislation in question."[182]

Popenoe earlier had praised the Nazi government for basing its sterilization legislation on science rather than ideology. "The policy of the present German government is . . . to gather about it the recognized leaders of the eugenics movement, and to depend largely on their counsel in framing a policy which will direct the destinies of the German people," he wrote in 1934. Conceding that "mistakes will be inevitable," Popenoe insisted that "the Nazis seem, as this scientific leadership becomes more and more prominent in their councils, to be avoiding the misplaced emphasis of their earlier pronouncements on questions of race, and to be proceeding toward a policy that will accord with the best thought of eugenists in all civilized countries."[183] Popenoe also reported with satisfaction that "Hitler . . . has long been a convinced advocate of race betterment through eugenic measures," and pointed out that in *"Mein Kampf . . .* he bases his hopes of national regeneration solidly on the application of biological principles to human society."[184]

Some American eugenists were envious of the Nazi sterilization program; it was much more comprehensive than patchwork American efforts. "The Germans are beating us at our own game," complained Virginia doctor Joseph DeJarnette in the mid 1930s.[185] Superintendent of the Western State Hospital in Virginia,[186] DeJarnette wanted to implement more aggressive eugenic restrictions in the United States. He was not alone. Harry Laughlin in 1930 was already expressing the hope that state governments would take eugenics to the next level. Sterilization of "extreme cases" was praiseworthy, to be sure, but "in the future the several states may well look toward the establishment of a still higher biological standard for the legalization of parenthood."[187]

Laughlin hoped in vain. Instead of becoming more aggressive, the American eugenics crusade gradually dissipated over the next two decades as opposition mounted among religious traditionalists, doctors, and scientists.

The Decline of Eugenics

Although endorsed by many among the intellectual elite, eugenics never escaped public controversy. When the Nebraska legislature passed a forced-sterilization law in 1914, for example, Governor John Morehead vetoed the bill, charging that "it seems more in keeping with the pagan age than with the teachings of Christianity. Man is more than an animal."[188]

Nebraska eventually enacted a sterilization statute, but the national controversy over eugenics persisted, fed in large part by the opposition of evangelical Protestants and traditionalist Roman Catholics. While many liberal and modernist clergy offered fervent support for eugenics, conservative religious leaders were much more skeptical.[189] Evangelical firebrands William Jennings Bryan and Billy Sunday both condemned eugenics, with Bryan dismissing it as a program for "scientific breeding . . . under which a few supposedly superior intellects, self-appointed, would direct the mating and the movements of the mass of mankind—an impossible system!"[190]

Roman Catholics probably provided the stiffest resistance. Pope Pius XI strongly condemned forced sterilization in an encyclical in 1930, criticizing eugenists for calling on "the civil authority to arrogate to itself a power over a faculty which it never had and can never legitimately possess."[191] In a number of states Catholic clergy actively challenged the drive for sterilization. In Louisiana, the Catholic Church played a key role in frustrating eight years of fierce agitation for a statewide sterilization law.[192] Catholic priests could also exert powerful influence at the local level, as a hapless welfare official in North Dakota discovered during the summer of 1937.

The imbroglio started when Joseph Boucher of Bottineau County, North Dakota, went to the local office of the County Welfare Board seeking additional support for his family of seven children.[193] A "respectable farmer" before going on relief during the Depression, Boucher had part-time work through the Works Progress Administration. But "$40 a month was not quite enough to get along," Boucher told local welfare official R. H. Mehus. He asked Mehus to visit his family and see for himself. Instead of dispensing more support, Mehus allegedly told Boucher that he had too many children and should not have any more. After Boucher replied that "it could not be helped," Mehus allegedly said that Boucher could be sterilized. He recommended a doctor who could do the operation and promised that the County Welfare Board would pay for it.

Troubled by Mehus's proposal, Boucher consulted his priest, who took the issue to his superior, Father Joseph Andrieux. Andrieux acted swiftly. He had Boucher swear out an affidavit about his meeting with Mehus, and he asked the County Welfare Board to hold a hearing to investigate what had happened. In the meantime, Andrieux spoke to the state's attorney in the area and learned that others had made similar complaints against Mehus. He also talked with Wilfred Dore, who allegedly told him that he was sterilized with government money after being advised to get the operation by Mehus. Father Andrieux subsequently had a heated confrontation with Mehus. According to the priest, Mehus confessed to him his misdeeds and agreed to resign. Mehus did not resign, however, and at the Welfare Board hearing the facts of the situation quickly became a muddle. Wilfred Dore now denied that Mehus had advised him to be sterilized, and Mehus denied the charge as well.[194] Regarding Boucher, Mehus claimed that it was Boucher who first raised the subject by telling Mehus "that he did not want to have any more children if he could help it."[195] That was when "he advised him to go down and see a doctor." Unable to determine conclusively what happened, the Board declined to ask for Mehus's resignation. But it did pass a resolution stating that "it is not, and never has been, the policy of our County Welfare Board to advocate, advise or discuss this subject [of birth control] in any way."[196]

Not satisfied with the result, Father Andrieux wrote an article in the local newspaper denouncing what had happened to Boucher and urging the removal of Mehus from his position. Father Andrieux railed at what he viewed as an attempt to prevent poor people from having "any children without a special permission from a bureaucrat."[197] "Is this Russia, or is it still America?" he asked. "Even in Russia and Germany they advocate sterilization only of the feebleminded and diseased, but not of men of sound mind and good health." Andrieux ended with an appeal for people to petition the governor to remove Mehus. Eugenists responded by defending Mehus, and Margaret Sanger sent a letter to the Welfare Board urging that the board retain him: "If the facts are as stated . . . we feel that Mr. Mehus should be commended for suggesting that no more should be born to an individual who already had more than he could feed and clothe . . . [S]terilization was certainly indicated in this case."

Sanger went on to complain that North Dakota's existing sterilization law was too narrow because it only applied "to certain types of socially inadequate inmates of state institutions . . . It is a pity that the law is not broadened to require sterilization for certain indications of social inadequacy among the population at large."[198] Sanger's letter arrived too late to influence anything. Responding to Father Andrieux's call to arms, two thousand citizens had already signed petitions to the governor urging that Mehus be sacked, and Andrieux claimed to have gathered additional affidavits supporting the original charges.[199] Three days before the petitions were presented to the governor by Father Andrieux, the County Welfare Board met again and unanimously asked for Mehus's resignation.[200] Several days later, Andrieux wrote a second article for the newspaper replying to an essay by a local eugenist. "National advocates of sterilization regard those on relief as a distinct class of undesirables," he wrote. But "the thousands on relief in North Dakota do not constitute a class of undesirables. Most of them are respectable citizens who are suffering from misfortunes, brought about through crop failures and unfair prices." He insisted that if farmers had received fair prices for their crops in recent years, most of those on relief "would be taking care of their families today. We would have no public relief officials with fat salaries dictating the private lives of respectable citizens."[201]

Opposition to eugenics was not limited to traditionalist clergy. From the start, there were also dissenters within the medical community. In 1915, a local newspaper reported that there was "much opposition" to the Wisconsin eugenics law on the part of the doctors in Dane County, Wisconsin.[202] The issue was of such interest that the Dane County Medical Society invited the governor of the state to address the group about the law. Eventually some biologists began to raise questions about the eugenics crusade as well, although the extent of their opposition was in many cases limited.

Thomas Hunt Morgan, who would win the Nobel Prize in 1933, was one of the few leading geneticists to express open skepticism about the rationale for eugenics by the 1920s.[203] In the 1925 edition of *Evolution and Genetics*, Morgan emphasized just how little scientists knew about the hereditary basis of human defects such as "feeblemindedness." Given this lack of knowledge, he concluded that "social reforms might, perhaps, more quickly and efficiently get at the root of a part of the trouble," and he cautioned that "until we know how much the environment is responsible for, I am inclined to think that the student of human heredity will do well to recommend more enlightenment on the social causes of deficiencies rather than more elimination."[204] Morgan nonetheless agreed that a limited program of negative eugenics might be useful: "I will . . . venture to go so far as to suppose that the average of the human race might be improved by eliminating a few of the extreme disorders, however they may have arisen."[205] Despite his reservations, Morgan did not crusade against eugenics, and by the early 1940s he apparently had agreed to become a member of the Human Betterment Foundation, one of the most strident pro-sterilization groups in the nation.[206]

Historian Daniel Kevles also identifies biologists Edwin Conklin and Herbert Jennings as among the critics of what he calls "mainline" eugenics.[207] That is true, but only to a point. Conklin was skeptical that eugenics could usher in Utopia, but he continued to attack modern society for encouraging "the perpetuation of the worst lines through sentimental regard for personal rights, even when opposed to the welfare of society," and he urged society "to substitute intelligent artificial selection for natural selection," since natural selection was "in so far as possible nullified by civilized man."[208] Whatever his doubts about the ultimate efficacy of eugenics, Conklin continued to embrace the movement's major public-policy goal, which was the prevention of propagation by the unfit.[209]

Jennings held similar views. In 1931, well after he was supposed to have become a critic of "mainline" eugenics, Jennings wrote that "in spite of difficulties and aberrations the better considered aims of eugenics . . . may be held as desirable and their realization to some degree not out of the range of possibility."[210] He added that "in ridding society of certain classes of seriously defective individuals negative eugenic measures in the present state of knowledge can do more than can be done in any other way," and he expressed hope "that with an increase in knowledge they may become much more discriminating and effective." In lectures published in 1933, Jennings made even more clear that his disagreement with other eugenists was more practical than principled. He thought that breeding better humans was "a difficult task in our present state of knowledge and organization," but he did not doubt that "the promotion of the future fulness and adequacy of life" would include not just "the improvement of the conditions of life" but also "improving the living material itself." Jennings added that the latter technique "requires some control of the matings that are made" as well as "the prevention of the propagation of individuals that are imperfect in constitution." Looking to the future, Jennings declared that "there is an unlimited field for interest, for ideals, for effort."[211]

The halfhearted criticisms offered by scientists like Conklin and Jennings did little to stop the push for eugenics. It took the Nazis to finally cause many people to reconsider the wisdom of the eugenics crusade.[212] As the enormity of Nazi crimes in the name of both negative eugenics and racial superiority became clear, many Americans, including those in the medical and scientific communities, recoiled. Nazi genocide effectively poisoned public discussions of eugenics for decades to come. Despite this odium, true believers in the eugenics gospel continued to propagate it as best they could.[213]

Margaret Sanger of Planned Parenthood still championed sterilization of the unfit in the 1950s, although she lamented that because of the Nazi program, "the word has acquired some unpleasant connotations which it does not deserve."[214] In a bizarre effort to disentangle eugenics from the Nazis, Sanger tried to blame such outrages as "concentration camps" on too little population control rather than too much. She told the Planned Parenthood Federation in 1950: "Human beings are herded into concentration camps, into vast slave labor prisons. Whole nations are made homeless and displaced. These manifestations are symptoms of a complete lack of population policies and of political foresight as to the value and meaning of dignified human living on this earth."[215] That is, the Nazis had sterilized too few people. At the end of her speech, Sanger urged the Planned Parenthood Federation to support the creation of a federally funded sterilization program targeting "the feebleminded and the victims of transmissible, congenital diseases" around the nation.[216]

Although eugenics seemed to disappear as a mainstream movement after World War II, it would be more accurate to say that it was repackaged in order to make it more palatable to the public. One of the main architects of the repackaging was Frederick Osborn, sometime president of the American Eugenics Society.[217] Osborn is typically portrayed as a reformer who had tried to purge eugenics of its unsavory connections to racism, class prejudice, and compulsory sterilization as early as the 1930s.[218] There is some truth to this portrayal, but there is also plenty of fiction. As late as 1937, Osborn voiced appreciation for the Nazi eugenics program. Summarizing a presentation on the German program at the Conference on Eugenics in Relation to Nursing, Osborn concluded approvingly: "Taken altogether, recent developments in Germany constitute perhaps the most important social experiment which has ever been tried."[219] Osborn did note that "the existence of scientifically valid bases for the ban on interracial marriages was greatly questioned in a discussion following the reading of the paper," but he added in the Nazis' defense: "It is however worthwhile to remember that any country may, to be useful to the best interest of its people, enact legislation and initiate measures to correct existing internal conditions."[220]

More than most eugenists, though, Osborn understood that the eugenics movement needed to reformulate its message. Speaking to eugenists in Great Britain in 1956, Osborn frankly acknowledged that previous eugenics efforts had failed: "The eugenic movement is nothing but a few small handfuls of men in various countries," and "the very word eugenics is in disrepute in some quarters." Thus, "we must ask ourselves, what have we done wrong?" Osborn's answer was revealing:

> I think we have failed to take into account a trait which is almost universal and is very deep in human nature. People simply are not willing to accept the idea that the genetic base on which their character is formed is inferior and should not be repeated in the next generation. We have asked whole groups of people to accept this idea and we have asked individuals to accept it. They have constantly refused, and we have all but killed the eugenics movement.[221]

To deal with this problem, according to Osborn, eugenists needed to downplay their true views:

> My own feeling is that if eugenics is to make any progress in the foreseeable future, we will not only have to drop the idea of assigning genetic

superiorities to social or racial groups, but we will even have to stop trying to designate individuals as superior or inferior. To many eugenists
this would seem a radical step, almost the abandonment of eugenics. But
a little consideration will show that there are means of selection which do
not require that we humiliate one half of the individuals who comprise the
human race by telling them that they are not as fit as the other half to pro-
create the next generation.[222]

Chief among the "means of selection" advocated by Osborn was the
wide availability of birth control. In 1953, the American Eugenics Association issued a document titled "Freedom of Choice for Parenthood: A Program of Positive Eugenics," which declared that "there is only one method
which would be appropriate and acceptable in a democracy, and which fits
within the limits of our present knowledge: it is the method of free choice
by parents."[223] Under Osborn's careful guidance, American eugenists
threw their wholehearted support behind the voluntary family-planning
movement, eventually being all but subsumed by it. As a consequence,
the eugenic arguments for birth control were increasingly replaced in the
public debate with social and libertarian arguments, which included the
claims that "children have the right to be born to parents who will care
for them properly" and "the spread of voluntary parenthood is desirable,
together with equal availability of birth control."[224]

Osborn was in a key position to promote eugenics in the guise of family planning. In the early 1950s, he helped found the Population Council, which soon became one of the preeminent groups advocating family-
planning programs around the world. From 1957–59, Osborn served as the
organization's president.[225]

Despite the new slogan of "freedom of choice," the motives and goals
of postwar eugenists were strikingly similar to those of earlier eugenists.
Darwinist materialism remained at the core of their ideology, which is why
Osborn continued to complain that civilized societies were counteracting
evolution's law of natural selection to their own detriment. "Unless we can
substitute some other and equally effective means of natural selection,
man will not continue his long advance," he insisted.[226] Osborn also cast
his promotion of voluntary birth control in Darwinian terms, suggesting
that it would more closely approximate the workings of natural selection
than would compulsory sterilization. When aviator Charles Lindbergh expressed doubts to Osborn about the ability of the human intellect to replace
the workings of natural selection, Osborn replied that the eugenics program
he was now advocating "comes near the old forms of natural selection."[227]

Whether the "new" eugenics promoted by Osborn was apprecia-
bly more voluntary than the "old" eugenics is questionable. At the same

time Osborn claimed that decisions to limit family size based on heredity "should be made by the parents on a wholly voluntary basis," he also
made clear that such "voluntary" choices could be achieved by exerting
appropriate pressure: "It should be possible to develop such strong social
and psychological pressures that in the great majority of cases known carriers of serious defect would entirely refrain from having children [and]
. . . carriers of less serious defect . . . would have less than the usual number
of children." Osborn believed that "the pressures necessary to bring about
these results are in the field of public health."[228] Accordingly, in a pamphlet
titled "Preparing for Parenthood," he urged doctors, nurses, and welfare
workers to use their influence to guide parents of inferior genetic stock to
eugenically correct decisions.[229]

In addition to such indirect coercion, Osborn still endorsed compulsory sterilization for certain classes of people. "There will remain the problem of those carriers of defect who are themselves so defective that they
are unable to control their own actions . . . It seems inevitable that in their
case the community will have to use some form of compulsion." Although
Osborn implied that he was discussing only the most seriously mentally
handicapped individuals, he later predicted that "in many instances cus-
todial cases can be released on parole if they are first sterilized."[230] But if
these candidates for compulsory sterilization could live successfully outside an institution, then it is at least debatable whether they were seriously defective to begin with. Osborn's reasoning was precisely the kind of
double talk used by an earlier generation of eugenists to justify the forced
sterilization of the "feeble-minded": On the one hand, they claimed that
the feebleminded were so grossly defective that they had to be incarcerated in institutions and sterilized; on the other hand, they insisted that
many of the feeble-minded were functional enough to be released and live
productive lives after their operations, thus saving taxpayers money.[231]

All things considered, the policy differences between eugenists in the
1950s and '60s and those in the 1920s and '30s were considerably smaller
than is usually supposed. Both groups supported curtailing childbearing
by those they deemed genetically unfit, and both supported compulsory
sterilization, although the later eugenists downplayed the coercive parts
of their agenda in their public pronouncements.

Even in the area of race relations, the differences between the new and
old eugenists were less than razor sharp. Although Osborn repeatedly dis-
associated himself from the racial views of earlier eugenists,[232] his own
record on race was uneven. In a lecture to the American Philosophical Society in 1952, he warned that non-European peoples were reproducing "at
an unprecedented rate," which meant that "non-Europeans will inevitably
comprise an even more preponderant majority of the worlds [sic] people

than they do at present." He then suggested that the "survival of our particular type of civilization" may be at stake and that people needed to begin asking themselves "whether the next fifty years will see an improvement in the ability and character of our people sufficient to offset their relative numerical decline."[233]

Two years later, Osborn wanted to publish a "Eugenics Credo" that would have attacked the beliefs that "all men are 'equal,' [and] . . . all races are 'the same.' " The draft credo also condemned racial intermarriage: "Eugenics will encourage the marriages of like to like [and] . . . oppose miscegenation, believing that each race, whose evolution has taken untold ages, has its own contribution to make to the future, and that a single mixed race would endanger further evolution."[234] Osborn ultimately abandoned the credo as bad science,[235] and he continued to declare that "there is no scientific evidence for 'superior' races."[236] At the same time, however, he expressed concern that blacks were reproducing at too high a rate: "Negroes as a group have been increasing in numbers more rapidly than the whites, with a probable trend . . . towards producing relatively fewer able individuals. This is the kind of dyagenic [sic] situation which eugenics aims to remedy." He further claimed that "marginal Negroes have more children than the corresponding whites. This would be a dangerous trend if it should continue for long."[237] Despite Osborn's disavowals, it is hard to deny the racist implications of such comments.

In private correspondence, Osborn made clear that his new emphasis on voluntary birth control was a change in tactics, not fundamental principles. He saw his efforts as a continuation of his work in the 1930s to curtail breeding by the unfit and increase childbearing by the eugenically superior. Acknowledging the importance of reducing childbearing among "the very large group at the bottom," he warned a correspondent in 1965 that

> it would do no good in the present state of public understanding to urge it for eugenic reasons. The public will accept it, and have accepted it, for entirely other reasons. The situation is so urgent that I am willing to let the program of eugenic education for birth control be secondary to the other reasons of birth control. In doing this we have been very successful: Witness the change in public attitudes towards birth control since I became head of the Population Council 10 years ago, and in which the Planned Parenthood people, the Population Council and all others interested were successful because they went along with the main currents of contemporary thinking.[238]

In another letter, Osborn likewise cautioned that "I am not sure that the name eugenics would just now be a help in practical programs" to improve the distribution of births between the fit and unfit. "I think the other reasons for liberalizing abortion laws, fighting for sterilization, and the other things you suggest might be retarded if a eugenic argument was used in their behalf."[239]

In retrospect, Osborn's reformulation of the rhetoric of the eugenics movement was a brilliant move, and it continued to shape the public debate over both family planning and abortion in the ensuing decades. The old Darwinian justification for birth control was almost entirely supplanted in the public arena by the rhetoric of "freedom of choice." However, echoes of the past reverberate, especially when it comes to abortion.

The efforts of postwar eugenists to enlist doctors in their movement to root out hereditary defects certainly bore fruit. Prenatal genetic screening has become the rule, and many doctors regularly encourage their patients to abort defective babies. Frederick Osborn was right that the "social and psychological pressures" exerted on patients could render legal coercion unnecessary. Today's expectation by doctors and medical staff that a woman will abort a defective baby is no less powerful because it is not found in any statute book. "The right to abort a disabled child . . . is approaching the status of a duty to abort a disabled child," writes George Neumayr, warning that "it is the parents with disabled children who must justify themselves to a society that tacitly asks: Why did you bring into the world a child you knew was disabled or might become disabled?"[240]

Anyone who doubts the coerciveness of the present situation should read the poignant first-person accounts collected in *Defiant Birth: Women Who Resist Medical Eugenics* (2006). This book recounts the hurdles faced by women who decide to bring to term babies who are considered to be less than perfect. According to editor Melinda Tankard Reist,

> one American woman, pregnant at 46 with triplets, was rejected for care by twelve doctors. Others were abandoned by medical practitioners when they declined "the standard of care" on offer: termination. Still more were disparaged and treated as pariahs for departing from accepted medical wisdom about becoming pregnant at all.
>
> A disturbing number of women . . . were given grave diagnoses for their babies—regaled with a litany of abnormalities and "life-threatening" conditions. But their babies were born without the predicted problems or with lesser difficulties. This raises questions about the accuracy of screening procedures and the clearly ill-placed faith in their veracity. How many women are being forced to make agonizing decisions on the basis of inadequate—even inaccurate—information?[241]

Outside the doctor's office, meanwhile, certain eugenic-tainted questions that used to be verboten have been put back on the table for open discussion. During the summer of 1999, a fractious public debate broke out over research by University of Chicago economist Steven Levitt and Stanford law professor John Donohue purporting to show that "legalized abortion appears to account for as much as 50 percent of the recent drop in crime," possibly by reducing the fertility of subpopulations most likely to produce criminals.[242] Levitt and Donohue speculated that much of the decline in the murder rate could be attributed to the prevalence of abortion among blacks: "Fertility declines for black women are three times greater than for whites . . . Given that homicide rates of black youths are roughly nine times higher than those of white youths, racial differences in the fertility effects of abortion are likely to translate into greater homicide reductions."[243] Although Levitt and Donohue insisted they were not recommending abortion as a method of fighting crime, many people worried that their research sounded like a frightening justification for eugenics-based—and possibly racist—population controls. (The empirical basis of Levitt and Donohue's conclusions was later seriously challenged by other scholars.)[244]

It was only a matter of time before someone resurrected an explicitly Darwinian defense of family planning, and in 2004, Alexander Sanger—Margaret Sanger's grandson—did precisely that. Chair of the International Planned Parenthood Council, Sanger published a provocative book arguing that the opponents of abortion ignored the facts of evolutionary biology, because they fail to recognize that human beings have evolved through natural selection the capacity to control their own reproduction. In his words, "humanity has evolved to take conscious control of reproduction and has done so in order to survive . . . We cannot repeal the laws of natural selection. Nature does not let every life form survive. Humanity uniquely, and to its benefit, can exercise some dominion over this process and maximize the chances for human life to survive and grow."[245] Claiming to disavow the old eugenics because of its junk science and coercion, Sanger nonetheless made what is essentially a eugenics-based argument: Abortion on demand is biologically justified because it aids the human race in its struggle to survive. Thus, "abortion is good,"[246] and "we must become proud that we have taken control of our reproduction. This has been a major factor in advancing human evolution and survival."[247]

"We live in a Darwinian world, like it or not," Sanger later explained in a speech to prochoice Republicans in Florida. "The rules of evolution and the rules of natural selection apply to us like any other species . . . And natural selection favors women who take control of their reproduction. When they do, they are more likely to survive child-bearing and have their children survive."[248]

Perhaps we are not so far removed from the eugenic horror as we like to think.

Understanding the Real Lessons of Eugenics

The eugenics movement is typically remembered today as an example of how politicians can hijack science for their own ideological ends. But this lesson is misleading. The leaders of the eugenics crusade were largely university-trained biologists and doctors, not politicians, and they pushed for eugenics because they thought it was fully justified by biological science. "Eugenics is a branch of biology—social biology—and its study has been cultivated chiefly by the biologists," insisted Charles Davenport.[249] "The biologist . . . demands cures instead of first-aid," added Edward East, who condemned most social-service programs as "unsound biologically" and justified eugenic birth control as the appropriate scientific alternative.[250] Rather than an effort by politicians to twist science for their own ends, eugenics is more accurately described as an effort by scientists to dictate government social policy based on their presumed scientific expertise.

Nor can advocates of eugenics be easily dismissed as being on the fringes of science. They represented mainstream science. Prominent eugenists were affiliated with institutions such as Harvard, Princeton, Columbia, and Stanford.[251] They were leaders in America's premier scientific organizations, such as the American Association for the Advancement of Science (AAAS) and the American Museum of Natural History in New York (which sponsored an extensive exhibit promoting eugenics during the Third International Congress of Eugenics in 1932). In sum, eugenists were members in good standing of the scientific establishment, and their views became so dominant for a time that eugenics was for all practical purposes the "consensus" view of the scientific community. Lectures and courses on eugenics were taught at many American colleges and universities, and the topic was included as a standard part of high school and college biology textbooks, as will be seen in chapter 11.[252] Daniel Kevles, who makes much of the eventual scientific opposition to eugenics, concedes that it took "some years" for biologists critical of eugenics to start speaking out. "Many geneticists, both British and American, either were themselves caught up in the mainline creed or were reluctant . . . to offend their pro-eugenic colleagues."[253]

Historically speaking, the eugenics movement is important because it was one of the first—and most powerful—efforts to use science to expand the power of the state over social matters. Eugenists claimed that their superior scientific knowledge trumped the beliefs of nonscientists, and so they should be allowed to design a truly scientific welfare policy. Critics of eugenics disputed this claim early on, pointing out that good public policy requires more than just scientific expertise. When Pennsylvania's governor Samuel Pennypacker vetoed a forced sterilization proposal in 1905, he announced: "Scientists, like all other men whose experiences have been limited to one pursuit . . . sometimes need to be restrained. Men of high scientific attainments are prone, in their love for technique, to lose sight of broad principles outside their domain of thought."[254]

Although the eugenists' attempt to fuse science and social policy ultimately unraveled, it paved the way for a new generation of scientists and "social scientists" to engage in similar efforts in the future. As will be seen in the next chapter, eugenics also had long-term consequences that spread beyond welfare and into the American business community and its policies toward workers and customers.

8

The Science of Business

He had a suave manner and a winning personality. Respected by both colleagues and employers, the up-and-coming business executive was a pillar of the community and active in his church. "Like a bolt out of a clear sky therefore came the revelation that he had robbed his employers of more than a hundred thousand dollars."[1]

It was a disaster that did not have to happen, according to personnel consultants Katherine Blackford, M.D., and Arthur Newcomb. Writing in their book *Analyzing Character: The New Science of Judging Men* (1916), they explained why the scandal should have been easy to prevent. All the man's employers had to do was recognize the telltale signs of corruption: his "curly blond hair," "his large, somewhat fleshy nose," and "the small and retreating contour of his chin." Such plain "external evidences" should have been enough "to determine at a glance . . . that he was selfish by nature, grasping, extravagant, too hopeful, too optimistic, too fond of money, too self-indulgent."[2]

During the early decades of the twentieth century, Blackford and Newcomb lobbied for reform in management-labor relations. Attacking traditional methods of employee selection as based on personal prejudice and superstition, they urged America's businesses to reinvent their employment policies by drawing on the discoveries of modern science, especially Darwinian biology. Employment selection, in short, needed to be based on the facts of natural selection. It was the application of eugenic principles to business.

According to Blackford and Newcomb, "Every bone, every muscle, every nerve, every feature of the body, as well as the general physical conformation, colour, texture, and consistency, are the result of this ages-long process of selection and survival." As a result, "every feature of [a man's] body, as well as every little twist and turn of his mental abilities, his morals, and his disposition, are the result of heredity and environment of his ancestors extending back into antiquity . . . plus his own environment and experiences."[3]

Moreover, "every mental and psychical state and activity is accompanied by its particular physical reaction." Therefore, to determine a person's moral and mental characteristics, one merely needed to examine the corresponding physical manifestations of those moral and mental traits. Promoting a system of scientific "character analysis" that might be described as a cross between phrenology and eugenics, Blackford and Newcomb identified nine physical traits that they said provided the keys to unlocking a potential employee's inner secrets, including skin color, form (e.g., the shape of the nose, chin, and mouth), physical size, and the structure of the muscles, the brain, and the digestive system.[4]

Blackford and Newcomb insisted their ideas were based on sound science, not quackery, and they claimed to have conducted a scientific study of more than twelve thousand employees to justify their conclusions. Thus, their system of character analysis was "fundamentally sound from the standpoint of evolution, heredity, environment, biology, physiology, and psychology, and has been verified by thousands of careful observations."[5]

Whether or not it was sound science, it was certainly popular. What became known as the "Blackford Plan" for employee selection was adopted by companies throughout the country, and books by Blackford and Newcomb went through multiple editions. In addition to general books on the topic, they published a home-study course to train employers how to use the techniques they developed for character analysis.[6] In 1924, they published *The Right Job*, a two-volume, 603-page treatise "for parents, guardians, teachers and vocational counselors" on "How to Choose, Prepare for, and Succeed in" the right occupation.[7] Dr. Blackford was a celebrated and well-connected businesswoman of her day. She spoke on behalf of the Republican National Committee during the election of 1924 and was featured at the "fourth annual exposition of women's art and industries" in New York City in 1925.[8] The introduction to *The Right Job* was supplied by American industrialist Charles M. Schwab, chairman of the Bethlehem Steel Corporation.[9]

Blackford and Newcomb believed that skin color was one of the most important traits in understanding someone's capabilities; they divided all human beings into "blondes" and "brunettes"—"those with white skins and those with dark skins."[10] "The albino is the most extreme blond; the black negro the most extreme brunette. Those fairer than halfway between the two are blond; those darker, brunette," they wrote. "[P]rimitive man was brunette," while the blond races were the products of more recent evolution among the peoples of Northern Europe. "Under [harsh] Northern climatic conditions . . . only the largest, strongest, healthiest, most intelligent, most hopeful, most courageous, and most aggressive individuals would survive. The natural result would be the evolution of a race of men and women endowed with robust physical, mental, and psychical characteristics."[11]

"Brunette" races, meanwhile, occupied milder climates, which allowed them "to survive without the exuberant health, vigour, intelligence, resourcefulness, and aggressiveness required of blonds."[12] As a consequence, "Brunettes require less food and breathe less oxygen than blonds . . . They do not become ill so quickly, but are more subject to chronic diseases."[13] Brunettes are "more conservative, more constant" and do not "seek the limelight" or the "dominating position." Instead, they tend "to be resigned and religious and imitative."[14] It was only natural that the blond races had "become the conquerors and rulers of brunette races less aggressive, less bold, less domineering, less vigorous, because their more kindly environment had not necessitated the evolution of these rugged traits."[15] Not that a brunette minded such domination: "Not having a genius for organization and government, he is usually willing to permit the domineering blond to take this burden off his hands."[16] So concluded two blonds.

Blackford and Newcomb did acknowledge that brunettes excelled in some areas, especially culture. "The dark races of the earth seem to have a genius for art, for literature, for religion, and for conservatism."[17] The "dark man" can even be considered "highly intelligent,"[18] although that very intelligence may make him more dangerous to society. While blondes are more likely to be guilty of "crimes of passion and impulse," brunettes are "more likely to commit crimes of deliberation, specialization, detail, such as murder, counterfeiting, forgeries, conspiracy, etc. . . . [T]he brunette is far more liable to harbour resentment, to cherish a grudge, to plan revenge, to see the dark side of life, and often to be melancholy and pessimistic" than the naturally "optimistic" and "good-humoured" blond.[19] These racial differences had an important "commercial application" according to Blackford and Newcomb. "The active, restless, aggressive, variety-loving blonde is found in large proportions amongst speculators, promoters, organizers, advertising men, traveling salesmen; while the more stable and constant brunette predominates amongst the plodders, the planners, the scientists, the administrators, and the conservators."

Just as race provides important information about the capabilities of potential employees, so too does the shape of body parts like the nose, face, forehead, chin, and mouth. "We find that the low, flat nose is everywhere the nose of indolence and passivity, while the large nose, high in the bridge, is everywhere an indication of energy and aggressiveness."[20] Body size also mattered. "Other things being equal, the small man is more excitable and becomes angry more easily than the large man. He also cools down more quickly."[21]

The scientific character analysis of Blackford and Newcomb did not even require personal interviews. Photographs were enough for a trained analyst to draw scientifically justified conclusions about the subjects under study. "In fact, some analysts are far more reliable in making analyses from photographs than in personal interviews. In dealing with the photograph they apply the principles and laws of the science relentlessly and almost mathematically, while, in a personal interview, they are irresistibly influenced by their sympathies, their likes and their dislikes."[22]

Although this pair of charlatans may have represented the extreme end of the spectrum when it came to applying "science" to employee selection, their underlying assumptions were widely shared at the dawn of the twentieth century. Progress was in the air. Many predicted that the same scientific method that enabled man to manufacture better products out of raw materials would also enable him to manufacture better employees.

"As in the physical world we select first the material suited to our purpose, then turn the iron into steel and temper the steel for the knife," wrote J. M. Cattell in the journal *Science*, "so in the world of human action we must learn to select the right man, to educate him and to fit him for his exact task."[23] Most of the attention paid to applying science to business during this period was directed toward the "scientific management" school championed by Frederick Taylor.[24] Emphasizing such practices as time and motion studies (popularized by the work of Frank and Lillian Gilbreth of *Cheaper by the Dozen* fame), "scientific management" had little to do with the insights of either biology or eugenics. The field of employee selection was a different matter.

The interest in applying biology to employee selection could be seen in the "vocational guidance" movement that emerged in the early part of the century. The man credited with founding that movement, Frank Parsons of the Vocation Bureau in Boston, urged that vocational choices ought to be made "in a careful, scientific way."[25] For Parsons, this meant that job seekers must list and analyze their traits, their abilities, their heredity, and their weaknesses. "You must learn what you are best adapted to do, and get started in that line."[26] Much of Parsons's advice was simply common sense and not particularly dependent on "science," let alone the insights of biology. For the most part, he eschewed the fanciful physiological and evolutionary explanations championed by Blackford and Newcomb. But not entirely. When a young man sought vocational guidance from him, Parsons typically asked the youth to think of himself as a plant or an animal being classified by a biologist: "If all the boys in Boston were gathered here together and a naturalist were classifying them as he would classify plants and animals, in what division would you belong?"[27] Moreover, Parsons's interview techniques were not exactly free from physiological speculations. In *Choosing a Vocation* (1909), he explained that while interviewing someone he would "carefully observe the shape of his head" as well as "the relative development above, before, and behind the ears."[28] The purpose of such observations? "If the applicant's head is largely developed behind the ears, with big neck, low forehead, and small upper head, he is probably of an animal type, and if the other symptoms coincide he should be dealt with on that basis."[29]

Another vocational expert, the exquisitely named J. Adams Puffer, was more pronounced in his reliance on biological determinism. In *Vocational Guidance: The Teacher as Counselor* (1913), Puffer warned vocational counselors of their duty to prevent men from "attempting a higher grade of work than one's natural ability warrants. The business of the counselor is to guide youth in exercising its powers, not in pressing beyond them."[30] In understanding the natural limits of different individuals, Puffer advised that "racial psychology is often a defense against some of the commoner errors." He then explained that "the abler races of the world, such as the Scots and north-country English, the 'first families' of the South, the old New England stock and its derivatives in the West, are likely to come to maturity slowly."[31] Therefore, vocational counselors should be careful not to underestimate the potential of the children of these "abler races," for they will likely blossom over the long term. "They are as likely . . . to exceed their early promise as children of the less evolved races are to fall short of them."[32]

The implied corollary, of course, was that vocational counselors should not overestimate the abilities of children from "less evolved races," no matter how promising their early track record. Puffer added that vocational counselors should learn about "the new science of human heredity," because "most of the qualities in which we are interested are strongly inheritable; especially is this true of high talents and peculiar gifts."[33] Indeed, "present-day scientific opinion is that virtually all the native qualities of the young are but recombinations of identical elements in the parents. The stream, proverbially, does not rise higher than its source. Neither does it commonly fail to attain its level."[34] Puffer tempered his biological determinism by acknowledging that "there is always the exception that tries the

168 ♦ Darwin Day in America

rule," and even suggesting that "one must keep in mind the black sheep which may turn up in any flock, and, on the other hand, the single boy or girl who stands apart from and above the family group."[35] Elsewhere he cautioned about underestimating the native abilities of "recent immigrants," noting that "the children of an oppressed people are often of a higher natural quality than the social position of their parents would lead one to suspect."[36]

Nevertheless, Puffer preached that biology is destiny: "A reasonably safe working rule is this. If the parents are thoroughly commonplace, mediocre persons, repeating accurately the qualities of their social type, then usually all the children will be like unto them. The grade of their vocation is predetermined."[37] The same page that offered that guidance also reprinted a photograph of more than a dozen young boys sitting on the hillside of a coal pit. The caption read: "Coal picker. The grade of their vocation is largely predetermined."

Teachers were increasingly urged to sort students into their mental "types" from their earliest years. In a booklet published in 1912 by the Central Committee on Vocational Guidance in New York City, Gustave Blumenthal asserted that "there are three types of school children": "[t]he coarse type," "[t]he fine type," and "[t]he medium type."[38] The coarse type of children "are the draught-horses of life, producing all the time." But "they are seldom able to study for a profession as their ideas are always practical." The fine type of children "are the most brilliant, but seldom the most successful . . . If negative, they become parasites and prey on the other two classes."[39] Children of the medium type are a mixture of the first two types, and these children "are by far the most intelligent and represent the middle class in life."[40] "The females [of the medium type] often make good housewives and also shine in social matters." According to Blumenthal, the first job of a teacher is to classify students according to these types, a task that "any teacher of average intelligence can do . . . easily."[41]

Speaking to a group of kindergarten teachers in 1915, Columbia University sociologist Franklin Giddings similarly encouraged teachers to "distinguish between the children that are natural imitators and followers . . . and the children that are natural leaders."[42] He claimed that "these two kinds of children can be distinguished one from the other before they are three years of age." "If you are a teacher, if you are a parent, if you are interested in the child, the sooner you discover to which of the two classes your child belongs, the sooner you will be on the right track in guiding the further development of his character and his mind."[43] Perhaps not surprisingly, Giddings served on the advisory council of the American Eugenics Society.

Fortunately, this kind of biological reductionism met with caution and even opposition from other proponents of vocational guidance. In the textbook *Plant Management* (1919), Dexter Kimball wrote respectfully that "the background" of the theories of Blackford and others "is sound," yet "it is questionable as yet whether they have been advanced to the stage where they can be applied intelligently to the problems of industrial selection."[44] Indeed, "it will probably be some time before we shall be able to . . . measure men and women as we now measure inert material. Nevertheless, this phase of industrial progress should be carefully studied by the progressive manager."[45]

Other critics were sharper. In *The Vocational-Guidance Movement: Its Problems and Possibilities* (1919), John Brewer devoted a chapter to what he called "Pseudo-Guidance," skewering efforts to categorize children according to predefined mental and physical "types" as well as the tendency to overestimate the importance of physical characteristics and the results of psychological testing. Commenting on the classification of some students as "the manual type" (fit only for manual labor), Brewer responded: "Without doubt the student of vocational guidance wishes to foster the introduction of more handwork into the schools; without doubt also, however, he must protest at assigning a thirteen-year-old boy to a class which will become for him an educational blind alley."[46]

Brewer pointed out that the abilities and interests of children can change over time, and so it is unfair as well as unwise to only provide manual training to some students and intellectual training to other students. "How can we know that the boy who is just now disinclined toward Scott's *Lady of the Lake* . . . should therefore be guided into a mechanical career?" asked Brewer. "Perhaps he should cultivate his manual powers, but at the same time have an education so broad that the way may be kept open for him to become a teacher of manual work . . . a chemist, a surgeon, a patent lawyer . . . or, indeed, a professor of philosophy."[47]

Brewer was equally dismissive of claims that character traits could be ascertained through a person's physiognomy, complaining that "we have . . . grown into a materialistic centering of the thought on bodily characteristics," which results in "the injustice of misjudging men." He chided vocational counselors who relied on such teachings for "going back to methods no better than astrology and phrenology."[48] Similar criticism came from Herman Schneider, dean of the College of Engineering at the University of Cincinnati. Criticizing the use of physical characteristics as a guide to character traits, he reported that a study of the physical characteristics of business executives showed that such a "system was not reliable."[49] Psychologist Hugo Münsterberg of Harvard University likewise dismissed Blackford's explanatory framework as one of many "arbitrary pseudo-

theories," although he also praised her effort to get businesses to central-
ize their employee-selection process in a single office staffed by experts.[50]

Defenders of character analysis by physical traits responded that their
critics were attacking the progress of science itself. "Every new scientific
discovery has always been rejected by many recognized authorities after
what they considered to be careful and convincing tests," wrote Blackford
and Newcomb.[51] "Harvey nearly died in trying to maintain his theory of
the circulation of the blood; Darwin's theory was insistently repudiated
and rejected by many scientific men of his day." Galileo, Columbus, even
"the inventor of the steam locomotive engine" all "failed to convince the
recognized authorities of their times."

Blackford and Newcomb could also have pointed out that there was
more common ground between themselves and some of their detractors
than their critics might want to acknowledge. Dean Schneider, for example,
dismissed the use of some physical characteristics to judge employees, but
he drew on the insights of biology to justify his own version of biological
determinism. Contrasting what he called the "deliberate" and "impulsive"
types, Schneider contended that "the northern races are usually deliberate,
the southern impulsive; one controls its passions, the other is frequently
controlled by them."[52] According to Schneider, "the working of the principle of evolution is at once evident" in the development of such biological
differences. "Every distinct people possesses certain characteristics, the
result of the thousands and thousands of years of conditions peculiar to
it. Thus the Chinese are settled, the Arabs are roving, the Sicilians are
impulsive, the Hindus are deliberate, the Japanese are manually accurate,
[and] the Persians possess a refined color sense."[53] Read today, it is difficult
to see a sharp difference between these comments and some of the views
expressed by Blackford and Newcomb.

Although controversial within the vocational-guidance movement itself, efforts to use evolutionary biology as a means of determining who
was "fit" for certain jobs became widespread in the early part of the century, especially in schools and other government agencies. As will be discussed in chapter 10, some high school biology texts explicitly provided
vocational guidance based on the teachings of Darwinian eugenics. The
federal government's agricultural policy, meanwhile, was profoundly influenced by academic speculations about which farmers were most fit in
evolutionary terms.

Writing in 1918, agricultural economist Charles Josiah Galpin of the
University of Wisconsin contrasted the "primitive muscular type" of
farmer with the "new cerebral type."[54] According to Galpin, the "primitive
muscular type" was the "hoe farmer" who relied primarily on his own
physical labor and rudimentary tools for his success. His "constant use [of]

the large muscles of the legs, thighs, trunk, and shoulders . . . resulted in both a massive development of the muscles themselves and indirectly in an evolution of vigorous viscera, lungs, and heart."[55] Unfortunately, while the muscles of the "hoe farmer" bulged through constant use, his mind atrophied: "Nature denies to the mind of the peasant lifter and carrier the privilege of having his big muscle movements summon into the stream of his ideas and thinking the rich and varied ideas stored up in latent form in his higher brain areas."[56] By contrast, the "new cerebral type" of farmer relied increasingly on machines and science, not brute strength. This "cerebral type" of farmer exhibited "smaller, finer muscles" that "have their controls higher up in the nervous system, much nearer the brain cells of the cortex," resulting in "a stream of mental associations" and "a large range of intellectual intercourse."[57]

Galpin did not claim that the "primitive muscular type" and "cerebral type" were fixed by heredity. Instead, they were bred largely by environment. He wrote dismissively of the social environment of rural communities, claiming that "the few small, simple institutions of rural life . . . have failed, broadly speaking, to relate the rural community to the dynamic side of the thought-world."[58] In his view, rural life instilled in farmers unhealthy traits of "animism," "self-reliance," "individualism," and "conservatism," all of which he believed were unhealthy.[59] Galpin hoped that the spread of "scientific farming" practices would eventually overcome such traits and generate greater numbers of "cerebral" farmers.

But what about existing farmers of the "primitive muscular type"? Could they be retrained to become a "cerebral type" of farmer? Or were they forever doomed to their inferior state? Galpin seemed skeptical that the "primitive muscular type" could be reformed. Although he admitted that such farmers had once had the biological potential for higher mental achievements, he thought that their potential for mental growth once they attained adulthood was probably small: "the adult peasant, it may well be believed, has lost, through disuse of his associational mechanism, the ability thoroughly to revise the habits of his isolated life."[60]

Galpin's colleague at the University of Wisconsin, Henry Taylor, was similarly concerned about the drag of inferior farmers on American agriculture. In *An Introduction to the Study of Agricultural Economics* (1905), he warned that "the labor of all those who are engaged in agriculture is not equally productive." One segment of the "agricultural population" in particular was "markedly inefficient, both from the standpoint of the quantity and the quality of their work."[61] Taylor was referring to black farmers, and he attributed their inefficiency to bad breeding, quoting another scholar's comments that "the negro is decidedly lacking . . . Something is *holding him back* . . . [I]t is his inheritance of thousands of years in Africa."[62] Taylor seemed to be of two minds about the impact of natural selection on the quality of farm labor. On the one hand, he trusted that the Darwinian process would eventually weed out inferior farmers:

> the Darwinian idea of evolution through the struggle for existence and the survival of the fittest when viewed in the light of our present knowledge of the variations in the economic productivity of men leads to the view that it is those who are less capable as producers who are in danger of not being able to make a living in normal times. This means that there is a process of natural selection going on which tends to eliminate the less efficient, and thus lift the average of human efficiency.[63]

On the other hand, Taylor also worried that the economic enticements of city life would lead to dysfunctional selection, depleting rural America of "its best elements" and installing "a dependent tenantry of inferior stock in place of the independent tillers of the soil who own the farms they cultivate and vote as they see fit."[64] The question was how to keep these "best elements" from emigrating to the cities.

Taylor eventually became chief of the Bureau of Agricultural Economics at the U.S. Department of Agriculture, while Charles Galpin was appointed "head of the Bureau's division of farm population and rural life studies."[65] During their tenure, concerns about the propagation of inferior farmers came to the forefront. According to agricultural historian Harry McDean, federal officials worried that economic woes such as "the collapse of farm prices in 1920–1921" would spur even more emigration to America's cities, leaving behind the "primitive, muscular types" in rural areas who "were certain now to produce even more children."[66] Federal officials searched for ways to identify and aid the fittest farmers in order to stave off the degeneration of the farming population. By the mid-1920s, efforts were underway to develop "tests to measure the aptitude of individuals for farming. Many suspected that there might be a correlation between success in farming and testable technical and mechanical skills."[67] Some also believed that characteristics such as "race" and "genealogy" could be used to identify successful farmers. "If such measures proved accurate the government might develop a policy that would both test aptitudes of potential farmers and help those whose test scores indicated they should be successful," wrote Harry McDean. The Department of Agriculture conducted a national study of "pathological farming areas" in America, which would "provide a profile of the unprogressive farmer."[68]

Taylor's patronizing attitude toward American farmers created political problems, especially when he testified to congressional committees early in his government career. Congressmen from farm states weren't amused at hearing about the racial degeneration of their constituents. Another official in the agriculture department rebuked Taylor privately for speaking to Congress about "the inferiority of farm people," noting that such comments were "a very dangerous thing, especially before a [congressional agriculture] committee." Not that the official thought Taylor's point was wrong. He agreed that "it is true but I doubt the wisdom of saying it before the committee . . . when we are trying to get funds."[69]

Taylor was fired by President Calvin Coolidge in 1926, [70] but he continued to sound the alarm about the degeneration of America's farmers. In the revised edition of *Outlines of Agricultural Economics* (1931), he warned that the migration of "the best elements of the rural population" to the cities would be "fraught with great danger to the nation," and he implied that this danger was in fact being realized.[71] "Even now one may see, here and there in the United States, the emergence of a genuine peasant type," he wrote.[72] "In some instances this is due to the displacement of the original American stock by immigrants," while in other "places it is due to the gradual withdrawal of the better elements of the rural population, leaving behind a residuum with a lower standard of living, a lower standard of efficiency and a lower standard of citizenship." The result would be "the degradation of the rural population."

It may have been reasonable for Taylor to fear the negative consequences of a brain drain in American agriculture, but his concerns led him to embrace the utopian scheme of trying to breed better farmers through eugenics. About the same time Taylor issues his dire predictions about rural America, he was at the helm of the Vermont Commission on Country Life, which was conducting a eugenics-tinged survey of Vermont's rural population.[73]

The centrality of eugenics to the commission's work was plainly evident in Taylor's introduction to its 1931 report, which opened with a farming analogy: "For more than a century, Vermont has been one of the most reliable seedbeds of our national life . . . How may the fertility of this seedbed be maintained and how may the quality of the human stock be conserved are questions which rightfully command the attention of the leaders of today in the Green Mountain State."[74] Later in the introduction Taylor listed various questions addressed by the survey. They had a decidedly eugenic flavor: "How may the best elements and qualities in the basic human stocks be conserved and improved?" "Are there 'pockets of degeneracy' hidden among our hills?" How can "the blind, mentally defective and other handicapped persons . . . be kept at a minimum in the future?"

Not everyone agreed with Taylor's doomsday warnings about rural migration. Fred Yoder, professor of rural social science at the State College of Washington, insisted that the "selective effect of rural migration has been greatly exaggerated," pointing out that failed farmers as well as capable ones migrated to the cities.[75] Hence "it is not safe to make any sweeping generalization about this matter until we have more first-hand evidence from field studies." Despite his skepticism about the supposed negative effect of rural migration, Yoder shared the concern about biological defectives in rural areas. He seemed convinced that rural America already had a significant proportion of biological misfits. He noted that "a survey made in New Hampshire . . . found that the proportion of persons feeble-minded increased with the ruralness of the counties" and "the selective draft [found] that the mental deficiency rate was nearly twice as high among draftees from rural sections as from the urban."[76] But in his view, feeble-mindedness was only the tip of the proverbial iceberg: "above the class of the actually feeble-minded is a much larger class on the borderline between the subnormal and the normal. They are the never-do-wells. In some sections, among the white population, they are called 'poor white trash.' They are shiftless, indolent, improvident, and stupid." Yoder conceded that "it was impossible to say whether natural inheritance or generations of unfavorable social heritage accounts for their backwardness." But no matter what the original cause of their mental retardation, "the mentally deficient and backward class of farmers cannot attain land ownership." He added his belief that "a thoroughgoing mental survey" of sharecroppers in the South, white as well as black, would show that their "economic backwardness . . . cannot be understood without a knowledge of their mental conditions."[77]

The idea that chronically poor farmers were either biologically or socially unfit for modern farm life apparently exerted a profound influence on the agricultural policies of the New Deal. On the positive side, the desire "to save rural America from genetic degeneration" laid the groundwork for "a myriad of programs to benefit farmers whom social scientists found to possess aptitudes for success in farming."[78] On the negative side, the belief that many poor farmers were inherently inferior spawned plans—largely unrealized—for a massive resettlement program to move these defective farmers to new communities where they would be expected to become factory workers, an occupation considered more suited to their limited capabilities.[79]

The crudest forms of biological selection in vocational guidance and personnel management were eventually debunked. In 1922, for example, the *Journal of Personnel Research* published a study exploding the myth that there were significant innate mental differences between so-called "blondes" and "brunettes."[80] More sophisticated forms of biological ranking persisted, however, fueled by the rise of IQ and other psychological tests. Harvard's Hugo Münsterberg, who had sharply criticized the

Blackford Plan, had no problem recommending in its stead a battery of psychological tests that were also supposed to measure a person's innate capacities.[81] For Münsterberg, the primary problem with Dr. Blackford's proposals was not that innate biological differences did not exist, but that Dr. Blackford had chosen the wrong method of identifying them. Münsterberg was surely correct that some important individual differences in ability are innate. But, like Blackford, he intended to inflate the extent and importance of such differences, and he was overconfident in the ability of science to measure them accurately. After writing that "the decisive tendencies of our mind are inherited and cannot be fundamentally changed," Münsterberg claimed that psychology alone was capable of helping people "to discover the capacities and dispositions with which an individual has to make his struggle for existence."[82]

Modern psychology turned out to be the key to unlocking the inherent capacities of workers. But it promised much more. If psychology could unlock the deepest secrets of workers, why stop there? Why couldn't it help the businessman better understand the needs and wants of consumers?

The Psychology of Selling

When Kellogg needed advice about Tony the Tiger, Seagrams wanted to know more about whiskey, and Samsonite wanted to understand the deeper meaning of luggage, they all called one man: Clotaire Rapaille, Boca Raton marketing guru extraordinaire.[83] A native of France, Rapaille has parlayed a master's degree in psychology and a doctorate in medical anthropology from the Sorbonne into a lucrative career in the high-stakes world of corporate advertising. Featured by such news outlets as CNN, the *New York Times*, and *Newsweek*, Rapaille has assembled an elite client list straight from the Fortune 100.

Rapaille specializes in blending biology, psychoanalysis, and cultural anthropology to tell corporate executives how their products connect with consumers' deepest yearnings. He is especially popular among auto makers, having worked on various projects for General Motors, Chrysler, and Ford. According to Rapaille, "a car is a message . . . It has eyes, a mouth, a chin. It has a face, and that face speaks to you." In his view, people are attracted to big SUVs because of the primal urge to survive that has been programmed into human beings down through the ages. "That's the reptilian response," he says. "If you're big and strong, you will survive." He adds that "the reptilian brain always wins." Auto executive Bryan Nesbitt swears by Rapaille, although he admits not everyone is sold on him. "He talks about motivations and calibrates that with biology, which some people find hard to connect with product development." But not Nesbitt, who says that "when you design a vehicle, understanding its emotional value becomes important."[84]

While Rapaille's claims may seem off-the-wall, he stands in a long line of psychologists and other scientists who have tried to fuse advertising with what they saw as the insights of modern science, especially efforts to view human beings as nothing but the passive products of their biology and environment.

The attempt to create a truly "scientific" advertising was already widespread by the early 1900s. The "scientific method has been the secret of modern progress," wrote Herbert Casson in *Ads and Sales* (1911). "It first revolutionized botany, geology, astronomy, chemistry, physics, etc. . . . And now the next great step, in the general swing from metaphysics to science, is to apply the principles of Efficiency to the selling and advertising of goods."[85]

Psychology was regarded as the preeminent means of applying the scientific method to advertising. "I have never seen or heard any reference to anything except psychology which could furnish a stable foundation for a theory of advertising," explained Walter Dill Scott, director of the Psychological Laboratory of Northwestern University, in 1903.[86] "Scientific advertising follows the laws of psychology," echoed an article in the trade journal *Printer's Ink* in 1907. "The successful advertiser . . . must understand how the human mind acts. He must know what repels and what attracts. He must know what will create an interest and what will fall flat."[87] Lest advertisers have any doubts about the efficacy of psychology, they were told to "remember that psychology is a science . . . Science is impartial; it searches for truth; it proceeds by reliable and carefully formulated methods."[88]

The psychology offered up to advertisers was usually based on a blend of materialism and reductionism. As Harvard's Hugo Münsterberg put it in *Business Psychology* (1922): "Today the psychologist is not concerned with philosophical ideas about the soul, but he examines the facts of our inner experience and analyzes and describes them with the same scientific calmness with which the naturalists study chemicals."[89] He went on to insist that man's mental life could be reduced completely to the physical processes of the brain. "Whatever man is doing is done by muscle action, and these muscle actions are caused by the foregoing brain processes. This is the reason why it makes hardly any difference for our practical purposes whether we speak about the mental experiences of a man or about his bodily behavior."[90]

The ultimate goal was to use science to exert better control over consumer behavior. "The aim of applied psychology is the prediction and control of human behavior in every sphere of activity," wrote Columbia University psychologist Albert T. Poffenberger in 1925. Thus, "the aim of advertising is the prediction and control of human behavior in a specialized field of activity; namely, the purchase of goods."[91]

Early psychologists writing about advertising often emphasized the importance of human "suggestibility," claiming that customers could be virtually compelled to act in a certain way if advertisers applied the right scientific techniques. "All men and probably all the lower animals are suggestible," wrote Walter Dill Scott, pointing to animal stampedes as an example. "One animal becomes frightened and starts to escape. The others see the one fleeing, and the idea of flight and escaping from danger is suggested to them by this action. The suggestion is so overpowering that in many cases the most sedate and steady animals go wild with fright."[92] Scott drew a parallel between such stampedes "among dumb brutes" and the "stampede among intelligent business men, when making a run on a bank."[93] He also pointed to the prevalence of copycat crimes. "We do things that we don't want to, simply because the thought of it has been suggested to us, and we feel compelled to carry it out."[94]

Hypnotism was held up as an example of just how far human beings were open to manipulation from without, and advertisers were encouraged to think that they could exercise the same sort of power over consumers by adopting the right scientific techniques. "The suggestibility can certainly be strongly influenced from without," wrote Hugo Münsterberg. "We can make a person more suggestible and thus bring him into a mental state in which he is more willing to accept our propositions on account of the decreased resistance by opposing ideas. The strongest possible intrusion of this kind is hypnotization."[95] While it may not be possible (or ethical) to hypnotize customers, "this does not mean . . . that the merchant has to abstain entirely from such suggestive influences."[96]

According to Münsterberg, "true hypnotism is secured by monotonous sounds, slight strain of converging the eyes, soft touch sensations, and so on. The effect probably results from a contraction of blood vessels in certain parts of the brain." Münsterberg thought that this was "quite similar" to the "way the mild, insistent, somewhat monotonous words of the clerk can affect the listener. The comfortable surroundings produce moreover a general relaxation which also strengthens the suggestibility."[97] Perhaps realizing that such open advocacy of the mental manipulation of customers was unseemly, Münsterberg added that such tactics could be misused and even noted that understanding how the techniques of suggestion operate could also be used to show people "how to strengthen the mind against the temptation which the displays and the suggestive effects of the stores [sic] offer to the population."[98] Of course, Münsterberg was writing a textbook for business students, who presumably wanted to use the techniques he described to sell products, not to inoculate customers against their appeals.

The discussion of suggestibility by proponents of scientific advertising exemplified their view of human beings as fundamentally nonrational creatures who could—and should—be manipulated in the same way as the lower animals. "Man has been called the reasoning animal," commented Walter Dill Scott, "but he could with greater truthfulness be called the creature of suggestion. He is reasonable, but he is to a greater extent suggestible."[99] Scott and others were obviously correct that human behavior is the product of more than unadulterated reason, and that a significant portion of human behavior can be explained by the same mechanisms that drive the behavior of other animals. The problem was that the "scientific" approach to advertising increasingly treated human beings as if they were *wholly* nonrational beings, despite occasional asides to the contrary. This penchant to focus almost exclusively on the subrational aspects of human nature could be seen even more starkly in the ad men's preoccupation with human "instincts."

Early boosters of scientific advertising spent considerable time enumerating and elucidating the various "instincts" that were supposed to drive human choices. The idea was that once advertisers understood these instincts, they could use them to find new ways to promote their products. "The advertiser should study human nature to discover these hidden springs of action," wrote Walter Dill Scott. "If he can find a method whereby his efforts are seconded by some of the most powerful of the human instincts, his task will be simplified to the extreme."[100] Scott conceded that reason might allow a man to "control his instinctive actions and thus obliterate their instinctive appearance," but he insisted that "such actions" were nevertheless "fundamentally instinctive."[101]

The Advertising Handbook (1921) basically agreed. Although a particular person "may get so far away from primitive things as to lose" a particular instinct, "nevertheless, the advertising man or woman needs to take account of the existence of all of these pronounced instincts, motives, emotions and tendencies, for they are such an intimate part of mankind that he is sure in his general appeals to be able to get a point of contact with many of them."[102] Advertisers should try to find ways to use human instincts to guide consumer choices.

To this end, Scott launched into an extensive listing of the instincts he thought advertisers could manipulate. These included the "Instinct to Preserve and Further the Material Possessions," "Food Instincts," the "Instinct of Bodily Preservation," the "Hoarding" instinct, and the "Hunting Instinct."[103] Explaining the hoarding instinct, Scott compared the human proclivity to collect more goods than necessary to "the hoardings of a wood-rat in California" and to "the squirrel [who] is said to collect nuts and store them away simply because that is the very action which is in itself more delightful than any other possible action."[104]

Regarding the hunting instinct, Scott said that "we like to hunt and to fish because we have inherited the hunting instinct from remote ancestors. For the civilized man such an instinct is often worthless, but to our ancestors it was necessary for the preservation of life."[105] Scott added that "the annual sale of rifles, revolvers, fishing tackle, fishing boats, etc., is beyond anything which could be attributed to their practical need. The hunting instinct shows itself in our fiendish desire for conflict."[106] Concluding his discussion of instincts, Scott wrote that "we have seen that to secure action along these lines it is not necessary to show the value of such action or the necessity of it, but merely to present the proper stimulus, and the action is forthcoming immediately."

The ad men's fixation on human instincts was dehumanizing in that it offered a truncated view of the human person, downplaying to the point of irrelevancy the part of man's nature that used to be regarded as his crown: his rationality. The view of consumers as little more than animals was widespread among advertising experts. "Strictly speaking, a sweetheart is a secondary want," advised Edward Strong in *The Psychology of Selling and Advertising* (1925). "She is desired because she satisfies certain bodily cravings."[107] "Take a more conservative view, the proper one for the student of advertising, and you find in the sex motive and the self-assertion motive two springs of behavior which may be effectively appealed to by advertising," counseled Albert Poffenberger during the same year.[108] Poffenberger added that "in most instances . . . advertising simply presents new means of satisfying wants that already exist." These "natural desires may be thought of as directing the individual toward the fulfilment [sic] of some fundamental biological need, either of the individual or of the group to which he belongs."[109]

The fascination with manipulating human instincts through advertising lasted for decades. Melvin Hattwick's text *How to Use Psychology for Better Advertising* (1950) contains an extensive discussion of the role of "basic wants" in human beings that could have been written in the early 1900s. According to Hattwick, human drives for things like food and sex "go back thousands of years" to a time when "[m]en killed each other over food, drink, women, comfortable caves, the desire to be chief, and so on." While "we still want the same things our ancestors wanted in the Stone Age," we have "spread a social veneer over their primitive methods" and "go after things we want with greater finesse." Nonetheless, "our 'drives' to get what we want are as strong today as they were hundreds of years ago."[110] Advertisers should try to exploit those basic wants. For example, humans clearly "want to live forever. That's right, forever. Self-preservation is probably the strongest, and it is certainly the longest-lasting, of all wants." That is why "a household convenience that saves you work or a gown that makes you look youthful, each has a commercial value when presented as a way to live longer."[111]

Some early psychologists dismissed efforts to uncover human beings' underlying drives and instincts. Chief among the critics was John B. Watson, founder of the behavioral school of psychology. Watson opposed the preoccupation with instincts not because he thought it was dehumanizing or placed human beings on the same level as animals, but because he thought it still treated human beings as if they were qualitatively different from animals. For Watson and his followers, the focus on instincts and drives smacked of the old idea that human beings had a mental life apart from their physical reactions—in short, it seemed reminiscent of traditional claims that human beings (unlike animals) possessed a soul.

"The new psychology . . . abandons the attempt to define the mind in terms of mental faculties, thought processes, feelings, and motives," wrote Watson disciple Henry Link in *The New Psychology of Selling and Advertising* (1932). Instead, behavioral psychology reduces all human action to a mechanical process of stimulus and response: "How, under certain circumstances (technically known as stimuli) does the person act (technically called reaction)."[112] Watson criticized scientists who reverted "outside of their laboratories" to what he called "folklore . . . crystallized into religious concepts," which included the belief "that every individual has a soul which is separate and distinct from the body."[113] He complained that "no one has ever touched a soul, or seen one in a test tube . . . Nevertheless, to doubt its existence is to become a heretic."[114]

Watson defined behaviorism as "an attempt to do one thing—to apply to the experimental study of man the same kind of procedure and the same language of description that many research men had found useful for so many years in the study of animals lower than man."[115] He compared opposition to behaviorism to the "resistance that appeared when Darwin's 'Origin of species' was first published." In his view, the root of the resistance to Darwin and behaviorism was the same: "Human beings do not want to class themselves with other animals." Watson attributed the rejection of behaviorism by some psychologists to their unwillingness to accept "the raw fact" that "to remain scientific" they "must describe the behavior of man in no other terms than those [they] . . . would use in describing the behavior of the ox [they] . . . slaughter."[116]

Watson conducted a series of hideous behavioral experiments on human infants that today would be regarded as child abuse. Seeking to identify the source of the "fear response" in children, he would place an infant in a bare room, "the walls of which were painted black," then let loose a cat or another animal to see if the infant would panic.[117] He also would make sudden loud noises near babies, drop them, and jerk blankets out from under them.[118] Loud noises, Watson reported clinically, could provoke "crying, falling down, crawling, walking or running away."[119] Watson also tried to generate a "rage" response by holding the babies' heads "lightly between the hands," pressing their arms "to the sides," and holding their "legs . . . tightly together."[120] Perversely, he tried to spark what he called a "love" response in babies by stroking their "nipples . . . lips and . . . sex organs."[121]

Although Watson is remembered today for his role in founding behaviorism, he also had a direct influence on the advertising industry. Dismissed from his post at Johns Hopkins University after leaving his wife for a graduate student, he subsequently joined the J. Walter Thompson Company, one of the nation's leading advertising agencies.[122] He rose to the position of vice president.[123]

The behavioral approach to advertising was arguably less reductionistic in practice than some other psychological approaches. At least in the early years, behaviorists in advertising tended to study consumer choices as brute facts, deferring to consumers' self-expressed desires and beliefs rather than trying to deconstruct them as the products of various subrational "instincts" and "drives." The result was the acceptance of consumer wants more as a given than as something to be manipulated. Behaviorists advocated surveying consumers as a way of finding out what they wanted already so that companies could create products better tailored to consumers' existing wants. Watson contrasted this new approach based on market research with previous efforts to use psychology to provide the "advertiser a tool by which to make people buy what they might otherwise not buy."[124] He declared, "The supremacy of the consumer, as a major premise, constitutes an entirely new platform for the psychology of selling and advertising."

Behaviorism also undercut the view that some consumers should be classified as biologically inferior to other consumers. As late as 1950, one proponent of scientific advertising still appealed to the IQ tests of World War I draftees to demonstrate the supposed fact that there is "a wider gap between the idiot and the genius than between the idiot and the chimpanzee!"[125] The lesson to be drawn? Keep ads simple so that the "millions . . . in the average and below-average intelligence group" will understand them. "That's why the advice of the advertising man is sound who said: 'Tell it in language the Smiths can understand; the Smythes will get it too.' "[126]

Although he acknowledged that heredity could produce mental defectives in limited cases, he argued that with the proper environment nearly every human baby could thrive. In his typically flamboyant manner, Watson even issued a notorious boast about human malleability: " 'Give me a dozen healthy infants, well-formed, and my own specified world to bring them up in and I'll guarantee to take any one at random and train him to become any type of specialist I might select—doctor, lawyer, artist, merchant-chief, and yes, even beggar-man and thief."[127]

Of course, the equality preached by Watson was not the same as traditional conceptions of human dignity. In terms of behavioral psychology, the fundamental equality shared by human beings was an equal openness to manipulation through environmental conditioning. That is why behaviorism is justly linked by some to all sorts of later coercive political and cultural schemes, and why it ultimately fed into an even more virulent strain of reductionism in advertising. According to Watson, science could show advertisers how to sell products by linking them through conditioning to the primal reactions of love, fear, and rage in consumers.[128] Determining the "appropriate stimuli" that would spark such reactions could be done in laboratory experiments and then applied to the larger population. The goal was to learn how to influence consumer preferences and purchases through environmental conditioning. Watson himself probably conducted few experiments in this area, but his views certainly helped lay the groundwork for what came to be known as "motivational research," which fused the insights of behaviorism with a renewed emphasis on subconscious drives that Watson likely would have detested.[129]

By the 1950s, the result was increased interest in probing consumers' physical and even subconscious responses to advertising messages in order to find more sophisticated ways to condition their behavior. For example, laboratory experiments were conducted with something called a "psychogalvanometer," "a device for measuring the very small changes in the amount of perspiration secreted by the skin during any kind of emotion." The object of the research was to determine which words carried "more emotional 'kick' than others" so that advertisers might frame their advertising copy accordingly.[130]

Others turned to Freudian psychoanalysis and "depth psychology" to gain insight into how advertisers could tap into the subconscious. One of the pioneers of this field was Viennese psychologist Dr. Ernest Dichter, who regarded advertising agencies as "one of the most advanced laboratories in psychology" and argued that a successful agency "manipulates human motivations and desires and develops a need for goods with which the public has at one time been unfamiliar—perhaps even undesirous of purchasing."[131]

Dichter emigrated to the United States in the late 1930s and built a career fusing psychoanalysis and marketing. One of his first jobs was a market study of why people buy milk.[132] "People were being asked through questionnaires why they were buying milk . . . and I just couldn't swallow that. It was almost comparable to asking people why they thought they were neurotic or to a physician asking a patient whatever diseases he thought he had."[133] In Dichter's framework, consumers were ignorant "patients" while market researchers were the expert "doctors" who needed to scientifically determine the real root causes of consumer behavior. The self-understanding of the consumer was of minimal importance: "When I want to find out why somebody does something . . . the last thing I should do is to let the person who behaves in one form or another interpret his or her behavior."[134]

Louis Cheskin of the Color Research Institute was another proponent of psychoanalyzing consumers to uncover their deepest reactions to ads. One of his experiments involved showing consumers "two packages that differed only in one respect, one had an illustration of the product on the front panel, the other had a crest."[135] When directly asked about which package they preferred, participants claimed to prefer the product with the product illustration "because they could see what was in the package." But in "association tests conducted on an unconscious level, the results were exactly the opposite. The package with the crest . . . was associated with high quality and the package with the product illustration was associated with an inferior product."

The Chicago advertising agency of Weiss and Geller also championed the focus on the subconscious. One of its more notorious studies focused on "women's menstrual cycle and the emotional states which go with each stage of the cycle." The agency found that at one stage women are "likely to feel creative, sexually excitable, narcissistic, giving, loving, and outgoing," while at another stage they are "likely to need and want attention and affection . . . and [to] have everything done for [them]."[136] Edward Weiss, president of the ad agency, explained how these findings could be applied to selling cake mix. One ad might be calculated to reach a woman "in her creative mood" by encouraging her "to try something new," while at the same time appealing "to another woman whose opposite emotional needs at the moment will be best satisfied by a cake mix promising 'no work, no fuss, no bother.'"[137]

The heyday of psychoanalytic advertising was the 1950s and '60s, after which it fell into disrepute among academics, although not necessarily among advertising practitioners and their corporate clients.[138] As the elite client list of Clotaire Rapaille attests, tapping into subconscious urges never went out of style for many executives among top companies. At the same time, attempts to correlate advertising messages with physical responses have grown increasingly sophisticated in the past few decades. Instead of just measuring sweat droplets (as with the "psychogalvanometer"), advertising researchers can now measure eye dilation, saliva production (in response to food ads), voice stress, and even brain waves.[139] Significant efforts have also been made to understand how the brain recalls and responds to television ads.[140]

Perhaps the best-known effort to influence the consumer subconscious was a complete flop. Starting in the 1950s, fears were raised in the public mind about the power of "subliminal advertising" to dictate consumer choices. Subliminal advertising was the idea that certain messages could be hidden in audio or video signals that could not be detected by the conscious mind. The claim was that the "unconscious" would detect the message and obey it. Probably the most infamous example was the story of the movie theatre that flashed the words "Eat popcorn" on the screen, supposedly generating a dramatic increase in popcorn sales. The story turned out to be a hoax.[141] Recent research has all but repudiated the claims for subliminal advertising.[142]

That doesn't mean advertising researchers are no longer interested in using subrational appeals to shape consumer responses—far from it. The quest to find new ways to manipulate consumers through their biological or physiological responses has continued unabated. One recent study, for example, advises how to use neurobiological differences between men and women to increase the effectiveness of direct-mail appeals. ("Women respond well to bright colours, photographs and images and men respond well to bold headlines, bullet points and graphs."[143]) Another study recommends the use of sex in advertisements when trying to persuade a target population that "is likely to be antagonistic toward, or counterargue, the advocated position."[144] According to the researchers in that study, "persuasive messages employing sexual information inhibit elaboration or systematic processing," which they viewed as "a desirable outcome for topics of appeals likely to evoke counterarguments."[145] In other words, sex should be used precisely because it diminishes a person's ability to process rationally the message conveyed by the ad. The last thing an advertiser wants is for the target of the ad to respond with "counterarguments"! Additional research is being conducted to determine how businesses can use music, color, and even odor to induce shoppers to buy more.[146]

Other experts propose drawing on evolutionary psychology to create more effective ads. According to psychologist and ad consultant Mark Cary, "evolutionary psychology suggests we have a 'Stone Age' brain inside our modern skulls—a brain composed of mental organs designed for the specific adaptive tasks of our ancestors."[147] Thus, "advertising that appeals to both the civilized needs and the underlying 'Stone Age' motivations should be more effective."[148] As an example, Cary says that when trying to evoke fear in an advertising campaign, an advertiser should ignore modern "dangers like electrical outlets" and focus instead on some fear "that touches our evolutionary past and evokes a deeper gut reaction," such as spiders or snakes.[149] Cary also advises advertisers to exploit what he calls "biological blind spots." "In simple animals a single stimuli can trigger an innate releasing mechanism," he explains, citing a certain fish that will try "to mate with a ridiculously crude model of any female fish as long as the belly is swollen." Cary asserts that just "like these animals," human beings "also have biological blind spots that can be exploited by advertisers."

Perhaps the most powerful lingering influence of scientific materialism on the advertising industry is the centrality of "branding" in the modern economy. Corporations can spend hundreds of millions of dollars not to show why their products are objectively superior to other products but merely to implant in consumers' brains a positive feeling about a certain brand name. The idea is to train consumers to associate the brand name with a positive emotional response through repeated exposure, making it more likely that consumers will purchase products which carry the brand name. "Repeated ad nauseam, the ads may drive you nuts," writes Alan Mitchell in *Marketing Week*, "but that's also a part of what makes them work: endless repetition makes the association stick."[150] At its core, corporate branding efforts are little more than the conditioning process championed by the behaviorists, which is why in the 1980s Coca-Cola executive Joel S. Dubow dubbed famed Russian behaviorist Ivan Pavlov "the father of modern advertising."[151] Dubow explained, "Pavlov's unconditioned stimulus (UCS) was a spray of meat powder which produced salivation . . . But if you think what Pavlov did, he actually took a neutral object and, by associating it with a meaningful object, made it a symbol of something else; he imbued it with imagery, he gave it added value. And isn't that what we try to do in modern image advertising?"

Defenders of modern advertising are apt to point out that advertising is wedded to scientific materialism for the most practical of all reasons: It works. While this statement is undoubtedly true, it does not make modern advertising any less dehumanizing. By treating consumers as animals to be conditioned rather than as rational beings to be persuaded, the scientific ad men essentially reduce human beings to the level of lab rats. Instead of appealing to the better angels of human nature, they target people's lowest material appetites. To be sure, in one sense there is nothing new about efforts to manipulate human behavior by means of the subrational. Persuasion has always included appeals to the irrational part of human nature. Aristotle, for example, provided a detailed examination of the uses of emotion in his work on rhetoric.[152] Yet, in its traditional sense, the art of persuasion preeminently involved rational appeals. Simply playing on an audience's emotions was regarded as an affront to human dignity and morality.[153] The problem with modern advertising is that it rejects the primacy of rational persuasion. Built on the assumptions of scientific materialism, modern advertising treats rational appeals as beside the point. Why seek to persuade through reason, when one can induce through science?

Scientific materialism shapes our mental imagery and interior life through the omnipresence of modern advertising. But it has exerted just as profound an influence on our exterior environment and physical surroundings. As the next chapter will show, the very buildings we inhabit have been shaped by materialist ideology.

9

Building the World of Tomorrow

Chicago was abuzz with excitement. In the midst of the Great Depression, the city had decided to mount an ambitious world's fair. Dedicated to the theme "A Century of Progress," the fair was set to open in 1933 and would celebrate the amazing achievements of the age of modern science.

"In scope, plan and architecture, it will be such an exposition as the world has never seen before," promised the *Chicago Tribune*. "Its keynote is the dramatization of the scientific revolution; its external expression, its architecture, a bold breaking away from unreasoned tradition—and for the builders of the future, perhaps . . . a prophetic exemplar."[1]

As buildings for the fair started going up, however, misgivings were voiced. Writing in the *Tribune* in 1931, Eleanor Jewett commented acidly on the new buildings "which at present disfigure the [Chicago] lake front." She worried that fair planners were adopting the Swedish example of modernist ugliness: "Last summer Sweden had an arts and crafts exhibition housed in modernist buildings . . . The buildings were like rows of factories. There was nothing inspiring or beautiful about them. They exemplified utility brought to the nth degree." She urged those behind the fair to do something to fix the situation "before it was too late." "The present specimens on the lake front could not, by any stretch of the imagination, be cherished by any one," she spat. "Typical, ugly, utilitarian buildings, graceless and awkward, your only reaction towards them is the shocked ejaculation (as it dawns on you that these are soberly being built for 1933): 'How ghastly!' "[2]

Many letter-writers to the *Tribune* agreed. "What nonsense is babbled over the modernistic type of architecture of the 1933 World's Fair," wrote Loren Knox. "This day, especially in America, has no soul for architecture or for any other fine art. Science, by its materialism and overintellectualization, has killed soul in the age, the folk, and the artists."[3]

Learning that the U.S. Congress had declined to appropriate requested money for the fair, F. W. Fitzpatrick joked that perhaps this was because "some of our national solons have had a glimpse of what is being done architecturally at the exposition." Describing one building as "a glorified gas tank," he wrote that another building reminded him "of the hull, upside down, of a sunken wreck being raised from the deep." All of it was "silly, desperately ugly stuff."[4]

The controversy soon spread beyond Chicago. A *New York Times* profile in 1932 carried the headline, "Bizarre Patterns for Chicago's Fair: Cubist Lines, Shining Steel and Many-Colored Walls Proclaim a New Tempo in Architecture." The article quoted an unnamed Chicago businessman calling the fair buildings "metallic monstrosities."

"At least we shall not bore the public with a monotonous repetition of past patterns," retorted Daniel Burnham, the director of works for the fair. Burnham explained that the fair's planners had "frankly discarded the classical style with its balanced grouping of buildings around a court . . . And we turned our architects loose." In Burnham's view, new times required a new style of building. "After all, the century and the culture we are celebrating are as far removed from classical Rome and Greece as the electric lamp of today is distant from the smoking torches of Homeric times—and our celebration calls for an architecture in step with the tempo of the modern age."[5]

Others remained unpersuaded. President Franklin Roosevelt reportedly called the architecture of the fair "crazy looking,"[6] and a letter writer to the *Times* from Kennebunkport, Maine, remarked that "the one comforting thought about the Century of Progress washboiler architecture, in Chicago, is that fair buildings eventually fall into decay and are taken down."[7]

Even those who were inclined to be open-minded about the architectural design of the fair acknowledged that many people might find it hard to take. After praising the fair for being "a veritable laboratory of modern developments in architecture," J. Wendell Clark conceded that "some of the effects are shocking to the uninitiated and their first impulse [will be] . . . to condemn the whole movement as ugly and repellent." Nevertheless, Clark urged tolerance: "Change of form dawns slowly on the human affections and perhaps we should be as slow in our judgment of what the next generation will consider useful and beautiful."[8] Daniel Burnham allowed that "the uniqueness of the thing architecturally may give [some people] a mental jolt," but he remained hopeful "that most of them will be interested, challenged, and attracted."[9]

They weren't. Instead, many Chicagoans yearned for the past, comparing the new fair unfavorably to Chicago's first world's fair a mere forty years earlier.

The World's Columbian Exposition of 1893 had been one of the great triumphs of American art and architecture.[10] Dubbed "the White City," the Columbian Exposition's white-stucco buildings were resplendent examples of neoclassical design, frankly emulating the styles of ancient Greece and Rome and traditional European architecture. The impressive buildings were set in the midst of grand plazas, fountains, canals, and lagoons, and at night they were lit up with thousands of brilliant lights.

The overall effect was staggering to contemporary observers. It was as if the greatest cities of Europe had been magically transported to the lakeshore of Chicago and reassembled. Indeed, according to some observers, it was as if ancient Greece herself had been reborn in the American Midwest. In the words of one poet:

Say not, 'Greece is no more' . . .
O! happy West—
Greece flowers anew, and all her
Temples soar![11]

Most visitors to the 1893 fair found its buildings and grounds stunning in their beauty—and some continued to think so decades later. Architecture critic Thomas Tallmadge asserted in the 1930s that "the Exposition . . . furnished a spectacle unequalled in the history of the world for the magnificence of its beauty. Imperial Rome in the third century might have approached but surely did not surpass it."[12]

The lingering memory of the 1893 exposition made the boast "A Century of Progress" seem hollow to many critics of the new fair. Instead of progress, the new exposition looked like a regression. "What a contrast between these harsh, stony, 'original' hulks of 1933 and those unoriginal adaptations of classic grace which made the Columbian exposition of 1893 a joy forever!" complained Loren Knox. "Visitors, even though they know nothing of true art, can feel these harsh edges, angles, and masses of this 'sincere' style beating at them and hacking at their eyes, and will hate them."[13]

The clashing architecture of the two world's fairs starkly symbolized the transformation that had occurred in the American building industry by the early decades of the twentieth century. While the World's Columbian Exposition paid tribute to traditional standards of architectural design and beauty, the Century of Progress was a glorification of the modern style of architecture that was beginning to remake America's cityscapes, industrial areas, and neighborhoods.

That modern style drew inspiration from nineteenth-century scientific materialism as well as new techniques in construction, and it had begun making inroads in America shortly after Chicago's first world's fair.

Louis Sullivan and the Birth of the Modern Building

"The damage wrought by the World's Fair will last for half a century from its date, if not longer. It has penetrated deep into the constitution of the American mind, effecting there lesions significant of dementia."[14] So groused architect Louis Sullivan (1856–1924), reflecting on the World's Columbian Exposition.

Often considered the father of modern architecture in America, Sullivan was demoralized by what he regarded as the Exposition's slavish obeisance to the architecture of the past. Sullivan believed that architecture must reflect the values and needs of the present rather than be frozen in the styles of the past, and he rejected the notion that architects should design their buildings according to some independent standard of beauty sanctioned by either philosophy or tradition.

Adopting the credo "form . . . follows function," Sullivan argued that the shape and style of a particular building must reflect its actual function as a building.[15] If a building is supposed to be a department store, its structure and style should make that fact plain rather than seeking to follow some extrinsic standard of beauty. In Sullivan's eyes, the beauty of a building came from its honest expression of its function. By seeking to disconnect architecture from a reliance on any independent standard of aesthetics, Sullivan's standard pushed architecture toward a view of buildings as merely the sum of their material components.

One can see this tendency at work in Sullivan's own buildings, including the Transportation Building he designed for the 1893 World's Fair he so despised. As Thomas Tallmadge later observed, the Transportation Building was a sterling illustration of Sullivan's injunction that form follow function. Looking neither like a "magnificent palace" nor a "Roman bath," the Transportation Building openly acknowledged its purpose of housing railroad cars and other forms of transport. But according to Tallmadge, Sullivan went even further by having his building frankly reveal its building materials. "He said, in effect, 'The building must tell the truth about its construction and material. Instead of the magnificent shams of . . . [other fair buildings] which pretend to be of marble but are in reality built of plaster, my building shall tell the world that it is plaster, and glory in the fact.' "[16]

Here was the triumph of materialism over artistic expression. A building needed to be true to its materials by openly admitting the physical components out of which it was made. If a building is made of plaster, then it shouldn't have a facade that looks like marble. If a building is made of steel, it shouldn't be dressed up by a masonry front that makes the building look like it was actually built from stone, for "a steel frame function in a masonry form" is as nonsensical as "[f]rog-bearing pea vines" or "[t]arantula-potatoes."[17] For a building to be honest, the physical functions of its constituent parts must be open for everyone to see rather than covered up with false ornamentation. The building where every part of the structure is allowed to display its mechanical function approached Sullivan's ideal of "organic architecture."[18]

Sullivan's outlook was profoundly shaped by "the emergence and growth of science," which he credited with liberating mankind from the dogmas and superstitions of its "feudalist" past.[19] Sullivan declared that "the sciences Physics, Biology, and Psychology" are the "pioneers" that "have opened and are preparing the way"[20] to a new world of democracy and human freedom. He chided his fellow architects for ignoring "the towering fact that modern science . . . has placed freely at our service the most comprehensive, accurate and high-powered system of organic reasoning that the world has known."[21]

Sullivan gleaned special insights from the theory of evolution. Retelling the story of his life in the third person for *The Autobiography of an Idea* (1924), Sullivan observed that "in Darwin he found much food. The Theory of Evolution seemed stupendous."[22] The importance of evolution on Sullivan's thinking can be seen in his account of the development of man's mental and spiritual qualities, which could have come straight from Darwin's *Descent of Man*. According to Sullivan, man's "highest thoughts, his most delicate yearnings," developed "through an imperceptible birth and growth, from the material sense of touch." Likewise, "from [physical] hunger arose the cravings of his soul." And "from urgent [sexual] passions have the sweetest vows of his heart arisen." Finally, "from savage instincts came the force and powers of his mind."[23] The development of man's spiritual nature he thus ascribed to an evolutionary process originating in man's physical cravings for touch, food, sex, and "savage instincts" that presumably included self-preservation.

Sullivan thought that all of Nature was in a similar process of endless flux and development.[24] Nature's evolution supplied the basis for Sullivan's insistence that architecture must be kept forever up-to-date. Instead of following the architectural styles of another era, buildings—like the rest of the material universe—must be continuously adapted to meet current exigencies. In short, architecture must follow the natural law of adapt or die. To be static—or worse, to retrogress by following tradition—was to commit the unpardonable sin against Nature herself.

The influence of modern science on Sullivan's architectural views can also be detected in the terminology he imported into his discipline from the natural sciences. By advocating such concepts as "organic architecture" and "the anatomy and physiology of architectural planning and design," Sullivan talked as if architecture were somehow an extension of modern biology.[25] Moreover, his belief that each part of a building must arise naturally from its function, and that "each part must so clearly express its function that the function can be read through the part," was clearly derived from his understanding of the operations of organisms and inorganic substances in the natural world.[26]

In nature, it is the physical function that determines the form of something. "This law is not only comprehensive, but universal. It applies to the crystal as well as to the plant, each seeking and finding its form by virtue of its working plan, or purpose or utility."[27] According to Sullivan, there is such a complete identification between form and physical functions in the natural world that it might be said that "all is form, all is function" and "all is function, all is form.[28]

Despite the influence of modern science on his views, Sullivan likely would have bristled at being labeled either a scientific materialist or a reductionist. Although he praised Darwin and Spencer, he condemned "the survival of the fittest" mentality in business and the petty materialism of businessmen.[29] And even while he seemed to lay the groundwork for a reductionist architecture, he continued to insist that architecture was preeminently a spiritual process involving the imagination of the architect. "I value spiritual results only," he proclaimed. "It is for this reason I say that all mechanical theories of art are vanity."[30] Sullivan also acknowledged the danger that "a building may have its functions of plan and purpose expressed in a literal mechanical way that tends to repel," which was not his goal.[31]

Perhaps the most striking evidence of Sullivan's apparent dissent from materialism was the multiplicity of passages in his writings that rhapsodically appealed to "Spirit," "soul," "the One Infinite Spirit," "the Infinite Creative Spirit," "the Eternal Spirit," "the All-Spirit," and similar concepts.[32] "There is no matter," he declared in one passage. "[T]he power of man's spirit is at one with the All-Spirit . . . [M]an is first, last and all the time a spiritual being."[33]

What Sullivan meant by spirituality was considerably different from the traditional definition of the term, however. For Sullivan, "the physical, the mental, and the spiritual are in essence one,"[34] and he eschewed any form of dualism that might suggest an independent existence of the spiritual realm apart from matter. Lashing out at various manifestations of dualism, Sullivan wrote that "it is assumed the day is passing, wherein men seriously believe . . . in the fetish of mind as an active something apart from another passive fetish called Nature; in the fetish of soul as apart from a fetish called body." To Sullivan,

> such conceptions are but survivals . . . of the feudal notions of good and evil, of the phantasmal notions of sin and redemption—Survivals of the notion of an external God—as though there could be anything external—external to what? Survivals of the notion that man is apart from nature, exterior to nature—where could man go to get away from nature?[35]

In one particularly blunt passage, Sullivan made clear that "the imagination, intuition, reason, are but exalted forms of the physical senses, as we call them. For Man there is nothing but the physical; what he calls his spirituality is but the most exalted reach of his animalism."[36] As for God, when Sullivan wrote about the "Infinite Creative Spirit," he was not speaking about anything a traditional monotheist would recognize. According to Sullivan, human beings created God and the gods, and the notion of a personal deity was simply retrograde. "Throughout the history of men's thoughts, no conception has been so tragically obscuring and paralyzant as the notion of a Personal God."[37]

Nevertheless, Sullivan saw modern materialism itself as providing a footing for a new spiritualism. "Human development . . . has now arrived at a materialism so profound, so exalted as to prove the fittest basis for a coming era of spiritual splendor."[38] The goal of this new spirituality would be for man to recognize that "Gods are for slaves—Spirit is for Man . . . For man is Creator. No conceivable god can exist without man's consent."[39] In free-flowing prose poetry, Sullivan sketched the implications of his view by offering up an eclectic mixture of materialism, spirituality, Nature-worship, and utopianism. Reminiscent of "Leaves of Grass" by Walt Whitman as well as the philosophy of Friedrich Nietzsche, Sullivan's speculations can be challenging to read and even more challenging to comprehend.[40]

The same sort of mystical materialism animated Sullivan's most famous American disciple, Frank Lloyd Wright (1869–1959). Like Sullivan, Wright spoke at length about the spiritual aspects of architecture, even as he tried to define both spirituality and architecture in material terms. He lamented the failure of Christianity to see "matter and spirit as the same thing," and he too saw God as one of the creations of Man. Wright further criticized Christianity for not inspiring "the nature-worship of the creative mind" and for inviting man "to become a parasite upon the Lord."[41] Again like Sullivan, Wright placed materialism near the heart of his conception of architecture:

> Now a chair is a machine to sit it in. A home is a machine to live in. A human body is a machine to be worked by will. A tree is a machine to bear fruit. A plant is a machine to bear flowers and seeds. And, as I've admitted before somewhere, a heart is a suction pump. Does that idea thrill you? Trite as it is, it may be as well to think it over because the least any of these things may be, is just that. All of them are that before they are anything else. And to violate that mechanical requirement in any of them is to finish before anything of higher purpose can happen.[42]

Yet, also like Sullivan, Wright did not end with mechanism. While it might be "either sentimentality or the prevalent insanity" to ignore the material basis of all things, Wright conceded that "were we to stop with that trite acknowledgement, we should only be living in a low, rudimentary sense."[43]

Reluctant materialists in the end, Sullivan and Wright tempered the reductionism of modern science with their romanticism and their worship of Nature with a capital "N." Whether their view was ultimately coherent can be questioned, and it proved but a temporary resting place on the journey toward a truly modern architecture. By the opening years of the twentieth century, a zealous new generation of European designers was pushing for an architecture that was much more consistently materialistic.

Biological Architecture

"'Oh, our concepts: space, home, style!' Ugh, how these concepts stink!" proclaimed German architect Bruno Taut in 1920. "Destroy them, put an end to them! Let nothing remain! . . . Death to the concept-lice! Death to everything stuffy! Death to everything called title, dignity, authority! Down with everything serious!"[44]

Taut's diatribe against traditional architecture captures the radical spirit that prevailed among young European architects in the early part of the twentieth century. Iconoclasts who wanted to demolish the established order, they wanted to wipe the slate clean and start anew.

"Smash the shell-lime Doric, Ionic and Corinthian columns, demolish the pinheads!" shrieked Taut. "Down with the 'respectability' of sandstone and plate-glass, in fragments with the rubbish or marble and precious wood, to the garbage heap with all that junk!"[45]

What was to replace traditional architecture after it had been reduced to rubble? The new alternative went by different names—"the New Architecture," "Futurist Architecture," "the International Style," or simply "Modern Architecture"—but its attributes were the same: Flat roofs. Bare walls shorn of ornamentation. Buildings designed as cubes. Exposed concrete. Steel cages enclosed by glass. Anything but classical forms of design and ornamentation. While the various schools of modern architecture had their differences, most followed Sullivan in viewing architecture as an evolutionary discipline as well as embracing a reductionist view of buildings as the sum of their functions and building materials. But some of the new firebrands pushed these articles of faith with a vengeance that even Sullivan might have found disconcerting. Not confined to any one country, the revolutionaries of the new architecture soon claimed that they were forging an "international style." The claim was much more than bluster. In country after country in Europe, the modernist rebels sought to overthrow what they regarded as the outdated trappings of the past.

In Vienna, Adolf Loos crusaded against ornamentation on buildings. Having come under the spell of Louis Sullivan while living in America during the 1890s, Loos went back to Austria determined to propagate the new faith. In his essay "Ornament and Crime" in 1908, Loos explicitly drew on evolutionary biology to make his case, beginning his essay with German Darwinist Ernst Haeckel's theory that "the human embryo in the womb passes through all the evolutionary stages of the animal kingdom."

According to Loos, "when man is born, his sensory impressions are like those of a newborn puppy. His childhood takes him through all the metamorphoses of human history."[46] Loos then claimed that ornamentation—whether it be tattooing one's skin, or painting a complicated design on the surface of a piece of china—is a throwback to a previous chapter in human evolution. The urge for such ornamentation is essentially erotic, and while the practice might be understandable in savage cultures, the civilized artist or architect who pursues ornamentation today is akin to "a criminal or a degenerate" who "smears the walls" of a lavatory "with erotic symbols."[47] According to Loos, *the evolution of culture is synonymous with the removal of ornament from utilitarian objects.*[48]

In Italy, the modernists were known as Futurists, and they preached a full-blooded architectural materialism befitting the age of machines. A manifesto on "Futurist Architecture" issued in 1914 castigated the "astounding outburst of idiocies and impotence known as 'neo-classicism'" and complained that "the new beauty of concrete and iron is profaned by the superimposition of carnival decorative incrustations justified neither by structural necessity nor by our taste."[49]

In the Futurists' view, a building really was reducible to its construction materials, and therefore the most honest building was one where the materials rather than aesthetic considerations were allowed to determine everything about the building. That is why the development of construction materials such as iron and reinforced concrete was considered so revolutionary by the Futurists. Iron and concrete made many devices of traditional architecture such as arches, columns, and stone facades unnecessary. Since the physical structure of a building was the only reality according to the Futurists, this evolution of the means of construction meant that the accoutrements of classical architecture must go. In fact, the new building materials

> rule out "architecture" in the classical and traditional sense. Modern building materials and our scientific ideas absolutely do not lend themselves to the disciplines of historical styles and are the chief cause of the grotesque appearance of buildings a la mode, in which an attempt is made to force the splendidly light and slender supporting members and the apparent fragility of reinforced concrete to imitate the heavy curve or arches and the massive appearance of marble.[50]

Because material reality was paramount, it was dishonest to put an arch or a column in a building if its only function was to serve an architectural tradition an independent standard of beauty.

For Futurists, the most appropriate model for the architecture of the future was the machine, the example par excellence of an entity whose structure is completely prescribed by its requisite physical operations. But not only should modern architecture find its "inspiration in the elements of the immensely new mechanical world," it should also strive to be "the finest expression, the most complete synthesis, the most efficacious artistic integration" of that world of machines.[51] Consequently, the house of the future must be conceived of explicitly as "an enormous machine . . . of concrete, glass, and iron, without painting and without sculpture, enriched solely by the innate beauty of its lines and projections, extremely 'ugly' in its mechanical simplicity."[52] To those who might rebel at such "ugliness," the Futurists asserted that what was needed was "a new ideal of beauty."[53]

In Russia, the modernist impulse in architecture came under the banner of "Constructivism." The banner may have been different, but the proposals were the same. Architecture, to be "true," must be stripped down to its material reality, allowing the building materials to stand for themselves without adornment or even paint. "We reject decorative colour as a painterly element in three-dimensional construction," read a manifesto outlining the principles of constructivism issued in Moscow in 1920. "We demand that the concrete material shall be employed as a painterly element." Likewise, "we reject the decorative line. We demand of every line in the work of art that it shall serve solely to define the inner directions of force in the body to be portrayed."[54]

In France, the architect Le Corbusier similarly advocated the view of buildings as machines. "If we eliminate from our hearts and minds all dead concepts in regard to the house, and look at the question from a critical and objective point of view, we shall arrive at the 'House-Machine' . . . beautiful in the same way that the working tools and instruments which accompany our existence are beautiful."[55] Le Corbusier regarded "American grain elevators and factories" as "the magnificent FIRST FRUITS of the new age" of modern architecture.[56]

But it was probably in post–World War I Germany where the new style of architecture gained its most influential expression in the school for the arts known as the Bauhaus. Established by Walter Gropius (1883–1969) shortly after the First World War, the Bauhaus and its students had become so influential that by 1938 Alfred Barr, director of the Museum of Modern Art in the United States, could claim that "Bauhaus designs, Bauhaus men, Bauhaus ideas, which taken together form one of the chief cultural contributions of modern Germany, have been spread throughout the world."[57]

The Bauhaus sought to offer a radically streamlined architecture fit for the new age of science and machines. A few years before founding the Bauhaus, Walter Gropius had written that "false historical nostalgia can only blur modern artistic creation and impede artistic originality." His goal was to teach "organic design," replacing "the old, discredited method" of "stick[ing] unrelated frills on the existing forms of trade and industrial products."[58] After establishing his new school, Gropius declared that the "entire 'architecture' . . . of the last generation" was "with very few exceptions, a lie." Rather than looking to architects, the builders of the future should look to engineers, because it was "the engineer . . . unhampered by esthetics and historical inhibitions . . . [who] has arrived at clear and organic forms."[59] The goal of the Bauhaus, in Gropius's view, was to "develop a new attitude toward design" that included "a resolute affirmation of the living environment of machines and vehicles" and "the organic design of things based on their own present-day laws, without romantic gloss and wasteful frivolity."[60]

Applying his theories to the creation of residential housing, Gropius declared that "the dwelling is an industrial-technical organism whose unity is composed organically of numerous individual functions" and suggested that housing design should be dictated by the same principles used for industrial design. Just as engineers sought "optimal solutions for factories" by "making use of the minimum possible expenditure of mechanical and human energy and the least amount of time, material, and money to gain the maximum result," so too should they "adopt the same goals in the building of houses." The ugly monotony produced by such a prescription was apparently unimportant. As long as the houses fulfilled their material functions, there was no need to worry about variety or individuality:

> For the majority of individuals the necessities of life are the same. It is therefore logical and consistent with an economic approach to satisfy these homogeneous needs uniformly and consistently. Hence it is not justifiable for each house to have a different floor plan, a different shape, different building materials, and a different "style." To do this is to practice waste and to put a false emphasis on individuality.[61]

In later years Gropius would go to great lengths to try to rebut the "portrait of the early pioneers of the modern movement as men of rigid, mechanistic conceptions, addicted to the glorification of the machine and quite indifferent to intimate human values." As he dryly put it, "being one of these monsters myself, I wonder how we managed to survive on such meager fare. The truth is that the problem of how to humanize the machine was in the foreground of our early discussions and that a new way of living was the focus of our thoughts."[62]

However, it is rather difficult to uncover the humanizing tendencies in Gropius's own building designs from the Bauhaus period. For instance, his houses for the Weissenhof suburb in Stuttgart look like white cardboard boxes. Whatever the validity of Gropius's later explanations, the designs as well as the rhetoric coming out of the Bauhaus during the 1920s was preoccupied with a narrow utilitarianism growing out of a rigid scientific materialism. That focus became even more pronounced after Gropius left the directorship of the Bauhaus in 1928 and was succeeded by socialist Hannes Meyer (1889–1954). It was during Meyer's tenure that the Bauhaus provided its most pristine articulation of an architecture based on scientific materialism.

Meyer and his followers at the Bauhaus were avowed materialists who made the connection between their architecture and their scientific beliefs explicit. Meyer stated that his "aim was design founded on science." And what were the relevant teachings of science? "To understand the only reality that can be controlled, that of the measureable, visible and ponderable"; and to understand that "all life is a striving for oxygen + carbohydrates

+ sugar + starch + protein. All forms of construction must therefore be anchored in the world." This meant that "building is a biological, not an aesthetic process,"[63] and a house should be "not only a 'machine for living', but also a biological apparatus serving the needs of body and mind."[64] Thus, the "only motives when building a house" should be to provide for physical needs such as sex, sleep, personal hygiene, and weather protection.[65] Standards of beauty and ugliness were beside the point.

The new architecture and its boosters did not go unchallenged. In England, Geoffrey Scott supplied a stringent critique in his book *The Architecture of Humanism* (1924). Ostensibly a study of the architecture of the Renaissance, Scott's book was actually a blistering indictment of modern architecture's devotion to what he termed the "mechanical fallacy" and the "biological fallacy." The mechanical fallacy was the assertion that "architecture is construction" and nothing more.[66] In Scott's view, such a claim represented an absurd attempt to subordinate "aesthetics . . . to the categories of materialistic and mechanical science"[67] and was part of a general crusade to use science to explain everything in life in purely mechanical terms. The "biological fallacy" was the effort to treat the development of architecture as a process akin to biological evolution. Scott argued that the evolutionary approach not only failed to do justice to the actual facts of architectural history, but it was analytically shallow because it never had to confront the question of whether architectural styles were actually good or bad. It replaced the study of individual architectural styles with the study of the sequence in which those styles came to be, and "in such a view there is no place for praise or blame."[68] The simple fact that architecture had changed over time told one nothing about whether the changes were good. Hence, "the question is no longer what a thing ought to be, no longer even what it is; but with what it is connected."[69]

Even architects who tried to keep an open mind about the new architecture raised doubts about its unrelenting utilitarianism. In America, Thomas Tallmadge inveighed against what he termed "the Puritanism" of modernists such as Gropius and Le Corbusier. "Cotton Mather denounced no more sternly the pretty brooch on the white throat of a New England girl than do these reformers condemn ornament on the fair surface of a modern building."[70]

Despite its detractors, the modernist tide was not about to be stemmed. In the United States, where homegrown Sullivan and Wright had received a chilly reception in earlier years, the new style was winning disciples and gaining influence.

The Bauhaus Comes to America

As early as 1922, Henry-Russell Hitchcock and Philip Johnson (1906–2005) had expounded the case for the new architecture to America in their book *The International Style*. "The architect who builds in the international style seeks to display the true character of his construction and to express clearly his provision for function," they stated baldly.[71] They heaped high praise on one skyscraper for coming "nearest to achieving aesthetically the expression of the enclosed steel cage."[72]

Hitchcock and Johnson found little to admire in existing American architecture, with one notable exception. Like Le Corbusier, who was enthralled by "American grain elevators and factories," Hitchcock and Johnson swooned over American industrial architecture. "Only the acceptance of a thoroughgoing aesthetic discipline by our architects would make it possible for our skyscrapers to be finer than our factories," they wrote.[73]

By the 1930s, the work of the Bauhaus was being toasted at exhibitions held by the Museum of Modern Art in New York,[74] and the principles of the new architecture were beginning to reach the general public. At Chicago's 1933 World's Fair ("the Century of Progress"), average Americans were treated to buildings built in the Bauhaus style, and in books like *The New World Architecture* (1935) by Sheldon Cheney they were presented with a popular case for both architectural evolution and functionalism. Articulating the now-standard idea that architecture must keep up with the times, Cheney complained that most people lived in "inefficient" houses "stylistically smelling of the past"[75] and excoriated nineteenth-century architects for being "servants of other men's minds, imitative professionals ... [I]n light of the creative new spirit abroad in the world, they were timid, impotent, slave-minded."[76] His model for ideal housing was the sleek functionalism of the automobile, and he wondered "why our dwelling-place couldn't have been conceived and built as cleanly, as efficiently—and as beautifully—as our automobile: that has just the combination of mechanical efficiency and comfort, of cleanliness and pleasurable brightness, of mechanically perfect shelter and of beauty out of proportioning and structure, that we would relish in a house."[77] Cheney intended praise when he analyzed the modern office building as

> simply a series of cubicles piled thirty stories high, with . . . electric elevators up and down, scientifically calculated halls and aisles; steel frame sheathed with baked clay; concrete floors, tile-and-plaster walls, metal doors and window-frames . . . And the architects have mercifully left out all Gothic, Greek, and Baroque ornament . . . This business building brings

us back to civilization and to what is coming after . . . here is a new sort of universality, a stripping clear of all the historic styles.[78]

The modern style in America also received an unexpected boost from National Socialism in Germany. The Nazis cracked down on leftists at the Bauhaus, and Hannes Meyer was forced out as director of the school in 1930.[79] He was replaced by Ludwig Mies van der Rohe (1886–1969), a political chameleon who had been sympathetic to the left in earlier years but who now did his best to curry favor with fascists as he tried to obtain commissions for government building projects.[80] Mies van der Rohe's efforts to placate the new government went nowhere, and the Nazis ultimately shut down the Bauhaus in 1933, leading the director to emigrate to the United States. From the Bauhaus's demise arose the twin myths that Mies van der Rohe had been a valiant opponent of Nazism and that architectural modernism was somehow the architecture of freedom while classical architecture was the architecture of totalitarianism.

There was just enough truth in the second myth to make it plausible. Adolf Hitler and Albert Speer, his favorite architect, did have a mania for neoclassical architecture, wishing to exploit the aesthetic legacy of the Roman Empire on behalf of their thousand-year Reich.[81] Yet the situation was more ambiguous than Hitler's architectural tastes implied. High-level Nazis such as Hermann Goering and Joseph Goebbels took a more positive view of modernist architecture than their Führer, and Italian Fascists embraced the modernist buildings of the Futurists as an appropriate expression of the Fascist movement in Italy.[82] The equally totalitarian Soviet Union welcomed former Bauhaus director Hannes Meyer from 1930–36.[83]

Whether or not modern architecture was in fact the "architecture of freedom," the Nazi persecution of Bauhaus leaders certainly improved those leaders' positions when they emigrated to America. Walter Gropius arrived in 1937 and became chairman of the architectural department at Harvard University. Ludwig Mies van der Rohe came the following year and went on to teach at the Illinois Institute of Technology in Chicago. Hannes Meyer's friend László Moholy Nagy (1895–1946) arrived during the same period and went to Chicago, where he founded the "New Bauhaus" and then the "Institute of Design."[84] (Moholy Nagy shared Meyer's hard-core materialism, once writing someone, "Do you really still believe in the old story about man's soul? What people call the soul is nothing more than a bodily function."[85])

Bauhaus expatriates propagated a two-pronged gospel of evolutionary architecture and architectural materialism in their new country. "There is no finality in architecture—only continuous change," proclaimed Walter Gropius, who explicitly founded his architectural relativism on modern science.[86] "Science has discovered the relativity of all human values and that they are in constant flux. There is no such thing as finality or eternal truth according to science. Transformation is the essence of life."[87]

Both Gropius and Mies van der Rohe were more nuanced in their defense of modernism than radicals such as Hannes Meyer, especially in later years as modernism's harsh utilitarianism became more controversial. Mies van der Rohe began to temper his views by the time he took over the Bauhaus under the Nazis, and Gropius in later years proclaimed that "the creation of beauty and the forming of values and standards" moved man "more deeply and more lastingly than the satisfactions of comfort."[88] Despite the moderation in rhetoric, however, the actual design of modern buildings—and the underlying principles of modern architecture—remained largely the same.

Still, modernists occasionally succeeded in making their vision of the future alluring to the masses, perhaps most notably at the 1939 World's Fair in New York City. Dedicated to "Building the World of Tomorrow," this gleaming utopia celebrated the industries and technologies of the future (including a newfangled invention called television). Befitting its focus on the future, the New York fair was constructed almost entirely according to the "modern style" in architecture, meaning that its exhibition halls tended to be boxes of concrete, steel, and plaster, unadorned by unnecessary embellishments. "There was an absolute conviction that buildings must be made to look what they are—temporary exhibit structures," proclaimed the fair's official guidebook. "No imitations either of historic architecture or imitations of permanent materials were permitted" except in pavilions operated by the various American states.[89] In an effort "to emphasize the frankly temporary nature of the buildings," fair structures "were constructed with large blank wall surfaces and without the superposition of meaningless architectural forms."[90]

But unlike the tacky cubist structures of the Century of Progress in 1933, the modernist concoctions for the New York World's Fair were sleek, striking, even entrancing. The modernists had successfully created their version of Shangri-La. The fair's signature structures were the Perisphere and Trylon, a dazzling white sphere two hundred feet in diameter positioned next to a seven-hundred-foot-high white obelisk.

It was only after fairgoers entered the Perisphere that they saw that the modernists' dream of the architectural future was more like a nightmare. Inside the sphere they looked down from the sky on an elaborate model City of Tomorrow. "Democracity," as this hideous vision of the future was called, featured streets of identical box-like buildings fashioned from concrete and steel and located on sprawling city blocks. Cold, utilitarian, and numbingly standardized, the City of the Tomorrow of 1939 seems far from appealing today, perhaps because many Americans have had to live in communities designed on its architectural principles.[91]

It is not my intention to claim that architectural modernism was wholly sterile, or that its creations could never be defended on grounds other than a reductive materialism. Modernism's yearning for simplicity, for example, clearly predates scientific materialism and can be defended on aesthetic grounds apart from a crude reductionism. Frank Lloyd Wright was not wrong to preach that "simplicity is a spiritual ideal."[92] The problem is that modern architecture picked one or two principles from the canons of traditional design and turned them into unquestioned dogmas. There was no moderation in modernism, no sense of proportion, no balance. One thinks of Aristotle's dictum that to pursue any goal to the exclusion of all others is the very essence of tyranny. Modernism led to architectural tyranny. Whereas simplicity was one of a hierarchy of goods in traditional architecture, it became the all-consuming goal of modern design. The general effect was deadening, even if there were notable exceptions.

The catastrophe of modern architecture in America has been ably chronicled in such books as *From Bauhaus to Our House* (1981) by Tom Wolfe and *The Geography of Nowhere* (1993) by James Howard Kunstler.[93] As contemporary architecture critic Catesby Leigh puts it, "the glass and steel-building-box was the direct result" of the progenitors of the modern style, "and so was the ubiquitous irruption of sterility and ugliness in America's downtowns and suburban office parks."[94] The fruits of modernism could also be seen in America's public-housing projects. As Donald Drew Egbert points out, the residential designs of Gropius and his colleagues in Germany supplied "the model for much of the urban housing built in the United States under the New Deal."[95]

Eventually there was a backlash against the banality of modernism's uniformity and reductionism. "The material uninhabitability of the slums is preferable to the moral uninhabitability of functional, utilitarian architecture," declared one architectural manifesto in 1958. "In the so-called slums only man's body can perish, but in the architecture ostensibly planned for man his soul perishes."[96] "Architecture loses its expression when technological, functional methods are employed," insisted two other European architects in 1960. "The result is apartment blocks that look like schools, schools like administrative buildings and administrative buildings like factories . . . The resulting 'architecture' is the expression of a materialistic social order whose principles are the primacy of technology and equalization."[97]

The "postmodern" architecture that subsequently became fashionable was not much better than what had come before it and was arguably just a final phase of modernism. By the 1990s, however, traditional principles of design seemed to be reviving. In the United States, the "new urbanism," with its "traditional neighborhood developments," argued for architecture rooted in the traditions and history of particular communities. The new urbanists advocated town planning that emphasized the cultural and aesthetic needs of people as well as their material needs, and which deemphasized modern culture's subservience to machines such as the automobile.[98] New urbanist developments invoked both the architecture and the town-planning principles of an earlier era.[99] At the same time, there was renewed interest in renovating—rather than razing—historic buildings in urban neighborhoods to adapt them to new uses, and in applying new urbanist principles to the redevelopment of existing urban and suburban communities.[100]

Meanwhile, the few remaining architects still around from the "golden age" of modernism struggled to stay true to their old faith. They did not always find it easy. In 2001, a party was held in New York City to toast the ninety-fifth birthday—and the latest design—of noted architect Philip Johnson (since deceased), who had been one of the first to promote the merits of the "International Style" back in the early 1930s. While Johnson's own pristine modernism had long since devolved into "postmodernism," it retained many of the signature elements of the "International Style."

Johnson's new building was described as "an apartment tower inspired by cubist painting," and the twenty-six-story skyscraper was nearly as ugly as many of the "modernist" products of the past century. Johnson announced plans to duplicate the same structure in half a dozen cities around the world. But the neighborhood chosen to host the building in New York City wasn't properly impressed. Politely acknowledging Johnson's reputation (and even offering that "people like the general concept"—whatever that meant), a local official indicated that the neighborhood wished to decline the honor. "It's like putting the Empire State Building in Levittown," he blurted out. Johnson, who proudly insisted that his ugly structures were "sculptures" akin to the Venus de Milo, wasn't pleased and indicated that he was searching for another neighborhood that would be more welcoming.[101]

America's detour into modernist architecture highlighted again the key role played by intellectual elites in applying the faith of scientific materialism to the rest of the culture. Many of the businessmen who provided the capital to build modernist skyscrapers and housing developments were probably traditionalists who preferred older styles of art and architecture. John D. Rockefeller Jr., for example, loved medieval art and spent a large part of his life restoring colonial Williamsburg to postcard-perfect beauty. But when it came to building Rockefeller Center in New York City, he apparently was persuaded by his academically trained architects to build in the ugly modernist style.[102] Just as many politicians followed the lead of biologists and social scientists when they embraced eugenics, many businessmen followed the lead of academic architects and planners when it came to creating housing developments and putting up office buildings.

Given that scientific materialism was an idea embraced primarily by the intellectual classes, it was natural that its strongest champions would come from academic institutions. It was equally natural that the educational system would be one of the most fertile grounds for further propagating the scientific materialist faith. During the twentieth century, America's schools were charged with educating new generations about the scientific worldview and the contributions of modern science to industry and culture. In the next section, we will explore how educators fulfilled this charge in the teaching of evolution and sex.

Schools and Scholars

10

Darwin Day in America

At Ennis High School in Texas, students participated in a "Darwin look-alike contest." In San Francisco, people attended "Evolutionpalooza," featuring a Darwin impersonator and a game called "Evolutionary!" Not to be outdone, the University of Louisiana at Monroe offered "a lunch of Primordial Soup and Birthday cake," and the California Academy of Sciences hosted "Natural Selection: A Party Dedicated to Darwin and Dancing," featuring "Lavay Smith and Her Red Hot Skillet Lickers."

Students of a New Jersey City University sociology professor may have obtained the best deal: Their professor cancelled classes. "Last year when I announced to my classes that I wouldn't be in class . . . I was extremely disturbed to discover that none of the five classes I met that day (about 120 mixed undergraduates) knew who Charles Darwin was," said the professor. "And my university is located about five miles from the Statue of Liberty!" What Charles Darwin has to do with the Statue of Liberty the professor did not explain.[1]

February 12 used to be known in schoolrooms across the nation as Abraham Lincoln's birthday, but in recent years schools and community groups have started to use the day to celebrate the birthday of the father of evolution instead. The movement to establish February 12 as "Darwin Day" is spreading, promoted by an evangelistic nonprofit group with its own website (www.darwinday.org) and an ambitious plan "to build a Global Celebration of Science and Humanity that is intended to promote a common bond among all people of the earth."[2]

Promoters of Darwin Day insist they are not promoting the worship of Charles Darwin.[3] That may be true, although the Darwin Day website speaks of Darwin in glowing language usually reserved for hagiography. According to Amanda Chesworth, the original executive director of the group promoting Darwin Day, Charles Darwin "cast a bright, explanatory light on reality, our self-knowledge, and on the world of which we are a part. He contributed to our understanding at the deepest level and forever changed the way we see ourselves in this vast, impersonal universe." Whether or not backers of Darwin Day promote uncritical adoration of Charles Darwin, they do seem to be encouraging the uncritical worship of science.

Chesworth describes science as the guiding star of both civilization and personal meaning: "Through science we use our intelligence to illuminate the mysteries of the universe, explore the unknown, and probe the very fabric of reality. We find . . . a guide in answering the hows, whys and wherefores that have intrigued our species for as long as we've been around." Science also "offers a tool kit for use in solving the practical enigmas of human life, from the small scale, day by day, personal problems that face us, through the grander ones of civil society, to the immense concerns of managing a planet." In Chesworth's words, science is the one true tie that binds: "Through science, in spite of our relatively superficial differences in culture, nationality, ethnicity or religion, we discover our kinship as one species—the human species. The knowledge we share in science serves to shrink our differences and smooth the path to our collective future. In this sense we have the necessary ingredients to build a just society."[4]

Promoters of Darwin Day deny that their activities are antireligious, despite the fact that they trace the origins of their movement to an event sponsored by the Stanford Humanists and the Humanist Community in 1995.[5] Their denial is hard to square with the way they champion modern science as offering a universal philosophy of everything, or with their quasi-religious justification for the celebration itself: "In Darwin Day we recognize and pay homage to the indomitable minds and hearts of the people who have helped build the secular cathedrals of verifiable knowledge."[6] "Secular cathedrals"? This is Science with a capital "S," not science as the mere acquisition of neutral facts about the natural world.

The supposed religious neutrality of the celebration is further undermined by the antireligious agendas of its official endorsers. The first "honorary president" of the Darwin Day program was biologist Richard Dawkins, author most recently of *The God Delusion* (2006).[7] Dawkins is best known for such statements as "faith is one of the world's great evils, comparable to the smallpox virus but harder to eradicate," and "Darwin made it possible to be an intellectually fulfilled atheist."[8] By 2005, the group's ad-

visory board had expanded to include not only Dawkins but such prominent atheists and agnostics as Eugenie Scott of the National Center for
Science Education (an original signer of the "Humanist Manifesto III"),
philosopher Daniel Dennett (who praises Darwinism as the "universal
acid" that eats away traditional religion and morality), and *Scientific American* columnist Michael Shermer (an atheist who writes that "Science Is My
Savior" because it helped free him from "the stultifying dogma of a 2,000-
year-old religion").[9]

The list of groups sponsoring Darwin Day events also betrays an antireligious bias. Over the past decade, the list has been top-heavy with
organizations bearing such names as the "New Orleans Secular Humanist
Association," the "San Francisco Atheists," the "Gay and Lesbian Atheists
and Humanists," the "Humanists of Idaho," the "Central Iowa Skeptics,"
the "Southeast Michigan Chapter of Freedom from Religion Foundation,"
the "Long Island Secular Humanists," and the "Atheists and Agnostics
of Wisconsin" (this last group has sponsored an "Annual Darwin Day
March" in the Wisconsin state capital).[10]

Given the groups sponsoring Darwin Day events, it should be no surprise that the events often have an avowedly antireligious bent. At New
Rochelle High School in New York, students wore shirts emblazoned with
messages proclaiming that "no religious dogmas [were] keeping them
from believing what they want to believe," while in California a group
calling itself "Students for Science and Skepticism" hosted a lecture at the
University of California–Irvine on the topic, "Darwin's Greatest Discovery: Design without a Designer." In Boston, a group sponsored a similar
event on "Biological Arguments Against the Existence of God."[11]

A musical group calling itself "Scientific Gospel Productions" mocks
gospel music by holding annual Darwin Day concerts featuring such songs
as "Ain't Gonna Be No Judgment Day," the "Virgin of Spumoni" (satirizing
the Virgin Mary), and "Randomness Is Good Enough for Me," the lyrics
of which proclaim: "Randomness is good enough for me. / If there's no
design it means I'm free. / You can pray to go to heaven. / I'm gonna try
to roll a seven. / Randomness is good enough for me." The same group's
website offers a CD for sale titled "Hallelujah! Evolution!"[12]

Perhaps in an effort to clean up the image of Darwin Day as merely
Christmas for atheists,[13] a professor from Wisconsin is now urging churches to celebrate "Evolution Sunday" on or near Darwin Day.[14] But enlisting
an assortment of liberal churches to spread the Darwinist gospel cannot
cover up the antireligious fervor that pervades grassroots Darwinian fundamentalism.

The Darwin Day movement is fascinating because it reveals a side
of the controversy over teaching evolution that is rarely covered by the

mainstream media. Although journalists routinely write about the presumed religious motives of anyone critical of evolution, they almost never explore the metaphysical baggage carried by many of evolution's staunchest defenders. Yet the ideology of some of evolution's proponents has always been a key reason why teaching evolution has been controversial in the United States. Indeed, the teaching of evolution in American schools has been intertwined with theological, social, and even political agendas from the very beginning.

Civic Biology

Most people have never read, nor even heard of, George William Hunter's *A Civic Biology* (1914). Yet this textbook was at the center of one of the most notorious education controversies of the twentieth century. In 1925, high school teacher John Scopes was prosecuted in Dayton, Tennessee, for teaching evolution from *A Civic Biology*. Although the Scopes controversy is widely remembered through the highly fictionalized account offered by the film and play *Inherit the Wind*, Hunter's textbook has been consigned to history's proverbial dustbin.[15] That is unfortunate, because the book has much to tell us about the political and cultural context in which teaching about evolution has occurred during the past century.

If any teacher in America tried to teach evolution using *A Civic Biology* today, the teacher would be stopped just as surely as John Scopes, and not merely because the book's science is outdated. The very title supplies a clue to the modern reader that the book is about something more than a neutral quest for knowledge. The goal was to equip students to be good citizens, and it is the social and political agenda of the textbook that most dates it. For example, the textbook enlists evolutionary science to justify white supremacy. In his discussion of human evolution, Hunter informs students that there are "five races or varieties of man. . . the Ethiopian or negro type . . . the Malay or brown race . . . the American Indian . . . the Mongolian or yellow race . . . and finally, the highest type of all, the Caucasians, represented by the civilized white inhabitants of Europe and America."[16]

Even more pronounced than the book's racism is its advocacy of eugenics in the chapter on "Heredity and Variation." After discussing how the power of natural selection can be harnessed through domestic breeding, Hunter makes an explicit application to human society. "If the stock of domesticated animals can be improved, it is not unfair to ask if the health and vigor of the future generations of men and women on the earth might not be improved by applying to them the laws of selection."[17] Hunter's answer to this question is an unambiguous "yes." Although he gives passing notice to the improvement of the race through better hygiene and environment, he makes clear that biological inheritance is the key. Under the subheading "Blood Tells," he declares that "brilliant men and women" are the products of "the good inheritance from their ancestors," while under the subheading "The Jukes" he attempts to show that bad biology is the cause of criminality, disease, alcoholism, and immorality.[18] Hunter stigmatizes the families produced by such bad biology as "parasites":

> Just as certain animals or plants become parasitic on other plants or animals, these families have become parasitic on society. They not only do harm to others by corrupting, stealing, or spreading disease, but they are actually protected and cared for by the state out of public money . . . They take from society, but they give nothing in return. They are true parasites.[19]

What should be done about these "parasites"? Hunter is blunt: "If such people were lower animals, we would probably kill them off to prevent them from spreading." But the author retreats from this radical suggestion, acknowledging that "[h]umanity will not allow" such a policy. So instead, "we . . . have the remedy of separating the sexes in asylums or other places and in various ways preventing intermarriage and the possibilities of perpetuating such a low and degenerate race."[20] Preventing marriage between the unfit is a major theme of this part of Hunter's text. "When people marry there are certain things that the individual as well as the race should demand. The most important of these is freedom from germ diseases which might be handed down to the offspring. Tuberculosis . . . epilepsy, and feeble-mindedness are handicaps which it is not only unfair but criminal to hand down to posterity."[21] In the lab manual accompanying *A Civic Biology*, students are asked to study charts "showing heredity of feeble-mindedness, alcoholism, epilepsy, etc." and answer the question, "Should feeble-minded persons be allowed to marry?"[22]

After the Scopes trial, the Hunter textbook was reissued in a revised edition (1926). Hunter's *New Civic Biology* was changed in some interesting ways. The reference to the Caucasian race being the "highest" was removed, as was the inflammatory observation that if defectives "were lower animals, we would probably kill them off."[23] Indeed, the word "evolution" itself was removed from the textbook, indicating increased sensitivity to public controversy.[24]

However, the practical application of evolutionary theory through eugenics was actually expanded in the revised edition. Statistics were now presented to prove just how widespread and costly to society the biologically unfit were. Students were told that one defective family line alone "cost the state of New York more than $2,500,000, besides immensely lowering the moral tone of the communities which the family contaminated." Moreover, "between 25% and 50% of all prisoners in penal institutions" were estimated to be "feeble-minded," and "at the lowest figure there are 600,000 feeble-minded persons in this country, most of whom are free to breed their kind."[25] Although the evils of the biologically unfit were emphasized more strongly in the new edition, so too was the idea that greatness in any field was dependent on heredity. Indeed, "no one becomes great unless he or she has a nervous system of a superior capacity."[26] Students learned that abilities in music, science, literature, and politics were all subservient to the laws of inheritance.[27]

Hunter's *New Civic Biology* urged students to apply eugenics to their own lives. "Two applications of this knowledge of heredity stand out for us as high school students. One is in the choice of a mate, the other in the choice of a vocation."[28] Regarding marriage, students were told that they should recognize that a strong marriage is based in large part on good heredity. Moreover, they had a responsibility to produce children who strengthen the race: "You should be parents. Will you choose to have children well born? Or will you send them into the world with an inheritance that will handicap them for life?" Concerning their choice of a career, students were instructed that they should let heredity be their guide, because although learning and environment may be important, "success, after all, depends on our inheritance."[29]

The suggested activities in the revised textbook's lab manual were even more genetically determinist. One proposed activity used materials supplied by the Eugenics Record Office and asked students to create a personal eugenic history of physical and mental traits going back at least three generations. The concluding question of the activity was, "What family traits have I inherited, and from whom?"[30] Another suggested activity asked students to consider how their heredity and environment affected their choice of vocation. "Remember that in choosing a life work it is not always what you want to do that counts, it is what you are best fitted to do."[31] The teachers' manual for the textbook told instructors that the material on eugenics was "of more lasting value than any other part of the course" and urged them to "make this chapter count, especially in its implications in vocational guidance."[32]

Hunter's advocacy of eugenics in his biology textbook was not unusual.[33] A study of secondary-school biology texts used in the United States from 1900–1977 showed that the use of Darwinian theory to promote eugenics was a standard feature of biology textbooks from 1920 to 1939, with less frequent treatments of the topic continuing at least until 1949. In the 1920s, "concerns often were expressed that natural selection was no longer allowed to operate on humans, and as a result, the population of defective individuals was increasing."[34] In the 1930s, "concern still was expressed that human beings were interfering with the process of natural selection and eugenics needed to be practiced more intensively."[35]

College biology texts could be just as outspoken in their advocacy of eugenics.[36] In *Fundamentals of Biology* (1928), University of California botanist Arthur Haupt complained that educational, religious, and charitable agencies were essentially wasting their time by trying to improve environmental conditions as a method of improving society. "While these influences are indispensable, it should be realized that they affect only the individual, not the race. Man cannot be improved racially by improving the environment any more than a farmer can create prize winners from scrub stock by giving them the best of food and care."[37]

Under the subheading "Elimination of Defectives," Haupt warned that "a great many defectives" are left "at large, free to propagate their kind," doing serious harm to society.[38] Haupt used the term "defective" to describe "not only the feeble-minded and insane, but criminals, paupers, tramps, beggars, and all persons who are a burden to society." Conceding that "many defectives may be the victims of a poor environment," he still insisted that "there is no question but that most defectives are mentally deficient and simply lack the capacity for improvement." Indeed, he claimed that "mental tests performed on juvenile criminals in state 'reformatories' have shown 50 to 90 per cent to be feeble-minded."[39]

The problem, explained Haupt, was that such defectives had been treated with far too much compassion and respect, as if they were somehow normal. "The tendency has always been to protect and care for the unfit, to give them every opportunity to improve, and except in extreme cases to grant them the same 'personal rights' as normal individuals. The result has been a constant multiplication in numbers."

In a later chapter on the evolution of man, Haupt made explicit the connection between Darwin's theory of natural selection and eugenics, citing another biologist who attacked "charity which fosters the physically, mentally, and morally feeble" because it is "contrary to the law of natural selection" and therefore "must . . . in the long run have an adverse effect upon the race."[40] In Haupt's view, the "most satisfactory" solution to the problem of the unfit was "sterilization."[41] But he lamented that "there are few sterilization laws in operation, chiefly due to the fact that public sentiment is not yet sufficiently enlightened to the necessity of enacting and enforcing them." His discussion was clearly designed to build support for the eugenists' political agenda.

This advocacy of eugenics as a Darwinian imperative was not a case of nonscientists imposing their agenda on the field of biology. The promotion of eugenics was done in biology textbooks written by biologists for students in biology classes. At the college level, eugenics courses were taught in biology and zoology departments in colleges and universities across America, including Boston College, Adelphi College, the University of Oregon, the University of Mississippi, the University of Texas, the University of Wisconsin, the University of Utah, and the University of Washington.[42] By the early 1940s, the pro-eugenics Human Betterment Foundation was still distributing roughly one hundred thousand copies per year of its pro-sterilization pamphlets for use by students in college classes across the country.[43] The widespread embrace of eugenics by science educators during the first half of the twentieth century is an example of just how difficult it can be to disentangle science from its political and social implications, even for scientists.

The temptation to mix science education with politics continues into the present, although today's issues are different from those of the past. In an article in the *American Biology Teacher* in 2000, three science professors urged biology teachers to redefine "biological literacy" to include information about the supposed genetic origins of sexual orientation. The purpose of presenting such information was to convince students that discrimination against homosexual behavior is wrong because homosexuality is genetically determined. "If more people understood that homosexuality is as 'natural' as left-handedness," wrote the authors, "it is likely that the climate of fear and hate that is fostered by ignorance would be reduced considerably with a corresponding reduction in the negative feelings that lead some homosexuals to commit suicide."[44]

The authors placed their proposal within a larger context of "taking Darwin seriously," which they defined as "relying more fully on naturalism" as the explanation for human behavior.[45] They even referenced the book *Darwin's Dangerous Idea* (1995) by atheist philosopher Daniel Dennett.[46] In a separate editorial, one of the article's authors added that he "can think of no more important area of study than human sexual orientation for high school biology teachers and social studies teachers to collaborate on and discuss with their students," although he conceded that "many persons may resist the idea, just as Darwinian evolution has been resisted by those predisposed to other ways of 'knowing.' "[47]

The *American Biology Teacher* is the leading journal for secondary-school biology teachers. In other words, it is an organ of "mainstream" biology education. The fact that the journal would publish naked appeals to use biology curricula to lobby students in favor of gay rights reveals how blind many science educators remain to the misuse of science curricula for political purposes. The blindness in this case is all the more egregious since the scientific data are ambiguous. Despite the assurances offered in the

American Biology Teacher, the biological origins of homosexuality remain controversial and hotly contested.[48]

But social issues like eugenics and homosexuality are not the only areas where science educators have blurred the lines between science and the rest of the culture. Despite commonly repeated claims that teaching evolution has nothing to do with religion, the content of biology textbooks over the past century tells a different story.

Darwinian Metaphysics

In *Fundamentals of Biology* Arthur Haupt articulated what soon became a standard textbook explanation of the relationship between evolution and religion. Science and religion are separate spheres, Haupt wrote in 1928. Science focuses on the facts, while religion focuses on "ideals." Hence "there can be no real conflict because each properly occupies a different sphere . . . It is only when one takes upon itself the functions of the other that difficulties arise."[49]

The idea that faith and science are two completely separate realms has become so common in popular discussions of evolution that it can be regarded as a truism (for a recent expression of this view, see the late Stephen Jay Gould's *Rock of Ages*).[50] Yet this truism is not based on any empirical scientific investigation. It is, in fact, a philosophical claim about the proper dividing line between science and theology. Defenders of teaching evolution often maintain that philosophy (and theology) is inappropriate in the science classroom. But if that is the case, then teaching that religion and science are two separate spheres in biology class also should be inappropriate. The separate-spheres approach is only one of a number of competing philosophical/theological positions in the science-theology debate. Another view is that religion and science are complementary realms that overlap in certain areas.[51]

Haupt, at least, attempted to be consistent in following the framework he proposed, stating that "science does not deny the existence of a creator" and "to say . . . the world has developed in accordance with natural laws does not deny the existence of an unknown guiding influence, call it what we may."[52] In other words, Haupt left open the question whether evolution is guided in some way (which a traditional theist would maintain) or whether it is driven only by an undirected process of chance and necessity. Many other biology textbooks have not been so careful or evenhanded.

Consider *Biology: The Story of Living Things* (1937), coauthored by George William Hunter. Like *Fundamentals of Biology*, Hunter's 1937 text also seemed to stake out a separate-realms position: "The religious person and the evolutionist both approach the citadel of truth with equal reverence, but from somewhat different directions. There is nothing to prevent their harmonious meeting within the portals."[53] Unlike Haupt in *Fundamentals of Biology*, however, Hunter and his coauthors filled their textbook with a variety of philosophical, theological, and moral pronouncements made in the name of evolutionary science. For example, they put evolution forward as a principle that explains everything in the universe, including religion:

> The evolutionary principle is everywhere observable, even in other than strictly biological fields . . . Even our idea of God has evolved from that of the originally exclusive individual household god, up through tribal gods, and the more inclusive national gods, until finally there has been accepted the idea of universal human brotherhood with one God over all.[54]

The authors give a reductionist account of the development of religion, implying that religion is a human creation that developed over time through completely natural processes. But it is incoherent to claim on the one hand that religion and science are separate and harmonious, yet assert on the other hand that evolution can explain religion as the product of natural (rather than supernatural) forces. The monotheistic religions of the Bible put forward the claim that monotheism was the original faith of the human race, revealed by God to a faithful remnant. The point here is not to argue whether the claim made by biblical religions is true, but only to note that if science and religion in fact are separate realms that do not challenge each other, then biology textbooks have no business offering a naturalistic account of the origins of religion.

But Hunter's textbook went still further in challenging religion. He and his coauthors seemed to foreclose the possibility that any intelligence was involved in guiding either evolution or the origin of life. In discussing the origin of life, the authors first disparaged the concept of "special creation" and then observed that "probably the theory which has the most hope of ultimate solution is the belief that at some time life arose by a chance combination of chemical elements of which the earth is made."[55] But to assert that life originated by "a chance combination" is to make a metaphysical claim, not merely an empirical one—a metaphysical claim that flatly contradicts traditional theism. In another part of the book, the authors contended that the natural selection that drives the development of life must be understood as an impersonal and undirected process,[56] making hash out of the claim that religion and science are somehow different but harmonious.

Hunter's textbook also invoked natural selection to deconstruct the idea of moral absolutes. In the section titled "Struggle for Existence," the authors described how when house spiders hatch from their eggs "there is nothing for the young spiders to eat except brothers and sisters, which they proceed to devour . . . The worse a spider is ethically, according to human standards, the better that spider is as a spider." The lesson to be drawn from this example according to the textbook authors is "that not only the movements of the heavenly bodies are subject to Einstein's law of relativity, but . . . ethics are also."[57] Now science becomes a justification for moral relativism.

In *The Science of Biology* (1959), by Paul Weisz, one finds a more sophisticated kind of double talk about science and religion. Weisz wrote that science is only one of many different "languages" that allow us to understand the world, and "it should be clear that no one language is 'truer' or 'righter' than any other. There are only different languages, each serving its function in its own domain." Hence "in one language man was created by God; in another man is a result of chance reactions among chemicals and of evolution. Again, neither the scientific nor the religious interpretation is the truer . . . The impasse is permanent, and within their own systems of communication the scientist and the theologian are equally right."[58] Elsewhere Weisz emphasized that "God is outside the domain of science, and science cannot legitimately say anything about him."[59] Yet despite the apparent humility of these statements, the rest of Weisz's book repeatedly made claims against God and religion.

Although Weisz stated that the question of whether there is purpose in the universe cannot be answered by science, he proceeded to answer the question anyway by offering scientific evidence purporting to show that evolution was purposeless and nondirected.[60] When discussing the origin of life from nonlife, Weisz dropped all pretense and bluntly asserted that "nothing supernatural was involved."[61] In later editions of the book (1963, 1967, 1971), that statement was moderated to "nothing supernatural appears to be involved," but the implication that science had somehow disproved any supernatural role in the origin of life remained.[62] Weisz also added a section in later editions emphasizing that science is required to embrace a full-blooded materialism (or what he called "the mechanistic philosophy"): "On the basis of the total experimental experience, the mechanistic philosophy holds that if all physical and chemical phenomena in the universe can be accounted for, no other phenomena will remain. Therefore, the controlling agent of the material in the universe must reside within the material itself. Moreover, it must consist of physical and chemical events only."

At this point, Weisz finally acknowledged "a conceptual conflict between religion and science." But he argued that "the conflict is not necessarily irreconcilable"—as long as a theist concedes that once God set up the universe with its natural laws "any direct influence of God over the universe must have ceased."[63] In other words, the type of God not allowed by science turns out to be the traditional conception of God held by the vast majority of Jewish, Christian, and Muslim theists.

As for Weisz's insistence that religion and science are equally legitimate (but different) ways of understanding reality, that claim too turns out to be less than meets the eye. Elsewhere in his text, Weisz made clear that science is the only method of inquiry that provides any real knowledge about the world. True, science cannot provide us with knowledge about final truths, values, or the purposes of the universe. But that is because such knowledge is apparently unobtainable. "After centuries of earnest deliberation, mankind still does not agree on what truth is, values still change with the times and with places, and purposes remain as unfathomed as ever. On such shifting sands it has proved difficult to build a knowledge of nature. What little of nature we really know, and are likely to know . . . stands on the bedrock of science."[64] So it turns out that religion and science are not equal routes to understanding the world after all.

In addition to antireligious metaphysics, Weisz supplied students with a sanitized account of the history of Darwin's theory. By the time Weisz wrote, eugenics and Social Darwinism had long been out of favor, and so Weisz sought to distance Darwin and nineteenth-century biology from these unsavory movements. Writing that Darwin's theory was "widely misinterpreted," Weisz blamed the confusion on "[s]ocial philosophers of the time and other 'press agents' and disseminators of 'news,' not biologists." These philosophers and journalists "thought that the essence of natural selection was described by the phrase 'struggle for existence.' They then coined alternative slogans like 'survival of the fittest' and 'elimination of the unfit.' "[65] Weisz failed to acknowledge that many who invoked these terms were in fact biologists, and even Darwin employed some of the very terms that Weisz found misleading. Apparently Darwin misrepresented his own theory.[66]

While many biology textbooks have avoided the sort of extensive propagandizing found in Weisz's book, the slippage of biology texts into metaphysics has been far more common than is usually recognized, especially at the college level. Examples are plentiful. According to *Biology: An Introduction to Medical and Other Studies* (1956), the evolutionary process must be regarded as purposeless, and scientists "are forced to assume" the "non-existence" of any supernatural element operating in life.[67] The introductory college biology text *A View of Life* (1981) asserts that Darwin's theory refuted human beings' "cosmic arrogance" in thinking that they have special status because they were "created in the image of an all-powerful God." The authors write: "In arguing that we are but one product of a natural process without purpose or inherent direction, Darwin forced us to seek meaning within ourselves, not in nature."[68] *Biology: Discovering Life* (1992, 1994) claims that "Darwin knew that accepting his theory required believing in *philosophical materialism*, the conviction that matter is the stuff of all existence and that all mental and spiritual phenomena are its by-products." As a result, "humanity was reduced to just one more species in a world that cared nothing for us. The great human mind was no more than a mass of evolving neurons. Worst of all, there was no divine plan to guide us."[69] In the words of *Evolutionary Biology* (1998), "by coupling undirected, purposeless variation to the blind, uncaring process of natural selection, Darwin made theological or spiritual explanations of the life processes superfluous."[70] The "profound, and deeply unsettling, implication" of this "purely mechanical, material explanation for the existence and characteristics of diverse organisms is that *we need not invoke, nor can we find any evidence for, any design, goal, or purpose anywhere in the natural world*, except in human behavior."[71] And *Life: The Science of Biology* (2001), accepting "the Darwinian view . . . means accepting not only the processes of evolution, but also the view that . . . evolutionary change occurs without any 'goals.' The idea that evolutionary change is not directed toward a final goal or state has been more difficult for many people to accept than the process of evolution itself."[72]

Textbooks that put forward such metaphysically loaded statements typically offer the obligatory caveats that of course Darwin's theory and religion are not necessarily incompatible.[73] But these caveats are impossible to take seriously. Asserting that the development of life is "purposeless," that supernatural explanations are "superfluous," and that embracing Darwin's theory "require[s] believing in philosophical materialism" are very peculiar ways of demonstrating to students that Darwinism and traditional theism are harmonious.

Professional organizations in science education have championed the same God-absent understanding of evolution. In 1995, the National Association of Biology Teachers (NABT), a professional association for biology teachers at the secondary and college levels, adopted an official statement about how evolution should be taught. Although the statement declared, "Evolutionary theory . . . neither refutes nor supports the existence of a deity or deities," it defined evolution as "an unsupervised, impersonal, unpredictable and natural process."[74]

There is a coherence problem here. If evolutionary theory truly takes no position on God, how can evolution be defined as "unsupervised" and "impersonal"? By definition, the NABT statement flatly denies that evolution could be guided by a higher intelligence. In 1997, philosopher Alvin Plantinga and religion professor Huston Smith, both distinguished scholars, wrote to the NABT urging it to drop the terms "unsupervised" and "impersonal" from its definition of evolution. According to Plantinga and Smith, the NABT statement stepped over the line into metaphysics, because "[h]ow could an empirical inquiry possibly show that God was not guiding and directing evolution?"[75] When the NABT board met in October 1997, it initially rejected the request by Plantinga and Smith, but then just a few days later it reversed itself after lobbying from Eugenie Scott, executive director of the National Center for Science Education.[76]

The NABT decision to modify its statement on teaching evolution sparked an outcry. More than one hundred evolutionists signed an open letter decrying the change, charging that "it represents the first wedge of a movement intended to surreptitiously introduce religious teachings into our public schools." According to signers of the open letter, there was nothing wrong with the words "unsupervised" and "impersonal." While they agreed that "evolutionary biology . . . cannot prove or disprove the existence of some kind of god," they were clear that "evolution denies many attributes of various forms of [the] Christian god." In particular, "all we know so far about the evolutionary process tells us that there is no supervision except for the action of natural selection," and so "evolution indeed is, to the best of our knowledge, an impersonal and unsupervised process." The NABT's change was wrong, they charged, because it "leaves open the possibility that evolution is in fact supervised in a personal manner. This is a prospect that every evolutionary biologist should vigorously and positively deny."[77] Signers of the open letter included the president-elect of the Society for the Study of Evolution, the editor of the *Journal of Ecology*, and Harvard zoologist Richard Lewontin.[78]

Eugenie Scott, who had pushed for the change, responded that something had to be done because the NABT's original wording had "generated an unanticipated public relations problem."[79] The NABT "was accused of making antireligious statements, and it is obvious that such accusations would make it more difficult for teachers to teach evolution." Scott insisted that the change only dropped "two nonessential words" and "would not affect the statement's accurate characterization of evolution."[80] Scott emphasized in particular that evolution was still described as a "natural process," and natural selection was still defined as having "no specific direction or goal."[81] In other words, the slight change adopted by the NABT solved a public-relations problem without changing the actual meaning of the NABT statement. The NABT still endorsed the view that the development of life was purposeless and undirected, but it did so in a manner that was less inflammatory. Scott did state that "one cannot make a scientific statement that the universe is in any absolute sense 'impersonal' and 'unsupervised,'"[82] but given her defense of evolution as a purposeless process, it is unclear what she meant by this statement.

Massimo Pigliucci, the evolutionary biologist who drafted the open letter, suggested that Scott was being "intellectually dishonest" because she seemed to be arguing publicly that modern science has no problem with a belief in "a personal god [who] intervenes in every day events."[83] In Pigliucci's view, this is precisely the sort of religious belief that science does refute, and he implied that Scott was being less than candid in her public comments by not saying so. Pigliucci noted that Scott says that she is a philosophical naturalist who does not believe in God but embraces that position for "personal" rather than scientific reasons. Relying in part on personal correspondence with Scott, Pigliucci found Scott's explanation wanting because Scott's "personal reasons" turned out to be "her deep understanding of science and of evolution in particular."[84]

Further evidence of Scott's disingenuousness comes from a document she signed called the "Humanist Manifesto III." This document celebrates "the inevitability and finality of death" and proclaims that "humans are . . . the result of unguided evolutionary change."[85] By specifically citing "unguided evolutionary change" as part of its case for "a progressive philosophy of life . . . without supernaturalism," this manifesto clearly suggests that evolution properly understood contradicts belief in a personal God. Did Scott fail to understand the document when she signed it along with such antireligious zealots as Richard Dawkins and Michael Shermer?

Scott's hypocrisy aside, the dispute over the NABT statement illuminates how central the idea of purposeless evolution is to many who advocate Darwin's theory. It is certainly central to the scientists who signed the open letter criticizing the NABT. But it seems nearly as important to the defenders of the NABT (such as Eugenie Scott) who insisted that the changes in wording were minor and did not compromise NABT's support for undirected evolution.

Scott was surely right, for despite the changed wording, the NABT has remained a champion of impersonal and unsupervised evolution. Anyone who doubts this should read a remarkably candid article that appeared in the *American Biology Teacher,* the NABT's journal, in August 2001. Titled "Effect of a Curriculum Containing Creation Stories on Attitudes about Evolution," the article was written by Dorothy Matthews, who teaches college-level biology courses in New York.[86]

Citing public-opinion surveys showing that nearly half of Americans are creationists, Matthews lamented that "our current approach to evolution is failing."[87] To solve the problem, she advocated having students read "creation stories" from various religions so that they could compare the reasonableness of these nonscientific accounts "with the scientific explanation of the origin of life."[88] The idea was to allow students to openly discuss their nonscientific views of creation so that these beliefs could then be refuted more effectively.

What was most interesting about Matthews's article was how she measured the success of her new method of teaching evolution. She administered a series of surveys to her students that were designed to elicit whether they agreed or disagreed with certain "non-scientific" statements. These included "human beings are different from nonliving things because they possess a soul," "it seems reasonable that the universe was created by God," and "all events in nature occur as part of a predetermined plan."[89] Matthews's goal in teaching evolution was to get students to abandon such "non-scientific" viewpoints. She believed that teaching evolution should convince students that human beings do not possess souls, that it is unreasonable to believe that "the universe was created by God," and that events happen only because of chance and necessity.

Another article published in the *American Biology Teacher* (this one in 2003) urged teachers to use their positions to criticize religion as incompatible with science: "When young children are indoctrinated into believing that for which there is no evidence (God, Heaven, Hell, etc.), a habit of mind is being developed that is inconsistent with the open, inquiring mind needed for scientific study. The habits of mind are not merely different, they are incompatible between science and religion."[90] "[T]he story of evolution" presented in the classroom, writes the author, "must include the history of the struggle between religion and science, not just the facts of evolutionary science."[91]

The fact that such articles could undergo peer review by the *American Biology Teacher* and still be published indicates how deep-seated the adherence to materialistic philosophy remains among some science educators. If evolutionary theory is truly neutral on the subject of religion, then why is the *American Biology Teacher* publishing articles that tie the teaching of evolution to erasing students' beliefs in the supernatural?

Nor is the *American Biology Teacher* the only science journal whose editors and reviewers apparently have no problem linking modern science to an explicitly antitheistic agenda. Consider these opening sentences from an article in the prestigious journal *Cell*:

> The greatest scientific advance of the last 1000 years was providing the evidence to prove that human beings are independent agents whose lives on earth are neither conferred nor controlled by celestial forces. Although it may be more conventional to measure scientific progress in terms of specific technological developments, nothing was more important than providing the means to release men and women from the hegemony of the supernatural.[92]

Is this claim—which appeared in an article on the otherwise arcane topic of membrane traffic—science or metaphysics?

Despite typical protestations that modern biology is neutral on the subject of religion, many biologists are not in fact neutral. The strong antireligious sentiment within the biology community has been confirmed by multiple surveys. According to a poll of scientists listed in *American Men and Women of Science,* 43.5 percent of the biologists who responded disbelieved in God and 44.5 percent disbelieved in personal immortality. Another 14 percent were agnostic about the existence of God and 14.9 percent were agnostic about life after death. The nation's most elite biologists were even more atheistic. According to a 1998 survey of members of the National Academy of Sciences, 65.3 percent of the NAS biologists disbelieved in God, while an additional 29.1 percent were agnostic. Similar percentages rejected life after death.[93] A 2003 survey of leading scientists in the field of evolution, meanwhile, indicated that 87 percent denied the existence of God, 88 percent disbelieved in life after death, and 90 percent rejected the idea that evolution is directed toward an "ultimate purpose."[94] By contrast, the vast majority of Americans continue to believe both in God and in personal immortality.[95]

Given the religious beliefs of most Americans, the linking of evolution curricula to an avowedly (or even implicitly) antireligious agenda can only invite continued cultural strife. Perhaps that is why efforts are now underway—as a public-relations strategy—to invoke rather than refute theological arguments in the biology classroom. Instead of claiming that biology undermines theology, public defenders of evolutionary theory increasingly argue that theology supports Darwin's theory—and that public-school students ought to be taught this.

Promoting Darwin's God

Brown University biologist Kenneth Miller is a man with a mission. A fierce critic of creationism, Miller says he also wants to rescue Darwinism from some of its defenders—in particular, those like Richard Dawkins and Daniel Dennett who insist that evolution makes belief in God irrational. As a Roman Catholic, Miller wants to preserve his belief in God. At the same time, he wants to defend a robust view of evolution as purposeless and unguided. "Evolution is a natural process," he says, "and natural processes are undirected."[96]

In *Finding Darwin's God: A Scientist's Search for Common Ground between God and Evolution* (1999), Miller attempts to square his two beliefs. Whether Miller's theological apologia for Darwin's theory is successful is open to debate. In order to defend evolution as an unguided process, he must venture far afield from traditional Christian theology. Historic Christian teaching asserts that man has an immaterial soul, but Miller appears to deny this, stating that "although we might like to think there is a divine spark within us, in fact, our lives and the lives of every animal, plant, and microorganism of this planet have a physical basis in the chemistry of cells and tissues . . . [O]ur origins as individuals come entirely from the materials of life."[97] Traditional Christian teaching also claims that the natural world as a whole results from God's specific intentions, and that human beings in particular were specifically designed by God in his image. But Miller flatly denies that God guided the evolutionary process in order to produce any particular result—even the development of human beings. In fact, he says he agrees with the view "that mankind's appearance on this planet was *not* preordained, that we are here not as the products of an inevitable procession of evolutionary success, but as an afterthought, a minor detail, a happenstance in a history that might just as well have left us out."[98]

Given this view, how can Miller also insist that "the final result of the process may nonetheless be seen as part of God's will"?[99] What Miller apparently means by this is that once God set up the undirected and unpredictable process of evolution, he could know that "given evolution's ability to adapt, to innovate, to test, and to experiment, sooner or later it would have given the Creator exactly what He was looking for—a creature who, like us, could know Him and love Him."[100]

But Miller is engaging in double-talk. He plainly does not believe that human beings represent an "exact" intention of God. In Miller's view, God wished for some sort of rational creature to develop in the universe, but he assigned the job to an undirected process that need not have produced human beings. At a 2007 conference, Miller admitted as much. "If you let the videotape of life run again," he stated, the intelligent creature produced by evolution "might be a big-brained dinosaur" or even "a mollusk with exceptional mental capabilities."[101]

Miller's view is a radical departure from traditional Christian (or Jewish or Muslim) teaching that human beings are created as the result of God's specific plan. The point here is not to argue whether Miller's view or traditional theology is correct, but to point out that Miller's theological defense of unguided evolution is open to significant dispute within traditional religious communities.[102]

Miller's theological defense of Darwinism is especially relevant because efforts are now being made to import religious arguments for Darwinism into public education under the guise of science education. This was a prime objective of an eight-hour PBS series titled *Evolution*, which appeared in 2001. Underwritten by billionaire Paul Allen and created by his film company, *Evolution* was targeted largely at the nation's schools. Thousands of glossy teacher's guides were sent to instructors around the country, and the companion website offered an array of tools to be used in the classroom. The creators of the series hoped to transform the way evolution was taught by reaching up to 10 million students over ten years. Although the documentary series was supposed to be about science, much of it was about religion—to be more specific, about how religion allegedly supports Darwin's theory.

Kenneth Miller had a starring role in the series. "I'm an orthodox Catholic and I'm an orthodox Darwinist," he declared in the first episode.[103] Miller also refuted as bad theology the idea that God is the designer of the natural world. Focusing on the human eye, he argued that the eye should not be viewed as reflecting God's design because it is imperfect, and this imperfection provides proof that the eye originated in "the blind process of natural selection" rather than in a purposeful plan.[104] As at least one critic pointed out, Miller was making a theological argument, not a scientific one, which rested on the assumption that God only creates perfect things.[105] Further doses of theology came in the last episode of the series, "What about God?" which told how evangelical Christian students at Wheaton College in Illinois came to accept the truth that their religious beliefs can be made compatible with Darwinian evolution.

The message that religion is consistent with Darwin's theory was presented even more strikingly in a six-and-a-half minute video segment titled "Why Is Evolution Controversial Anyway?" Designed for use in the classroom,[106] the segment argued that "most Jews, Protestants, Catholics, and other religious people embrace evolution." The only people who don't are a handful of fundamentalist zealots, represented in the film by a crazed-looking fundamentalist minister shown attacking evolution. While organ music played softly in the background, Kenneth Miller concluded the segment by giving testimony that God and Darwin are in fact harmonious:

> So not only was Darwin right about the origin of species, and not only was Darwin right about the mechanisms of evolutionary change, but there's nothing about those origins or that mechanism of change which goes against religious belief, and therefore I sort of find this absolutely wonderful consistency with what I understand about the universe from science and what I understand about the universe from faith.

Along the same lines, the teacher's guide for *Evolution* advocated telling students "that many religions support the teaching of evolution,"[107] and it implied that all criticism of evolution stems from a "fundamentalist viewpoint about the interpretation of Biblical literature."[108] Teachers were also instructed that "intelligent design" is just another version of "creation science," and that while they should familiarize themselves with such "creationist" arguments, "don't introduce this examination into your classroom because introducing religion into a science classroom is inappropriate."[109]

It is simply untrue that the only religious criticism of evolution stems from fundamentalism, and it is likewise untrue that "most Jews, Protestants, Catholics, and other religious people embrace evolution." Public-opinion surveys consistently show that Americans overwhelmingly reject Darwin's theory of unguided evolution. Indeed, a significant percentage of Americans continue to reject evolution even when it is described as being directed by God.[110] Additionally, the lumping together of "intelligent design" with "creation science" was misleading. Unlike "creation science," "intelligent design" is not based on an interpretation of the Bible.[111]

Groups that usually cry "foul" when anyone wants to inject a religious critique of evolution into the schools were notably silent in criticizing *Evolution's* religious defense of evolution—even though the *Evolution* series was clearly designed for use in public education, science classes in particular. Indeed, the major national group seeking to keep religion out of science classrooms helped to frame *Evolution's* discussion of religion and science. Eugenie Scott of the National Center for Science Education served as both a consultant for the *Evolution* series and as one of its official spokespersons, and NCSE materials are referenced throughout the educational materials created for *Evolution*. Moreover, Scott appeared on screen to make the argument that religion is harmonious with Darwinism. This was rather incongruous, given that her organization steadfastly insists that religion has no place in science education. For Scott and the NCSE, religion must not be used to critique evolution, and even scientific critiques of evolution are to be shunned because someone might derive an indirect religious implication from the discussion. But it is perfectly all right—no, desirable—to invoke religion in *support* of evolution.

In 2004, the NCSE took this approach a step further when it unveiled a website for teachers titled "Understanding Evolution," jointly developed with the University of California Museum of Paleontology.[112] Funded in part by more than a half-million dollars in federal money from the National Science Foundation, the website was intended to instruct teachers how to teach evolution.[113] But in addition to science information, the website included a module encouraging teachers to use religion to endorse evolution. Teachers were told that nearly all religious people, theologians, and scientists who hold religious beliefs endorse modern evolutionary theory, and that such a view "actually enriches their faith."[114] In addition, the website directed teachers to the NCSE's website in order to read statements by a variety of religious groups giving their theological endorsement of evolution.

For example, teachers could read a statement from the United Church of Christ that "modern evolutionary theory . . . is in no way at odds with our belief in a Creator God, or in the revelation and presence of that God in Jesus Christ and the Holy Spirit."[115] Needless to say, statements from religious groups and scholars who critique Darwinism because of its claim that the development of life was an unguided process were not included.

Elsewhere on the website, Eugenie Scott encourages biology teachers to spend class time having students read statements by religious leaders and theologians supporting evolution. Scott even suggests that students be assigned to interview local ministers about their views on evolution—but not if the community is "conservative Christian," because then the lesson that "Evolution is OK!" may not come through.[116]

Taxpayers might wonder why it is the government's business to tell them what their religious beliefs about evolution should or shouldn't be. Presumably the National Science Foundation grant for the "Understanding Evolution" website was supposed to be spent on teaching science, not on convincing people that evolution comports with "the revelation and presence of . . . God in Jesus Christ and the Holy Spirit." It is difficult to see how the website's presentation of religion even comes close to following Supreme Court precedents on the establishment clause of the First Amendment. Imagine the outcry if some group advised opponents of Darwin's theory to invite clergy into schools to criticize evolution. How long would it take before the ACLU was on the scene?

This effort to use religion to endorse evolution is part of a larger public-relations strategy devised by the NCSE to defuse public skepticism of neo-Darwinism. On its website, the group urges evolution supporters to invite ministers to testify before school boards in favor of evolution,[117] and it has created a curriculum to promote evolution in the churches.[118] The NCSE even has a "Faith Network Director" who claims that "Darwin's theory of evolution . . . has, for those open to the possibilities, expanded our notions of God."[119] Other evolutionists have followed the NCSE's lead, collecting signatures from liberal clergy in support of evolution as part of "The Clergy Letter Project" and urging churches to celebrate "Evolution Sunday" on the Sunday closest to Darwin's birthday.[120] Whether NCSE staff members actually believe their own claims about the compatibility of religion and evolution is questionable.

The attempt to put a religious face on modern evolutionary theory is an effort to deal with what might be called Darwinism's "Dawkins problem." Oxford biologist Richard Dawkins is one of the world's foremost boosters of Darwinian evolution. Unfortunately for evolutionists, Dawkins zealously expounds the antireligious implications of the theory and regularly denounces religion.[121] By highlighting the religious defenders of evolution, the NCSE undoubtedly hopes to depict Dawkins as a fringe figure whose views are not representative of Darwinists as a whole. Unfortunately for them, he is hardly on the fringe when it comes to evolutionary biologists or even grassroots supporters of Darwinism. As noted earlier, the vast majority of biologists in the elite National Academy of Sciences identify themselves as atheists or agnostics.

Although the effort to promote a theological justification for evolution in public schools represents a new wrinkle in the controversy over teaching evolution, it is far from surprising. Throughout American history, from crusades over eugenics to more recent antireligious polemics, biology education has been entangled with politics and theology. Once one recognizes this fact, it becomes tougher to justify a blanket exclusion of religious and philosophical critics of Darwin's theory from a place at the table in public science education. Biology classes are already addressing religious and philosophical issues, and so as a matter of simple fairness, it would seem that students ought to be able to learn about the full spectrum of religious and philosophical views of evolution.

But metaphysics masquerading as empirical science is only part of the problem with how evolution is currently taught in American schools. According to a growing number of critics, the real problem with modern biology education is not that some religious and philosophical viewpoints are shut out, but that modern biology's commitment to neo-Darwinism discourages students and teachers from honestly addressing the scientific shortcomings of evolution. These critics allege that neo-Darwinism has become an ideology that no longer tolerates real debate and that science teachers and scientists question at their peril. We will explore these charges in the next chapter.

11

Banned in Burlington

Burlington, Washington, is a sleepy town located a couple of hours north of Seattle. Surrounded by farmland, it is the sort of quiet place someone might go to get away from it all. But in the late 1990s, Burlington became embroiled in an emotional, bitter battle over science education that attracted national attention. At the center of the conflict was high school biology teacher Roger DeHart.[1]

Popular and well respected before the controversy, the soft-spoken De-Hart suddenly found himself the target of strident criticism for his teaching of evolution. To some observers, DeHart's plight seemed like an updated version of *Inherit the Wind*, the fictionalized drama about the Scopes trial of the 1920s. Like the protagonist in that drama, DeHart was an idealistic teacher who faced ostracism for teaching a controversial scientific theory.

Yet the story had a twist: The protagonist in *Inherit the Wind* was trying to promote Darwin's theory of evolution. DeHart wanted to criticize it.

For one day out of a two-week unit on evolution, DeHart would tell students about scientists who were raising questions about the validity of Darwin's theory of unguided evolution, including scientists who advocated the emerging theory known as "intelligent design."[2] Scientists who advocate intelligent design in biology claim that Darwin's mechanism of natural selection is insufficient to explain the development of life. They believe that parts of biological life—such as the amazing molecular motors within the cell, or the information content embedded in DNA—are so intricate and highly ordered that they are best explained as the products of an intelligent cause rather than chance and necessity.

Similar arguments about the evidence for design in nature have been made outside biology in physics and cosmology for decades. As Jonathan Witt points out, "beginning with Fred Hoyle's discovery of the carbon-12 resonance in the early 1950s, physicists began uncovering a number of ways the universal constants of physics and chemistry (gravity, the strong and weak nuclear forces, etc.) were fine tuned for complex life. Reviewing these developments in 1982, leading theoretical physicist Paul Davies described the fine-tuning of the universe as 'the most compelling evidence for an element of cosmic design.' "[3]

Critics of intelligent design often lump it together with biblical creationism, but that assessment is inaccurate as well as unfair.[4] The intellectual roots of the idea reach back to ancient Greece and Rome,[5] and the term itself was employed as an alternative to blind evolution by Oxford scholar F. C. S. Schiller in the 1890s.[6] More recent usage of the term dates to a book by James Horigan published by the Philosophical Library in 1979, the agnostic Hoyle in 1982, and chemist Charles Thaxton in the late 1980s.[7]

Whatever its weaknesses as a scientific theory, intelligent design is not based on a reading of Genesis nor is it promoted primarily by Christian fundamentalists. Instead, it is advocated by scholars from a variety of religious backgrounds who hold standard academic credentials, such as biochemistry professor Michael Behe at Lehigh University; microbiology professor Scott Minnich at the University of Idaho; neuroscientist Jeffrey Schwartz at the University of California, Los Angeles; Cambridge-trained philosopher of science Stephen Meyer; and mathematician and philosopher William Dembski, formerly of Baylor University.[8]

Strictly speaking, intelligent design is not "anti-evolution." It does not challenge the idea that living things "change over time," nor does it deny that Darwin's mechanism of natural selection can produce changes in living things.[9] It is not incompatible with the idea that all living things arose from a universal common ancestor, although scientists who support intelligent design differ on whether the scientific evidence substantiates such a conclusion.[10] Intelligent design does oppose the central claim of Darwinian evolution that all of the highly specified complexity in nature can be accounted for through an undirected process such as natural selection acting on random variations.

Although critics often label intelligent design a "purely negative" argument,[11] the primary argument for intelligent design is positive. Based on our own extensive experience in the natural and social worlds, we know that intelligent causes are habitually capable of producing certain kinds of highly functional complexity (what mathematician William Dembski defines as "specified complexity.")[12] Thus, whenever we find highly functional complexity we have prima facie evidence that it was produced by an intelligent cause—because we know from our own experience that intelligent causes habitually produce this kind of complexity.

This argument is based on our knowledge of the cause-and-effect operation of the world around us, not on our ignorance. So when we see Mt. Rushmore, or a laptop computer, or even a beaver's dam or a bird's nest, we are acting rationally when we conclude that they were all the result of purposeful action rather than a blind process of chance and necessity.[13]

It might be added that the positive case for intelligent design is based on the same scientific methodology employed by Darwin. He embraced what is known as "uniformitarianism," an approach popularized by his friend Charles Lyell, the founder of modern geology. According to uniformitarianism, science seeks to explain past events by invoking causes that regularly operate in nature today rather than by hypothesizing unique or ad hoc causes that may have existed in the past.[14]

This was precisely the scientific approach followed by Darwin when he drew on his knowledge of the current effects of breeding or "artificial selection" to provide evidence for his view that "natural selection" was the primary mechanism of evolution in natural history. The same uniformitarian approach is employed by the modern proponents of intelligent design who infer intelligent causation in the past based on what we know intelligent causes are capable of doing in the present.

Of course, conclusions reached by applying uniformitarianism can be mistaken. It is logically possible for similar effects to be the products of different causes. That is why the positive argument for intelligent design is strengthened by a negative argument: At the same time intelligent causes habitually produce certain kinds of highly functional complexity, nonintelligent causes (chance and necessity) do not seem capable of generating these same kinds of highly ordered complexity. This is the point of biochemist Michael Behe's argument about "irreducible complexity."

A biological system is "irreducibly complex," says Behe, if it "is composed of several interacting parts that contribute to the basic function," and if "the removal of any one of the parts causes the system to effectively cease function." Irreducibly complex systems cannot be generated through a direct evolutionary route by natural selection "since any precursor to an irreducibly complex system is by definition nonfunctional. Since natural selection requires a function to select, an irreducibly complex biological system, if there is such a thing, would have to arise as an integrated unit for natural selection to have anything to act on."[15]

Taken together, the positive and negative arguments offered by intelligent-design scientists reinforce each other: If we know that intelligent causes habitually generate systems of highly functional complexity, and we also know that nonintelligent causes typically do not do so, then the best explanation under the circumstances ("the inference to the best explanation")[16] is that an intelligent cause was involved whenever we see highly functional complexity. Intelligent design as a scientific hypothesis is thus a probabilistic argument based on our current knowledge, which means that it is also subject to testing and possible refutation based on new evidence of how nature operates.[17]

Although intelligent design is first and foremost a positive argument, it is certainly true that the negative argument is quite important. If nonintelligent processes are typically capable of producing new highly functional complexity, then it brings into question whether intelligent causes are the best explanation for any given case of such complexity. At the same time, the less capable nonintelligent causes are of producing highly functional complexity, the more likely intelligent causation is the best explanation, given what we know about the capabilities of intelligent causes.

In his Burlington High School biology classes, Roger DeHart did not plan to delve into great depth about intelligent design, but he did want to inform students about Michael Behe's hypothesis that some biochemical systems are so "irreducibly complex" that they are better explained as the products of intelligent design rather than a Darwinian process of blind, unguided natural selection. At the end of the evolution unit, DeHart asked his students to prepare papers describing what they thought the best evidences were for either evolution or for intelligent design. Students were also asked to critically evaluate their own views—so students who wrote in favor of Darwin's theory had to describe if there were any problems with evolutionary theory, and students who wrote in favor of intelligent design had to describe any possible drawbacks of that theory. Those students who chose not to write papers were asked to prepare a classroom debate on the subject of Darwinism v. design.

DeHart says he always made sure that each side had the same number of students and was given the same amount of time to state its views before the class. After the debate, students were able to give their own views during a class discussion. That concluded the unit on evolution.

DeHart maintains that he was evenhanded and impartial in his presentation of material in class: "I think that before the controversy broke . . . most students . . . would never know what side of the fence I was on; whether I was a design advocate or an evolutionist. I taught the unit as a neutral person."[18] In fact, says DeHart, students frequently asked him to reveal his own views and he would tell them that his opinion wasn't important because he worried that the minute he stepped in and gave his opinion he would be squelching student discussion.

In 1997, a parent of one of DeHart's former students filed a complaint with the American Civil Liberties Union, and the ACLU subsequently tried to get the school district to curtail DeHart's discussion of intelligent design. Later a group of local citizens formed an anti-DeHart group called "The Burlington-Edison Committee for Science Education," and the National Center for Science Education (NCSE) also began lobbying the school district against DeHart. After first upholding DeHart's right to teach about intelligent design as part of the larger unit on evolution, the school district reversed itself and told DeHart he could no longer mention intelligent design to students.

It soon became clear, however, that DeHart's opponents would not be satisfied with merely silencing DeHart's discussion of intelligent design. They didn't want DeHart to present any information that might cast any doubt on Darwinian theory, even if that information came from mainstream science journals. In their view, telling students about recognized problems in evolutionary theory was tantamount to "creationism" or at the very least would confuse high school students by implanting doubts about evolution. "The role of a high school teacher is not to be on the cutting edge of research," insisted Eugenie Scott of the NCSE. "It's not doing the students any service to confuse them about some of the esoteric elements of a scientific discipline."[19]

Initially, DeHart was told by district officials that he could still inform students about scientific criticisms of evolutionary theory, but he soon learned otherwise. He tried to assign articles from mainstream science publications correcting various misstatements in his high school biology textbook, but the articles were vetoed by his principal. Things got worse. "They went from not allowing me to introduce any supplemental materials, then to telling me that I could not verbally share any of this information with my students," recalls DeHart. "Basically I was to stick to and just teach what the textbook said"—even when the textbook was wrong.[20]

DeHart was forbidden to show students an article by Stephen Jay Gould, in which the noted evolutionist criticized textbooks for using outdated drawings of vertebrate embryos based on a diagram created by nineteenth-century German Darwinist Ernst Haeckel.[21] Haeckel's diagram purported to show that all vertebrates come from a common ancestor, since in their earliest stages of embryonic development they look almost identical. But Haeckel's pictures made vertebrate embryos look much more similar than they are in reality, and Haeckel may well have faked the drawings. Although Gould strongly advocated the theory of evolution, he criticized textbook use of Haeckel's bogus drawings. Even though the textbook DeHart used reprinted the drawings, the teacher wasn't allowed to tell his students that they were no longer considered good science.

DeHart's critics agreed with the school district's decision. "The issue of Haeckel's embryos is a tiny little historic anomaly in terms of the big picture of high school biology," said Eugenie Scott of the NCSE. "Why does Mr. DeHart feel so strongly to have these articles read? Well it's because he wants to call into question the whole issue of whether evolution happened. I don't think this is good science education for the students in that district. And apparently neither did his colleagues nor his superintendent."[22]

The DeHart controversy reversed the usual cultural stereotypes. In the popular mind, evolutionists are expected to be champions of academic freedom, while religious fundamentalists play the role of inquisitors who persecute teachers. But in the Burlington conflict it was the defenders of evolution who seemed to be the inquisitors, and their rhetoric didn't do anything to allay that impression. "Fanatics like DeHart and his cohorts will never be satisfied," fumed one DeHart critic in the local newspaper.[23] "Mr. DeHart . . . stands exposed as the fox in the henhouse from which he should be ejected," cried another, who added that it was "unfortunate that youth are the target of the pseudo-science mullahs who would preach blind belief in the supernatural to explain what they do not comprehend about complexity."[24] "Mr. DeHart . . . now wants to teach about 'problems' in the textbook," complained a third critic. "Every year there is a new name for creationism. When the materials, tactics, and it's [sic] originators are so continually intellectually dishonest, the district and community should say, 'no, go away.' "[25]

DeHart's critics portrayed the controversy as an epic battle between local citizens and a fundamentalist conspiracy that was seeking to fuse church and state. The Burlington-Edison Committee for Science Education (BCSE) even set up a website devoted to attacking DeHart and denouncing the supposed conspiracy. The website sought to expose DeHart as "a fundamentalist" who "was working in concert with fundamentalist foundations worth millions of dollars who make it their business to generate questionable evidence and fund local anti-science initiatives." Chief among these groups was "the big bad Discovery Institute," a nonprofit think tank based in Seattle that supports scholars skeptical of materialistic Darwinian theory (including this writer). According to the BCSE, the goal of Discovery Institute was to "move our world back into the dark ages where scientific observations were allowed to be explained by supernatural miracles." Indeed, the institute had a twenty-year master plan "to have religion control not only science, but also everyday life, laws, and education." Moreover,

Discovery was tutoring DeHart "on how to subvert science in subtle ways that appeared to be legitimate biology." To be confronted by such an evil agenda was nothing short of "frightening" to the BCSE, but defenders of true science were told to take heart: Discovery Institute "is being defeated in our town, and they can be defeated in yours as well."[26]

The attempt to depict DeHart as the puppet of a fundamentalist conspiracy was ridiculous on several levels. Discovery Institute, the supposed mastermind of the conspiracy, is in fact a secular organization that has sponsored programs on a wide array of issues besides intelligent design, including mass transit, technology regulation, the environment, and foreign and defense policy. At the time of the DeHart controversy, the chairman of its board was Jewish, its president was Episcopalian, and its various fellows represented a range of religious views ranging from Roman Catholic to agnostic.[27] Far from advocating a union between church and state, it sponsored a small program on religion and civic life that advocated religious liberty and the necessity for religious believers in the public square to base their proposals on the common ground they share with all citizens rather than on sectarian religious teachings.[28]

The effort to portray Discovery Institute as the mover and shaker of the Burlington controversy was also incongruent with the facts. The institute came to the defense of DeHart only after the Burlington controversy was underway. It was the ACLU and the NCSE that lodged the original complaints launching the controversy, not Discovery. Since DeHart's opponents were the ones who had enlisted the intervention of these outside groups, it was odd for them later to suggest that the conflict was somehow generated by DeHart and Discovery Institute.

The strident campaign against DeHart had an effect. After fifteen years of teaching biology in the Burlington-Edison school district, DeHart was informed during the summer of 2001 that he was being reassigned to teach earth science.[29] Rick Jones, the district superintendent, flatly denied that the reassignment was designed to punish DeHart for his views, but even the local newspaper (which had been critical of DeHart previously) had a difficult time believing that claim.[30] The new biology teacher was fresh out of college and had majored in physical education rather than biology, like DeHart.[31]

No longer allowed to teach his chosen field in Burlington, DeHart resigned and took a position in an adjoining school district in order to teach biology there. However, local newspapers discovered his new position and published stories playing up his controversial background.[32] That made the new school district skittish, and it also promptly reassigned DeHart to teach earth science, even though he had been hired to teach biology. Only after DeHart brought in an attorney did the new district relent. But the victory was temporary. Later in the school year, DeHart was again reassigned from biology to earth science—against the wishes of both his principal and the physics teacher, who was now expected to teach biology.[33]

Neither Burlington nor DeHart's new district alleged that DeHart had been insubordinate or continued to teach things not allowed by the districts. Indeed, in Burlington DeHart had been given glowing evaluations and was regarded as an exemplary instructor.[34] Worn out by years of hectoring, DeHart resigned from his new job in 2002 and moved to another state to teach at a private school.[35]

In many ways, Roger DeHart represents the new face of the controversy over evolution in American education. It is no longer teachers and professors who advocate evolution who face intimidation and censorship. Instead, it is those who criticize Darwin.

At San Francisco State University, tenured biology professor Dean Kenyon was removed from teaching introductory biology classes in the early 1990s. Once an influential proponent of Darwinian evolution, Kenyon had come to doubt key parts of Darwin's theory and expressed those doubts to students in class, including his belief that some biological features exhibited evidence of intelligent design. Kenyon was luckier than most academic critics of Darwin. After his plight was publicized by an article in the *Wall Street Journal*, his university was shamed into reinstating him.[36]

In Minnesota, high school teacher Rodney LeVake was removed from teaching biology in 1998 after expressing doubts about Darwin's theory. LeVake, who holds a master's degree in biology, agreed to teach evolution as required in the district's curriculum but also said he wanted to "accompany that treatment of evolution with an honest look at the difficulties and inconsistencies of the theory."[37]

At Baylor University in the fall of 2000, mathematician William Dembski was fired as head of an academic center he had founded to explore intelligent design as a scientific paradigm. Several years later, his faculty contract was not renewed after it had expired. Dembski, who holds two Ph.D.s (one from the University of Chicago), had exemplary academic credentials and publications, but his research center had been strenuously opposed by Baylor's biology faculty.[38]

Oregon community college instructor Kevin Haley was terminated in 2000 after it became known that he criticized evolution in his freshman biology classes.[39] Haley's college refused to state why his contract was not renewed, but some of Haley's colleagues were upset that students who took his biology class were starting to challenge evolution in their classes. Before the controversy over evolution, Haley had been regarded as an excellent teacher. His former department chair had praised him in glowing terms, saying that students "perceive that he is interested in them. He generates curiosity and stimulates their thinking. Those are things that I think are not always there in a professor."[40]

In 2003, chemistry professor Nancy Bryson was removed from her post as head of the science and math division of Mississippi University for Women after she delivered a lecture to honors students about some of the scientific weaknesses of chemical and biological evolution. "I was harshly attacked by Darwinist colleagues," she explained later. "Students at my college got the message very clearly, do not ask any questions about Darwinism."[41]

In 2005, biology professor Caroline Crocker at George Mason University was "barred by her department from teaching both evolution and intelligent design" after committing the crime of mentioning intelligent design in a course on cell biology. "It's an infringement of academic freedom," she told the journal *Nature*.[42] Subsequently her contract was not renewed.[43]

Also in 2005, Ohio State University doctoral candidate Bryan Leonard had his dissertation defense put in limbo after three pro-Darwin professors filed a bogus complaint attacking Leonard's dissertation research as "unethical human subject experimentation." Leonard's dissertation project looked at how student beliefs changed after students were taught scientific evidence for and against modern evolutionary theory. The complaining professors admitted that they had not actually read Leonard's dissertation. But they were sure it must be unethical. Why? Because there is no valid evidence against evolutionary theory. Thus—by definition—Leonard's research must be tantamount to child abuse.[44]

In 2004, evolutionary biologist Richard Sternberg faced retaliation by officials at the Smithsonian Institution's National Museum of Natural History (NMNH) after accepting for publication a peer-reviewed article supportive of intelligent design by philosopher of science Stephen Meyer in a biology journal he edited. A research associate at the museum, Sternberg said that after the article was published he was told to vacate his office space and was shunned and vilified by colleagues. Efforts were also made by administrators to discover Sternberg's personal religious and political beliefs.[45] Investigators for the U.S. Office of Special Counsel concluded that "it is . . . clear that a hostile work environment was created with the ultimate goal of forcing [Dr. Sternberg] . . . out of the [Smithsonian]."[46]

Smithsonian officials denied any wrongdoing, but in late 2006 Sternberg was demoted from a research associate to a "research collaborator" without explanation.[47] A seventeen-month investigation by subcommittee staff of the House Committee on Government Reform confirmed and elaborated on the previous findings of the U.S. Office of Special Counsel. In a detailed report issued in December 2006, subcommittee investigators concluded that they had uncovered "substantial, credible evidence

of efforts to abuse and harass Dr. Sternberg, including punitively targeting him for investigation in order to supply a pretext for dismissing him,
and applying to him regulations and restrictions not imposed on other
researchers."[48]

Congressional investigators accused NMNH officials of conspiring
"on government time and using government emails . . . with the pro-
evolution National Center for Science Education (NCSE) . . . to publicly
smear and discredit Dr. Sternberg with false and defamatory information."[49] The NCSE even provided a set of "'talking points' to . . . NMNH
officials on how to discredit both Sternberg and the Meyer article." In addition, the NCSE was asked by senior museum administrator Dr. Hans
Sues "to monitor Sternberg's outside activities . . . The clear purpose of
having the NCSE monitor Dr. Sternberg's outside activities was to find a
way to dismiss him."[50] Congressional investigators concluded that "the extent to which NMNH officials colluded *on government time and with government resources* with the NCSE to publicly discredit Dr. Sternberg's scientific
and professional integrity and investigate opportunities to dismiss him is
alarming."[51]

When asked about Sternberg's plight by the *Washington Post*, Eugenie
Scott of the NCSE replied: "If this was a corporation, and an employee did
something that really embarrassed the administration, really blew it, how
long do you think that person would be employed?"[52]

Nor is biology the only scientific field where litmus tests are being applied. At Iowa State University, pro–intelligent design astronomer Guillermo Gonzalez was denied tenure in 2007 despite the fact that his work on
design focused on physics and astronomy and did not challenge biological evolution.[53] An outstanding scientist whose research has been featured
in *Science, Nature,* and on the cover of *Scientific American*, Gonzalez was
rejected for tenure even though he had produced 350 percent more peer-
reviewed publications than needed to demonstrate research excellence in
his department, and even though his articles have been cited more frequently by his peers than the work of all but one of his astronomer colleagues (according to a key measure of scientific citations).[54] University
president Gregory Geoffroy insisted that Gonzalez's denial of tenure had
nothing to do with his views on intelligent design.[55] But members of Gonzalez's department admitted otherwise, including one colleague who published a newspaper article highlighting Gonzalez's views on intelligent
design as the only reason he voted to deny him tenure.[56] Gonzalez's most
vocal opponent on campus was atheist religion professor Hector Avalos,
who has argued that the Bible is worse than Hitler's *Mein Kampf*.[57] The very
semester Gonzalez was turned down for tenure at Iowa State, Avalos was
promoted to full professor.

Even scholars outside the natural sciences have faced mistreatment for suggesting that there might be a serious intellectual debate over Darwinism. At Baylor University, philosopher and legal scholar Francis Beckwith initially was denied tenure in the spring of 2006, despite an outstanding record of academic research and publications.[58] Although Professor Beckwith was well known for his pro-life views, he was most controversial for his law-review articles and an academic book defending the constitutionality of teaching about intelligent design.[59] It is important to note that Beckwith did not advocate that intelligent design should be taught in public schools—only that it was constitutional to teach it in an appropriate manner. But that nuanced position was too much for some Darwinist colleagues to accept. Fortunately for Beckwith, after a public outcry the president of Baylor granted him tenure later in the same year.[60]

In the popular mind, science is often thought of as a disinterested search for empirical truth. However, philosophers and historians of science know well that science, just like any other human profession, is often shaped by egos, ideology, partisanship, and struggles for power. "What science says the world is like at a certain time is affected by the human ideas, choices, expectations, prejudices, beliefs, and assumptions holding at that time," write philosophers of science Peter Machamer, Marcello Pera, and Aristides Baltas.[61] When one scientific theory is enshrined as the reigning paradigm, dissenting views are often silenced for reasons other than lack of evidence. Dissenting views represent a threat to the ruling paradigm, and so those who have earned power and prestige from advancing it are loath to let their authority be eroded. For this and other reasons, scientists who have spent their lives working within one paradigm may have a difficult time acknowledging problems within that paradigm no matter how much contrary evidence accumulates.[62]

Some scientists are now complaining that this is exactly what has happened with Darwin's theory of evolution. They claim that the theory has taken on nearly dogmatic status with many in the profession, and that anyone who dares to criticize it faces ostracism—or worse. As one Berkeley-trained biologist has learned, scientists who express public doubts about Darwinian theory should be prepared to have their integrity challenged and their personal lives exposed.

Berkeley's Heretic

Jonathan Wells may be a natural-born heretic. He certainly is unafraid to buck the Establishment. During the 1960s, he was active in the antiwar movement in Berkeley, and he spent more than a year in federal prison because he refused to return to active duty in the Army because of his op-position to the Vietnam War.[63] Refusing to wear a military uniform again, Wells spent the first four months of his prison sentence wrapped in a blan-ket in solitary confinement. But the violence of some antiwar protesters (especially campus Marxists) ultimately disillusioned him, and after he was released from prison and finished his undergraduate degree in physi-cal sciences, he "built a cabin in the woods and lived like Thoreau for a year."[64]

Up to that point, Wells says he fully accepted Darwin's theory of evo-lution. But as he began a wide-ranging exploration of nature, philosophy, and world religions, he started to have doubts. "As I got more in touch with the wild, with living things, I just decided that Darwin's mechanism was not adequate to explain what I was seeing," he explains. "I still be-lieved that everything descended from a common ancestor, but I just ques-tioned whether natural selection and random variations could account for everything I saw."[65]

At the same time, Wells underwent a spiritual awakening, an expe-rience that ultimately led him to make another heretical—or at least un-conventional—choice. He joined the Reverend Sun Myung Moon's Unifi-cation Church, impressed by the way Moon's followers were willing to stand up to Marxists on college campuses.[66] Wells eventually earned a Ph.D. in religious studies at Yale University, writing his dissertation on nineteenth-century theologian Charles Hodge's critique of Darwinism.[67] At Yale, Wells became increasingly disenchanted by the way many scien-tists and philosophers had used Darwin's theory to promote materialism, and he wanted to oppose such efforts to join evolutionary theory with an-tireligious metaphysics. He decided to pursue a second Ph.D.—this time in biology—to better equip himself to challenge what he viewed as the con-tamination of evolutionary biology with metaphysics. At this point Wells says he still accepted Darwin's theory of common ancestry.[68]

It was only after he began to study evolutionary theory at the gradu-ate level that Wells started to question not only Darwin's mechanism of unguided evolution but the claim that all life developed from a univer-sal common ancestor. In graduate school at the University of California at Berkeley, Wells says he discovered to his surprise that much of the stan-dard textbook evidence offered to support evolutionary theory was in fact no longer accepted as good science by many biologists. In time, that discovery inspired him to write *Icons of Evolution* (2000), a book expos-ing what he regards as widespread factual inaccuracies in many textbook treatments of evolution that overstate the evidence for Darwin's theory.[69]

Chapters in Wells's book were peer-reviewed by other scientists in various subfields of biology, and the book's claims were exhaustively documented with seventy pages of references to the primary scientific literature. Wells claimed that the ten examples most commonly used by biologists (and textbooks) to corroborate evolution were highly misleading. "Some of these icons of evolution present assumptions or hypotheses as though they were observed facts," he wrote. "Others conceal raging controversies among biologists that have far-reaching implications for evolutionary theory. Worst of all, some are directly contrary to well-established scientific evidence."[70]

The examples criticized by Wells included Haeckel's fraudulent embryos, which are supposed to show that "amphibians, reptiles, birds and human beings are all descended from a fish-like animal"; "the evolutionary tree of life," which is supposed to be based on "a large and growing body of fossil and molecular evidence"; similar bone structures between animals, which are supposed to "indicate their evolutionary origin in a common ancestor"; "peppered moths on tree trunks," which supply "the most famous example of evolution by natural selection"; and "fruit flies with an extra pair of wings," which purportedly demonstrate "that genetic mutations can provide the raw materials for evolution."[71]

Because *Icons of Evolution* was written by someone with academic credentials in biology from a leading research university, it was a hard book for evolutionists to ignore, and it was widely reviewed in the nation's top science journals, including *Nature, Science, BioScience,* and the *Quarterly Review of Biology.*[72] Although some reviewers grudgingly acknowledged Wells's technical mastery of his subject, the reviews in mainstream science journals were mostly negative. Given that Wells launched a frontal assault on the evidence for Darwin's theory, that reaction was not surprising. What was unexpected was the overwhelmingly personal tone of many of the criticisms—which frequently had little to do with science.

Perhaps the most outlandish example was the review penned by University of Chicago evolutionary biologist Jerry Coyne for the prestigious journal *Nature.*[73] More than 40 percent of its text was devoted to outing Wells as a member of the Unification Church and ridiculing him as a follower of Reverend Moon. As a rhetorical strategy, this approach may have been effective. The Unification Church is certainly an unpopular religious movement in the United States, and focusing attention on Wells's religious affiliation was an easy way of marginalizing him without having to answer his arguments. As a form of scientific discourse, however, Coyne's attack on Wells's religion—in a book that dealt with science, not religion—was despicable. Here was a scientific journal unashamedly using religion as an argument in a scientific dispute.

Appalled by what he regarded as Coyne's "ad hominem" attacks, one scientist complained to *Nature* that Coyne's tactics reminded him of the

Nazis' effort to delegitimize music written by composers of Jewish ancestry. "I find no difference between the rejection of that music on those irrelevant religious grounds, and Dr. Coyne's dismissal of Wells' book on the basis of his religion," wrote the scientist. "If the editors of *Nature* cannot understand this, then scientists are going to doubt the validity of anything published in the magazine."[74]

Coyne's effort to stigmatize Wells as outside the religious mainstream was twinged with irony, because Coyne's own metaphysical beliefs happen to be far from the American mainstream. Coyne endorses the idea that humans are "random blind products" of natural selection and contends that "we're not the special products of any creative process." Although "people don't like that because . . . they think it deprives them of having a purpose in a universe and it makes them feel alone . . . I'm perfectly content, myself, to live with that."[75] Had Coyne articulated his own metaphysical beliefs in his *Nature* review rather than simply pillorying those held by Wells, readers might have concluded that it was Coyne whose beliefs are most out of the religious mainstream.

Of course, the deeper issue is not whose religious beliefs are more mainstream but whether pointing out a supposed religious motivation constitutes a refutation of someone's scientific views. One of the more troubling aspects of the contemporary debate over evolution is the largely unargued presumption that critics of evolution must be motivated by religion, whereas defenders of evolution are supposedly the disinterested pursuers of truth.

Yet evolutionary biologists are human beings just like everyone else, which means that they are subject to the same influences of passion, pride, and even prejudice. If scientists start to dismiss their colleagues' work based solely on motivation (especially supposed motivations), not much would be left of science. Some scholars have argued that Darwin's work was influenced by his rejection of God after the death of one of his daughters. Does that mean Darwin's theory should be rejected because of a presumed antireligious motivation? For that matter, what about Jerry Coyne's motives for defending evolution? Might Coyne be so vehement because of his own metaphysical commitment to the idea that human beings are "not the special products of any creative process"?

The focus on motives in science quickly gets one nowhere. Science is supposed to focus on analyzing evidence for scientific theories rather than unraveling the personal motivations of scientists. But preoccupied with Wells's religion, Coyne spent very little space actually analyzing the evidence in Wells's book. He only discussed one of Wells's "icons" in any detail—Haeckel's embryos—and his analysis mostly ignored rather than engaged the material Wells offered on the topic. The only other "icon" mentioned by Coyne was Wells's chapter on human evolution, which he dismissed by stating that "Wells can only mumble about the Piltdown Man hoax, and imply that the vigorous scientific debate about the course of human evolution proves that humans did not evolve."[76] Wells's twenty-page chapter did much more than that, but Coyne didn't attempt to offer a real critique.

Coyne's review was not the only one that relied more on invective than a serious review of the evidence. "The Talented Mr. Wells" in the *Quarterly Review of Biology* was in some ways even more gratuitous and ad hominem. Written by Kevin Padian and Alan Gishlick, both officials with the National Center for Science Education, "The Talented Mr. Wells" compared Wells to the protagonist in the film *The Talented Mr. Ripley*.[77] During the course of the movie, Mr. Ripley is revealed to be a habitual liar who misrepresents his background and stops at nothing to get what he wants, including murder, so this comparison amounted to character assassination. Padian and Gishlick also impugned Wells credentials by claiming that after he obtained his Ph.D. in biology, he "followed this with a [five]-year postdoctoral position . . . during which time he seems to have performed no experiments" and "no peer-reviewed publications resulted."[78] Both claims were false. Wells had performed experiments in his postdoctoral position, and those experiments had generated peer-reviewed publications.[79]

The scientist who complained to *Nature* about Coyne's "ad hominem attacks" on Wells and his failure to deal with the substance of Wells's book received a letter back from an editor defending the attack on Wells's religion and arguing that *Nature*'s book reviews "are not intended to list every point made in the book. Nor do they necessarily have to be objective."[80] Nor did *Nature* allow Wells an opportunity to respond to Coyne's critique, even though Wells sent the journal a letter challenging (with appropriate scientific citations) some of Coyne's scientific claims.[81] The *Quarterly Review of Biology* exhibited similar defensiveness. When one of Wells's colleagues at Berkeley sent in a letter verifying that she and Wells had in fact performed experiments together and that these experiments had resulted in peer-reviewed publications, editor Albert Carlson wrote back: "The *Quarterly Review of Biology* does not print retractions."[82]

Even when Wells's critics did challenge the substance of his case, they were determined to defend traditional Darwinian theory no matter what the evidence. Consider the peppered moth. Peppered moths come in lighter and darker varieties. During the industrial revolution in England, the darker variety gradually overtook the lighter variety. For decades, most biology texts cited this fluctuation as a prime example of natural selection in action. According to the standard textbook account, increased pollution during the industrial revolution killed off the light-colored lichens that grew on trees, making tree trunks darker. In the new environment, darker moths were able to survive better because they were harder for birds to spot on the darker trunks.

This account was based largely on experiments conducted during the 1950s by Bernard Kettlewell that demonstrated that lighter moths resting on tree trunks were more likely to be eaten by birds. Additional confirmation of the camouflage-predation hypothesis seemed to come after pollution was reduced during the latter part of the twentieth century. The lichens returned to the trees (making the tree trunks lighter), and the lighter moths once again overtook the darker moths in the overall moth population. The changes seemed to provide striking confirmation of how Darwin's theory of natural selection actually works in the wild.

However, Jonathan Wells pointed out that the traditional textbook story has major flaws.[83] Field studies have demonstrated that pollution and tree lichens are not always correlated with a greater proportion of darker moths. In one place, for example, the number of darker moths increased after pollution decreased.[84] In another area, the number of darker moths "began decreasing before lichens returned to the trees."[85] Perhaps the most serious flaw of all was that "since 1980 . . . evidence has accumulated showing that *peppered moths do not normally rest on tree trunks.*"[86] Instead, the moths likely rest in unexposed places among tree branches, where they are not easily observed by birds. But if the moths usually rest in places where birds cannot easily see them, then the bird-predation hypothesis is not supported by the empirical evidence.

Given the new evidence, Wells criticized textbooks for publishing photos showing the moths resting on tree trunks. Like many wildlife photos, these were invariably staged, sometimes using dead moths glued to the trunks. Wells argued that such "staged photos may have been reasonable when biologists thought they were simulating the normal resting-places of peppered moths," but "by the late 1980s . . . the practice should have stopped" because biologists had come to know that the moths do not normally rest on tree trunks in the wild.[87] In Wells's view, textbooks that continue to use the staged photos of peppered moths are publishing "faked photographs."[88]

Wells accepts the role of natural selection in promoting microevolution, and he does not claim that the lack of proof for the traditional account of peppered moths refutes the reality of natural selection. His claim is more nuanced: Peppered moth fluctuations may still result from natural selection, but they should not be offered as proof for natural selection or the camouflage-predation hypothesis until the research confirms this to be the case. Nor should students be shown staged pictures of moths resting on tree trunks without being told that the pictures were staged and that moths do not normally rest on tree trunks. If textbook writers want to continue to include the peppered moth, they should tell students the whole story. They should introduce them to the genuine complexities of scientific research—an endeavor where evidence does not always neatly confirm one's theories, where credible scientists often disagree, and where all scientists need to be open to new data. Wells's plea for full disclosure on the peppered moth would seem to be a fairly moderate proposal to which both evolutionists and their critics should be able to agree. Nevertheless, some evolutionary biologists refused to admit that Wells's criticisms had any merit, especially in public discussions of educational policy.

Kenneth Miller was one of the most vocal defenders of the standard peppered moth story, which he had included in his own textbooks. At a meeting of the Ohio State Board of Education in March 2002, Miller accused Wells of engaging in repeated misrepresentations and even fraud. Wells's critique of the peppered moth story was exhibit number one in Miller's indictment: "In his book, Dr. Wells made the claim, quote 'Peppered moths don't rest on tree trunks.' But he didn't present any data. When you do look at the data, what you discover is that the major observations that have been made of peppered moths in the wild most frequently shows that they rest on tree trunks—and therefore, that claim is incorrect."[89] As for the photos of peppered moths resting on tree trunks that appear in biology textbooks like his own, Miller insisted that "those 'faked' photographs aren't faked at all; they're real moths, on real trees, in the real positions that moths have actually been found in the wild." Miller later wrote that he challenged Wells's account of the peppered moth story to demonstrate how Wells "misrepresents, distorts, or simply lies about the facts."[90]

Readers of Wells's book, however, might have concluded that it was Miller who was engaging in misrepresentation. Contrary to Miller's claim that Wells "didn't present any data" in his book to back up his arguments, Wells in fact provided a detailed examination of the scientific research showing that peppered moths do not normally rest on tree trunks.[91] When Wells responded to Miller's accusations with a careful rebuttal reciting the evidence for his view, Miller posted an essay on his website with the self-pitying title "Paying the Price." Although Miller had previously accused Wells of being a liar and a fraud, he now portrayed himself as an aggrieved victim: "What is the price of publicly correcting the mistakes and distortions of the Reverend Jonathan Wells . . . As I have just found out, it is to be the recipient of page after page of personal attacks on one's honesty and integrity." According to Miller, Wells's factual rebuttal to Miller's previous attack was an effort "to smear me." Miller also played the religion card, deriding Wells as "the Reverend Jonathan Wells" and supplying a link to a Unification Church website.[92] For someone so loudly complaining about smears and "personal attacks," it was a performance of giddy chutzpah.

Despite Kenneth Miller's vigorous public defense of the peppered moth story during the first half of 2002, it was deleted from the next edition of one of his own biology textbooks.[93] The change was just in time. Later that year a devastating book-length critique of the conventional peppered moth story was published by science journalist Judith Hooper.[94] Hooper's *Of Moths and Men* suggested not only that the standard peppered moth account was unsupported by more recent research, but that the original experiments by Kettlewell were full of holes. Wells was fully vindicated, but no apologies were forthcoming from his critics.

The vehement reaction provoked by Wells indicates that the debate over Darwin's theory is something more than a merely academic dispute over scientific evidence. Wells has been treated as if he were a religious heretic questioning a sacred religious dogma. Wells thinks the analogy is an apt one. "Darwinism has all the trappings of a secular religion," he says.[95] Surprisingly, some evolutionists agree. Leading Darwinist Michael Ruse argues that "evolution is promoted by its practitioners as more than mere science. Evolution is promulgated as an ideology, a secular religion."[96] In Wells's view, that is why he has been the target of such personal abuse. He says evolution's "priests forgive a multitude of sins in their postulants . . . but never the sin of disbelief."[97] Is Wells right? Is Darwinian theory regarded as so sacrosanct in the modern scientific community that any questioning is treated as heresy?

At one level, there is considerable debate among biologists over certain parts of evolutionary theory. "The science journals right across the sub-disciplines of biology are full of criticisms of the Darwinian approach to the history of life," notes philosopher of science Stephen Meyer. "The Darwinian mechanism is held by many evolutionists to be insufficient to produce the new form and function in the history of life."[98] Dozens of recent peer-reviewed articles from mainstream science journals and books challenge, question, or raise problems for different aspects of Darwinian theory.[99]

When it comes to public discussions of evolutionary theory, however, the freedom for scientists to raise hard questions evaporates. "There's a feeling in biology that scientists should keep their dirty laundry hidden because the religious right are always looking for any argument between evolutionists as support for their creationist theories," explained evolutionist and computer scientist W. Daniel Hillis in the mid-1990s. "There's a strong school of thought in biology that one should never question Darwin in public."[100] As a consequence, the public has rarely heard about mainstream scientific disagreements over evolutionary theory. That situation seems to be changing in the latest battles over science-education policy, especially those sparked by the No Child Left Behind Act of 2001.

The New Battle over Evolution in the Schools

Under the No Child Left Behind Act, every state had to adopt science standards and statewide assessments of student knowledge in the sciences.[101] As state governments moved to fulfill this new federal mandate, they confronted head-on the issue of how to teach evolution. The result was a series of high-profile policy battles around the country.

Ohio was the first state to adopt science standards requiring students to know about scientific criticisms of Darwin's theory. After months of intense discussion and debate, the Ohio State Board of Education approved language in 2002 stipulating that all students must know "how scientists continue to investigate and critically analyze aspects of evolutionary theory."[102] The Ohio Board followed this action by adopting a model curriculum in 2004 that included a lesson on the "Critical Analysis of Evolution."[103] New Mexico adopted similar science standards in 2003 requiring students to "critically analyze the data and observations supporting the conclusion that the species living on Earth today are related by descent from the ancestral one-celled organisms,"[104] and Minnesota approved science standards in 2004 directing that students "will be able to explain how scientific and technological innovations as well as new evidence can challenge portions of or entire accepted theories and models including . . . [the] theory of evolution."[105] In 2005, the Kansas State Board of Education adopted science standards "call[ing] for students to learn about the best evidence for modern evolutionary theory, but also to learn about areas where scientists are raising scientific criticisms of the theory."[106]

Opponents of these efforts to present scientific criticisms of evolution in the classroom frequently argued that there are no genuine scientific disputes over evolutionary theory for schools to cover. When the Texas State Board of Education in 2003 considered requiring textbook discussion of the scientific weaknesses of evolution as well as the strengths, many scientists and citizens strenuously insisted that evolution had no weaknesses. Geologist Steven Schafersman, head of Texas Citizens for Science, claimed that current biology textbooks were already completely "factually accurate and free of errors concerning evolution . . . nor do they omit scientific information critical of evolution, because there isn't any such information."[107] An adjunct biology instructor declared that "if textbooks need to be modified in any way, it would be to provide stronger support for the absolute certainty of evolution, natural selection and the common descent of all life."[108] Nobel-laureate physicist Steven Weinberg told the Texas Board that "there is not one thing that is known to be inexplicable through evolution by natural selection."[109] The strongest assertion came from philosopher Robert Pennock, who insisted that "we have even more and better evidence for Darwin's discovery than we do for a view that the earth goes around the sun."[110]

This rhetoric about the supposed scientific invincibility of modern evolutionary theory is being challenged by a growing minority of scholars within the scientific community who believe that evolution should be taught more critically. During the debate over science standards in Ohio in 2002, fifty-two Ohio scientists signed a statement urging that the "science curriculum should help students understand why the subject of biological evolution generates controversy" and opposing "the censorship of scientific views that may challenge current theories of origins."[111] Many signers of the statement were on the faculties of Ohio universities, including several from the state's major research institution, Ohio State University. When the Ohio Board of Education ultimately adopted a model lesson plan on the "Critical Analysis of Evolution" that examined mainstream scientific criticisms of evolutionary theory, a similar group of scientists endorsed the lesson, and two state university biologists, Daniel Ely and Glen Needham, publicly testified in favor of it.[112]

Needham and Ely served as members of the Science Advisory Committee that helped draft Ohio's model science curriculum. Needham, a professor at Ohio State, argued that the "Critical Analysis of Evolution" model lesson "provides a corrective to the overly simplistic presentations that one often finds in high-school biology textbooks." In Needham's view, the lesson plan was simply a matter of full disclosure to students of the current findings of biology. "The new lesson is good science because it encourages students to apply critical-thinking skills to analyze the evidence. Good students know there is a growing body of published criticism of evolutionary theory," he argued.[113]

The level of personal animosity experienced by scientists who come "out of the closet" with their doubts about Darwin's theory is hard to overstate. A leaked e-mail from one of the leading opponents of Ohio's "Critical Analysis of Evolution" lesson plan, Ohio State University anthropologist Jeffrey McKee, later revealed with embarrassing and brutal clarity the burn-them-at-the-stake attitude harbored by some Darwinists. McKee compared Glen Needham and another scientist at his university who publicly defended the lesson plan to "parasitic ticks hiding in the university's scalp . . . I learned in Boy Scouts to twist the ticks when taking them out, so their heads don't get embedded in the skin. Others prefer burning them off. What fate awaits OSU's ticks remains to be seen."[114] Ironically, the genial and soft-spoken Needham is an entomologist who is one of the nation's leading authorities on ticks. He later pointed out to McKee that burning or twisting the head off a tick would not be a very effective (or safe) way to get rid of it.

Despite having to face Darwinist anger and abuse, a growing number of scientists are going public to encourage more open discussion of the central claims of modern evolutionary biology. By 2007, more than seven hundred doctoral scientists had signed their names to "A Scientific Dissent from Darwinism," which announced that they were "skeptical of claims for the ability of random mutation and natural selection to account for the complexity of life" and stated that "careful examination of the evidence for Darwinian theory should be encouraged."[115] Signers of the declaration included members of the national academies of science in the United States, Russia, Poland, the Czech Republic, and India (Hindustan), as well as faculty and researchers from a wide range of universities and colleges, including Princeton, MIT, Dartmouth, the University of Idaho, Tulane, and the University of Michigan.

"Some defenders of Darwinism embrace standards of evidence for evolution that as scientists they would never accept in other circumstances," said signer Henry Schaeffer, director of the Center for Computational Quantum Chemistry at the University of Georgia.[116] Other signers expressed similar concerns. "The ideology and philosophy of neo-Darwinism, which is sold by its adepts as a scientific theoretical foundation of biology, seriously hampers the development of science and hides from students the field's real problems," declared Vladimir L. Voeikov, Professor of Bioorganic Chemistry at Lomonosov Moscow State University.[117] Microbiologist Scott Minnich at the University of Idaho complained that Darwinian theory was "the exceptional area that you can't criticize" in science, something he considered "a bad precedent."[118] Evolutionary biologist Stanley Salthe argued that biology students needed to be exposed to the weaknesses as well as the strengths of Darwin's theory:

> Darwinian evolutionary theory was my field of specialization in biology. Among other things, I wrote a textbook on the subject thirty years ago. Meanwhile, however, I have become an apostate from Darwinian theory and have described it as part of modernism's origination myth. Consequently, I certainly agree that biology students at least should have the opportunity to learn about the flaws and limits of Darwin's theory while they are learning about the theory's strongest claims.[119]

Signer and U.S. National Academy of Sciences member Philip Skell says that Darwinian theory is largely "superfluous" in biological explanations of how things work. Skell, the Emeritus Evan Pugh Professor at Pennsylvania State University, argues that "Darwinian evolution . . . does not provide a fruitful heuristic in experimental biology" and "the claim that it is the cornerstone of modern experimental biology will be met with quiet skepticism from a growing number of scientists in fields where theories actually do serve as cornerstones for tangible breakthroughs." Skell reports that he "recently asked more than 70 eminent researchers if they would have done their work differently if they had thought Darwin's theory was wrong. The responses were all the same: No." Moreover, when Skell examined "the outstanding biodiscoveries of the past century," such as "the discovery of the double helix" and the "mapping of the genomes," he "found that Darwin's theory had provided no discernible guidance, but was brought in, after the breakthroughs, as an interesting narrative gloss."[120]

The rising tide of scientists who are publicly skeptical of neo-Darwinism has made it harder for Darwinists to deny outright the existence of real scientific controversies involving evolution. Such denials become even more difficult when policymakers are presented with articles from the peer-reviewed science literature raising questions about such issues as the role of microevolution in explaining macroevolution, the efficacy of natural selection, or the origin of animal body plans during the Cambrian explosion.[121]

When defenders of evolution are forced to respond to specific criticisms, they sometimes modify their rhetoric. During the 2003 debates over biology textbooks in Texas, Texas Citizens for Science leader Steven Schafersman started out by asserting that were no genuine scientific controversies over evolution for textbooks to cover. But as public debate progressed and critics of the textbooks enumerated specific examples, Schafersman's rhetoric began to, well, evolve.

Schafersman soon conceded that there are indeed some legitimate scientific controversies involving evolution, but he asserted that these "disagreements and controversies are found at the frontiers of research and graduate education, not at the level of introductory biology textbooks."[122] By the time the textbooks were actually voted on by the Texas Board of Education, Schafersman was acknowledging that there are in fact "many disagreements among scientists about the correct nature or explanation of the evolutionary process."[123] These legitimate controversies included "the sufficiency of microevolution to explain macroevolution," "disagreements about the primacy of natural selection," and "the extent to which evolutionary theory can explain or account for human morality, religion, behaviors, self-awareness, free will, etc." Notably, these were some of the very controversies that critics of the textbooks had identified as worthy of discussion. Schafersman was reduced to arguing that high school students are too immature to appreciate them.

"Scientific theories are too massive and established to expect any high school student to critique or question," he insisted. "The vast majority of high school students would not be able to perform such critiques in a scientific way. Scientific theories should be accepted as reliable knowledge in K-12 classes, and not made the object of questioning until they have the educational training necessary to do so, which consists of years of graduate study at universities." That is why even "real scientific problems, controversies, etc., should not be included in introductory science textbooks, because they are almost always too difficult to understand and their presence would only lead to student confusion and frustration."

Whatever the merits of this line of argument, it was entirely different from Schafersman's initial claim that there were no scientific controversies about evolution for students to know about. Schafersman's logic was also suspect. If high school students are mature enough to understand the evidence and arguments in favor of Darwin's theory, why aren't they mature enough to be exposed to evidence and arguments on the other side? High school students are expected to exercise critical thinking in social studies and literature; why not in biology?

Most defenders of Darwin's theory likely recognize the inherent weakness of this type of argument, which may explain why they try to shift the focus of public debates about science education away from science and onto religion. This was a common tactic in Texas in 2003, as critics of biology textbooks were denounced as the "Texas Taliban,"[124] as "anti-science zealots" who were "uninterested in the truth or science or facts and only interested in promoting a religious and political agenda,"[125] and as part of the frightening "Radical Religious Right" that wanted to "convince Texas to teach the Bible as fact, in addition to (or even in the place of) science."[126] Of course, there was no effort in Texas to insert either the Bible or creationism into biology textbooks. In Ohio, supporters of requiring the critical analysis of evolution were similarly denounced as "religious terrorists" and "our local Ohio 'Taliban'" in a publication distributed by Ohio Citizens for Science, the state's pro-evolution lobbying group.[127] Referring to critics of Darwin's theory as "Taliban" has become a common rhetorical ploy by Darwinists across the United States.[128]

Attempts to stigmatize critics of evolution based on their religion—or supposed religion—have not been confined to grassroots activists. When Ohio State professor Glen Needham publicly endorsed his state's "Critical Analysis of Evolution" lesson plan, he was attacked in the *Columbus Dispatch* by a university colleague who falsely labeled Needham a creationist and ludicrously claimed that students in one of his classes "were required to interpret the Bible and specifically to hold it in opposition to evolution" in order "to earn a grade."[129] The local newspaper printed the defamatory charges against Needham but refused to allow Needham to correct the record.

Biologist Paul Gross and philosopher Barbara Forrest have pressed the religious motives argument to the point of hysteria. Gross derides scholars critical of neo-Darwinism as "crackpots," "bogus scientists," or "scientific illiterates" who are driven by religious fanaticism and who are part of a "vast right-wing conspiracy" against the separation of church and state.[130] Gross and Forrest take particular aim at scholars supportive of the idea of intelligent design in biology, people whom Forrest describes as "part of a much larger Religious Right network which has plans for the country to undermine church/state separation."[131] "At heart," claims Forrest, "proponents of intelligent design are not motivated to improve science but to transform it into a theistic enterprise that supports religious faith."[132] Gross and Forrest seem to believe that religious motivations disqualify scientists from being heard in the public square—no matter what their qualifications and no matter what the quality of the evidence they present.

In a republic whose founding document guarantees free speech and religious liberty, the effort to muzzle citizens based on their presumed religious views is curious, to say the least. All citizens, regardless of their religious beliefs, have an equal right to be heard in the public square. The religious motives of participants in public discourse ought to be irrelevant. It is one thing to claim that a specifically religious public-policy proposal by a religious adherent may be illegitimate in a secular democracy. It is quite another to suggest that a secular policy proposal by religious people is suspect merely because of the personal religious beliefs of its proponents. In the case of science education, most current proposals for reform do not involve the teaching of biblical creationism or even the scientific theory of intelligent design. They focus instead on the presentation of secular scientific criticisms of modern evolutionary theory that are well developed in the peer-reviewed scientific literature. Yet these proposals are invariably attacked as illegitimate simply because the private motives of those who make the proposals are supposedly religious. For example, some critics claimed that Ohio's model lesson plan on the "Critical Analysis of Evolution" violated the establishment clause of the Constitution because of the motives of some of its supporters. Developed by a committee of professional scientists and educators, the lesson plan did not discuss religion or creationism in any way, nor did it teach about intelligent design. Yet, according to critics, the motives of those who supported the lesson plan made it constitutionally impermissible. Such efforts to exclude religious citizens from equal participation in the development of secular educational policies is antithetical to the principles of a free society.

Many of those who are most vocal in trying to deny religious citizens a voice in science-education policy are far from disinterested on the subject of religion. Barbara Forrest serves on the board of directors of the New Orleans Secular Humanist Association, which describes itself as "an affiliate of American Atheists, and [a] member of the Atheist Alliance International."[133] Physics professor Lawrence Krauss, one of the chief opponents of Ohio's "Critical Analysis of Evolution" lesson plan, received the "Humanism Award" in 2003 from the Free Inquirers of Northeast Ohio,[134] and he thinks that the small number of members of the National Academy of Science who believe in God are suffering from "delusions."[135] Chris Mooney, author of the anti-ID tome *The Republican War on Science*, was "copresident and a founding member of the Yale College Society for Humanists, Atheists and Agnostics" and even helped write a "Bill of Rights for Unbelievers."[136]

Steven Schafersman, head of Texas Citizens for Science, describes himself as a "secular humanist"[137] and maintains that "[s]upernaturalistic religion and naturalistic science . . . are and will remain in eternal conflict."[138] He also insists that "Scientists have been obliged by the evidence" to accept the "conclusion that the process of evolution is blind, mechanistic, purposeless, goalless, unplanned, and completely natural and material." He further claims that according to "evolutionary scientists . . . the universe and life is [sic] devoid of immanent meaning and purpose."

Physicist Steven Weinberg, who championed Darwinism before the Texas State Board of Education in 2003, believes that delegitimizing religion is probably "the most important contribution" science can make to the world. In his own words, "I personally feel that the teaching of modern science is corrosive of religious belief, and I'm all for that! One of the things that in fact has driven me in my life, is the feeling that this is one of the great social functions of science—to free people from superstition."[139] Lest there be any doubt about what Weinberg means by "superstition," he goes on to say that he hopes "that this progression of priests and ministers and rabbis and ulamas and imams and bonzes and bodhisattvas will come to an end, that we'll see no more of them. I hope that this is something to which science can contribute and if it is, then I think it may be the most important contribution that we can make."

Should Weinberg's views on evolution be dismissed simply because he is an atheist? Of course not. But neither should the scientific views of religious scientists be dismissed simply because they happen to be theists. Unfortunately, a double standard regarding motives currently prevails in discussions about evolution, a double standard reinforced by both popular culture and reporters.

Inherit the Spin

More than forty years ago, the film and stage play *Inherit the Wind* depicted the famous Scopes trial of the 1920s as a battle between bigoted stick-figure fundamentalists who champion a literal reading of Genesis and an open-minded science teacher who is jailed simply for teaching the facts of biology. In one especially over-the-top scene, a crazed preacher calls down the wrath of God on anyone who defends the poor persecuted teacher. In another scene, angry townspeople threaten to run the defense attorney out of town. As Pulitzer Prize–winning historian Edward J. Larson (*Summer for the Gods* [1997]) and others have pointed out, *Inherit the Wind* is more fiction than fact. Teacher John Scopes was never jailed; the prosecution of Scopes was contrived by the ACLU, not creationist bigots; and Dayton, Tennessee, the friendly town where the trial occurred, welcomed Scopes and his attorneys.[140] Nevertheless, *Inherit the Wind*'s caricature of the debate over evolution has supplied the prism through which many journalists view conflicts over evolution in the schools. As a result, the public can have a difficult time obtaining accurate information about science-education battles from the media.

A typical example of misreporting took place in the fall of 2002 when the board of the Cobb County school district in Georgia unanimously approved a policy encouraging teachers to discuss "disputed views" about evolution as part of a "balanced education."[141] Critics of the policy cried foul, asserting that the policy was a thinly disguised attempt to smuggle the Bible into science class. Much of the national news media followed suit, setting up the Cobb County conflict as a classic battle between science and biblical creationism. CNN declared that the school board had voted "to allow creationism in class," and the Associated Press made a similar claim.[142] *Science* magazine proclaimed that "[t]he forces of creationism scored a victory in Georgia last night."[143]

The only problem with this reporting was that the school board said exactly the opposite. "We expect teachers to continue to teach the theory of evolution," said school board chair Curtis Johnston at the meeting during which the policy was adopted. "We do not expect teachers to teach creationism . . . Religion has no place in science instruction." Lest Mr. Johnston's comments leave any room for ambiguity, the entire board then adopted a statement expressing its intent that the new "policy not be interpreted to restrict the teaching of evolution" or "to promote or require the teaching of creationism."[144]

By clinging to old stereotypes, the national news media missed the real controversy in Cobb County, which was significantly more interesting than a cardboard battle between religious fundamentalists and the defenders of science. National news accounts played up the fact that some scientists from Georgia universities opposed the Cobb County policy. What the national news media failed to report was that a group of twenty-eight scientists from the very same educational institutions (places like the University of Georgia and Georgia Tech) wrote the Cobb County board expressing their skepticism of Darwinism and urging "careful examination of the evidence for Darwinian theory."[145]

Sometimes poor reporting is the result of sloppiness, but other times it results from bias. After the Ohio Board of Education voted to adopt its model lesson plan on the "Critical Analysis of Evolution," for example, the local Associated Press reporter filed a story accurately describing what took place, but the national AP desk rewrote the opening to state that the Ohio Board of Education had "approved a lesson plan . . . that includes a religious theory for the creation of life."[146] This statement was false. The lesson plan adopted by Ohio did not include either creationism or intelligent design; it focused only on mainstream scientific criticisms of evolutionary theory.

Sometimes journalistic sloppiness and bias veer into fabrication. In July 2003, CNN reporter Ed Lavandera filed a report about a hearing held before the Texas Board of Education on new biology textbooks.[147] Lavandera framed the hearing as a conflict between defenders of evolution and those who wanted to insert creationism or intelligent design into textbooks. There were two major problems with the CNN report. First, Lavendera conflated creationism with intelligent design, even though the two concepts are not the same. Second, and more importantly, there was no effort in Texas to insert either creationism or intelligent design into any biology textbooks. Critics of evolution at the textbook hearing had more limited objectives: They were seeking corrections of documented factual errors and the inclusion of certain scientific criticisms of evolutionary theory. Because not a single person testifying at the hearing advocated either biblical creationism or intelligent design to the board, Lavandera was unable to show anyone on screen defending these causes. So what did he do? He showed soundless video of people testifying at the hearing and simply asserted in his voice-over that people were proposing to insert intelligent design into textbooks. Lavandera made up the key part of his story.

Nearly as fictitious was a 2004 story by CNN reporter Denise Belgrave, who claimed that Missouri and other states were considering bills that would fire teachers who did not teach the theory of intelligent design.[148] Such a bill had been proposed months earlier in Missouri, but by the time of CNN's report the sponsor of the bill no longer supported the teacher-penalty provision and had already submitted a revised bill without it. In addition, contrary to CNN's claims, no other state had considered a similar measure. Although Belgrave made the Missouri bill the centerpiece of her story, she did not bother to contact the legislator who sponsored the bill, which is why she failed to discover that the teacher-penalty provision had been dropped. Belgrave's shoddy reporting crossed the line into invention after CNN steadfastly refused to correct the record once the errors in its original report were brought to its attention.[149]

Similarly inaccurate reporting dominated news coverage of the Kansas science-standards debate in 2005. The Kansas State Board of Education's decision to adopt science standards encouraging the critical analysis of evolution was widely—but wrongly—reported as an effort to insert intelligent design into the state curriculum. Even the usually thoughtful Charles Krauthammer, who should have known better, condemned the Kansas board for "forcing intelligent design into the statewide biology curriculum."[150] In fact, the new science standards contained an unequivocal statement to the contrary: "We also emphasize that the Science Curriculum Standards do not include Intelligent Design."[151] Somewhat amusingly, CNN reporter Ed Lavandera staged a repeat performance in Kansas, recycling an old clip of someone testifying in Texas and leaving the clear impression that the clip depicted the Kansas debate two years later.[152]

But the Kansas coverage was a model of journalistic probity compared to reporting about the celebrated *Kitzmiller v. Dover* intelligent design case.

Inherit the Wind for the Twenty-First Century?

In October 2004, a misguided school board in Dover, Pennsylvania, adopted a policy that contained the words "intelligent design." Although the policy was soon portrayed as being about the "teaching" of intelligent design, in reality the Dover board had required only that students hear a brief statement telling them that "Intelligent Design is an explanation of the origin of life that differs from Darwin's view"—a superficial declaration that supplied no meaningful information about the substance of the theory. Students were further notified that "the reference book, *Of Pandas and People*, is available for students who might be interested in gaining an understanding of what Intelligent Design actually involves."[153]

In retrospect, it is clear that most Dover board members knew little, if anything, about intelligent design when they enacted their policy. The instigators of the policy were supporters of biblical creationism, not intelligent design, and after the policy's adoption, board members continued to find it difficult to define intelligent design or summarize its key tenets.[154]

The Dover board adopted its policy in the face of stiff opposition from Discovery Institute, which represents most of the scientists and scholars who favor intelligent design. Although the institute supported the right of teachers to voluntarily discuss intelligent design in an objective and pedagogically appropriate manner, it opposed efforts to mandate intelligent design (even, as in this case, when the mandate was so inconsequential). The institute's preferred approach to the teaching of evolution—which it had advocated previously in Ohio, Minnesota, and other states[155]—was for schools to teach the scientific evidence for and against Darwin's theory without even getting into alternative theories.[156] The idea was to include mainstream scientific criticisms being raised even by some evolutionists. After all, if scientists could read such criticisms in their journals, why couldn't students study them in their biology classes? Institute leaders believed that this limited effort to "teach the controversy" represented a common-ground approach. It also had the benefit of not politicizing the debate over intelligent design, for such politicization would prevent ID supporters from gaining a fair hearing for their views in the scientific community and academia.[157]

States and school districts that embraced the approach recommended by the institute did not end up in court. Indeed, during the very time that the Dover board was rejecting Discovery's approach, a school district in Grantsburg, Wisconsin, adopted it and found that the controversy over its policy largely disappeared. During the summer of 2004, the Grantsburg board passed a motion directing its "science department to teach all theories of origins." Predictably, supporters of evolution were incensed, but the board also faced opposition from Discovery, which advised it to adopt a teach-the-controversy approach instead. The board eventually did so, enacting a new policy in December 2004 stating that "students shall be able to explain the scientific strengths and weaknesses of evolutionary theory." The board made clear that its new policy did "not call for the teaching of creationism or intelligent design."[158] The revised policy was immediately praised by Discovery, while the reaction of evolutionists was mixed.[159] Some thought the revised policy was acceptable—except for the fact that Discovery Institute had favored it.[160] Others, such as University of Wisconsin–Oshkosh dean (and biologist) Michael Zimmerman, condemned the policy as "a smoke screen for a school board that supports a policy of discrediting evolutionary theory and, despite the qualification, would like to allow other theories—namely creationism and intelligent design—to be introduced."[161] But no lawsuit was filed, and an effort to replace board members who were supportive of teaching the controversy over evolution failed in 2005.[162] The Grantsburg policy was still in effect as of 2007, and by that time even Michael Zimmerman was insisting in print that he had never opposed the revised policy![163]

Discovery attempted to convince the Dover school board to follow the same approach as Grantsburg, but to no avail.[164] Discovery's efforts to convince the board to change course were thwarted by the Thomas More Law Center, which wanted to litigate a test case on intelligent design. Frustrated by the board's intransigence, Discovery ramped up its public criticism. On the same day the ACLU and Americans United for Separation of Church and State filed their lawsuit against the school district, the institute issued a press release raising questions about the Dover policy's wisdom and constitutionality and urging that it "be withdrawn and rewritten."[165]

The fact that the major institutional supporter of intelligent design strongly opposed the Dover policy should have been an important news story. It wasn't. Except for one story by the Associated Press in November 2004,[166] reporters at national publications either refused to mention or downplayed the institute's objections to the Dover policy. For example, *Washington Post* reporter Valerie Strauss wrote an article that discussed efforts to insert intelligent design into schools.[167] Although I was quoted in the article, nowhere did the story mention the fact that Discovery opposed the Dover policy, even though this point was stressed in my interview with Strauss. When challenged on the accuracy of her story, Strauss responded with a technicality: "The story does not say you or your institute favors teaching intelligent design . . . It doesn't say anything about your position on it at all. It sounds to me as if you are misrepresenting my story."[168] But suppressing any mention of Discovery Institute's opposition to mandating intelligent design in a story that described efforts to teach intelligent design certainly constituted a material misrepresentation of the facts. As a result of such pervasively poor reporting, the erroneous belief that Discovery somehow encouraged the Dover board to adopt its intelligent design policy has been hard to correct, even among those who should know better.

Despite the fact that the Dover board did not speak for the scientists and other scholars who support intelligent design, the ACLU attorneys in the case were determined to turn their lawsuit into a broad referendum on the intellectual and scientific validity of the theory. Unfortunately, federal judge John E. Jones obliged them, issuing a 139-page ruling in December 2005 that purported to refute in detail the scientific claims made by ID proponents.[169]

Jones's ruling was quickly lavished with the kind of praise usually reserved for prize-winning books and outstanding scholars. University of Chicago geophysicist Raymond Pierrehumbert called Jones's ruling a "masterpiece of wit, scholarship and clear thinking."[170] Lawyer Ed Darrell described it as "a model of law . . . a model of argument . . . a model of legal philosophy . . . [and] a model of integrity of our judicial system."[171] *Scientific American* editor John Rennie said that Jones's opinion constituted "an encyclopedic refutation of I.D."[172] Bloggers at the pro-Darwin *Panda's Thumb* website hailed Jones's decision as an important work "of both scholarship and history,"[173] and they lauded Jones as "a top-notch thinker,"[174] "an outstanding thinker,"[175] someone who "is as deserving of the title 'great thinker' as someone who writes a great mathematical proof or a great work of music criticism."[176] Not to be outdone, *Time* put a picture of Jones on its cover and honored him as one of "the world's most influential people" in the category of "scientists and thinkers."[177]

Those who read Judge Jones's opinion in *Dover* without previously having studied the evidence and arguments in the case may well be impressed by his seemingly authoritative command of the facts. But such an impression would be unwarranted. Judge Jones repeatedly misrepresented both the facts and the law in his opinion, sometimes egregiously. When crosschecked against the evidence and arguments presented in the court record, many of Judge Jones's key assertions turn out to be false, contradictory, or beside the point.[178]

Part of the problem was undoubtedly Judge Jones's excessive and uncritical reliance on plaintiffs' attorneys. It turns out that the key section of his "masterpiece of wit, scholarship and clear thinking" was copied wholesale from a document prepared for him by ACLU lawyers. To be precise, 90.9 percent (or 5,458 words) of Judge Jones's 6,004-word section analyzing intelligent design as science was copied verbatim or virtually verbatim from the proposed "Findings of Fact and Conclusions of Law" submitted to Judge Jones by ACLU attorneys nearly a month before his ruling. Jones essentially cut-and-pasted the ACLU's wording into his ruling to come up with his analysis.[179]

Although such extensive verbatim copying from one party's legal documents is frowned upon by appellate courts,[180] it is not considered "plagiarism" in judicial circles, and Jones was not forbidden from doing it. Nevertheless, it offers an explanation for why he repeatedly misstated facts in his opinion. In numerous instances, he apparently cited claims made by the ACLU attorneys without double-checking the trial record and verifying them for himself.

For example, Judge Jones claimed that biochemist Michael Behe, when confronted with articles supposedly explaining the evolution of the immune system, replied that these articles were "not 'good enough.' "[181] In reality, Behe said the exact opposite at trial: "it's not that they aren't good enough. It's simply that they are addressed to a different subject."[182] The answer cited by Judge Jones came from the ACLU's proposed "Findings of Fact," which misquoted Behe, twisting the substance of his answer.[183]

Judge Jones also copied verbatim the ACLU's false allegation that intelligent design "is not supported by any peer-reviewed research, data or
publications."[184] Again, the court record shows otherwise. University of
Idaho microbiologist Scott Minnich testified at trial that there were between "seven and ten" peer-reviewed papers supporting ID, and he specifically discussed Stephen Meyer's explicitly pro–intelligent design article
in the peer-reviewed biology journal *Proceedings of the Biological Society of
Washington*.[185] Additional peer-reviewed publications, including William
Dembski's monograph *The Design Inference* (published by Cambridge University Press), were described in an annotated bibliography submitted in
an amicus brief accepted as part of the official record of the case by Judge
Jones.[186]

Jones's uncritical copying of the ACLU also led him to wrongly assert
that ID "requires supernatural creation," that "ID is predicated on supernatural causation," and that "ID posits that animals . . . were created
abruptly by a non-natural, or supernatural, designer."[187] Although Jones
insisted that the "defendants' own expert witnesses acknowledged this
point,"[188] they did nothing of the sort. When asked at trial "whether intelligent design requires the action of a supernatural creator," biochemist Scott
Minnich replied, "It does not."[189]

When Judge Jones's extensive copying was revealed in December 2006,
many defenders of Darwin's theory were incensed—not at Judge Jones, but
at the critics who would dare to raise such an issue. The main defense
of Judge Jones seemed to be that "everybody does it" and that wholesale
copying from documents supplied by the parties to a case is perfectly acceptable in legal circles. Some even claimed that it was far better for judges
not to write their own opinions. According to one Darwinist blogger,

> it's not "acceptable," it's better. We don't want judges writing all the elabo-
> rate technical stuff. Some judges might be able to understand the issues at
> hand, and from following the case it's clear Judge Jones was one. However
> it's better when possible to have the actual final materials written by some-
> one with, or even a group of people with, deep background knowledge . . .
> [A]ny judge ruling on scientific matters would be bound to make minor er-
> rors, errors which both would not be acceptable (you can't go back and just
> correct a judicial decision later with a red pen) and which in a worst-case
> scenario might even result in a needless appeal. This is not what we want.
> This is not a judge's job.[190]

A few lonely voices among the Darwinists did express disillusionment
about the whole affair. Biochemist Larry Moran noted plaintively that the
comments of fellow Darwinists had "seemed to confirm that Jones had

written this decision all by himself and deserved full credit for his brilliant analysis. As it turns out, this isn't true and I feel deceived." For his honesty, Moran was roundly condemned by other Darwinists.[191]

Even before the revelations that Judge Jones's critique of intelligent design had been copied wholesale from the ACLU, the partisan tone of his opinion had attracted severe criticism. Distinguished University of Chicago law professor Albert Alschuler, for one, chastised Judge Jones for trying to smear ID proponents as biblical fundamentalists:

> If fundamentalism still means what it meant in the early twentieth century . . . accepting the Bible as literal truth—the champions of intelligent design are not fundamentalists. They uniformly disclaim reliance on the Book and focus only on where the biological evidence leads. The court's response—"well, that's what they say, but we know what they mean"—is uncivil, an illustration of the dismissive and contemptuous treatment that characterizes much contemporary discourse. Once we know who you are, we need not listen. We've heard it all already.

According to Alschuler, in Judge Jones's eyes "*Dover* is simply *Scopes* trial redux. The proponents of intelligent design are guilty by association, and today's yahoos are merely yesterday's reincarnated."[192] There may have been more truth to that observation than Alschuler suspected: During the trial, Judge Jones told a reporter that he planned to watch *Inherit the Wind* for "historical context."[193] Alschuler countered that "proponents of intelligent design deserve the same respect" as evolutionists in the evaluation of their arguments, something they did not get from Judge Jones. Their ideas should be evaluated on their merits, not on presumed illicit motives. As Alschuler put it, "freedom from psychoanalysis is a basic courtesy."[194]

Even if Judge Jones's indictment of intelligent design had been exemplary, however, there is a serious question about whether it should have been issued at all.

It is a standard principle of good constitutional interpretation that a judge should venture only as far as necessary to answer the issue before him. If a judge can decide a case on narrow grounds, then that is what he ought to do. He should not use his opinion to answer all possible questions.[195] In this case, Judge Jones found that the Dover school board acted for clearly religious reasons rather than for a legitimate secular purpose.[196] Having made this determination, the specific policy adopted by the Dover board was plainly unconstitutional under existing Supreme Court precedents.[197]

That should have been the end of the decision. There was no need for the judge to launch an expansive inquiry into whether intelligent design is science, whether there is scientific evidence for the theory, whether the theory is inherently religious, whether Darwinism has flaws, and even whether Darwinian evolution is compatible with religious faith. A judge bound by judicial restraint would not have ventured into these other areas, because they were unnecessary for the disposition of the case.

Judge Jones made a point of insisting in his opinion that his decision was not that of "an activist court."[198] But Jones's protestations to the contrary, his decision was judicial activism—with a vengeance. Judges who believe in judicial restraint refrain from employing their judicial power to decide divisive social questions unless it is legally necessary to do so. In this case, Judge Jones had narrow grounds on which to base his decision. But he chose not to stand on those grounds. Judge Jones exhibited the same kind of activism that led the federal courts to entangle themselves in a host of social conflicts (such as abortion) during the second half of the twentieth century. Far from resolving controversial issues, such activism disserves the democratic process and often leads only to further polarization. Unlike judicial activism that seeks to decide issues by judicial fiat, the democratic process promotes incremental solutions and compromise, and it tends to cool passions over the long term by giving everyone a stake in the decision-making process.

Even some legal scholars who oppose intelligent design have criticized the overreach of Jones's decision. Boston University law professor Jay Wexler, for example, says, "The part of *Kitzmiller* that finds ID not to be science is unnecessary, unconvincing, not particularly suited to the judicial role, and even perhaps dangerous both to science and to freedom of religion."[199]

The *Kitzmiller v. Dover* decision was not appealed because the school-board members who adopted it lost reelection shortly after the conclusion of the trial. As a result, Judge Jones's ruling is not a binding legal precedent outside of the Dover school district. That has not prevented Darwinists from trying to use the decision to shut down debate over evolution in other states.

The State of the Debate

Less than two months after the *Dover* ruling, the Ohio State Board of Education repealed its Critical Analysis of Evolution science standard and lesson plan. Repeal forces used *Dover* to play up fears about a potential lawsuit, despite the fact that the Ohio science standard had been in place since 2002 and the model lesson plan since 2004—with no lawsuits.[200] Later in 2006, voters in Kansas replaced two teach-the-controversy members of their state board of education, and in early 2007 the new board voted 6-4 to repeal the Kansas science standards presenting scientific evidence critical of Darwin's theory.[201] Astonishingly, the Kansas board members also repealed language that asked students to study not only the successes of science but also abuses, such as eugenics and the infamous Tuskegee syphilis experiment that targeted African Americans. Studying scientific abuses was apparently considered too depressing by Kansas Darwinists.[202]

Despite these high-profile reversals, other states and communities have continued to press for greater openness in how evolution is taught post-*Dover*. Six months after the *Dover* ruling, South Carolina approved a statewide science standard on the critical analysis of evolution virtually identical to the one Ohio had repealed.[203] School districts in California and Louisiana also adopted local teach-the-controversy policies, and a bipartisan majority of Oklahoma's House of Representatives voted in favor of an academic freedom bill protecting the right of teachers to teach scientific criticisms of Darwin's theory along with the evidence supporting the theory.[204]

Moreover, most Americans continued to express support for a more thorough presentation of evolutionary theory in schools. According to a national Zogby poll conducted two months after the *Dover* decision, only 21 percent of American adults believe that "biology teachers should teach only Darwin's theory of evolution and the scientific evidence that supports it," while an overwhelming 69 percent of Americans think that "biology teachers should teach Darwin's theory of evolution, but also the scientific evidence against it."[205] The level of support expressed for teaching both sides of the evolution controversy was virtually unchanged from an earlier national Zogby poll in 2001, and state surveys have shown similar results.[206]

Some Darwinists are expressing concerns that they are losing the war of ideas. According to the *New York Times*, at a 2006 gathering of eminent pro-Darwin scientists at the Salk Institute for Biological Studies in California, participants expressed "a rough consensus" that the theory "of evolution by natural selection" is "losing out in the intellectual marketplace."[207] That was a stunning admission.

Unfortunately, many Darwinists seem to believe they can resolve the current impasse over evolution through better public relations and spin control rather than a more robust exchange of ideas. A plea for greater public-relations skills was the main point of biologist and filmmaker Randy Olson's much-discussed documentary *Flock of Dodos* in 2006.[208] According to Olson, pro-Darwin scientists are losing the public debate over evolution not because of their lack of arguments, but because of their lack of basic communication skills. In Olson's view, if only evolutionists were more like the slick proponents of intelligent design, then they surely would be able to convince the public of the validity of their theory.

Flock of Dodos provided a revealing glimpse into the unsavory tactics Olson apparently believed should be used to win the public debate over Darwinism. Attempting to discredit biologist Jonathan Wells and his book *Icons of Evolution*, Olson seized on the single "icon" of Haeckel's embryo drawings discussed in the account of Roger DeHart at the beginning of this chapter. Olson conceded that Haeckel's diagrams were erroneous, but he sharply disputed Wells's claim that the diagrams had appeared in modern textbooks. Olson showed an ID proponent vainly flipping pages of a textbook unable to find the diagrams, and when Olson eventually located a textbook with the diagrams it turned out to be from 1914. The clear message communicated was that Wells and other ID proponents were perpetrating a hoax.

But if anyone was perpetrating a hoax, it was Olson. His assertion that modern textbooks had never used drawings derived from Haeckel contradicted the claims of many evolutionists.[209] In 2000, the late Stephen Jay Gould excoriated "the century of mindless recycling that has led to the persistence of these drawings in a large number, if not a majority, of modern textbooks!"[210] In 2001, *New York Times* science reporter James Glanz reported that Haeckel's "drawings were reproduced in textbook after textbook for more than a century," including a textbook coauthored by Bruce Alberts, then-head of the National Academy of Sciences, and Nobel Prize–winning geneticist James Watson.[211] Even Eugenie Scott of the National Center for Science Education did not have the audacity to maintain that Haeckel's embryo drawings had never appeared in modern textbooks. In an interview for the documentary version of *Icons of Evolution*, she explained that "the reason why the diagrams are reproduced is because they're easily available. There's no copyright on them. It's an easy way to illustrate a point."[212]

After being confronted with evidence that Haeckel's embryos had in fact been used in modern textbooks just as Wells had said, Olson reluctantly conceded the point at a public screening. But now he insisted that criticisms of his film's inaccurate claims about Haeckel's embryos were unfair because the issue was supposedly inconsequential—even though he was the one who had made the bogus embryo drawings the centerpiece of his attack on Jonathan Wells.[213]

Olson ended up discrediting himself in his clumsy effort at public relations, but he wrongly faulted Darwinists as a whole for their poor PR skills. In fact, many Darwinists have been all too effective in employing spin and public relations in order to avoid a serious exchange of ideas. Consider the well-publicized resolution condemning intelligent design enacted by the board of the American Association for the Advancement of Science (AAAS) in 2002.[214] The resolution was widely treated as an authoritative determination that intelligent design is not science. In reality, it was simply a brilliant public-relations stunt. When AAAS board members were surveyed about what articles and books they had actually read by proponents of intelligent design prior to reaching their decision, it became clear how little they really knew about the subject. Of the four board members who responded, none could identify even a single article or book.[215] Although one board member said vaguely that she had perused unspecified sources on the Internet, Alan Leshner, the head of the AAAS, did not even make that claim. He responded that the issue had been looked at for him by the group's "science policy staff."[216] In other words, board members had voted to condemn intelligent design without even investigating for themselves the claims made by intelligent-design proponents. That did not keep the resolution from being repeatedly cited as an authoritative determination of the scientific legitimacy of intelligent design.

More generally, the effort by Darwinists to shift the debate over evolution to the motives, character, and funding of their opponents represents a clever rhetorical strategy to prevent debate over the flaws of Darwinian evolution and the scientific evidence for intelligent design. In an article outlining "Tactics in Fighting Creationists and IDers," Internet activist Lenny Flank warns fellow Darwinists: "Don't focus on the science . . . This isn't a science symposium. Don't treat it as one." Flank recommends responding to ID proponents by attacking "them on every possible front. Don't let up for a second, come at them from every possible direction, and don't give them an instant's rest."[217]

Perhaps the strongest evidence of the Darwinists' effectiveness at public relations in recent years has been the continued stereotyping of the evolution debate as a battle between unthinking religious fundamentalists and open-minded defenders of science and rationality. In the past, the cultural memory of the Scopes trial was key to preserving this stereotype, but today many young people have scant knowledge of such events from history. So evolutionists and the media are now offering the *Dover* lawsuit as the new Scopes trial, and Hollywood is preparing the obligatory feature film, which undoubtedly will pump new life into the old science v. religious fundamentalism trope.[218]

Notwithstanding their effort to turn *Dover* into this generation's version of *Inherit the Wind*, Darwinists may yet look back at the case as a pyrrhic victory. Defenders of evolution used to style themselves the champions of free speech and academic freedom against unthinking dogmatism. But they have become the new dogmatists, demanding judicially imposed censorship and discrimination against teachers, scientists, and students who dissent from Darwin. These are tactics of intellectual suppression, not free inquiry, and they betray the Darwinists' insecurity about the defensibility of their own theory.

Those who think they can kill the debate over evolution by court orders or intimidation are deceiving themselves. Americans do not like being told there are some ideas they are not permitted to investigate. Try to ban an idea, and you are liable to generate even more interest in it. Efforts to mandate intelligent design are misguided, but efforts to shut down discussion of Darwin's theory through harassment or government action are likely to backfire. The more Darwinists resort to censorship and persecution, the clearer it will become that they are championing dogmatism, not science.

In any case, evolution may be the most obvious topic implicating scientific materialism in the classroom in recent decades, but it is not the only one, nor is it the most controversial. Human sexuality has been even more contentious, as we will see in the next two chapters.

12

Junk Science in the Bedroom

My friend recently had sex for the first time and she is a year younger than I am and I really want to have sex. Is it wrong to just want to get it over with or is that called being a slut?

You know, "slut" is a really subjective term. A lot of people use it out of spite, because they're afraid of sexuality, and because they're afraid of people who enjoy it, and on some level, secretly jealous. If someone calls you a slut for the choices you make, it's their problem, not yours.

Scarleteen.com—Help and Advice[1]

Founded in 1997 by Heather Corinna, who describes herself as a "queer writer" and "a pioneer of . . . sex-positive erotic art," *Scarleteen.com* dispenses sex information and advice to nearly two million people each year.[2] Teens are the target audience for the site, which in addition to providing information offers for sale such products as *The Penis Book* and lunch boxes emblazoned with male or female sex organs (the website promises that the "quality construction" of the lunch boxes "will make you the envy of everyone on the schoolyard").[3]

Scarleteen.com is one of several sex-ed websites recommended for use by "young people" by the Sexuality Information and Education Council of the United States (SIECUS), arguably the most influential private group promoting "comprehensive sex education" in America's schools.[4] Other recommended websites offer similar services. *Go Ask Alice*, developed by Columbia University's health-education program, supplies tips on how to engage in phone sex, sadomasochism, and group sex, as well as advice about how to order pornographic magazines for someone who doesn't "want to go through the embarrassment of buying them over the counter."[5] Planned Parenthood's *Teenwire. com* instructs teens on how to perform anal sex and tries to recruit them to fight for abortion rights and lobby their school districts against abstinence-only sex-education programs.[6]

America has come a long way from the days when filmmakers had to show married people sleeping in separate beds so as not to offend social conventions. This journey from supposed Puritanism to the sexual liberation preached by modern sex educators has deep roots in modern biology and scientific materialism. Darwin's evolutionary account of human mating practices in *The Descent of Man* encouraged a relativistic understanding of sexual morality. Other thinkers soon followed in Darwin's footsteps, perhaps most notably the Finnish sociologist Edward Westermarck (1862–1939), whose *History of Human Marriage* (1921) tried to provide a comprehensive account of the origin and development of human mating practices.[7]

Directly inspired by Darwin's *Descent of Man*, Westermarck believed that marriage, as well as other human sexual behaviors, had developed through natural selection.[8] Insofar as marriage was an institution grounded in biology, Westermarck was critical of efforts by radical sex reformers to abolish it.[9] He viewed it as a permanent feature of human society. At the same time, Westermarck embraced moral relativism and cast doubt on the validity of certain Judeo-Christian sex taboos, including homosexuality and even bestiality.[10] Westermarck predicted for the future "that in questions of sex people will be less tied by conventional rules and more willing to judge each case on its merits, and that they will recognise greater freedom for men and women to mould their own amatory life."[11]

In more recent years, the burgeoning field of "evolutionary psychology" has invoked Darwinian theory to offer biological explanations of such practices as rape and adultery. According to Randy Thornhill and Craig Palmer, "the ultimate causes of human rape are clearly to be found in the distinctive evolution of male and female sexuality."[12] The same is true for casual sex and extramarital affairs. Evolutionary-psychology proponent Robert Wright argues that Darwin's theory explains why husbands are much more likely to desert their wives than vice versa: "The husband can, in principle, find an eighteen-year-old woman with twenty-five years of reproduction ahead. The wife . . . cannot possibly find a mate who will give her twenty-five years worth of reproductive potential."[13] An evolutionary-psychology textbook, meanwhile, claims that casual sex is an evolutionary adaptation based not only on "obvious reproductive advantages . . . to men" but also "tremendous benefits to women."[14]

Of course, proponents of evolutionary psychology typically offer caveats that just because natural selection programmed a certain behavior does not make it morally right and that we are not necessarily slaves to our genes. Yet these caveats are, on their own terms, illogical. If fidelity and adultery both exist simply because they furthered the survival of the fittest genes, what objective basis do we have for preferring one trait over the other? And if human beings truly are "puppets" to their genes, puppets whose "emotions are just evolution's executioners" (to quote Robert Wright), in what sense can people be blamed if they act according to their deepest impulses?[15] One can't appeal to their free will, because "free will is an illusion, brought to us by evolution."[16]

In the end, "we cannot escape our animal origins" according to evolutionary accounts of sexuality.[17] The impact of this kind of explanation on popular culture could be seen during the scandal over President Bill Clinton's extramarital relationship with intern Monica Lewinsky. One newspaper article published during the controversy bore the headline: "Not Meant for Monogamy? Blame the genes: Evolutionary psychologists and biologists suggest that humans are naturally polygynous, with perpetuating the species the goal—which means that sticking with one woman is not an efficient reproductive strategy."[18]

As fashionable as evolutionary psychology has become over the past decade, the monumental shift in how both elites and ordinary people view human sexual behavior occurred much earlier, as did the transformation of American sex education. Perhaps the single most important influence on those changes was a celebrated Darwinian zoologist whose work revolutionized the understanding of human sexuality starting in the late 1940s.

The zoologist's name was Alfred Charles Kinsey (1894–1956).

Kinsey's Sexual Revolution

Alfred Kinsey's passion for the natural world was acquired early, as he rambled the countryside around his home in South Orange, New Jersey.[19] By high school his classmates half-jestingly prophesied that he would become a "Second Darwin," and at Bowdoin College he served as president of the campus biology club.[20] Kinsey earned his doctorate in science from Harvard, majoring in animal and plant taxonomy.[21] Caught up in the fads of the time, he became a booster of eugenics, although the high school biology textbook he authored touched on the subject only indirectly.[22] Much of Kinsey's early biological research focused on the classification of gall wasps, but by the late 1930s he had turned his attention to the study of human sexuality. By the 1940s he had obtained funding for an extensive study of human sexual behavior from the National Research Council (an arm of the National Academy of Sciences), which in turn received its money for the project from the Rockefeller Foundation.[23] It was this work on sex that would make Kinsey a household name.

In 1948, he released *Sexual Behavior in the Human Male*, a mammoth volume containing more than eight hundred pages of graphs, charts, and descriptions of nearly every conceivable sexual practice among white American males.[24] Unveiled with a publicity barrage that would have dazzled Madison Avenue, the book soon became the talk of the nation.[25] Many Americans were shocked by Kinsey's findings. Based on interviews with thousands of Americans, Kinsey claimed that half of all white married males had extramarital intercourse at some point in their marriages, and "about 69 per cent of the total white male population ultimately has some experience with prostitutes."[26] Kinsey further reported that "37 per cent of the total male population has at least some overt homosexual experience to the point of orgasm between adolescence and old age . . . This accounts for nearly 2 males out of every 5 that one may meet."[27] In addition, "10 per cent of the males are more or less exclusively homosexual . . . for at least three years between the ages of 16 and 55."[28] Kinsey even asserted that more than 17 percent of single males who had lived on farms participated in bestiality, and "for the rural boys who will ultimately go to college, about 28 per cent have animal [sexual] contact between adolescence and 15."[29] Despite conventional mores against premarital sex, Kinsey reported that 40 percent of single white males from adolescence to age fifteen had already engaged in premarital intercourse, and more than 70 percent of single white males from sixteen to twenty had done so.[30] The bottom line was that America's sex mores and the actual behavior of American men were at war. Time and again, Kinsey claimed that society's sex taboos did not match reality.

In *Sexual Behavior in the Human Male* as well as its sequel (*Sexual Behavior in the Human Female*, released in 1953), Kinsey treated the "human animal" as merely another type of mammal whose mating behavior could be fully explicated in terms of biology and conditioning. Kinsey believed that biology had made the "human animal" sexually omnivorous, arguing that the "general occurrence of the homosexual in ancient Greece" as well as modern cultures "suggests that the capacity of an individual to respond erotically to any sort of stimulus . . . is basic in the species."[31] Given this biological background, Kinsey thought that the truly interesting question was not why people practiced a particular sex act, but why they did not.

"Considering the physiology of sexual response and the mammalian backgrounds of human behavior, it is not so difficult to explain why a human animal does a particular thing sexually," he explained. "It is more difficult to explain why each and every individual is not involved in every type of sexual activity."[32] According to Kinsey, the reason why certain sexual acts were regarded as "abnormal" by various people was simple conditioning: "the scientific data which are accumulating make it appear that, if circumstances had been propitious, most individuals might have become conditioned in any direction, even into activities which they now consider quite unacceptable. There is little evidence of the existence of such a thing as innate perversity, even among those individuals whose sexual activities society has been least inclined to accept."[33]

Kinsey presented his research as a value-free endeavor that could neither condemn nor sanction different kinds of sexual behavior. "The social values of human activities must be measured by many scales other than those which are available to the scientist," he conceded near the end of his volume on male sex behavior.[34] Kinsey even questioned whether scientists should continue to employ such labels as "normal" and "abnormal" or "natural" and "unnatural," since these were often used as terms of moral approval or condemnation. "Whatever the moral interpretation . . . there is no scientific reason for considering particular types of sexual activity as intrinsically, in their biologic origins, normal or abnormal," he wrote.[35] Science could reveal the facts of human sexual behavior; it could not tell people which forms of sexual behavior were good or bad.

Kinsey's argument for value-free science was less than candid. In reality, he did not disavow the categories of "normal" and "abnormal." He merely redefined them to promote sex behaviors he found acceptable and discourage sex restrictions he deemed illegitimate. Kinsey repeatedly invoked the label "normal" to try to legitimize behaviors traditionally regarded as abnormal or immoral. His discussion of bestiality provides a good example.

Kinsey began by contending that the true biological mystery of such behavior was not why some animals engaged in sexual relations across species, but why more animals did not do so. ("Why should mammals mate only with mammals of their own kind? In the animal kingdom as a whole, is it to be believed that the sources of sexual attraction are of such a nature that they provide stimuli only for other individuals of the same species?"[36]) He then pointed to an "increasing number" of reports of "higher mammals . . . mating, or trying to mate, with individuals of totally distinct and sometimes quite remote species," and went on to describe his research findings that a substantial proportion of "rural boys" had "sexual contacts with animals to the point of orgasm."[37]

According to Kinsey, such encounters could generate loving relationships. "In some cases the boy may develop an affectional relation with the particular animal with which he has his contact, and there are males who are quite upset emotionally, when situations force them to sever connections with the particular animal."[38] Responding to readers who might deem such relationships "a strange perversion of human affection," Kinsey noted "that exactly the same sort of affectional relationship is developed in many a household where there are pets."

Kinsey's discussion of bestiality was not simply a neutral scientific description of a certain form of sex behavior. It was an apologia for why this behavior should be regarded as normal and acceptable. Kinsey dismissed as childish those who believed bestiality was immoral,[39] and he suggested that taboos against bestiality originated in "superstition."[40] He blamed modern psychology for inducing false guilt in rural youth over their sexual contacts with animals. "It is only when the farm-bred male migrates to a city community and comes in contact with city-bred reactions to these activities, that he becomes upset over the contemplation of what he has done. This is particularly true if he learns through some psychology course or through books that such behavior is considered abnormal."[41]

Kinsey encouraged doctors and psychologists to assure such young men of the normality of their behavior. "The clinician who can reassure these individuals that such activities are biologically and psychologically part of the normal mammalian picture, and that such contacts occur in as high a percentage of the farm population as we have already indicated, may contribute materially toward the resolution of these conflicts."[42]

Kinsey's treatment of other traditional American sex taboos was similar. He tended to regard all such taboos as illegitimate efforts to repress man's biologic nature. Male promiscuity, for example, was also biologically natural: "There seems to be no question but that the human male would be promiscuous in his choice of sexual partner throughout the whole of his life if there were no social restrictions. This is the history of his anthropoid ancestors, and this is the history of unrestrained human males everywhere."[43]

As for homosexuality, Kinsey criticized biologists and psychologists who "assumed that heterosexual responses are a part of an animal's innate, 'instinctive' equipment, and that all other types of sexual activity represent 'perversions' of the 'normal instincts.' Such interpretations are . . . mystical."[44] According to Kinsey, "sexual contacts between individuals of the same sex are known to occur in practically every species of mammal which has been extensively studied," and the available evidence "suggests that the capacity of an individual to respond erotically to any sort of stimulus, whether it is provided by another person of the same or of the opposite sex, is basic in the species."[45]

Sex behavior among children is likewise natural. "In the normal course of events, the primitive human animal must have started his sexual activities with unrestrained pre-adolescent sex play, and begun regular intercourse well before the onset of adolescence. This is still the case in the other anthropoids . . . [,] in some of the so-called primitive human societies which have not acquired particular sex taboos . . . and among such as the children in our society as escape the restrictions of social conventions."[46]

One sex taboo Kinsey treated with more circumspection was sexual relations between adults and children. Unlike bestiality, homosexuality, and intercourse before marriage, Kinsey did not claim that adult-child sex was normal or acceptable. Nevertheless, he did downplay its seriousness and undermined the reasons for punishing it. In his view, the emotional upset caused by a child's sexual contact with an adult was no more serious than the fright displayed by children "when they see insects, spiders, or other objects against which they have been adversely conditioned."[47] Kinsey implied that the trauma of child-adult sexual contacts did not lie in the molestation itself but in the social disapproval that surrounded it.

"If a child were not culturally conditioned, it is doubtful if it would be disturbed by sexual approaches of the sort which had usually been involved in these histories," he contended. "It is difficult to understand why a child, except for its cultural conditioning, should be disturbed at having its genitalia touched, or disturbed at seeing the genitalia of other persons, or disturbed at even more specific sexual contacts."[48] Kinsey suggested that the real blame for the trauma of child-adult sex should be assigned to "the emotional reactions of the parents, police officers, and other adults who discover that the child has had such a contact." Such reactions "may disturb the child more seriously than the sexual contacts themselves."

Kinsey concluded that "the current hysteria over sex offenders may very well have serious effects on the ability of many of these children to work out sexual adjustments some years later in their marriages." Kinsey made a similar argument regarding extramarital intercourse. While acknowledging that such activity could be disruptive to the marriage relationship, he implied that the disruption was caused more by intolerant sexual mores than the inherent wrongfulness of the behavior. "At lower social levels, where the most extra-marital intercourse occurs, wives rather generally expect their husbands to 'step out,' and some of them rather frankly admit that they do not object provided they do not learn of the specified affairs which are carried on," he argued.[49] Likewise, at "the upper social level . . . it is sometimes had with the knowledge of the other spouse who may even aid and encourage the arrangement." It was in the middle class, presumably with its stuffy adherence to antiquated moral codes, that "extra-marital intercourse is less often accepted" and "often leads to divorce." Kinsey offered a biological justification for male adultery, suggesting that men, unlike women, were naturally programmed to seek a variety of sexual partners.[50]

Although Kinsey was loath to classify anyone's sexual behavior as "abnormal," he did not show the same open-mindedness toward the voluntary suppression of sexual activity. In a discussion of whether human beings could sublimate their sex drive and channel it into some other area, Kinsey criticized those who curbed their sexual activity as "apathetic" or "timid or inhibited individuals" who were "afraid of their own self condemnation if they were to engage in almost any sort of sexual activity" and who became "paranoid in their fear of moral transgression, or its outcomes."[51] He suggested that such individuals might be susceptible to suicide and that their sexual inactivity was comparable to a physical defect like "blindness or deafness."[52] He also accused educated persons of offering spurious justifications for reducing their sexual activity, such as the idea "that pre-marital intercourse . . . unfits one for making satisfactory sexual adjustments in marriage," or the claim "that the homosexual is a biologic abnormality," or the belief "that extra-marital intercourse inevitably destroys homes."[53] Kinsey disparaged such justifications as little more than "excuses" and "rationalizations . . . clutched at in support of a sexual suppression that is too often mistaken." More generally, Kinsey suggested that sex mores—like mores dealing with food, clothing, or religion—"originate[d] neither in accumulated experience nor in scientific examinations of objectively gathered data" but "in ignorance and superstition, and in the attempt of each group to set itself apart from its neighbors."[54]

Kinsey's overall theme was that any sex behavior practiced by other human beings or lower mammals was by definition biologically normal and therefore should be accepted rather than condemned. He promoted the view of "sex as a normal biologic function, acceptable in whatever form it is manifested,"[55] and he reproved psychologists for classifying sexual behaviors traditionally considered immoral as psychologically damaging. "It is unwarranted to believe that particular types of sexual behavior are always expressions of neuroses. In actuality, they are more often expressions of what is biologically basic in mammalian and anthropoid behavior, and of a deliberate disregard for social convention."[56] Those who engaged in this "deliberate disregard for social convention" are among society's best and brightest, according to Kinsey. "Many of the socially and intellectually most significant persons in our histories, successful scientists, educators, physicians, clergymen, business men, and persons of high position in governmental affairs, have socially taboo items in their sexual histories, and among them they have accepted nearly the whole range of so-called sexual abnormalities."

Kinsey acknowledged that many Americans considered this view of sex as "primitive, materialistic or animalistic, and beneath the dignity of a civilized and educated people,"[57] and he complained that "the scientist who observes and describes the reality [of human sex behavior] is attacked as an enemy of the faith, and his acceptance of human limitations in modifying that reality is condemned as scientific materialism."[58] Nevertheless, Kinsey maintained that his view simply constituted the honest "acceptance of reality," whereas those who opposed the materialistic view of sex refused to face facts.[58] "They seem to ignore the material origins of all behavior."[60]

Kinsey regarded such a view as delusional, and he accordingly attacked traditional social and legal restrictions on sex behavior as unrealistic, when in fact they were based on a clear-eyed, startlingly realistic understanding of the dangers posed to society by unrestrained sexual passions, regardless of their basis. Dissecting restrictions on premarital sex, for example, Kinsey claimed that they were incompatible with the biological fact that males reach their peak sexual capacity as teenagers. Arguing for the "near universality of adolescent [male] sexual activity,"[61] Kinsey added that it was all but impossible to try to control "the imperativeness of the [young male's] biologic demands."[62] Given Kinsey's view, it is little surprise that he dismissed as biologically unworkable the "opinion that . . . youths should ignore their sexual responses and should abstain from sexual activities prior to marriage . . . There is no evidence that it is possible for any male who is adolescent, and not physically incapacitated, to get along without some kind of regular outlet until old age."[63]

Kinsey also crusaded for greater leniency when it came to sex offenders, including those accused of child molestation, exposing their genitalia to children, extramarital activity, and bestiality. "In many instances the law, in the course of punishing the offender, does more damage to more persons than was ever done by the individual in his illicit sexual activity," argued Kinsey.[64] Writing about older males who are convicted of the fondling or attempted rape of children, Kinsey claimed that they were often misunderstood; he equated their crimes with "such affectionate fondling as parents and especially grandparents are wont to bestow upon their own (and other) children." He blamed "public hysteria" for misleading the child victims into thinking that they were somehow harmed. "Many small girls reflect the public hysteria over the prospect of 'being touched' by a strange person; and many a child, who has no idea at all of the mechanics of intercourse, interprets affection and simple caressing, from anyone except her own parents, as attempts at rape."[65]

Kinsey's findings were presented to the public as sound, objective scientific research conducted by a scholar motivated by nothing more than the impartial pursuit of scientific truth. In his preface to *Sexual Behavior in the Human Male*, Dr. Alan Gregg of the Rockefeller Foundation lauded Kinsey's research findings for "their extent, their thoroughness, and their dispassionate objectivity." He went on to lavish praise on Kinsey's integrity as a scientist: "Dr. Kinsey has studied sex phenomena of human beings as a biologist would examine biological phenomena, and the evidence he has secured is presented from the scientist's viewpoint, without moral bias or prejudice derived from current taboos."[66]

As a rhetorical posture, this just-the-facts-ma'am approach was at once enticing and disarming. Just as disarming was Kinsey's carefully crafted public persona. Although his research may have been shocking, his public character definitely was not. Kinsey exuded conventionality as a Midwestern professor who wore bow ties and who had been married to the same woman for more than a quarter of a century. Publicity photos taken of him interviewing children made him look like a benevolent uncle, and his children even attended Sunday school. Kinsey's public image as the wholesome champion of objective science allowed him and his defenders to marginalize critics as mean-spirited prudes who were determined to thwart the march of science.

The theme of Kinsey as the prophet of scientific progress was accentuated time and again by reviewers who, in the words of *Time* magazine, "hailed [him] . . . as one of the greatest scientists since Darwin."[67] Others compared him to Galileo or Columbus. "The Kinsey report has done for sex what Columbus did for geography," wrote Morris Ernst and David Loth.[68] Kinsey himself compared the criticisms he faced with past opposition to "Kepler, Copernicus, Galileo, and Pascal."[69]

In retrospect, the uncritical embrace of Kinsey and his research was based more on fantasy than fact, as researcher Judith Reisman has extensively documented in her book *Kinsey: Crimes and Consequences* (2000).[70] Perhaps the most egregious falsehood was the public image of Kinsey as the typical American family man.

Kinsey's Unorthodox Family

Kinsey's personal life was anything but conventional. According to biographer James Jones, Kinsey expected his closest associates to engage in sex with each other and with each other's wives. Not only that—they were expected to perform sexual acts on film while Kinsey watched.[71] Kinsey set up a makeshift studio in the attic of his house, and he hired a photographer to film the various sexual encounters.

Although defenders of Kinsey insist that such sessions were voluntary, the recollections of Kinsey's associates at the time paint a starkly different picture. Some participants later made clear that they had not wanted to participate. One wife of a Kinsey colleague complained of "the sickening pressure" to have sex on film.[72] "I felt like my husband's career at the Institute depended on it," she recalled.

Kinsey associate Vincent Nowlis similarly recounted being pressured by Kinsey to engage in homosexual activity, which he refused. On a research trip with Kinsey and others, Nowlis was asked to come to Kinsey's hotel bedroom. There he found Kinsey and two other colleagues. Nowlis was told that he needed homosexual experience in order to be an effective interviewer of homosexual subjects, and they were prepared to help him. In other words, he was expected to have sex with Kinsey and his two other colleagues. Shocked, Nowlis declined and resigned instead.[73]

Kinsey himself pursued ever more destructive sexual practices. He had become addicted to masochistic masturbation in his teens, sticking larger and larger objects up his urethra and tying ropes around his scrotum.[74] At one point he circumcised himself (without anesthesia) with a pocketknife.[75] Kinsey also recruited a sadomasochist to be one of his sexual partners. This person found Kinsey's performance during rough sex disappointing, complaining that "he was kind of a punk when it comes to s/m" and that "he liked [for me] to beat him with a cat-o'-nine tails but not very hard."[76]

Kinsey's attacks on his critics as being motivated by their own sexual hang-ups were thus deeply hypocritical. He clearly had a personal stake in trying to justify his own private sexual demons to the world as something healthy and normal. Perhaps his need for social approval for his personal life partly explains the ruthlessness with which he pursued his research project. Female students at Indiana University complained that Kinsey tried to pressure them into giving him their sex histories, and the university's dean of women came to their defense, provoking Kinsey's wrath. Years later, the former dean fought back tears as she recalled Kinsey's bullying. When she continued to insist that participation in his interviews must be completely voluntary, he became enraged. He told her she was unsuited for her job and should give him her own sex history. "He went so far as to say I should have some treatment by a psychiatrist to correct my bad attitudes."[77]

Kinsey's religious views were no more mainstream than his sexual practices. Despite allowing his children to attend Sunday school, he hated "the Judaeo-Christian sexual tradition,"[78] and he told his associate Clarence Tripp that "[t]he whole army of religion . . . is our central enemy."[79] According to fellow sex researcher Wardell Pomeroy, "Kinsey began to lose his [religious] beliefs as a college student, when his study of science disclosed to him what he saw as a basic incongruity between it and religion."[80] When asked once by Pomeroy whether he "really believe[d] in God," Kinsey snapped, "Don't be ridiculous. Of course not." A thoroughgoing scientific materialist, Kinsey dismissed the idea that life could continue after death. "I believe that when you're dead, you're dead, and that's all there is." When his daughter Anne wanted to be confirmed as a member of the local Presbyterian church at age twelve, he forbade it. He also tried to discourage one of his lab assistants from teaching Sunday school.[81]

But from a public-policy standpoint the most serious problem with Kinsey was not his personal life. It was his shoddy scholarship. Although Kinsey wrapped himself in the mantle of scientific respectability, his research turned out to be classic junk science.

Kinsey's Science Fiction

The Achilles' heel of Kinsey's study was its unrepresentative sample. In *Sexual Behavior in the Human Male*, Kinsey extrapolated his data to explain the sexual behavior of the entire white male population. One of the primary goals of the book was to convince people that a large proportion of the population engaged in practices typically regarded as abnormal or immoral. Yet Kinsey's sample was in no way representative of the general male population. Kinsey made it difficult for other scholars to uncover how biased his sample was because he never presented a clear, systematic description of the total number of people he interviewed and their relevant demographic data. But over time he and his associates did provide enough information to piece together a damning picture. *Sexual Behavior in the Human Male* states that the Kinsey team had interviewed 6,300 males, but only 5,300 of the interviews (those with "white males") were actually used for the book.[82] And, according to University of Chicago statistician W. Allen Wallis (a former president of the American Statistical Association), the numbers included in the tables of Kinsey's book added up to no more than 4,120 men interviewed.[83]

Who were these men? Kinsey never provided thorough information. But from comments made by himself and others it has been learned that the sample included roughly "1,400 convicted sex offenders in penal institutions,"[84] 199 "sexual psychopath patients,"[85] 329 prisoners who were not convicted of sex offenses,[86] several hundred juvenile delinquents or otherwise "aberrant" boys,[87] an unspecified number of members of the "Underworld" (defined by Kinsey as including such people as "bootleggers, con men, dope peddlers, gamblers, hold-up men, pimps, prostitutes, etc."),[88] and more than 450 homosexuals (most of these were apparently recruited from what Kinsey described as "homosexual communities" in several large American cities).[89] There is, of course, no reason to assume that such a sample could be validly used to describe the sexual patterns of the general male population of the United States. Judith Reisman has concluded that various deviant populations probably account for approximately 86 percent of the 4,120 males who actually appear in the tables of Kinsey's book.[90] Even if the proportion of deviants and sexual minorities in Kinsey's sample was smaller, there is no question that the sample was blatantly unrepresentative of the population as a whole.

There were other flaws in the sample. Kinsey did not randomly select the people he interviewed. Many of his subjects volunteered to tell their sexual histories after having attended a lecture by him about sex. This presents two problems. First, those who attended sex lectures by Kinsey may not have been representative of the general population. Second, even if such groups were representative of the general population, the fact that Kinsey asked for volunteers from those groups to give him their sex histories further compromised the generalizability of his data. Such self-selecting samples are no longer used in public-opinion polling because they are known to grossly misstate the views of the entire population. Those who volunteer to be interviewed tend to have different characteristics than those who choose not to volunteer.

According to Kinsey's own figures, only about six thousand of the fifty thousand men and women who attended his lectures volunteered to give him their sexual histories, a response rate of about 12 percent.[91] There is no good reason to assume that the 12 percent of his audiences who volunteered to give him their sex histories were comparable to the 88 percent who did not. Kinsey had little excuse for not recognizing this problem. Concerned about volunteer bias, prominent psychologist Abraham Maslow conducted a study for Kinsey that clearly demonstrated how volunteer samples tended to overrepresent those who are most sexually active. "As I expected, the volunteer error was proven and the whole basis for Kinsey's statistics was proven to be shaky," Maslow wrote later.[92] But Kinsey didn't want to hear it. In *Sexual Behavior in the Human Male*, he not only disregarded Maslow's findings, he cited an earlier article by Maslow that made it appear that he supported Kinsey's faulty methodology.[93]

Kinsey did propose a technique to overcome the problem of nonrandom samples, something he called the "hundred percent sample."[94] That is, he was able to get 100 percent participation of people in certain groups, and his plan was to compare their responses to the responses of his self-selecting samples. But Kinsey did not provide a systematic breakdown of who made up these "hundred percent samples." The limited information he did provide indicated that these groups were just as unrepresentative of the general population as the rest of his subjects. Twenty-nine of the "hundred percent samples" came from college groups, including seven college classes.[95] Presumably these included some of Kinsey's own sexually explicit "marriage classes" at Indiana University. Those who attended Kinsey's classes were not necessarily representative of the general student population.

Other groups of "hundred percent samples" included "conscientious objectors," "hitch-hikers," "delinquent institutional groups," "penal institutional groups," a "mental institutional group," and unnamed "professional groups." From the sketchy information provided by Kinsey, it is doubtful that these groups were representative. Moreover, interviews with some of these individuals may have been skewed not by "volunteer error" but by what might be called coercion error. In order to get a "hundred percent sample," we know that Kinsey had to follow up aggressively with nonrespondents until they finally relented.[96] We further know that female students at Indiana University complained about the way Kinsey pressured them to give him interviews.[97] The validity of data extracted from such coerced interviews is questionable.

Years after Kinsey's death, Paul Gebhard (one of Kinsey's coauthors) and Alan Johnson attempted to "clean" Kinsey's original sample by separating out convicts and members of groups with known sexual proclivities (such as explicitly homosexual groups) and retabulating some of the data. The results were published as *The Kinsey Data* (1979), a book that is sometimes touted as having verified Kinsey's research.[98]

Gebhard and Johnson originally planned to write a chapter comparing their new tabulations with Kinsey's original published data, but they claimed that they "abandoned this intention as we realized that such comparison would necessitate not a chapter but another book." They further declared that "direct comparison [with Kinsey's original results] was usually impossible" and warned readers who might attempt such comparisons that they would find it "complex and often frustrating. This is particularly true of *Sexual Behavior in the Human Male*, which is one of the most difficult books to work with ever written."[99] Yet Gebhard and Johnson's failure to provide tabulations that could be systematically compared with Kinsey's original published findings did not prevent them from assuring readers that "the major findings of [Kinsey's] . . . earlier works regarding age, gender, marital status, and socioeconomic class remain intact" and "the important contributions of Dr. Kinsey will stand."[100]

Such assurances aside, Gebhard and Johnson's "cleaned" figures did nothing to eliminate the self-selection bias in the remaining sample. Even the supposedly "normal" participants in Kinsey's sample were probably not representative of the general population because they were not randomly sampled.

Just how radically skewed Kinsey's sample was finally became apparent as social scientists started to ask questions about sex practices in large, randomly sampled national surveys. While Kinsey claimed that "about 69 per cent of the total white male population ultimately has some experience with prostitutes,"[101] current survey data indicate that the proportion of American males seventy and over who have ever paid a partner for sex is 18.6 percent.[102] The highest incidence of the use of prostitutes by American males among any age group is 23.1 percent. Thus, at a minimum, Kinsey's statistic was off by a factor of three. Kinsey's estimate that half of all white married males had extramarital intercourse at some point in their marriages was equally off base.[103] According to recent survey research, the proportion of married males seventy and over who have ever had extramarital intercourse is 9.5 percent, and the proportion of married males 60–69 who have ever had extramarital intercourse is 16.5.[104]

But that was nothing compared to Kinsey's wildly overblown statements about homosexuality. Contrary to his claim that "10 per cent of the males are more or less exclusively homosexual . . . for at least three years between the ages of 16 and 55,"[105] recent research indicates "that only about 2–3% of sexually active men . . . are currently engaging in same gender sex" and as few as 1 percent of men over eighteen identify themselves as "gay."[106] As for Kinsey's finding that "37 per cent of the total [white] male population has at least some overt homosexual experience to the point of orgasm between adolescence and old age,"[107] only 4.7 percent of white males currently report having had a same-gender sexual partner since the age of eighteen.[108] The same proportion of males seventy and over report having had a same-gender sexual partner since eighteen.

Biased sampling was not the only weakness of Kinsey's research. A second major problem was his manipulative interviewing technique. Worried that interview subjects would try to conceal some of their sexual activities (especially if the activities violated social taboos), Kinsey and his colleagues abandoned any pretense of nondirective interviewing. Instead, sex researchers were supposed to act more like police detectives, grilling their subjects until they made a full confession. According to Kinsey,

> the interviewer should not make it easy for a subject to deny his participation in any form of sexual activity. It is too easy to say no if he is simply asked whether he has ever engaged in a particular activity. We always assume that everyone has engaged in every type of activity. Consequently we always begin by asking when they first engaged in such activity. This places a heavier burden on the individual who is inclined to deny his experience.[109]

Kinsey added that an interviewer should continue to press the issue if he comes to feel the person being interviewed is not telling the whole truth. Of course, since Kinsey's operative assumption was that "everyone has engaged in every type of activity," a person who denied involvement in certain activities was more likely to generate doubt than someone who admitted participating in them.[110]

The lengths to which Kinsey would go to bully an interview subject were on display in a practice he labelled "forcing a subject."[111] If an interviewer came to believe that the person he was interviewing was not telling the truth, the interviewer was supposed to "denounce the subject with considerable severity, and the interviewer should refuse to proceed with the interview," said Kinsey.[112] The purpose of this approach was to shock the interviewee into revealing the full truth about his or her sexual history. Kinsey proudly claimed that "no history has ever been lost as a result of such action, and the study has won a number of staunch friends because of our insistence on scientific honesty."[113] But whether such interrogative tactics elicited valid answers is debatable.[114]

A third major problem with Kinsey's research was the way in which he invoked science as the justification for revising traditional sex mores. Time and again, Kinsey seemed to suggest that his scientific research about how men and animals actually behave should decide the question of how men ought to behave. This was fallacious reasoning. Even if a majority of human beings (or mammals, for that matter) performed a certain sex act, that would not necessarily make it morally right. It would not even show that the act itself was biologically obligatory. Ever the materialist, Kinsey sought to reduce ethics to the facts of biology.

Many of the problems with Kinsey's research methodology were identified during his lifetime.[115] But he proved deft in deflecting efforts to scrutinize his scholarship. When an expert panel set up by the American Statistical Association prepared a scathing report on the validity of his study of male sexual behavior, Kinsey bullied committee members into watering down the wording of the report so that he could tout it as a vindication of his research.[116] Kinsey's messy personal life and unethical research practices also generally escaped scrutiny. Shortly after his death, a high-profile murder and sex-crimes trial took place in Germany that could have seriously tarnished Kinsey's reputation, but the American news media did not cover the story.

In 1957, German lawyer Fritz von Balluseck was tried in Berlin for child murder and for multiple sex crimes against children.[117] The sensational trial was front-page news throughout Germany. A serial child molester, Balluseck maintained detailed records of his sexual abuse of one hundred children, including his own daughter. Balluseck's abuse of children apparently included the victimization of young girls during World War II while he served as a Nazi official during the German occupation of Poland. But perhaps the most shocking revelation during Balluseck's trial was not the extent of his crimes. It was his long collaboration with Alfred Kinsey. Investigators discovered that Balluseck had maintained an extensive correspondence with Kinsey in order to supply Kinsey with data derived from his sexual contacts with children. According to one contemporaneous German newspaper account of the trial, Balluseck recorded the details of his crimes down to "the smallest detail" in his diaries. He then "sent the detail of his experiences regularly to the US sex researcher, Kinsey. The latter was very interested, and kept up a regular and lively correspondence with Balluseck."[118]

Letters were discovered in Balluseck's papers in which Kinsey thanked Balluseck for supplying him with data and encouraged Balluseck to continue his research on Kinsey's behalf, although Kinsey also warned Balluseck, "Be careful."[119] At trial, Balluseck testified that he had approached children to molest them because "Kinsey had asked me to do so."[120] Kinsey never notified any police authorities about Balluseck's criminal activities with children. Indeed, he rebuffed the FBI when it tried to collect information about Balluseck on behalf of Interpol.[121] The judge presiding over Balluseck's case expressed shock that Kinsey had shielded Balluseck from prosecution. "Instead of answering his [Balluseck's] sordid letters . . . the strange American scholar [Kinsey] should rather have made sure that Mister von Balluseck was put behind bars," declared the judge.[122]

The Balluseck trial exposed a dark secret at the heart of Kinsey's *Sexual Behavior in the Human Male*. In chapter 5 of that book, Kinsey presented extensive data about the sex behavior of children. In Table 31, Kinsey purported to supply data on "orgasms" in 317 boys from two months to fifteen years of age.[123] In Table 32, he purported to supply data about the time it took selected boys to reach orgasm (the data was supposedly based on "observations timed with [a] second hand or stop watch").[124] In Table 33, Kinsey purported to present data about multiple orgasms in 182 boys.[125] And in Table 34, he claimed to present data about the time it took to reach multiple orgasms in twenty-four boys from five months to fourteen years of age.[126] How could Kinsey have collected such highly sensitive intimate data on children as young as two months old?

Kinsey wrote that the research came from "adult males who have had sexual contacts with younger boys."[127] To be more specific, the data came from observations by "9 of our adult male subjects" who observed preadolescent boys in "self-masturbation, or who were observed in contacts with other boys or *older adults* (emphasis mine)."[128] Even from Kinsey's own sketchy explanation, it was apparent that these "9 . . . adult male subjects" may have been pedophiles, and that Kinsey protected them rather than notifying the police. Amazingly, no one seems to have publicly questioned Kinsey during his lifetime about the ethics or legality of the child sex data he collected for his chapter on early male sexuality.[129] Nor did American journalists follow up the revelations about Kinsey in the Balluseck trial. Had the news media actively pursued the Balluseck-Kinsey connection, they might have been more willing to raise questions about the adequacy of Kinsey's research sample and interviewing techniques. But they did neither.

Not until 1981 did Kinsey's use of child molesters to obtain data on child sex behavior finally become the subject of intense public scrutiny and debate. In July of that year, communications scholar Judith Reisman delivered a paper on possible ethics violations in Kinsey's research at an international meeting of sexologists in Jerusalem. Based on a close reading of the tables in chapter 5 of *Sexual Behavior in the Human Male*, Reisman charged that Kinsey relied on pedophiles to collect his child sex data and therefore, as the title of her paper put it, he was "a contributing agent to child sexual abuse." Reisman later documented in detail Kinsey's faulty research techniques (including his use of unrepresentative samples) in her books *Kinsey, Sex and Fraud* (1990) and *Kinsey: Crimes and Consequences*.[130]

Reisman's allegations put the Kinsey Institute on the defensive for much of the 1980s and '90s. The institute's responses were neither candid nor complete. Replying to criticisms that Kinsey's sample population included "sex deviants" and "sex criminals," institute director John Bancroft assured people that "the large majority of the 18,000 interviews carried out in Kinsey's research involved ordinary men and women."[131] Yet statements by Kinsey and his fellow researchers demonstrate that the interviews used by Kinsey were filled with individuals unrepresentative of "ordinary men and women." Bancroft also insisted that "the only research method used by Kinsey was the interview,"[132] ignoring testimony from Kinsey's associates that in addition to interviews Kinsey arranged for the direct observation and filming of sex acts.[133] Bancroft conceded that Kinsey obtained some data from pedophiles, but the institute's statements on this point are less than forthright. The institute's website maintains that Kinsey "did not . . . collaborate, or persuade people to carry out experiments on children."[134] Whether or not Kinsey had to persuade his pedophile correspondents to cooperate with him in supplying data on their activities, it is hard to see how their cooperation was not "collaboration."

The sort of data Kinsey asked for required detailed record keeping and scientific measurements. For example, Kinsey's correspondents timed with watches the number of seconds it took for a boy to reach orgasm. They also observed the number of orgasms in various boys during a twenty-four-hour period. It is incredible to believe that this sort of detailed information was not collected at the behest of Kinsey himself. Moreover, even if such data were not collected at Kinsey's specific request, Kinsey was the recipient of the data and used it for his research. That clearly constituted "collaboration" with pedophiles for research purposes under any usual understanding of the term. Bancroft also tried to minimize the pedophile issue by asserting that although Kinsey talked with several pedophiles, the data he cited in his books only came from one pedophile. (Other sources have identified him as Rex King.) "Without any doubt, all of the information reported in Tables 31–34 came from the carefully documented records of one man," Bancroft claimed.[135] However, Bancroft's lone-pedophile theory flew in the face of Kinsey's comment that he derived the data in *Sexual Behavior of the Human Male* from "9 . . . adult male subjects" who "had sexual contacts with younger boys."[136]

Bizarrely, Bancroft acknowledged no ethical problem with using pedophiles to collect research data on child sexuality. In his view, Kinsey could "be criticized for making use of information about children's sexual responses obtained from individuals who were criminally involved with those children, not because it was improper to do so, but because of the uncertain validity of such information."[137] So although Bancroft conceded that the accuracy of such data can be questioned, he clearly implied that there is nothing ethically wrong with using pedophiles to collect data about sex behavior in children.

Bancroft insisted that "Kinsey did not employ criminal acts, against children or any one else, in his research," adding the caveat that "talking to criminals about their experiences and observations is a different matter" than a criminal act.[138] Here Bancroft fudged a crucial distinction. Talking with a convicted criminal about his past criminal activities is not a criminal act; but gathering data from a sex offender who has not yet been apprehended by the police and who is continuing to engage in the sexual abuse of children may well be a crime. The problem with Kinsey's use of pedophile correspondents was that Kinsey knew they were engaging in the abuse of children and yet did not report that abuse to the police. Kinsey shielded these pedophiles from prosecution in order to further his own research goals.

The Kinsey Institute has also been coy about Kinsey's relationship with particular pedophiles. For example, the institute's website acknowledges that a former Nazi who had "sexual experiences with children" (presumably Fritz von Balluseck) corresponded once with Kinsey and that while Kinsey's reply was "non-judgmental" it also "point[ed] out how strongly society condemned such behavior."[139] This answer does not square with contemporaneous German news reports of the trial of Balluseck, which make clear that Kinsey and Balluseck had an extensive correspondence (not just a single letter) and that Kinsey encouraged Balluseck to collect data for Kinsey's research. These accounts are corroborated by a past director of the Kinsey Institute, Paul Gebhard (who had worked personally with Kinsey), who has stated that Kinsey wrote Balluseck "questions . . . and they carried on quite a correspondence."[140] Because access to the

Kinsey Institute archives is restricted, it is difficult for independent investigators to verify the institute's current response regarding Balluseck. Nor can they verify the institute's response regarding a woman named Esther White who came forward in the 1990s with detailed, credible accusations that her father recorded his sex abuse of her in order to supply Kinsey with information.[141] The Kinsey Institute states that it has not been able to locate correspondence that would corroborate White's story.[142] But the public simply must take the institute's word for this. Even though the institute is located at taxpayer-funded Indiana University, it is a separate, private organization and is therefore immune to the normal forms of accountability associated with a public institution.

Either unable or unwilling to refute the charges against Kinsey with convincing evidence, the Kinsey Institute has fallen back on trying to stigmatize its critics as extremists. Institute director Bancroft has complained that allegations against Alfred Kinsey's use of pedophiles in sex research are a ploy by "those on the Religious Right who seek to discredit Kinsey" by drawing on society's "anxiety bordering on hysteria about child sexual abuse."[143] Such rhetorical posturing may be the best evidence that the original Kinsey data have been so thoroughly debunked in recent years that today they are beyond serious defense.

Unfortunately, the recent debunking of Kinsey's research has done little to undo its widespread impact on American public policy. For decades, Kinsey's data were largely accepted as good science, and as a result exerted a profound influence on the American legal and educational systems. In the field of criminal justice, Kinsey's ideas were cited to justify decriminalizing or reducing the penalties of a wide range of sex crimes, and Kinsey himself consulted with lawmakers in California, New Jersey, New York, Delaware, Wyoming, and Oregon about changes in sex laws.[144]

Kinsey's research was said to offer a scientific justification for abolishing or loosening outdated sex restrictions. "It is a cliché that we cannot legislate morality," wrote Fred Rodell in 1965. "Even less can we legislate effectively against conduct which a large part of the community has come to accept as not immoral."[145] Rodell cited Kinsey's research as proof that many illegal sex acts were in fact accepted by society, and he praised the legal community for finally beginning to recognize that "laws against common, voluntary [sex] acts by adults [are] more of a 'crime against nature' than the acts themselves."[146] Legal scholars compared America's sex laws with the ill-fated experiment of Prohibition. If Kinsey's statistics are accurate, wrote New York City magistrate Morris Ploscowe, "then it is obvious that our sex crime legislation is completely out of touch with the realities of individual living and is just as inherently unenforceable as legislation which prohibits the manufacture and sale of alcoholic beverages."[147] Morris Ernst and David Loth likewise observed that "the Kinsey Report gives us some idea of how common is the violation of laws which seek to regulate sexual behavior," and they asked, "how long can we keep our legal standards separated by such a wide gap from the actual practices of great masses of the population?"[148]

When the American Law Institute developed the Model Penal Code (MPC) in the 1950s, it repeatedly cited Kinsey's research to justify eliminating or reducing the penalties for various sex crimes.[149] As Kinsey biographer Jonathan Gathorne-Hardy puts it, the "Model Penal Code of 1955 is virtually a Kinsey document . . . At one point, Kinsey is cited six times in twelve pages."[150]

One area in which the drafters of the MPC followed Kinsey in spirit was sex crimes against children. The MPC raised the burden of proof in cases of both child rape and other child sexual assaults. Prior to the MPC, most states treated children as incapable of giving consent to sexual activity before the ages of eighteen or sixteen.[151] Any sexual contact with children under those ages were therefore presumed to be illegal. All the prosecutor had to show in such cases was that sexual contact had occurred. The MPC, however, proposed reducing the age of consent in rape cases to ten, which meant that child-rape victims age ten or older would have to prove not only that sexual intercourse had occurred, but that it had been compelled. Otherwise the rape would be treated as a lesser crime. Sexual assaults on children short of intercourse, meanwhile, were downgraded by the MPC to a misdemeanor. Unless it could be demonstrated that the sexual contact in question had been "offensive" to the child, the behavior was effectively presumed to be consensual—unless the child victim was under age ten, or unless the victim was ten to fifteen years of age and the perpetrator was "at least [four] years older" than the alleged victim.[152]

The rationale for making it more difficult to prosecute childhood sexual crimes was classic Kinsey. Especially in the case of sexual assaults short of intercourse, the MPC drafters argued that the new standards were needed because "it is imperative that normal adolescent sex play between males and females not carried so far as intercourse or attempted intercourse remain free of the taint of criminality."[153] Sexual activity by children as young as ten was regarded as healthy and "normal," and so the law must be careful not to penalize it. It should be emphasized that even in cases of severe age disparity (e.g., a thirty-year-old man and a ten-year-old girl), sexual molestation short of intercourse was only a misdemeanor under the MPC, not a felony. This lenient treatment of pedophiles was presumably justified in the minds of the MPC's drafters because they believed pedophiles were neither violent nor a continuing threat once caught. "Paedophilia is one of the most frequent types of sex offense but recidivism is low," they assured policymakers, drawing on questionable claims made by Kinsey's friend Dr. Karl Bowman.[154]

Legislators did not go as far as the drafters of the MPC wanted, but the age of consent for sexual intercourse was reduced in many states, especially in cases where the parties involved were within two to six years of each other in age.[155] The new standards had disastrous consequences for many girls. "Some states, in an effort to decriminalize consensual sex between teenagers, have drafted laws making it virtually impossible for a female under the age of eighteen to bring sexual assault charges against a nonrelated male in the same age group," writes Linda Brookover Bourque. "Yet young women are the most common victims of sexual assault, and young men the most frequent perpetrators."[156]

The reduced concern over sex involving youth exerted an impact even in those jurisdictions, such as California, where sexual contact with older children was still presumed, according to the laws on the books, to be illegal. In a particularly notorious 1993 case, a gang of teenage boys in Lakewood, California, competed to have sex with as many girls as possible. Despite clear evidence that many of the encounters involved coercion and intimidation, the district attorney's office ultimately dropped charges against all but one of the alleged assailants. The explanations offered for dropping the charges ranged from a claim that the sex was consensual (despite the fact that a number of the girls said otherwise) to worries that prosecutors could not prove that all of the encounters had been coerced.[157]

Old icons die hard. Despite recent withering critiques of Kinsey's research methods, his work remain culturally influential. Hollywood celebrated it in the 2004 film *Kinsey,* starring Liam Neeson, and Kinsey's research continues to be drawn on by many legal scholars, judges, and social scientists. From 1982 to early 2000, there were nearly 5,800 citations of Kinsey in law reviews and journals abstracted in the Social Science Citation Index and the Science Citation Index.[158] As Judith Reisman points out, that "is roughly double the citations for such luminaries as Sigmund Freud, Abraham Maslow and Margaret Mead."[159] Apparently unaware of the problems identified with Kinsey's methodology, one scholar writing in the 1990s maintained that Kinsey's "studies provide the best statistical evidence available, even four decades later, on the sexual activities of a broad cross-section of Americans—a tribute to the mass of data Kinsey was able to collect and the sophisticated statistical methods used to ensure that a representative sample was obtained."[160] As the next chapter will explain, nowhere is the continuing impact of Kinsey's research more pervasive than in the field of sex education.

13

Sex Miseducation

In *Sexual Behavior in the Human Male*, Alfred Kinsey excoriated existing sex-education efforts as too little, too late. He disparaged female teachers assigned to teach the subject for being mothers who were probably much less sexually experienced than their male students. "Many of these women, including some high school biology teachers, believe that the ninth or tenth grade boy is still too young to receive any sex instruction when, in actuality, he has a higher rate of [sexual] outlet and has already had a wider variety of sexual experience than most of his female teachers ever will have," Kinsey snapped.[1]

He then acknowledged that "whether there should be sex instruction, and what sort of instruction it should be, are problems that lie outside the scope of an objective scientific study"—but that did not stop him from offering advice. "It is obvious," he explained, "that the development of any curriculum that faces the fact [of adolescent sexual behavior] will be a much more complex undertaking than has been realized by those who think of the adolescent boy as a beginner, relatively inactive, and quite capable of ignoring his sexual development." Kinsey suggested that sex education begin no later than "ten or twelve, and in many instances at some earlier age."[2]

Kinsey's vision of a transformed sex-education system began to be realized after his death when, in 1964, Mary Calderone and Lester Kirkendall cofounded the Sex Information and Education Council of the United

States (SIECUS) to promote comprehensive sex instruction in kindergarten through high school.[3] (The title of the group was later changed slightly to the Sexuality Information and Education Council of the United States.) Early SIECUS board members included Kinsey associates Wardell Pomeroy and Harriet Pilpel. Pomeroy was coauthor of Kinsey's original published reports, and Pilpel was a lawyer who worked on a sex-law-reform project for Kinsey.[4] She later served as a vice president of the American Civil Liberties Union.

SIECUS forged ties with liberal members of the clergy, and one of its founding board members was a minister affiliated with the National Council of Churches. Like Kinsey, SIECUS cultivated a public image of probity and moderation, assuring parents and journalists that its goal was to educate people about the medical and scientific facts of sexuality so that they "may be aided toward responsible use of the sexual faculty."[5] Far from promoting permissiveness, SIECUS claimed to be on a crusade to protect the nation's children from the destructive effects of unwholesome sex. Its purpose was to champion "wholesome knowledge about sex . . . and to counteract the violent, dirty, threatening images of sex thrust at children from all sides."[6]

Executive director Mary Calderone reinforced her organization's image as an upholder of society's mores. A Quaker (a fact she did not hesitate to point out to reporters), Calderone spoke of her "profound belief that sex belongs primarily in marriage" and belittled as "blind alleys" such sex behaviors as "compulsive promiscuousness, homosexuality, 'fun and games,' and exploitation" (she later changed her mind about homosexuality).[7] Calderone emphasized that discussions of sex should not—indeed, could not—ignore the importance of values. "I do not see how it is possible to teach history, literature, social science or any form of science without discussing the moral values that underlie all human actions, past and present," she wrote.[8] She stressed: "There should be intensive consideration of the family and its relation to society, the institution of marriage, and the various factors that threaten the integrity of the family today—venereal disease, abortion, broken homes, divorce, etc."[9] Who could object to such appeals? Indeed, one liberal journalist later faulted Calderone for being too moralistic.[10]

SIECUS assisted local groups trying to establish sex-education programs in their communities, and it produced a series of booklets to help educators and administrators develop new sex-education curricula. By its own estimate, it had helped dozens of local groups establish sex-education programs by the end of the 1960s.

But success bred opposition, and by the turn of the decade sex education proved to be one of the most bitter and divisive issues within state and local education policy. The national news media portrayed these early battles largely according to the interpretive framework offered by SIECUS and its allies. In the minds of sex-education reformers, the conflicts pitted fundamentalist extremists against the enlightened forces of science. "I'm sure there are some well intentioned and responsible critics of sex-education programs," Mary Calderone conceded to one journalist.[11] "But we hear almost entirely from the extreme right, which is recruiting the timid and the conservative."

Sex-education critics featured in the news media tended to confirm the stereotype offered by Calderone. In an article bearing the tabloid-style title,"M[aryland] Mother Charges Sex Lecture Brainwashed Her 9th Grade Son," the *Washington Post* highlighted overwrought charges by a Maryland housewife about the supposed pernicious effects of a sex-education slideshow shown her son. Testifying before a state commission studying sex education, the housewife condemned the slide show in no uncertain terms as a plot to "brainwash established sources of guidance from the student's mind." She alleged that the slides "attack[ed] God, the Bible, the church, and parents." After watching for themselves, commission members concluded (according to the *Post*) that it did precisely the opposite of what the housewife asserted. In their view, the slides "upheld conventional morality rather than deriding it."[12] The foolishness of sex-education opponents was made abundantly clear by the article.

But sex-education critics were not merely dimwitted, according to reporters; they were also menacing. The *New York Times* chronicled the travails of a self-identified conservative school-board member in Wisconsin who faced anonymous phone calls branding him a "dirty Communist traitor" for supporting sex education.[13] The *Washington Post* highlighted accusations of terror tactics made by Maryland's state superintendent for health education. "They harass you all hours of the day and night," she complained about opponents of sex education. "They verbally abuse you. They use character assassination . . . They make us the target for crank calls. A den mother in Baltimore came out for [sex education] at a meeting and by the time she got home that night she had four calls, one of which threatened her children."[14] Sex-education opponents were depicted as being obsessed with conspiracy theories and notions that sex education was a communist plot. "To these critics," sex education "is manifestly the work of the [d]evil and the Communist Party," wrote a *Washington Post* reporter.[15] One opponent was quoted in the *New York Times* asserting that sex education was part of a "master plan" to create a "controlled, one-world society."[16] Another critic was quoted claiming that sex education was "part of a gigantic conspiracy to bring down America from within" and was described as "equat[ing] sex education with a Communist conspiracy to weaken the morals of youth."[17]

The depiction of sex-education critics as conspiracy-mongers was iron-ic, because the news media (following the lead of sex-education support-ers) put forward their own conspiracy theories. Some journalists claimed that opposition to sex education was masterminded by "ultraconservative groups." The *New York Times* implied that local associations opposed to sex education were "front groups" organized at the behest of the leaders of the John Birch Society.[18] The *Washington Post* was only slightly more nuanced. Admitting that "not all opponents of sex education are bumper-sticker conservatives or religious fundamentalists," a *Post* writer nonetheless in-sisted that "on the whole" opposition to sex education "is a crusade gen-erated by such organizations as the John Birch Society and the Rev. Billy James Hargis's Christian Crusade."[19] The *Post* then quoted an expert from an "organization studying extremism" to drive home the point that sex-education opposition was carefully orchestrated by right-wing extremists. Referring to a local sex-education battle in Maryland, the expert claimed, "It's been fanned professionally. It's not accidental and it's not even ca-sual."[20]

While national journalists were all too willing to portray critics of sex education as conspiratorial extremists, they tended to accept at face value the claims of moderation made by SIECUS and its supporters. Reporters were undoubtedly correct that some critics of sex education were bigoted, overbearing, and harsh (although some of the "ultraconservatives" dis-paraged by journalists were probably more thoughtful than they were given credit for). But nastiness could be found on both sides of the debate, even if the news media censored that fact in biased reporting. Consider the *New York Times* article highlighting the harassment of the Wisconsin school-board member who supported sex education. The article painted sex-education opponents in lurid colors as experts in smears and intimida-tion. This portrait was accentuated by the article's description of anti–sex education activists in a neighboring state. "Opponents of sex instruction in Minneapolis . . . have been on the offensive for some time," the *Times* writer reported grimly. "They have yet to suffer a setback and there is no sex education course in the city's schools." Mrs. Elsie Zimmerman, the leader of the opposition, was quoted by the *Times* as condemning sex-ed programs as "garbage" and threatening to use "every weapon" at her com-mand to stop sex education in Minneapolis. "We will be Paul Reveres in combat boots," Mrs. Zimmerman supposedly told her supporters.[21]

What the *Times* reporter neglected to mention was that sex-education critics in Minneapolis had been subjected to intense, even violent intimi-dation. Just a few months earlier, someone had used tear gas to disrupt testimony by Zimmerman and other sex-education skeptics at a legislative hearing. "This is what happens all the time," complained Zimmerman after gasping audience members had evacuated the auditorium. "They always try to disrupt us and to cause disturbances."[22] Since the tear-gas incident was reported in one of the state's main daily newspapers, and Mrs. Zimmerman had been one of the victims, it is difficult to believe that the *Times* reporter did not uncover it during his research. But violence directed at sex-education opponents did not fit the stereotyped story the reporter was trying to tell, so it was not newsworthy.

The media's suppression of facts extended an uncritical presentation of the beliefs of sex-education reformers at SIECUS and elsewhere. Reporters generally pooh-poohed accusations that SIECUS propounded radical ideas. Had they read the publications issued by the group they may not have been so dismissive.

Sex Education According to SIECUS

During the 1960s, SIECUS put out a series of study guides to help educators develop new sex-education curricula. These guides were later collected and published as essays in the book *Sexuality and Man* (1970).[23] Relying largely on Kinsey's research (which they cited repeatedly), the SIECUS study guides followed Kinsey in invoking the authority of science to legitimize as normal sex behaviors traditionally regarded as abnormal or inappropriate. In "Sexuality and the Life Cycle," SIECUS board members Lester Kirkendall and Isadore Rubin cited Kinsey's data collected from pedophiles to establish that children are sexual beings from infancy and that preadolescent sexual activities are perfectly natural among children.[24]

"Children learn about their sexuality in many ways," explained Kirkendall and Rubin. "Curiosity leads them to engage in exploratory activities involving themselves and other male or female children . . . They learn whether it is 'safe' to experiment with stimuli that adults call sexual and what sensations may come from experimentation."[25] By the time children reach the ages of nine to twelve, the authors wrote, "it is not uncommon . . . for exchanges of genital manipulation and exploration to occur." They added that "for most children probably the most traumatic part of such experiences is the horrified reaction of parents or other adults," because it implants the idea "that sex is dirty or evil." Kirkendall and Rubin claimed that "when they have an opportunity, children also engage in heterosexual play throughout prepuberty, but without the emotional commitment of love that comes later."[26] From puberty onward, "a good many youth (almost certainly more boys than girls) engage in erotic play, genital examination, and experimentation with members of their own sex."[27] Rather than consider such same-sex activities worrisome, the authors asserted that "[i]n the majority of cases, such experiences are likely to be simply a part of the development process."

In the SIECUS guide to homosexuality, Isadore Rubin similarly invoked science to dismiss the views that homosexuality was a mental illness, maladaptive, or contrary to nature. Echoing Kinsey, Rubin wrote that neither heterosexuality nor homosexuality was hard-wired into human biology. "It is worth noting that there is wide agreement today that man does not from birth possess an instinctive desire to achieve any specific goal in regard to sex, but that his sexual behavior is at any time the cumulative result of the learning and conditioning experiences he has had." Rubin cited Kinsey's research to establish the prevalence of homosexual behavior, asserting that "37 per cent of all males have some overt homosexual experience to the point of orgasm" and "over half of the males who remain unmarried at 35 have homosexual experience."[28]

In the guide to "Characteristics of Male and Female Sexual Responses," former Kinsey associates Cornelia Christensen and Wardell Pomeroy (also SIECUS board members) argued that "sexual deviations such as sadomasochism, fetishism, transvestism, voyeurism, and exhibitionism" among the male members of the species were merely the natural result of the male's stronger sex drive: "males in general are more inclined to want variety in sex than are females, both in regard to sexual techniques and in desire for change of sexual partners."[29] The SIECUS study guide on "Premarital Sexual Standards" maintained the impossibility of enforcing premarital abstinence as a social standard, especially for males: "during the past two thousand years, the Western world has never succeeded in bringing to adulthood as virgins the majority of even one generation of males."[30]

Unquestionably the most disturbing SIECUS study guide was "Sexual Encounters between Adults and Children" by SIECUS board member John Gagnon and William Simon. Gagnon and Simon consistently downplayed the negative consequences of child-adult sex, urged leniency for child molesters, and discouraged parents from reporting the sexual abuse of their children to police. The authors reported that "there is a wide range of sexual contacts between adults and children," but they insisted that "contacts that involve coercion or violence" are very rare.[31]

How could any sexual contacts between adults and children *not* be inherently coercive? Following Kinsey, the authors asserted that the real trauma from child-adult sex came from condemnation by uptight adults. "It may well be that the expression of outrage and anger by parents or other attending adults may place the child in greater jeopardy of psychic wounding than he or she faced during the commission of the offense itself."[32] Despite a tepid disclaimer that their comments did "not mean that offenses should not be reported to the police,"[33] Gagnon and Smith clearly encouraged parents to consider precisely that course of action. They argued that bringing molestation to the attention of police may damage the child, reported that "[m]ost parents do not report offenses to the police," and urged that "utilization of the police should be done with caution."[34] In short, the authors came as close as possible to discouraging cooperation with the police without actually advocating illegal behavior on the part of parents.

Gagnon and Smith repeatedly assured their readers that sexual abuse of children per se (apart from the negative reactions of adults) was no big deal. "The important fact that must be kept in mind is that for most children, in most situations, the victim situation is of short duration and of minimal effect."[35] Again, "the evidence suggests that the long-term consequences of victimization are quite mild . . . Indeed, there is evidence even among children who were the victims of father-daughter incest that the long-run outcomes were not . . . any worse than for a control population from the same social level." Even among "offenses that involve a great deal of violence . . . if the child does not assume from the reaction of his parents that he is the cause of the event, there is rarely a profound long-term effect."[36]

Not surprisingly, Gagnon and Smith urged leniency for those who committed sex offenses against children, asserting that a "large number of offenders who violate . . . sexual norms" do so merely "because of a maladaptive response to nonsexual stresses, such as work pressures that cannot be responded to in more direct ways."[37] The authors compared such offenders to "the abstainer or moderate drinker who suddenly 'goes on a binge.'" They added that "one of the mistakes made in dealing with these men . . . is the conventional punitive reaction, which insists on long sentences under repressive conditions."[38]

SIECUS study guides did not avoid the topic of sexual morality, but such discussions were presented within a framework of moral relativism. In "Sex, Science, and Values," SIECUS board member Harold Christensen acknowledged that some people took "morality" to mean "only those precepts of right and wrong that are intrinsic in the nature of things, and hence eternal or absolute." His study guide rejected that definition. Instead, it defined the term as "any system of propriety, regardless of the source . . . admit[ting] the possibility of relativistic as well as absolutistic formulations."

For followers of Judeo-Christian sexual ethics, wrote Christensen, "it is enough to say that 'God has spoken'; they do not feel obligated to test or prove anything, only to believe and obey . . . Thus the traditional Christian position is a dogmatic one . . . This is the old morality." By contrast, the new "relativistic" sexual morality was "a rational one, backed up by research."[39] This relativistic morality was the morality of the future, and it was the morality embraced by SIECUS.

Sex education must not be treated as "moral indoctrination," wrote Lester Kirkendall in another SIECUS study guide. "Attempting to indoctrinate young people with a set of rigid rules and ready-made formulas is doomed to failure in a period of transition and conflict such as ours."[40] Harold Christensen likewise proclaimed his acceptance of "the individual's right to make his own decisions . . . and so the attempt has been not to impose values, but to build perspective and encourage rational and responsible choosing."[41] Isadore Rubin offered advice about how to handle the values of "lower-class or minority cultures" in a value-neutral manner. Rubin wrote that "in both education and general thinking, deviations from white middle-class norms are viewed as pathological manifestations; defective variants or protests against middle-class norms; or examples of social disorganization." However, this judgmental attitude was fallacious because "the patterns of sexual behavior of lower-class or minority cultures are not necessarily examples of disorganization but interrelated aspects of an essentially adaptive way of life, rooted in long and different traditions." Rubin did not specify all of the value differences to which he was referring, but he provided a clue by explaining that a sociologist "has shown that many differences exist between white and black lower-class boys and girls in their attitudes toward premarital sex."[42] He quoted an expert from the federal Department of Health, Education and Welfare: "To imply that the cultural values that a student brings to school are inferior or inappropriate would seem to imply deficiencies in him and his parents . . . Values about sex which are different from those to which he has been exposed at home or in his neighborhood can be presented as another way of looking at the subject, not as the way, or a better way."[43]

Yet the relativism preached by SIECUS was ultimately contradictory and incoherent. Despite cultivating a posture of tolerance toward conflicting sexual values, SIECUS writers showed little concern about condemning values they did not share. For example, they had no problem dismissing traditional Judeo-Christian sexual standards as unrealistic and outdated. SIECUS writers were less intent on propagating relativism than on replacing the old moral absolutes based on religion and philosophy with a new set of absolutes based, allegedly, on science.

To be sure, SIECUS supporters were not always consistent in championing science as the new foundation for morality. In the SIECUS study guide "Premarital Sexual Standards," board member Ira Reiss disclaimed the idea that science could dictate moral choices. "The choice of a premarital sexual standard is a personal moral choice, and no facts or trends can 'prove' scientifically that one ought to choose a particular standard."[44] Harold Christensen similarly claimed to believe that "the ideal of science is that of value neutrality."[45] Nonetheless, he went on to use science to justify the superiority of moral relativism over other ethical theories. Because it is supposedly backed by scientific research, relativism "is the approach that seems to offer the most hope for consensus under modern conditions."[46] In short, relativism "is the most relevant" ethical position "to the conditions and the value orientations of this scientific age."[47] Christensen thought that "scientific research" could provide a grounding for relativism because it was "perhaps unique in offering us the possibility of reaching a common ground. Science is a method for dealing with data that permits the repeating of research and hence verification. It is objective rather than subjective, which makes its findings dependable—within the limits of its generalizations."[48]

Christensen's implication was that only values sanctioned by scientific research should be considered objective, and therefore those values should be used as the basis of social policy. Moral teachings based on anything else (such as religion) were inherently subjective and therefore should not be used as a basis for social legislation or mores. The near-deification of scientific expertise by SIECUS supporters explained why they frequently disparaged the idea that parents alone should direct the sex education of their children. Allowing parents to provide sex education "is both impractical and undesirable even if it were attainable," argued Lester Kirkendall. "It is impractical because parents are inadequately prepared to undertake the responsibility. No one or two persons can be adequately prepared in today's complex and varied world."[49] Hence, it was better for children to receive sex instruction from others, presumably those trained by SIECUS and its supporters.

Much of the advice supplied in the SIECUS study guides fell outside the American mainstream of the era. But the guides were tame compared to materials produced by other members of the sex-reform movement. In 1971, SIECUS cofounder and board member Lester Kirkendall coedited another collection of essays by some of the more radical proponents of sex reform. Titled *The New Sexual Revolution*, the book revealed just how extreme many of these people were and how clearly they were motivated by scientific materialism.[50]

In his preface to the book, humanist Paul Kurtz articulated with precision the central question posed by these reformers to American society: "Is man willing to take himself for what he is? Can he see that absolute moral codes of 'righteousness' that seem to go contrary to his most basic biological drives are profoundly immoral?"[51] Like Alfred Kinsey, these sex reformers viewed man as just another animal and believed that virtually every sex behavior exhibited by animals should be considered permissible. Traditional taboos against certain forms of sexual activity were the irrational products of superstition and outdated circumstances.

The book's chapter on "Cross Cultural Sexual Practices" by Herb Seal tried to show that anything and everything was acceptable so long as it gave pleasure. "If two people, be they two males, two females, male and female, parent and child, teacher and child, or whatever, are in a relationship of caring, any physical touch between them is a legitimate means of expressing tender recognition of mutual satisfaction in each other's presence." Seal claimed that extramarital sex was effectively allowed in many cultures and wondered "how long will it be before the surplus of American women grows enough for our laws to be revised to allow polygamous marriages?"[52] He claimed that sexual touching of children was part of the natural order of things, a claim that presented a problem when he tried to explain the origins of the incest taboo: "If the natural order of things is to allow stroking and touch, particularly of one's offspring, what occurred in the evolution of man to prohibit this phenomenon?"[53]

Seal didn't have a clear answer, although he did assert that many cultures throughout human history have practiced incest. While he doubted that American society would ever be able to get beyond the incest taboo, he certainly hoped so: "Perhaps a day will come when the behavior of a given person will be judged not according to arbitrary rules but on the nature and quality of the relationship that exists." He argued that incest could be beneficial to children. "That some children have been traumatized by brutal sexual assaults is a tragic truth, yet other children have looked back on early sexual experiences with high regard for the gentle and tender initiation to sexuality. We must avoid the judgments that speak of incest as 'universally forbidden, by even the most backward peoples,' when in fact it is not."[54] About the only practice relating to sexuality that Seal looked down upon was virginity; he claimed that there were cultures in which "a virgin was thought to be unfit to become a wife."[55]

In defining "immorality," another contributor to the book, Albert Ellis, argued that the concept "would better not be defined in terms of an individual's harming or acting unfairly toward another, but in terms of his needlessly or gratuitously injuring this other."[56] A prominent clinical psychologist (and board member of the American Association of Sex Educators, Counselors and Therapists),[57] Ellis was reluctant to condemn any type of sex behavior as always wrong, including abandoning a spouse and children for the thrill of having a sexual relationship with someone new. He believed it was moral for a man to seduce a woman with traditional sex scruples in order to "depropagandize her, induce her to surrender her sex-love hangups, and enable her to widen her potentialities for living."[58]

In Ellis's view, sex codes should be based on empirical evidence, and the empirical evidence showed that human beings "are quite varietist or non-monogamic in their sex desires," "are easily attracted to each other on very short notice, but their vital interest significantly wanes after prolonged sexual contact and shared domesticity," and "that large numbers of people can sexually desire and even be intensively, amatively attached to two or more members of the other sex simultaneously."[59] Thus, "certain ethical codes, such as once-in-a-lifetime devotion and sexual fidelity to a single member of the other sex, are, although advantageous in some respects, almost impossible for the average individual to achieve; and they would better be significantly altered or made preferential rather than mandatory."[60]

Given such biological imperatives, an appropriate query might be, "Is Monogamy Outdated?" Rustum and Della Roy tackled that question in another chapter.[61] The Roys purported to be supporters of monogamous marriage, but their views on how to save the institution were unorthodox. Their recommendations included "institutionalizing premarital sex,"[62] "expanding the erotic community" during marriage (in other words, allowing extramarital affairs),[63] "providing a relationship network for the single" (i.e., making room for mistresses in marriage),[64] and "legalizing bigamy." They went so far as to assert that "it is almost certainly true that contemporary-style monogamy would be greatly strengthened if bigamy (perhaps polygamy-polyandry) were legalized."[65]

Lester Kirkendall's contribution to *The New Sexual Revolution* was more provocative—liberated, he might say—than his essays in the official SIECUS volume designed for teachers and school administrators. He proclaimed that he "would like to drop the word 'morality,' certainly in the traditional sense," complaining that "[a]s we ordinarily speak of morals and moral behavior, particularly in the area of sex, our phraseology is judgmental, punitive, and endowed with the view that 'you have no one to blame but yourself' for your difficulties."[66] Among other things, Kirkendall's essay discussed "[o]ne of the common decisions which many boys . . . have to make." Not, mind you, whether to play baseball or football, but rather "whether to accept or reject homosexual approaches or what seem to them to be homosexual advances." According to Kirkendall, many boys are ill-prepared to decide whether to accept such propositions because "they may either experience a paralyzing fear or exhibit an irrational and violent anger. These reactions are typically much conditioned by the cultural attitudes which instill a frightened, recoiling, antagonistic attitude toward homosexuals." Kirkendall suggested that educators needed to rid boys of their "paralyzing fear" and "irrational . . . anger" toward homosexuality so that they could more rationally decide whether to accept homosexual advances.[67]

Remember that Kirkendall was writing here about the homosexual so-licitation of boys, not adult males. Since children cannot give legal consent to sexual relations, what Kirkendall was really describing was attempted rape or molestation. Yet in his view the problem was not that boys were being solicited for sex but that the boys reacted negatively (with fear or anger) toward attempts to rape or molest them. Such negative reactions were supposed to be unhealthy and irrational, and Kirkendall wanted to eliminate them so that boys would no longer reject homosexual advances out of hand. Kirkendall was not a fringe figure in the sex-reform move-ment. In addition to cofounding SIECUS, he served as vice president of the American Association of Sex Educators and remained a key figure in the sex-education movement until his death in 1991.[68] He also was a long-time professor at Oregon State University.

Other sex reformers proffered even more extreme views, such as trans-ferring the responsibility for child molestation to the child victims. In *Your Sexual Bill of Rights* (1973), Leonard Ramer argued that in child-adult sex encounters "many attractive children are the aggressors, using their charms to aid them in the role of the seducer. These children usually came from homes where they received little or no affection, or where there was an unnatural, fear-filled sexual atmosphere."[69] By the time young people reached their midteen years, they were the ones preying on older adults. "At that age, most of them know perfectly well what form of sexual plea-sure they desire, and they often take the initiative in homosexual advanc-es. As a matter of fact, unscrupulous youths often exploit their ability to tempt older men, sometimes for financial gain, fully realizing that if they get caught, the older man is blamed by the police and public."[70]

Seeking to establish the normalcy of child-adult sex, Ramer cited an-thropological studies purporting to document uninhibited sex in primi-tive societies, including adults teaching "masturbation to the children of the opposite sex because they believed it to be an important factor in sexual development."[71] The pioneering work of Alfred Kinsey made an obligatory appearance as well, Ramer lauding Kinsey for making "many of us aware of the impossibility of enforcing our sex laws when he discovered that if they were enforced as any law should be enforced, 95 percent of all men would be in prison, convicted of sex crimes, and only 5 percent would be left to act as jailers—if you could imagine such a ridiculous situation."[72]

Materials developed for young people reflected the new thinking about sex, including two books by former Kinsey coauthor Wardell Pomeroy, *Boys and Sex* (1968) and *Girls and Sex* (1969).[73] Pomeroy affirmed that "sex . . . cannot be denied until we are in our twenties or later."[74] Pomeroy's texts were largely recapitulations of Kinsey's tendentious findings, including claims that children are sexual beings from birth, that sex play among children is healthy, and that homosexual sex is normal mammalian behavior.[75] Although Pomeroy listed both the "pros" and "cons" of premarital sex, his discussion of the "cons" seemed designed to show that they were no longer relevant. Admitting that one "con" of premarital sex was the "danger of pregnancy," Pomeroy quickly assured readers that with the proper use of contraceptives "this possibility can be almost entirely eliminated."[76] In *Girls and Sex*, Pomeroy went so far as to suggest safe places where young couples could have sex without being interrupted.[77]

Like Kinsey, Pomeroy viewed human sexual behavior exclusively in terms of mammalian biology. "Man does, in fact, act very much like a mammal," he explained. "To speak in sexual terms, all species of mammals have in common the characteristic of self-play with the penis . . . all mammals perform intercourse by the male's thrusting his pelvis . . . [and] [a]ll species of mammals have homosexual relations," including cows, monkeys, and chimpanzees.[78]

Another influential sex-reform book was *Sex Before Marriage* (1969) by Eleanor Hamilton.[79] Although Hamilton claimed her book was written for "young people of college age," the book's subtitle promised "Guidance for Young Adults—Ages 16 to 20," clearly indicating that the intended audience included high school students.[80] Hamilton acknowledged that most high schoolers were not yet ready for intercourse, so she encouraged teenage couples to engage in mutual masturbation to the point of orgasm instead.[81] She heaped guilt on those who had scruples against masturbation, suggesting that they would be failures in marriage.[82] She tried to disabuse girls of the belief that they should stay virgins until marriage: "Today we know that most intelligent men consider virginity immaterial. It is about as useful in a prospective wife as an appendix."[83] She discussed how a boy could learn about sex by "try[ing] to find a sympathetic and experienced older woman with whom he can establish a tender and loving sexual relationship."[84] And she encouraged parents as well as college administrators to supply safe places for young people to go at it.[85] "Ninety-five percent of the dangers of premarital sex could be eliminated if parents would make their homes and their own knowledge available to their children as they would in all other areas of life," she wrote.[86]

In the short term, at least, these progressive sex-education books proved to be more the exception than the rule. By the end of the 1960s, early sex-education reformers had failed to remake the curricula in their own image. Public policy changed slowly, and few school districts were willing to adopt programs that fully implemented the ideas advocated by SIECUS or other sex-reform groups. It was difficult enough to get sex education adopted in many communities without broaching the sex reformers' unorthodox views on such topics as child-adult sex. To the dismay of reformers, many sex-education programs continued to affirm traditional Judeo-Christian moral teachings on such topics as premarital sex. Health educator Helen Manley from Missouri warned that a teacher must "avoid moralizing," but then listed the objectives of sex education as including the inculcation of "respect for social standards," "the reasons for proper behavior," "the need of controlling the sex urge by will power and self-discipline," the "dangers of promiscuity," and "continence during adolescence."[87] Manley's own sample answers for use by teachers provided the medical/scientific facts about sex within a context of natural theology and morality. For example, her description of the process of sexual intercourse for sixth graders used correct clinical language but also told students that the physical process of intercourse between a man and woman was designed by "our creator (God is used in some schools) . . . This is part of real love—this oneness (quote Bible if feasible) for two in real love . . . so when two people marry (stress this) they love each other and want to be as close as possible so mating is a fine part of marriage." Manley advised teachers that they should use their discussion of intercourse as "an opportunity . . . for [presenting] moral values—the wonder of life."[88] Other sex educators who produced materials for public schools in the 1960s were equally clear about the traditional moral point of view of their lessons, especially in their coverage of premarital sex. "We have tried to make clear the importance and desirability of chastity, always with the individual's future in mind," wrote one authority.[89] The issue of premarital sex "gives the teacher the opportunity to bring out not only the dangers involved, but also how rewarding the right way is," wrote another educator in the *American Biology Teacher*.[90] An article in the *Journal of School Health* rejected outright the claim that limits on premarital sex should be open to debate. "The wisdom of chastity and all that it implies need not be considered a debatable subject any more than monogamy or any other principle of behavior that represents 'the best experience of the human race.' "[91]

Yet the sex reformers did not give up, and their persistence wrought a transformation in sex education during the 1970s and '80s.

Achieving Sexual Maturity

"Adults and children dance together nude," reported the *Seattle Post-Intelligencer* matter-of-factly.[92] "[T]eenagers of both sexes are shown separately engaging in masturbation . . . live models turn for dramatic full-profile shots and then the camera zooms in to examine a variety of penises and vaginas." The explicit images came not from the latest X-rated film but from the award-winning sex-education documentary, *Achieving Sexual*

Maturity. Originally released 1973, the film gained nationwide notoriety in 1981 after it was apparently shown in a high school sex-education class in Washington state. A parent complained to the local chapter of the conservative group Moral Majority, which quickly demanded that the state library release a list of other schools that had ordered the film. No other school districts in the state were found to have used the film, but the controversy shed light on the new boundaries of propriety in sex education after the sexual revolution. Produced by a venerable New York academic publisher, *Achieving Sexual Maturity* was neither obscure nor aberrant. According to its distributors, "there were about 500 prints in circulation and . . . the Moral Majority complaint was the first they had heard of."[93] Presumably teachers somewhere were using those five hundred copies.

Sex education had come a long way from the 1960s, when filmstrips showing a cartoon of a man and woman in bed could be considered risqué. As often happens, the radicalism of one decade had become the mainstream of the next. By the end of the 1970s, a new generation of sex-education materials promoted nonjudgmental acceptance of a variety of sexual activities among children and teens. *Youth and Sex: Pleasure and Responsibility* (1973) by Gordon Jensen included photographs of nude couples and told readers that "several strong arguments can be made against virginity."[94] *The Sex Handbook: Information and Help for Minors* (1974) by Heidi Handman and Peter Brennan warned young people that "parents, schools, churches, and even the government try to make it as difficult as possible to enjoy your own body" and provided detailed information on how teens could lose their virginity and engage in oral and anal sex.[95] *Sex, with Love: A Guide for Young People* (1978) by Eleanor Hamilton recapitulated the ideas she expressed a decade earlier in *Sex Before Marriage*, except now there was no attempt to hide the fact that the book was targeted at teens, not college-age individuals.[96] This time around Hamilton gave teens instructions on how to ask parents for permission to engage in sex with their partners in their own bedrooms. *The Facts of Love* (1979) by Alex and Jane Comfort approved recreational sex among teens so long as each partner desired it,[97] assured young people of the harmlessness of pornography,[98] and promoted Kinsey's view of human beings as naturally bisexual ("Probably the ability to love anybody, of either sex, physically . . . is natural to humans.").[99]

Changing Bodies, Changing Lives (1980) contained sections on such topics as "exploring sex with yourself," "exploring sex with someone else," and "exploring sex with someone of your own sex."[100] Lauded by one sex-education curriculum specialist as "the most comprehensive, thorough, readable, realistic, and humane sex education book for older teens ever written,"[101] the book might be described more accurately as a manifesto of sexual license for kids. In addition to affirming adolescent and preadolescent sex play (one girl told about "spend[ing] hours . . . making out and fooling around" in a storage room with a boy in the fourth grade),[102] the book also perpetuated inaccurate factual information, most notably Kinsey's claim that 10 percent of the population is primarily homosexual.[103] *Changing Bodies, Changing Lives* probably went beyond many other texts in how hard it pushed the message of sexual self-determination by teens. Feeling guilty about engaging in sex acts not approved by your parents? Not to worry. The book quoted a counselor who explained: "Guilt may be a stage that [young] people have to go through as they stretch themselves past their parents' rules into their own new rules for themselves."[104] The book was replete with quotes from teens pleading for acceptance for sexual experimentation as a normal part of growing up. One teenage girl was quoted as saying that it "would be such a relief" if she could tell her parents about her lifestyle "and not get judged by them . . . If only they could understand that parties and staying out late and pot — and sex — are part of growing up for me."[105] "We've learned from the teens we've interviewed," concluded the book's authors, "that most want their parents to accept them as they are — people who are growing up and experimenting with life."[106]

The new philosophy of sexual liberation spread to children's fiction. Judy Blume's teen novel *Forever* (1975)[107] provided such detailed information about the first sexual encounters of a teenage couple that it was dubbed the "perfect sex education novel" by some progressive proponents of sex education.[108] In the new environment, earlier daring texts like Wardell Pomeroy's *Boys and Sex* and *Girls and Sex* became known as "serviceable middle-of-the-road classics."[109]

A good window into the mind-set of sex educators during the 1970s and early '80s can be found in *Discussing Sex in the Classroom: Readings for Teachers* (1982), published by the National Science Teachers Association.[110] The readings stigmatized traditional sex standards, promoted sexual relativism, and cultivated paranoia about parents critical of sex education. The themes came straight from the literature of the radical sex reformers of the 1960s and earlier. In "Adolescents, Sex, and Education," Adele D. Hofmann explained, in pseudo-anthropological terms, that "premarital chastity requirements" were the superstitious relic of "a patriarchal system":

> One cannot help but conclude that the asset of female virginity is strongly tied, first, to tribal economics wherein an 'unused' daughter brought the better bride price . . . and, second, to primitive magical thinking wherein there was inherent danger in menstrual and hymenal blood . . . Add in primitive man's animistic beliefs holding that evil spirits lurked everywhere, including body orifices, ready to harm if not properly propitiated or exorcised. It becomes relatively easy to contemplate how primitive patriarchal societies

saw the female as both "safe," as far as evil spirits were concerned, and of economic worth, only if she were kept premaritally virginal.[111]

Hofman branded as an "error" the effort "to find a single set of acceptable sex behaviors for all adolescents at all times." In her view, biological reality trumped the old morality. "Even if returning young people to the practice of non-marital sexual continence were a desirable goal, the analysis of trends in adolescent sexual behavior belies its feasibility."[112] What adolescents really needed was "an unbiased, open educational forum wherein they can freely explore their own concepts and come to their own conclusions within the context of that particular morality in which each was individually raised." Such "non-judgmental acceptance of an adolescent's behavior is essential. As already noted, teenagers are by nature experimenting . . . Blame should not be heaped upon those who transgress."[113]

In "Sex and Society: Teaching the Connection," Peggy Brick suggested breaking students into groups to create their own "Code of Sexual Behavior." Students were supposed to try to come to a consensus about the moral code, which seemed to suggest that morals were up for a vote.[114] Brick also believed in expanding sex education to encompass controversial political and social issues: "Speakers, debates, surveys, readings, films and journal entries all enrich our study of alternative lifestyles, the Equal Rights Amendment, abortion, prostitution and pornography."[115]

Parents who criticized sex education, meanwhile, were exposed as cranks and villains who would stop at nothing to get want they wanted. In "The New Opposition to Sex Education: A Powerful Threat to Democracy," Peter Scales deployed the old stereotypes of fundamentalist extremism to condemn continued criticism of sex education by parents. Ironically, Scales faulted sex-education opponents for exhibiting "paranoia over imagined conspiracies intent on radically changing society," not recognizing that he might be accused of the same failing.[116] Some of his statements seemed just as paranoid as those from the groups he was attacking. Citizen groups opposed to sex education "pose a threat, not only to the schools, but also to our basic democratic traditions," he warned ominously. Apparently, in his view, to use the political process to protest educational policies was subversive of democracy. Conceding that "not all the opponents to sex education are extremists and religious crusaders," he insisted that "it is the crusader . . . who is most likely to make his or her impact known." The "common unsavory tactics" of such crusaders "include quoting proponents out-of-context, spreading outright lies and fabrications, disrupting meetings and calling for demonstrations against 'controversial' speakers that sometimes succeed in getting them cancelled."[117] For Scales, the bullying and extremism were all on one side, and he accused critics of falsely smearing sex educators as pro-incest or favoring sexually explicit classroom materials.

Some of the specific attacks by sex-education critics cited by Scales may well have been inane, but Scales was ill-informed if he believed all such charges were baseless. Despite the impression he conveyed in his essay, fundamentalist crusaders did not hold a monopoly on extremism. In 1981, a distinguished group of sex-education scholars published *Sex Education in the Eighties: The Challenge of Healthy Sexual Evolution*.[118] Originally commissioned by the SIECUS board of directors as a festschrift for Mary Calderone,[119] the anthology provided an illuminating update of the agenda of the leaders of the sex-reform movement at the start of the 1980s. Their views had become more radical, particularly in the area of childhood sexuality.

In her own essay, Mary Calderone contended that "acceptance of the sexuality of children and infants" now should be made one of the movement's main goals. She complained that reaching "adolescents and pre-adolescents" with the message of positive sexuality was too late. Children from their earliest years had to be trained for sexual pleasure. Just as children are "taught how to use their bodies by . . . sports" and "use their minds in school and even earlier by attentive parents," they must also be taught "how to use their third human endowment, their sexuality." "[W]e must learn to speak with one voice on the sexuality of the infant and the child as normal, and the vital acceptance of this principle by parents as essential," she urged. "This must happen if 3-year-olds are to achieve ownership of their own bodies with all of their functions."[120]

Calderone was hazy about precisely what sex activities and training she was advocating for infants and children, but another contributor to the volume, Floyd Martinson of Gustavus Adolphus College in Minnesota, was happy to offer the details. In "The Sex Education of Young Children," Martinson nonchalantly discussed a range of "sexual learning experiences" that could help infants and children grow as sexual beings, including incest and sexual molestation.[121] Going far beyond the view of earlier sex reformers that incest might not be particularly harmful, Martinson supplied evidence that it could actually be a positive experience, quoting one scholar who affirmed incest by observing that "childhood is the best time to learn."[122] Martinson further reported research by another scholar who described families that practiced incestuous activities "in an 'educated, sophisticated, and carefully responsible manner.'" Parents of these families "felt that withholding expression of affection, even sexual expression, is damaging to the child." Martinson added that there is "a great deal of sex play among siblings" and "there is very little evidence that mutual sex play among siblings proves harmful." Indeed, "it can foster a robust, healthy, nonincestuous attitude later in life."[123]

Incest wasn't the only learning activity discussed as a positive experience by Martinson. He also spent considerable space legitimizing molestation as a natural part of sexual learning:

> In sex, as in most aspects of life, the older teach the younger. In the vast majority of cases such encounters are with someone close to the child: a family member, a relative, a neighbor, or a "baby-sitter." If the "teaching" is erotic, it is most often in the form of fondling or oral relations; attempted intercourse is relatively uncommon . . . If intercourse is attempted, it is often exploratory and becomes part of the initiator's learning experience rather than being purposeful, aggressive, or violent.[124]

Martinson lamented that "in the United States . . . '[i]ncest,' 'pedophilia,' 'exhibitionism,' and 'child sex abuse' have become pejorative terms," a state of affairs he condemned as an effort "to debase nonmarital sexual activity, especially that involving adults and children."[125] He went on to argue that "intimacy is a normal part of the maturational process of children, and pedophilia, if no violent aggression or physical harm accompanies the activity, need not create sexual trauma for the child. The child sometimes has pleasant memories of such encounters."[126] Once again, an appeal was made to anthropology. Martinson's vision of the future mirrored child sexuality in primitive cultures.[127]

Another contributing author to *Sex Education in the Eighties* revealed that SIECUS was trying to start "Sexuality Community Learning Centers" as part of its goal of "shifting the focus to younger children" and reaching them with the positive message of sexuality "before ignorance, fear, and guilt lead to poor judgment in sexual behaviors." "The initial target population for these centers will be the parents of children under 12 . . . The decade of the 1980s will need more such efforts at reaching the parents of younger children."[128] Given the views about childhood sexuality propounded by Calderone and Martinson, was this initiative supposed to be reassuring to parents?

The foray into child-adult sex was surely the most explosive part of *Sex Education in the Eighties*, but there were other issues that exposed the continued radicalism of the sex reformers. Lawyer and former Kinsey associate Harriet Pilpel coauthored an essay on "Sex Education and the Law" that took issue with statutes and regulations allowing parents to excuse their children from sex-education courses. Many of these exemptions had been enacted starting in the 1960s as a way of defusing parental opposition to sex-education programs: If the parents did not like the programs, their children did not have to take them. Pilpel argued that such exemptions might violate the constitutional right of privacy of students who wanted to take the courses.[129] In other words, if parents and students disagreed about whether to participate in a sex-education program, Pilpel thought that the state should trump parental rights.

Sex Education in the Eighties layed out an impressive agenda for the sex-education movement to go on the offensive in the 1980s and push society beyond remaining sex taboos. But political dynamics were already changing when the book came out, and the sex-education movement soon found itself on the defensive instead.

Turning Back the Sexual Revolution

By the early 1980s, the National Education Association counted more than three hundred groups in the United States opposed to sex education.[130] With the election of Ronald Reagan as president and Republican control of the United States Senate, sex-education critics found potent allies at the federal level. In 1981, conservative U.S. senators shepherded into law the Adolescent Family Life Act (AFLA), which provided federal funds for demonstration projects designed to discourage premarital sexual activity among teens.[131] The amount of money appropriated for AFLA was minuscule compared to the money spent by the federal government since 1971 on family-planning programs favored by the sex reformers.[132] Yet AFLA was a powerful symbol of a renewed effort to curb the excesses of the sexual revolution and reinstate traditional standards of sexual morality. Later in the Reagan administration, Secretary of Education William Bennett added his voice to those calling for an overhaul of permissive sex-education programs. "It is clear to me that some programs of sex education are not constructive," he wrote. "In fact, they may be just the opposite." Bennett faulted the supposedly nonjudgmental approach that typified many new sex-education programs, arguing that "a sex education course in which issues of right and wrong do not occupy center stage is an evasion and an irresponsibility."[133]

Mainstream news media were finally willing to investigate the radicalism of the sex-education lobby as well as their critics. In 1981, *Time* ran an article exposing the unconventional views about childhood sexuality championed by Mary Calderone, Wardell Pomeroy, Floyd Martinson, and other sex reformers. *Time* quoted other scholars who indicated that the views of Calderone et al. about child-adult sex were "full of crap."[134] The 1980s was also the decade of AIDS, which for many people seemed to dramatize the shocking consequences of sex conducted without any moral limits.

When Republicans won control of the U.S. House of Representatives in 1994, national policy shifted to the promotion of abstinence until marriage as the appropriate social goal. The landmark welfare-reform legislation enacted in 1996 established a federal entitlement (Title V, Sec. 510, 42 U.S.C. 701) to support abstinence-only programs in the states. Title V abstinence funds can only be used for programs that have as their "exclusive purpose, teaching the social, psychological, and health gains to be realized by abstaining from sexual activity." These programs must also teach "abstinence from sexual activity outside marriage as the expected standard for all school age children" and that "a mutually faithful monogamous relationship in context of marriage is the expected standard of human sexual activity."[135] Abstinence programs funded under Title V are "abstinence-only" programs and cannot provide information about the use of contraceptives. It should be noted that the Title V funding of abstinence programs was still dwarfed by the family-planning money provided under Title X by the federal government, which went to programs supported by SIECUS and Planned Parenthood. As of 2002, federal and state governments were still spending "$4.50 on promoting teen contraceptive use for every one dollar spent on teen abstinence."[136]

Despite the vast disparity in government funding of abstinence education and contraception, an independent evaluation of abstinence funds distributed under Title V found that they were "changing the local landscape of approaches to teenage pregnancy prevention and youth risk avoidance."[137] Whether or not that is true, it is clear that abstinence-only programs have been growing in popularity. "In 1988, only 2 percent of teachers responsible for sexuality education in public secondary schools reported teaching abstinence as the sole way to prevent pregnancy and STDs; by 1999 this figure had risen to 23 percent of secondary school sexuality education teachers."[138] Although most Americans now say premarital sexual activity in general is not morally wrong, a plurality of parents (47 percent) believe teens should be taught that "young people should not engage in sexual activity until they are married." Another 32 percent of parents favor teaching teens that "young people should not engage in sexual intercourse until they have, at least, finished high school and are in a relationship with someone they feel they would like to marry."[139]

Groups like SIECUS and Planned Parenthood have lobbied hard against additional money for abstinence-only programs. At the same time, SIECUS has continued to push for what it calls "comprehensive sex education," developing an influential set of guidelines for sex-education programs during the 1990s that it claimed became "the most widely recognized and implemented framework for comprehensive sexuality education across the country."[140] As in the past, SIECUS tried to describe its current approach as a mainstream method of teaching sex within a framework of values. Adapting to new political and social realities, SIECUS even tried to position itself as pro-abstinence by calling its favored approach "abstinence-plus." But a close inspection of its Guidelines for Comprehensive Sexuality does not support that description.

Sex Education for Five-Year-Olds

The SIECUS guidelines urge that, starting at age five, children be taught about such topics as vaginal intercourse, dating, masturbation, and homosexuality.[141] Starting at age nine, they should be taught about bisexuality,[142] homosexual fulfillment ("Homosexual love relationships can be as fulfilling as heterosexual relationships"),[143] the advantages of masturbation,[144] and the safety of abortions ("A legal abortion is very safe").[145] Starting at age twelve, children should be taught about the benefits of mutual masturbation ("Masturbation, either alone or with a partner, is one way a person can enjoy and express their sexuality without risking pregnancy or an STD/HIV"),[146] sexual fantasies (again, "alone or with a partner"),[147] the unchangeability of sexual orientation ("People do not choose their sexual orientation . . . Sexual orientation cannot be changed by therapy or medicine"),[148] and the normality of same-sex sexual experiences during childhood ("Many young people have brief sexual experiences [including fantasies and dreams] with the same gender").[149] By age fifteen, young people are supposed to be taught about the full range of sex techniques (including "massaging, sharing erotic literature or art, bathing/showering together, and oral, vaginal or anal intercourse"),[150] the uses of pornography to facilitate sex ("Some people use erotic photographs, movies, or literature to enhance their sexual fantasies when alone or with a partner"),[151] and more on sexual fantasies ("Many people's sexual fantasies include behaviors not actually acted upon or even desired in real life").[152]

As with past SIECUS pronouncements, these guidelines offer up a contradictory mixture of moral relativism and advocacy of liberal values. On the one hand, they call for tolerance ("In a pluralistic society, people should respect and accept the diversity of values and beliefs about sexuality that exist in a community").[153] On the other hand, they are dismissive of or actively oppose a variety of traditional moral standards. Starting in kindergarten, for example, children are supposed to be taught that "a family consists of two or more people who care for each other in many ways," a definition that is so broad as to be virtually meaningless.[154] On the issue of homosexuality, the guidelines glancingly acknowledge that "some religious groups" oppose homosexuality, but the overwhelming message is that homosexuality is positive, healthy, and normal.[155] The only empirical evidence presented about the issue is designed to reinforce this positive view of homosexuality. Regarding abortion, the guidelines profess neutrality by calling for students to learn that "some people believe abortion is morally wrong whereas others believe a woman has a right to choose abortion."[156] But, again, the empirical information presented is one-sided. Students are assured that "most women report no problems after having an abortion" and that "a legal abortion is very safe."[157] Assuming for the sake of argument that these statements are true, they tell only part of the story. A significant number of women do report consequences from abortions, and abortion, like any major medical procedure, can have serious side effects. Yet the guidelines do not require students to hear about the women who experience physical or emotional harm from abortion. Nor do the SIECUS guidelines require students to learn about how the fetus develops in the womb.[158] SIECUS wants students to learn that abortion is a "right . . . guaranteed by the Supreme Court," without requiring them to learn about those who criticize the Supreme Court for creating a right not actually protected by the Constitution.[159]

More skewed coverage comes in the unit on "Sexuality and Religion." While noting that "many religions teach that sexual intercourse should occur only in marriage," the SIECUS guidelines also require students to learn that "some people continue to respect their religious teachings and traditions but believe that some views are not personally relevant."[160] Strictly speaking, this statement is true. But without elaboration, it also borders on propaganda. Some people may think religious teachings are no longer personally relevant, but others continue to believe in their relevance. If students learn about those who reject the applicability of religious teachings, shouldn't they also have to learn about those who continue to accept their validity? If not, why not? This sort of presentation is neither objective nor fair. It is carefully slanted to manipulate the terms of debate. More bias occurs in the discussion of religion and homosexuality. The guidelines briefly acknowledge that "some religious groups oppose homosexuality" while highlighting the fact that "there are a growing number of congregations that welcome openly gay men and lesbians."[161] Both statements are misleading. It is not merely "some religious groups" that view homosexual behavior as morally wrong; it is *most* religious groups around the world. Likewise, although there may be a "growing number of congregations that welcome openly gay men and lesbians," the clear majority of congregations in most Christian denominations (including "mainline" Protestant groups) continue to disapprove of homosexual behavior.[162] This effort by SIECUS to downplay religious opposition to homosexuality is inconsistent as well as deceptive. When discussing sexual behaviors that SIECUS favors, its guidelines routinely stress that "many" or "most" people engage in or support the practices.[163] But SIECUS attempts to obscure the actual proportion of the religious community that continues to consider homosexual behavior illicit.

One notable area in which the SIECUS guidelines affirm traditional mores is child-adult sex. Apparently disowning the radical views of SIECUS leadership from the 1960s to the 1980s, the current guidelines maintain that "no adult should touch a child's sexual parts except for health reasons."[164] On the surface, the guidelines also appear to promote abstinence. Nine- to twelve-year-olds are supposed to be taught that "children are not ready for sexual intercourse." Twelve- to fifteen-year-olds are supposed to learn how "young teenagers are not mature enough for a sexual relationship that includes intercourse" and that "abstinence from sexual intercourse is the best method to prevent pregnancy and STDs/HIV."[165] But these statements are more ambiguous than they may first appear. SIECUS only encourages "abstinence from sexual intercourse," not all sexual activities. Another part of SIECUS's "abstinence" unit teaches twelve- to fifteen-year-olds that "there are many ways to give and receive sexual pleasure and not have intercourse."[166] The point is to frame "abstinence" in the narrowest terms possible so that teens can still consider themselves "abstinent" if they engage in sexual activities such as mutual masturbation and oral sex. It is doubtful that many parents would understand this sort of message as the promotion of "abstinence."

Whether many school districts actually follow SIECUS's complete formula for sex education is an open question. Many districts are likely to shy away from several of SIECUS's curricular guidelines, especially the ones advocating explicit sex instruction for children as young as age five. But such opposition is no longer the roadblock it used to be because of the Internet. By the end of the 1990s, SIECUS and other supporters of comprehensive sex education found they could go around parents and school districts altogether by putting sex education on-line. Because these on-line services do not have to be approved by elected school boards or parent groups, they offer pristine examples of the type of education sought by SIECUS and similar groups.

Sex Education on the Web

Although some of the sex-ed websites recommended by SIECUS pay lip-service to the idea that sex should be part of a relationship and not just a physical act, the nonsexual parts of their curricula are skimpy, to say the least, especially when compared to the exhaustive information presented about the techniques of sex play. Don't expect in-depth discussions of the central role of marriage, monogamy, or the true meaning of love and commitment in Western civilization. Some of the websites are more frank than others about their primary focus on sexual mechanics. Asked, "Why do you encourage sex without love?" the "Coalition for Positive Sexuality" website doesn't bother to deny the premise of the question. It merely responds: "Teaching about love is beyond the scope of what we do. We try to provide the information that teenagers need to make the sexual encounters that they do have free from serious consequences (pregnancy, STDs, AIDS, etc.)."[167]

Some sites do discuss the importance of "values," but the discussions usually are in the context of moral relativism, at least in the area of sexual relationships. As the Coalition for Positive Sexuality website advises, "Just Say Yes is about having a positive attitude towards sexuality—gay, straight, bisexual or whatever . . . You have the right to make your own choices, and to have people respect them."[168] Another site asks women, "Do you want losing your virginity to mean something symbolic to you, or don't you? Some women do, and some don't, and either approach is just fine if you're fine with it."[169] The same site declares that "no one can decide for you it if is okay for you to like pornography or not. Whether or not you like pornography and whether or not you decide to make pornography part of your sexual life is a decision that only you can really make."[170]

Moral relativism is perhaps most pronounced at Planned Parenthood's *Teenwire.com*, which pushes the message to the point of preachiness. "Only you can ultimately decide what is right for you and your body," says one article on the site.[171] "Everyone has different values about sexual experience. What's right for one person may not be right for another," says another.[172] "It's never OK to judge others," asserts a third article. "People's sexual activities and choices are their own business. Everyone's morality is different. What's right for one person may not be right for another."[173]

Planned Parenthood's site refuses to express outright disapproval of "hooking-up," the practice of engaging in sexual encounters for their own sake without the benefit of a long-term relationship. Although teens are told that "most people usually feel better about having sexual experiences when it's with someone they care about and trust," and even that "a healthy relationship requires more than just a desire to hook up," they are also assured that "hookups are common ways of exploring options" and are directed to a discussion-board section of the website where other teens discuss the question, "Do you think it's ok to have sex play with someone if you're not in a relationship with that person?"[174] More than 60 percent of posted responses indicated that it was fine.

These constant affirmations of sexual relativism are somewhat misleading, because there are in fact some things about which the on-line sex-ed sites are quite intolerant: getting or spreading a sexually transmitted disease, having an unwanted pregnancy, and, of course, judging others. There are even a few leftover sex taboos. One of the only sexual practices that does not elicit Planned Parenthood's "What's right for one person may not be right for another" mantra is incest. When queried by teens who wish to have sex with parents or siblings, Planned Parenthood urges them to "seek professional help" and provides a detailed list of the harms of incest.[175]

However, other than incest or "unsafe sex" it is difficult to find any sexual behavior that is unambiguously condemned on these sex-ed websites—even bestiality. Columbia University's *Go Ask Alice* suggests that someone attracted to bestiality might want to seek psychological help, but it also assures readers that they cannot get sexually transmitted diseases from having sex with animals, and it cites discredited statistics purporting to show that up to 8 percent of American males have had orgasm with an animal.[176] For a website that supposedly views bestiality as a psychological problem, it tries hard to present it as a normal sexual variation.

On the surface, abstinence is usually treated as a valid choice for at least some teens on these websites. But below the surface the message about abstinence is decidedly mixed. Planned Parenthood's site repeatedly affirms that it is okay to be a virgin and that one should never feel pressured to engage in sex, yet it also casts doubt on those who choose this option. An article titled "The Truth about Virginity Pledges" implies that they probably won't work and may even harm teens because they will not be as willing to learn about safe sex. It provides a checklist of questions to consider before making a virginity pledge, and portrays most of those who make the pledge as simply trying to please their parents. Teens considering virginity pledges are also told to "be real. If you think you may be in a situation where you end up having sex, by all means, protect yourself!" The clear implication is that teens who think that they can refrain from sex aren't being "real."[177] (In reality, most youth who take such pledges are significantly less likely to engage in intercourse.)[178]

There is also the slippery question of what type of "abstinence" is being offered as a legitimate option. Planned Parenthood officially defines abstinence as "not having sex play with a partner" on its teen website. But an article on the same site titled "Outercourse: Abstinence for Experts" tells teens that "there are two kinds of abstinence. In the first kind, partners have only very limited sex play—maybe you kiss, but there's no nakedness, no groping, no orgasms, nothing." By contrast, "the second kind [of abstinence] includes lots of sex play and is more open to possibilities, as long as partners don't have vaginal, oral, or anal intercourse. Planned Parenthood calles [sic] this type 'outercourse.' "

Although the article states that the first kind of abstinence "is the type encouraged by your parents, probably, and it's the right choice for most kids for a long time," it goes on to issue a virtual invitation to participate in the "second" type of "abstinence": "Hours of kissing sounds nice, huh? How about a little mutual masturbation that ends with orgasm? Get creative—talk to each other (before, during, and after you fool around)! Things can get pretty steamy even without having sex."[179] The article seems intended to convince teens that they can still consider themselves "abstinent" while engaging in all sorts of sexual encounters with others short of intercourse. Other articles on Planned Parenthood's teen site all but endorse "outercourse" as healthy and socially beneficial. A news brief states that teaching teens under sixteen about "other forms of sex play, like oral sex" could reduce the problem of teen pregnancies.[180] Another article in the "Ask the Experts" section lists the various kinds of "outercourse" ("body rubbing ['humping'], masturbation, deep kissing, erotic massage, oral sex play, role-playing, and sharing fantasies with a partner") and then assures readers that "outercourse offers nearly 100 percent protection against pregnancy" and "can also greatly reduce the risk of HIV and many other sexually transmitted infections."[181]

Despite unfettered access to American teens online, SIECUS and other groups are continuing their efforts to shape what schools teach about sexuality. One of their top priorities has been the elimination of abstinence-only funds under Title V.

The Battle over the Future of Abstinence Education

SIECUS and its allies criticize abstinence-only education as ineffective, unscientific, and focused more on scaring kids than helping them to build healthy relationships. They argue that their brand of "comprehensive sex education" (which includes detailed information about the use of contraceptives as well as abstinence) is far superior as well as favored by the overwhelming majority of the public.[182] According to a nationwide Zogby Poll in 2003, most parents do favor providing students with "basic biological and health information about contraception" along with abstinence (52 percent), but that does not mean that they favor the kind of comprehensive sex education advocated by SIECUS. Only about a quarter of parents (23 percent) favor the SIECUS approach of teaching abstinence but then supplementing it by providing teens with information on how to purchase and use condoms. A majority of parents also express doubts about teaching condom use as part of the same class in which abstinence is taught. A fifth of parents (21.7 percent) think condom use should not be taught at all, while another third (34.7 percent) favor keeping instruction in abstinence and contraception in separate classes.[183]

A similar Zogby Poll in 2007 found that 83 percent of parents believe it important for their children to wait until marriage to have sex, and that 78 percent of parents believe sex-education classes should place more emphasis on promoting abstinence than on contraceptive use.[184]

SIECUS's other attacks on abstinence-only curricula are also unpersuasive. SIECUS trumpeted a 2007 study of four abstinence programs that showed they had no impact on teen sexual activity, but three of the four programs did not continue into high school, when the pressure on teens to have sex is the most intense, and the remaining program stopped in the tenth grade.[185] At the same time, a growing number of peer-reviewed studies have demonstrated the effectiveness of abstinence programs.[186] SIECUS's demand for scientific proof that abstinence programs work is more than a little ironic, given its own history. During the late 1960s, SIECUS claimed social benefits for sex education beyond what the research demonstrated.[187] Even today, SIECUS advocates "comprehensive sex education," despite strong evidence that it does not produce the results claimed for it.[188]

As for the charge that abstinence-only programs are primarily fear-based and do not deal with healthy relationships, a 2004 Heritage Foundation study of sex-education curricula found exactly the opposite. What the authors described as "authentic abstinence curricula" (i.e., "abstinence-only curricula") devoted little time to trying to scare teens about the threats of sexually transmitted diseases and instead explained sex as part of healthy human relationships. The authors of the study concluded that "authentic abstinence programs take a more holistic approach to human sexuality: They are far more concerned with the social and psychological aspects of sex. Authentic abstinence curricula place a major emphasis on love, intimacy, and commitment. Young people are taught that human sexuality is not primarily physical, but moral, emotional, and psychological in nature." By contrast, "comprehensive sex-ed curricula largely depict human sexuality as a physical process. They focus on warning young people about the threats of teen pregnancy and sexually transmitted diseases that can result from unprotected sexual activity."[189]

The comprehensive sex-education curricula examined for the study apparently presented no substantive information on "healthy relationships and benefits of marriage,"[190] although it is possible that some information classified as dealing with "general behavioral skills" by the researchers could be reclassified as promoting healthy relationships. The authors of the study rejected the claim that comprehensive sex-education programs could be accurately described as "abstinence-plus," noting that only 4.75 percent of the content of these programs is devoted to abstinence-related material, compared to 53.7 percent of the content of "authentic abstinence curricula." According to the study's authors, comprehensive sex-education curricula devote nearly six times more content to promoting the use of contraceptives as they do to encouraging abstinence.[191] These curricula focus mostly on physiological pleasure and avoiding the negative physical consequences of sex. Human sexual relationships are essentially reduced to biology.

Despite criticism by SIECUS and other groups, Congress continued to authorize spending for abstinence-only programs into 2007, although Democrats vowed to stop the funding. Failing for the moment to eliminate abstinence-only programs directly, proponents of comprehensive sex education turned to urging states and the federal government to require sex-education curricula to be "medically accurate."[192] Although this proposal sounds unobjectionable (who could be against presenting information that is medically accurate?), it holds the potential to misuse science to censor information and viewpoints for ideological reasons.

SIECUS has objected to the abstinence-only criteria enacted by Congress in Title V by claiming that "much of [it] . . . has no basis in public health research."[193] But if one looks at SIECUS's specific objections to the Title V abstinence criteria it becomes clear that the real issue has little to do with accurate science. It has everything to do with conflicting views about morality.

SIECUS criticizes requiring abstinence programs to teach "that a mutually faithful monogamous relationship in the context of marriage is the expected standard of human sexual activity." But this statement does not make a scientific claim; it expresses the moral position of Congress that marriage is the best context for human sexuality. Because the statement expresses a moral position, not a scientific one, it is illegitimate to imply that it presents scientifically inaccurate information or that "science" somehow refutes it.

SIECUS complains that "the concept of chastity until marriage is unrealistic in an age when young people are reaching puberty earlier than ever before" and when research shows that most young people "have engaged in sexual intercourse" by the time they have graduated from college.[194] But the mere fact that most people may engage in sexual intercourse before marriage does not prove whether that behavior is morally right or wrong any more than current rates of assault and theft show whether societal standards on those behaviors are right or wrong. Whether theft is morally appropriate is not dependent on how many people currently steal. Similarly, whether the standard of sex within marriage is right or wrong is not dependent on scientific research into how many people currently fulfill the standard. SIECUS is perfectly entitled to argue that sex outside of marriage is morally acceptable. It is not entitled to present this view as value-neutral "science."

SIECUS's support for presenting medically accurate information is highly selective. In fact, SIECUS argues for the suppression of accurate health data that might undercut support for moral relativism vis-à-vis sexual behavior. Title V abstinence programs must also teach "that bearing children out-of-wedlock is likely to have harmful consequences for the child, the child's parents, and society." There happens to be a considerable amount of empirical research behind this statement,[195] but SIECUS doesn't want it presented. Why? Because SIECUS objects to presenting "one family structure as morally correct and beneficial to society." Such information might "cause negative feelings" among students whose parents are unmarried and is therefore inappropriate.[196] But this is another moral objection, not a scientific one. SIECUS is urging that schools censor accurate information about the consequences of bearing children out of wedlock because that information might cause hurt feelings. Regardless of whether SIECUS is correct in this moral judgment, no one should think it is somehow justified by science. As for the moral judgment itself, one wonders how consistently SIECUS is prepared to follow its own logic. Health-education classes invariably teach students about the harmful consequences of smoking. Presumably this message might "cause negative feelings" among students whose parents smoke. Is that sufficient reason not to tell students that smoking may kill them? Just how far is SIECUS willing to press this argument? Pretty far, it turns out.

SIECUS argues that hurt feelings provide a further reason not to teach about sex within marriage as the preferred social standard. It claims that the idea that sex is only appropriate in marriage "may prove particularly harmful to young people who are or have been sexually abused," because "it requires telling these students that the behaviors in which they have involuntarily participated go against society's 'expected standard.' Such statements are likely to produce additional feelings of guilt and shame." That's right: according to SIECUS, schools should not teach about reserving sex for marriage because it might make the victims of incest and molestation think that the sexual acts they were coerced into were somehow illegitimate.[197] What does SIECUS believe schools should do instead? Assure victims of sexual abuse that they shouldn't feel bad because the acts perpetrated against them were simply the expression of a different set of values? SIECUS's complaint is eerily reminiscent of claims by an earlier generation of SIECUS officials that the most traumatic part of molestation was that society disapproved of it.

The insistence by SIECUS and its allies that sex-education programs must only present scientifically accurate information is particularly ironic, not only because "comprehensive" sex-education programs are liable to contain numerous scientific and medical inaccuracies, but also because SIECUS and its associates have a long history spreading claims grounded in junk science.[198] Kinsey's research was cited repeatedly as authoritative by early SIECUS board members, and SIECUS in the 1980s provided a springboard for researchers like Floyd Martinson to offer pseudoscientific defenses of incest and child molestation. This track record of bogus claims made in the name of science does not exactly inspire confidence.

Recent trends in the sexual behavior of teenagers undercut the scientific credibility of the sex reformers even more. For years they routinely claimed that trying to curb sexual intercourse among teens was a useless endeavor because biology made widespread premarital sex a foregone conclusion. But by 2004, it was clear that the sex reformers had underestimated the capacity of human beings to make choices about their sexual behaviors.

It turns out that teens are perfectly capable of choosing to refrain from intercourse. According to the federal government's latest Youth Risk Behavior Survey, the proportion of high school students who have had sexual intercourse dropped from 54.1 percent in 1991 to 46.8 percent in 2005.[199] The causes of this drop in teen sexual activity are hotly debated, and whether the trend will continue is anyone's guess. But the most salient point is that the drop occurred. According to the biological reductionism of the sex reformers, it should have been impossible.

Sex is linked to fundamental questions about who we are as men and women and the meaning of our relationships with each other. Perhaps the only more fundamental questions we face deal with the very nature of life and death. When does human life begin, and when does it end? Scientific materialism has exerted a profound influence here as well, which is the subject of the final section of the book.

Life and Death

14

Redefining Life

The year was 1981. Ronald Reagan had just won a landslide victory against incumbent president Jimmy Carter in the previous autumn's election, and Republicans had taken control of the United States Senate for the first time in a quarter of a century. Social conservatives were widely credited with playing a key role in the Republican victory. As a result, they hoped that issues like abortion would soon win a place on the congressional agenda.

They did not have to wait long. A few months after Reagan took office, a subcommittee of the U.S. Senate Judiciary Committee held hearings on a proposed "Human Life Bill," which would have declared a congressional finding "that the life of each human being begins at conception." The subcommittee, chaired by Senator John East from North Carolina, heard from an array of scientists, lawyers, ethicists, and political activists on both sides of the abortion controversy for eight days in April, May, and June.

On the morning of Wednesday, May 20, the subcommittee took testimony from Dr. James Neel. Chairman of the Department of Genetics at the University of Michigan Medical School and a member of the National Academy of Sciences, Neel represented the elite of the American scientific and medical establishment. He had just been selected as president-elect of the Sixth International Congress of Human Genetics.

Launching into his prepared statement, Dr. Neel said that he found "it impossible to address . . . the issue of when, following conception, actual human life begins without some reference to the concepts of evolution."[1] He then offered senators a lesson in evolutionary biology.

"The early embryo appears to pass through some of the stages in the evolutionary history of our species," explained Neel. "The scientific dictum is: 'Ontogeny recapitulates phylogeny,' which translates into: during embryological development we repeat in abbreviated form many aspects of our evolutionary past." For example, "at about 30 days after conception, the developing embryo has a series of parallel ridges and grooves in its neck which are interpreted as corresponding to the gill slits and gill arches of fish." In addition, "It has a caudal appendage which is quite simply labeled 'tail' in many textbooks of human embryology. Traces of this phase of our development can even be found in the adult." Because of "these facts," Neel said that he found "it most difficult to state, as a scientist, just when in early fetal development human personhood beings, just as I would find it impossible to say exactly when in evolution we passed over the threshold that divides us from the other living creatures."

Although Neel inserted a few qualifiers in his presentation (e.g., "appears"), the implication of his testimony was clear. He was arguing that the value of human embryos could be discounted because for much of their development they were equivalent to earlier stages in man's evolutionary history. Neel's claim that "ontogeny recapitulates phylogeny" was popularized by Ernst Haeckel, the German Darwinist who created the spurious drawings of animal embryos. Haeckel argued that human embryos during their development supplied powerful evidence for the theory of evolution by replaying the evolutionary history of the species.[2] To be specific, human embryos were supposed to pass through past adult forms of man's evolutionary ancestors. At a certain point during pregnancy, for example, the human embryo supposedly went through a "fish" stage and even developed "gill-slits."

This is ridiculous. Human embryos merely exhibit folds in the neck area, not "gill-slits." While in fish embryos "pharyngeal folds" do eventually "develop into gills," explains embryologist Jonathan Wells, "in a reptile, mammal, or bird they develop into other structures entirely."[3] The resemblance of these pharyngeal folds to gills in fish has been termed "illusory" by British embryologist Lewis Wolpert.[4] By the 1960s and '70s, Haeckel's theory of the recapitulation of adult forms from an animal's evolutionary history had long been repudiated by embryologists. It was, in fact, junk science.

But that didn't stop some evolutionary biologists from trying to recast the idea of recapitulation so they could still claim it as evidence for Darwinian evolution.[5] In his book *Ontogeny and Phylogeny* (1977), Stephen Jay Gould acknowledged that a number of scientists had "tried to extend the concept of recapitulation" so that "if phylogeny is construed as a sequence of complete ontogenies, then any stage held in common by ancestors and descendants is a 'recapitulation.'"[6] In the view of such scientists, "since embryos do repeat the embryonic stages of their ancestors, why not call this recapitulation as well"?[7] Recent writers have focused more narrowly on the claim that vertebrate embryos are the most similar at their earliest stages of development, which is supposed to reflect a "conserved stage" from their evolutionary history. It turns out that this more limited claim is also untrue.[8]

Given the resilience of recapitulation as an idea, perhaps it should not be surprising that it was invoked by some abortion-rights advocates as "scientific" evidence that aborting a human embryo or fetus was no more problematic than destroying a fish.[9] In the nineteenth century, Haeckel had used recapitulation to justify abortion, so this sort of argument had a long pedigree.[10] Nevertheless, the number of reputable lawyers, doctors, and scientists who continued to cite the recapitulation argument for abortion long after it had been exposed as pseudoscience remains unsettling.

In a 1965 letter published by the *New York Times*, legal scholar Cyril C. Means Jr. argued that "embryological investigation beginning with Haeckel" justified the conclusion that a fetus does not become a human being until well into a woman's pregnancy. Means declared: "the Divine command imposes on man the duty . . . to destroy a fetus, still at the stage known to zoologists and embryologists as that of subhuman ancestral reminiscence, which, if allowed to pass beyond that stage, will predictably become neither an image nor a likeness of God, but only a grotesque caricature of man." In other words, there was nothing wrong in eliminating a fetus so long as it was still at a stage "of subhuman ancestral reminiscence," i.e., a prehuman stage of the evolutionary process.[11] Means was described as a "former legal adviser of the American Church Union" and was later a professor of law at New York Law School.[12] He was appointed by New York Governor Nelson Rockefeller to a commission to review New York's abortion law in 1968.

In a 1969 medical-journal essay, University of Washington medical professor Robert H. Williams similarly defended abortion by arguing that "the fetus has not been shown to be nearer to the human being than is the unborn ape." Williams went on to complain that too much was "made of 'quickening' of the fetus by many individuals as a time when 'life begins' . . . In reality, quickening symbolizes a very early stage of neuromuscular activity and of the recapitulation of phylogeny by ontogeny; and it takes man a relatively long time to attain the recapitulation."[13] Williams was the president of the Association of American Physicians when he made these arguments.

In 1980, another letter writer to the *New York Times* defended abortion and attacked the humanity of the fetus by appealing to recapitulation. According to Victor Eppstein, "If the ontogeny of the individual recapitulates the phylogeny of the race . . . the human fetus at various stages may be closer to a protozoan, a worm, a tadpole, a monkey, than to *homo sapiens.*"[14] This time, a retired professor from Seton Hall University responded with a withering rebuttal. "However superficially similar the embryos of various species may be," wrote Msgr. John M. Oesterreicher, "the human fetus does at no time pass through the stage of an amoeba, worm, fish or ape. Hence the *New Columbia Encyclopedia* calls the biogenetic law 'incorrect,' while the German embryologist E. Blechschmidt names it a 'catastrophic error in the history of natural science.' "[15]

Despite such occasional rebuttals, appeals to embryonic recapitulation by supporters of legalized abortion continued. During the 1981 Senate hearings on the "Human Life Bill," Dr. Neel was not the only one to use Haeckel's argument to defend the morality of ending fetal life. In a letter to the chair of the subcommittee, Dr. Milan Vuitch, medical director of the Laurel Clinic in Washington, D.C., castigated Congress for relying on what he regarded as outdated science. Vuitch had a long history in the abortion movement, having been prosecuted for performing abortions in Washington in 1969, pre–*Roe v. Wade*. His case made it all the way to the Supreme Court, which two years before *Roe v. Wade* upheld the District of Columbia's abortion statute against a claim that it was unconstitutionally vague.[16] According to Vuitch in 1981, the claim that human life begins at conception was based on junk science from "one or two centuries ago" when scientists still believed that the embryo was "a small human being" that simply needed time to grow. Thanks to Ernst Haeckel and his "law" of recapitulation, however, Vuitch said that modern scientists now know that "in the development of all Mammals [sic] each ontogeny must go through its phylogeny. To put it simply, this means that the development of a single organism must go through the evolutionary pattern of development of its phylum i.e. its 'basic division of animal kingdom.' " Hence, in its early stages the human embryo looks "very much like any developing zygote of any Primate [sic]." Only later does it "assume more and more human features." Vuitch claimed that Haeckel's "law" of recapitulation "is as valid and true now as it was at the beginning of this century and will be true in the future of Biology and Medicine."[17]

Dr. James Wilson, professor emeritus of research pediatrics at the Children's Hospital Medical Center in Cincinnati, Ohio, espoused a similar view in a letter to the subcommittee's ranking member, Senator Max Baucus (D-MT). "The embryo and fetus of man," wrote Wilson, "do not show any physiological activities not also displayed by similar developmental stages of other animals."[18] Although Wilson, unlike Vuitch, did not directly mention Haeckel, he clearly drew on the authority of the idea of recapitulation: "As existence *in utero* progresses, the fetus of man shows physical or anatomical differences from other species; but fetuses of all species begin to show their particular physical characteristics; and the magnitude of these differences is no greater, in fact may be less, between man and monkey than between monkey and mouse, for example."[19]

Writing in the *Medical Tribune* in 1985, Ohio surgeon George Crile argued that the determination of when human life begins should be answered "through the eyes of Darwin and evolution." Crile offered a particularly pristine appeal to recapitulation as a justification for abortion, claiming that "the product of fertilization goes through all the transitions of form that our remote ancestors have traversed in the course of evolution. Ontogeny (the development of the individual) repeats phylogeny (the development of the species), the saying goes." Crile added that the human embryo starts out "as formless as one of the ocean's floating flora. Then, since in the beginning all of us came from the sea, these develop the gill slits and the anatomy that characterize fish. Next comes the transition. The embryo begins to grow lungs and the organs necessary to breathe air and to live on land." According to Crile, these scientific facts undermined the belief that early human embryos are human beings: "It is hard to argue that before that time a human embryo is a human being. Until then it is only a recapitulation of the fish part of our evolution." Crile thought that the abortion controversy could be partly resolved if we were willing to look "at the embryo in the light of evolution." His solution: "Why not permit abortion in the first trimester, when the embryo is still a salt water creature . . . ?"[20]

In a 1989 article in *Perspectives in Biology and Medicine*, biochemist (and past board member of the National Center for Science Education) Rivers Singleton conceded that Haeckel's theory was no longer "strictly observed in biology." However, this didn't prevent him from suggesting that fetuses were throwbacks to man's evolutionary ancestors. "Early human fetuses, with their primitive gill slits and tails, more often than not resemble some primeval sea creature than a cuddly human baby," he wrote.[21]

In 1990, famed astronomer Carl Sagan and his wife Ann Druyan published a defense of abortion in *Parade* magazine that relied heavily on the idea of recapitulation without actually using the term. The article's account of development in the womb implied that the embryo retraces its evolutionary ancestry before becoming human; therefore there is nothing morally problematic about eliminating the embryo. "By the third week," the couple wrote, "the forming embryo . . . looks a little like a segmented worm." "By the end of the fourth week . . . something like the gill arches of a fish or an amphibian have become conspicuous, and there is a pronounced tail. It looks something like a newt or a tadpole." "By the sixth week . . . the reptilian face has connected slits where the mouth and nose eventually will be." "By the end of the seventh week . . . the face is mammalian but somewhat piglike." "By the end of the eighth week, the face resembles a primate's but is still not quite human." Only by the tenth week, according to Sagan and Druyan, does the embryo's face have "an unmistakably human cast."[22]

Junk science can be hard to correct, especially when it reinforces fashionable sociopolitical views. University of Arizona anatomy professor Clayton Ward Kischer found this out firsthand when he took it upon himself to try to correct the continued misuse of the recapitulation argument in the abortion debate. After the Sagan and Druyan article appeared, Kischer called the managing editor of *Parade* and complained about the bogus science presented in the article. Despite the fact that he was an embryologist at a state university, Kischer says his request for a correction was flatly rejected.[23] Kischer similarly tried to get the journal *Perspectives in Biology and Medicine* to accept an article of his that, among other things, countered the modified recapitulation argument offered by Rivers Singleton in the same journal. Kischer says that when he finally reached the editor of the journal on the phone, he was told his manuscript had not been sent out for review, and the editor acknowledged his own strong support for abortion rights.[24]

Underlying these repeated invocations of recapitulation was unabashed scientific materialism. According to many defenders of abortion, there was nothing sacred about a human embryo or fetus until it had actually developed the physical capacities that distinguished it from the lower animals. Until those capacities were realized, a human person by definition did not exist. The value of the embryo or fetus at any given stage of development was to be determined only by the physical capacities it had already realized. This view categorically rejected the idea that embryos and fetuses were intrinsically human because in their natural course of development they would become full-scale human beings, which was their preordained purpose. What something would become was beside the point. As social scientist Ernest van den Haag bluntly put it, "Things are what they are, not what they become."[25] Traditional debates about when a developing embryo acquired a "soul" were likewise regarded as irrelevant. The human embryo or fetus did not have the physical capacities of a full-grown human being; therefore, this entity was not to be regarded as human life, at least for most of its time in the womb.

The recapitulation argument was only one example of how the reductionistic interpretation of biological development surfaced in the abortion debate. Proponents of legalized abortion did not need to cite Haeckel to equate human embryos and fetuses with the lowest forms of animal life. Cornell University medical professor Jessica Davis maintained that the early human embryo was simply an "undifferentiated amorphous cell mass [that] must undergo millions of cell divisions before it assumes at best a primitive humanoid form."[26] University of California biological-science professor Clifford Grobstein similarly argued that "in early stages the embryo . . . is simply a collection of cells," and Rivers Singleton wrote that "the blastocyst and early embryonic forms are nothing more than poorly differentiated aggregations of cells."[27] Roy Lucas, who pioneered the argument that abortion is a constitutionally protected right, contended that a young fetus was "just a mass of protoplasm" that wasn't anything like a human being: "Physically and developmentally the fetus closely approximates . . . a collection of undeveloped growing cells infinitely more than it does a live infant. Certainly at the time most abortions could take place no one would confuse a fetus with anything physically human."[28] Lucas added that abortion laws interfered with a woman's right to believe "that life began in the 'prehistoric slime' and is not created but only passed along by conception."[29] A number of abortion-rights proponents believed that human fetuses were no more special than any other animal. "One need merely compare a human foetus with an ape foetus," wrote Michael Tooley. "What mental states does the former enjoy that the latter does not? Surely it is reasonable to hold that there are no significant differences in their respective mental lives."[30]

Not content to equate human fetuses with other forms of animal life, some defenders of abortion asserted that human fetuses were markedly *inferior* to lower animals. According to bioethicist Peter Singer, "On any fair comparison of morally relevant characteristics . . . the calf, the pig, and the much derided chicken come out well ahead of the fetus at any stage of pregnancy—while if we make the comparison with a fetus of less than three months, a fish, or even a prawn would show more signs of consciousness."[31] San Francisco State University philosophy professor Mary Anne Warren claimed that "even a fully developed" fetus "is considerably less personlike than is the average mature mammal, indeed the average fish." Hence, "if the right to life of a fetus is to be based upon its resemblance to a person, then it cannot be said to have any more right to life than, let us say, a newborn guppy."[32]

The more extreme abortion-rights supporters went beyond simply dismissing the humanity of the developing child and argued that embryos and fetuses should be regarded as parasites. Activist Lana Clarke Phelan argued in 1968 that human embryos were "endoparasitic growths, which if unchecked, threaten [women's] lives, their sanity, their existing families, their incomes and social futures."[33] Carl Sagan and Anne Druyen seemed to agree, writing that "the fertilized egg . . . destroys tissue in its path. It sucks blood from capillaries. It establishes itself as a kind of parasite on the walls of the uterus."[34] Joseph Pratt, emeritus professor of surgery at the Mayo Medical School, claimed that "a fetus is part of the woman, a parasite if you will."[35] Abortion, to these partisans, became no more morally problematic than fighting off a bacterial infection—or an assault by a stranger. After all, declared *Roe v. Wade* attorney Sarah Weddington, "the law presently allows no person . . . the right to use the body of another in a parasitic way, as does the fetus."[36]

The undercurrent in most of these comparisons was what one scientist has praised as "the trend toward the demystification of life that has been going on over the past century, in which life ceases to be a gift of God but, instead, becomes something we control."[37] A number of abortion-rights supporters repudiated the Judeo-Christian ethic of the sanctity of human life. That ethic holds that all human life originates with God, and no human being has the right to "play God" and usurp God's prerogatives. Each and every human life is regarded as having an immortal soul, making it of inestimable value. The development of human life in the womb was therefore to be regarded with sacred awe.

Nonsense, said some abortion-rights supporters. Speaking before the California Conference on Abortion in 1968, Lana Phelan urged the rejection of any sort of supernatural understanding of pregnancy and birth. "Men and women . . . must be taught there is nothing particularly spectacular in the act of conception, or even in giving birth. Any cow can do it."[38] Phelan said people should abandon the "timeworn" cliche that " 'life has begun and I cannot play god [sic],' " observing that a "doctor sees nothing unethical in excising a cancer . . . or using antibiotics to frustrate the will of god [sic] regarding life and death."[39]

The naturalness of the abortion procedure itself was sometimes justified by invoking Darwin's theory of evolution. A legal brief submitted to the Supreme Court in *Roe v. Wade* drew an analogy between "spontaneous abortions" (miscarriages) in nature and surgically induced abortions in society. The brief quoted a scholar who believed that "spontaneous abortion can be regarded as an important biologic mechanism which has evolved in viviparous animals to deal with the numerous embryologic errors arising during development." The implication was that since abortion was merely nature's way of weeding out the biologically unfit, there surely could be nothing wrong with having women employ the same method. "Both induced and spontaneous abortions amount to a rejection of pregnancy," explained the brief.[40] The primary difference between the two kinds of abortion was simply that one was dictated by nature, while the other was determined by human choice. The staying power of this evolutionary argument for abortion was demonstrated recently by Christopher Hitchens's *God Is Not Great* (2007), which lauded evolution's "creative destruction" in weeding out defectives in the womb and invoked (yet again!) embryonic recapitulation. (Hitchens assured his readers that "in utero . . . we begin as tiny forms that are amphibian, gradually developing lungs and brains . . . and growing and shedding that now useless coat of fur.")[41]

Of course, many scientists supportive of abortion rights publicly distanced themselves from scientific materialism, preferring to make a more limited argument: Science cannot determine when human life begins because this is a philosophical and religious question, not a scientific question. The "science cannot decide" argument was the one most frequently put forward by scientists and doctors when Congress took up the Human Life Bill in the 1980s and when the Supreme Court subsequently reviewed a Missouri statute which declared that "the life of each human being begins at conception."[42] "It is impossible scientifically to establish when personhood begins," said professor of medicine and genetics Arno G. Motulsky of the University of Washington.[43] "As a physician who has had some experience as an active scientist I consider it scientific nonsense to legislate that human life begins at the moment of conception," agreed Robert Ebert, former dean of Harvard Medical School.[44] The answer to the question of "when does a developing human embryo possess a soul . . . must spring from the philosophical tenets of the day, not from the laboratory," added Dr. Frederick C. Robbins, president of the Institute of Medicine of the National Academy of Sciences.[45] The membership of the National Academy of Sciences weighed in with a resolution declaring that the question of when human life begins was "a question to which science can provide no answer . . . Defining the time at which the developing embryo becomes a person must remain a matter of moral or religious value."[46]

On the surface, this line of argument was considerably more humble than efforts to enlist science to relegate human embryos to the status of fishes or pigs. After all, the new argument appeared to disavow the claim that everything could (or should) be explained in terms of scientific materialism. Indeed, it seemed to concede that nonscientific fields such as philosophy, ethics, and religion were the appropriate venues to determine the answer to the question of when life begins. On closer inspection, however, the humility of the argument was illusory. The concession that science could not determine when human life begins did appear to put limits on the reach of scientific authority. But the conclusion that followed undermined the concession: Because science cannot determine when life begins, it was argued that there are no other legitimate grounds for government to determine the answer to the question. What the scientists who invoked this argument most definitely did not say was that since science is silent, legislators were feel free to consult philosophy, ethics, or religion to come to an answer. In fact, they said quite the opposite.

"It is against the public interest to legislate in this realm since it will be impossible to arrive at scientific agreement," advised one medical professor, who added that "the interpretation of facts in this area depends upon philosophical, ethical, and religious attitudes" and "it is most dangerous . . . to legislate such attitudes."[47] Another professor insisted that since the question of when human life begins was "essentially a religious issue," it was wrong to impose an answer on others. "As a citizen I find it abhorrent to contemplate the force of law being given to one set of religious beliefs."[48] The assumption behind these comments was that science provides the only proper basis for public policy in this area, and since science was silent, the government must be mute as well. Far from acknowledging the limits of scientific materialism, the "science cannot decide" argument was merely a more sophisticated way to assert its primacy. The argument might be described justly as "arrogance cloaked in humility," to invoke a phrase coined by a Supreme Court justice in a different context.[49]

Yet the incoherence of the argument went even deeper than that. Its fundamental flaw can be seen in the majority opinion of *Roe v. Wade*, which similarly claimed that the court "need not resolve the difficult question of when life begins. When those trained in the respective disciplines of medicine, philosophy, and theology are unable to arrive at any consensus, the judiciary, at this point in the development of man's knowledge, is not in a position to speculate as to the answer."[50] Yet for all practical purposes the Supreme Court did decide when life begins. It had to decide this question by the very nature of the legal issue with which it was confronted.

Texas justified its statute criminalizing abortion by affirming that human life beings at conception.[51] If Texas's claim was correct—if human life really did begin at conception—then the state certainly had a compelling government interest that overrode any general right to abortion. Hence, the Court had to adjudicate the claim that human life begins at conception in order to dispose of the case. The result? In *Roe v. Wade*, the court flatly rejected Texas's claim about the beginning of human life, categorizing the unborn child as merely "potential life." The court decided that "the State's important and legitimate interest in potential life" only became " 'compelling' " at the point of "viability," which it defined as the time when "the fetus . . . presumably has the capability of meaningful life outside the mother's womb."[52] Only when the fetus was developed enough to survive outside the womb could the state prohibit his elimination. Postviability restrictions on abortion were acceptable because they have "both logical and biological justifications."[53] Biological justifications? The court seemed oblivious to the fact that it had just used science to determine that human life for all practical purposes does not begin until viability.[54] Scientists and doctors who claimed that science could not determine when life begins were caught in the same contradiction. They said that the beginning of life was not a scientific question, but they used science to deny the humanity of the unborn child. The ramifications of this denial of humanity soon spread far beyond abortion.

Fetuses, Guinea Pigs, and Defective Infants

Whilhamine Dick was finding it increasingly difficult to do her job. A nurse at the Magee-Women's Hospital in Pittsburgh, Dick was expected to participate in surgical abortions. What troubled her most was what happened to fetuses after the procedure. They were harvested for medical research. "It was repulsive to watch live fetuses being packed in ice while still moving and trying to breathe, then being rushed to a laboratory," she testified to the Pennsylvania Abortion Commission in 1972.[55] Dick asked to be excused from participating in abortions, but her request was denied, and she eventually quit.

The same year Dick testified, American medical researchers took part in a widely reported study of fetal-brain metabolism in Helsinki, Finland. Twelve fetuses ranging in age from twelve to twenty weeks were removed from their mothers via "abdominal hysterotomy" (essentially a C-section). After the fetuses' hearts stopped beating (but with their brain tissue still alive), their heads were cut off and attached to a pump that circulated a chemical mixture "through the internal carotid arteries."[56]

Other American researchers went to Finland to perform similar experiments. After injecting "a radioactive chemical into the fragile umbilical cords of fetuses freshly removed from their mothers' wombs," Dr. Jerald Gaull from New York cut out each fetus's "brain, lung, liver and kidneys" while its heart was still beating. The operation was performed without anesthesia, but according to the *Washington Post*, Dr. Gaull said that his first incision was to cut "the nervous connections that link the brain to the body 'to make sure the fetus will feel no pain.' "[57]

By the early 1970s, American researchers were regularly going abroad to gain access to intact fetuses from midpregnancy that had been aborted via hysterotomies. In Europe, such fetuses had become prime material for medical experiments. British researchers were even able to keep "months-old fetuses . . . alive for up to three or four days" outside the womb for various research purposes, although when this practice became public there was an outcry to curtail the experiments.[58] In Hungary, university researchers cut out the beating hearts of fetuses up to fifteen weeks for experimentation. The results of that study were reported in the *American Journal of Obstetrics and Gynecology* in 1974.[59]

But American researchers did not always have to go to Europe to experiment on fetuses outside the womb. In the 1960s, Robert Goodlin of Stanford University published research on submerging living fetuses in a saline solution designed to mimic the action of the placenta.[60] As part of the research, Goodlin sliced open the chests of the fetuses in order to directly observe the beating heart. "When the heart was beating, the fetus was returned to the chamber and the experiment was resumed."[61] Goodlin's research was funded in part by a grant from the National Institutes of Health.[62]

In May 1972, a group called the Student Pro-Life Federation organized picketing in front of the Stanford Medical Center to protest "abominable acts" by Dr. Goodwin during his research on fetuses. Acknowledging that the development of an artificial womb would be useful, protest organizer Mark Swendsen objected that the end did not justify the means. Dr. Goodlin's "experiments have involved cruel acts, such as slicing open the rib-cages of still-living aborted fetuses in order to observe their hearts," he charged. "We hold that the abortions which killed these children, as well as Dr. Goodlin's experiments on them in their dying moments, constitute violations of human rights. No human being should be made into an involuntary guinea-pig, no matter how much a doctor wants him as a subject."[63] Goodlin agreed to meet with leaders of the student protest, but neither side was satisfied. Goodlin allegedly told the news media, "I think they're nuts!" The student group issued a press release noting that "the opinion was apparently somewhat mutual."[64]

By the end of 1973, California banned medical experimentation on living aborted fetuses altogether after the state legislature heard testimony that California universities were experimenting on such fetuses up to six months in age. The legislation passed both houses of the state legislature "without a single dissenting vote."[65]

In addition to experiments outside the womb, fetuses inside the womb were used to test drug safety. Pregnant women intending to have abortions received live-virus vaccines for rubella in order to determine if the vaccine virus would cross the placenta and infect the fetus.[66] Clients of an abortion clinic in Arizona received the blood-pressure drug naldol to study whether it would harm their fetuses. According to a Yale University researcher, this kind of research was required "to develop a base of information on what are dangerous drug levels at different periods during pregnancy."[67] Of course, there were also experiments to improve the effectiveness of various methods of abortion.[68]

The controversy over fetal experimentation in America came to the forefront shortly after the Supreme Court handed down its decision in *Roe v. Wade*. Reports of the experiments in Finland and elsewhere sparked vigorous public criticism, and Congress swiftly adopted a temporary ban in 1974 on all fetal experiments, both inside and outside the womb, funded by the Department of Health, Education, and Welfare (now the Department of Health and Human Services).[69]

Many doctors and medical researchers were appalled—not by the experiments, but by the thwarting of what they saw as necessary scientific research. Some medical researchers only endorsed experimentation on pre-abortion fetuses still in the womb and postabortion fetuses once they were clearly dead. But others defended the experiments on postabortion fetuses whose hearts were still beating or whose brain tissue was still living.

Dr. Peter Adam, associate professor of pediatrics at Case Western Reserve University in Ohio, helped lead the Finnish study involving the decapitation of fetuses. Adam dismissed squeamishness about his research as "ritualistic absolution" and suggested that his critics ought to get over it: "People need to understand that the fetus doesn't have the neurologic development for consciousness or pain and that it also doesn't have the pulmonary system to survive."[70] In Adam's view, "Once society has declared the fetus dead and abrogated its rights, I don't see an ethical problem." Dr. Kurt Hirschhorn of Mount Sinai Hospital and Medical School in New York agreed. "I don't think it's an ethical problem. It is not possible to make this fetus into a child, therefore we can consider it as nothing more than a piece of tissue. It is the same principle as taking a beating heart from someone and making use of it in another person."[71] Hirschhorn endorsed keeping postabortion fetuses "artificially alive for a short period of time in order to investigate the effects of drugs, or environmental agents on fetal metabolic systems in the entire organism."[72] He added that scientists could multiply the number of fetuses available for this kind of research by using the right method of abortion to produce a "viable" fetus, "viable in the sense it can survive hours or a day." Joseph Fletcher disparaged opposition to experiments with the decapitated heads of fetuses as "entirely visceral" and "not rational."[73] In Fletcher's view, it was "justifiable . . . to make any use of live fetuses ex utero, previable or viable, if survival is not purposed or wanted, and if there is maternal consent."[74]

Dr. Gaull made the argument perhaps heard most frequently: "Rather than it being immoral to do what we are trying to do it is immoral . . . to throw these fetuses in the incinerator as is usually done, rather than to get some useful information."[75] Since the fetuses were doomed to die anyway, why not allow them to benefit the rest of mankind?

Willard Gaylin, president of the Institute of Society, Ethics and the Life Sciences, was willing to give up a right to experiment on living fetuses outside the womb. Drawing what he called "an arbitrary line between in utero and ex utero research," he acknowledged that "a whole set of new considerations and new moral dilemmas are created when we extend the life of a fetus outside of the womb for purposes of experimentation." He nevertheless defended fetal experimentation inside the womb on grounds similar to those espoused by more militant researchers. Gaylin wrote that "in the case of abortion the fetus cannot be 'helped' by being experimented upon since it is doomed to death anyhow, but perhaps its death can be ennobled because it served those more fortunate." In his view, fetal experimentation held the prospect of "endow[ing] . . . the process of abortion with human values it will not otherwise have."[76]

Dr. Andrew Hellegers, a professor of obstetrics at Georgetown University, responded that this view eerily echoed the rationale of German doctors who conducted experiments on prisoners in Nazi concentration camps. "It was the German approach, 'If it is going to die, you might as well use it.'"[77] Princeton bioethicist Paul Ramsey added that if it was truly "ennobling" to experiment on persons already condemned to die, why had society not experimented on death-row inmates? "After all, the condemned prisoner, like the living previable fetus, was going to die anyhow . . . Such experimentation could, you know, have been seen as 'ennobling' the death of the human subject by using him to make great contributions to mankind."[78] Society has not sanctioned such experiments, said Ramsey, because "we all know that would be a morally degrading use of the condemned for us, as a society, to make."

Closely related to the debate on fetal experimentation was one over harvesting fetal organs. If fetuses could be used for research, why not for organ donation? Could a wife become pregnant solely in order to grow an organ for transplantation into another family member? Bioethicist Mary Anne Warren approved such harvesting of fetal organs as perfectly legitimate: "While a fetus of five or six months may, perhaps, possess some flickering of sensation or some capacity to feel pain, this is equally true and probably even more true of creatures like fish or insects, which few would doubt the propriety of killing in order to save human lives."[79]

If late-term fetuses are not yet human and may be harvested, the use of tissue from early human embryos was even less objectionable. In the words of bioethicist Michael Lockwood, "I should have thought that, from any sane point of view, it was far preferable to experiment on a near-microscopic blob of unfeeling protoplasm than a feeling, caring being, albeit of a different species."[80] Actress Mary Tyler Moore urged Congress to permit the harvesting of early human embryos for stem-cell research:

"The embryos that are being discussed, according to science, bears [sic] as much resemblance to a human being as a goldfish."[81]

The 1974 congressional moratorium on federally funded fetal experimentation was lifted the following year and replaced with loose restrictions proposed by the National Commission for Protection of Human Subjects of Biomedical and Behavioral Research that allowed most experiments—including some on living fetuses outside the womb.[82] The commission had recommended permitting "ex utero" experiments on fetuses less than twenty-weeks old as long as "no intrusion into the fetus is made which alters the duration of life."[83] Keeping such fetuses alive several days for experimentation would be prohibited under the recommendation, as would harvesting their organs, as long as the fetuses lived. But decapitating the head of a pre-twenty-week-old fetus whose heartbeat had just ended (à la the Finnish study) and then attaching the head to a circulation system would be allowed. Presumably even invasive procedures like slicing open the chests of pre-twenty-week-old fetuses to observe their heart action (à la Goodlin's study) were permitted as long as the procedures did not shorten the lifespans of the fetuses. The commission also proposed establishing a double standard for research on fetuses still inside the womb. Research on fetuses to be brought to term was supposed to involve "minimal or no risk to the well-being of the fetus," while research on pre-twenty-week-old fetuses destined for abortion did not need to meet that standard.[84] The commission further recommended "that research on abortion techniques continue as permitted by law and government regulation."[85] Eventual federal regulations on fetal research funded by the Department of Health and Human Services (HHS) closely followed the commission's original recommendations.[86]

But the matter did not end there. During the Reagan administration, Congress enacted stricter statutory restrictions on live fetal research in 1985. Research on a live fetus outside the womb was now allowed only if it was designed to help the fetus or if it posed "no added risk of suffering, injury, or death to the fetus," regardless of fetal age. Moreover, in conducting fetal research inside the womb, the "risk standard for fetuses intended to be aborted and fetuses intended to be carried to term [had] to be the same."[87] It took more than fifteen years for HHS to finally issue revised regulations that were consistent with the new statutory language.[88]

Experimentation on fetuses outside the womb blurred the distinction between abortion and infanticide. Yet a number of leading abortion-rights supporters insisted that the distinction was essentially meaningless anyway. Said Peter Singer: "If the fetus does not have the same claim to life as a person, it appears that the newborn baby does not either, and the life of a newborn baby is of less value than the life of a pig, a dog, or a chimpanzee."[89] Joseph Fletcher agreed that the justifications for abortion and infanticide were closely intertwined, even observing that "it is reasonable, indeed, to describe infanticide as postnatal abortion."[90]

Although few abortion-rights supporters were as blunt as Singer or Fletcher in pressing their views to their logical conclusion, many openly admitted that the same rationale used to deny humanity to the fetus applied equally to the newborn. "Even the full-term infant must undergo many changes before attaining full status of humanity," acknowledged medical professor Robert Williams in 1969. "Only near the end of the first year of age does a child demonstrate . . . attributes that differentiate him significantly from other species."[91] Emeritus pediatrics professor James Wilson made the same point in a letter to Congress in 1981. "I believe that the condition or state of human personhood begins in fact when an individual begins to exhibit uniquely human traits, in distinction from the behavioral and mental traits which characterize other animals." These "traits which are distinctively human usually begin to appear a few weeks or months after birth." Wilson drew out the implications of his view for the mentally handicapped. He noted that "some unfortunate individuals with profound mental deficits . . . never achieve . . . [the] traits which characterize human beings. Strictly speaking such individuals do not attain human status." Only our sense of "compassion . . . has caused society to accord them legal status as human beings."[92]

Wilson was willing to continue treating mentally handicapped individuals as human beings even though they were not fully human in his view. Others, however, soon took the next step and advocated infanticide for mentally handicapped infants.

Nobel laureates Francis Crick and James Watson proposed that infants not be declared officially alive until three days after birth in order to allow the elimination of defective babies. As Watson explained in a 1973 interview: "If a child were not declared alive until three days after birth, then all parents could be allowed the choice that only a few are given under the present system. The doctor could allow the child to die if the parents so chose and save a lot of misery and suffering."[93] Michael Tooley made a similar suggestion: "everyday observation makes it perfectly clear, I believe, that a newborn baby does not possess the concept of a continuing self, any more than a newborn kitten . . . If so infanticide during a time interval shortly after birth must be morally acceptable."[94]

Peter Singer conceded that we are probably programmed by evolution to feel protective of young babies. ("It is true that infants appeal to us because they are small and helpless, and there are no doubt very good evolutionary reasons why we should instinctively feel protective towards them.")[95] But such emotions are insufficient to justify viewing newborns as human persons. Equally insufficient are beliefs about the sacredness of human life arising from religion. Singer grounded his case for infanticide in scientific materialism and its rejection of the Judeo-Christian world-view. "We can no longer base our ethics on the idea that human beings are a special form of creation, made in the image of God, singled out from all other animals, and alone possessing an immortal soul," Singer wrote in the journal *Pediatrics* in 1983. "Our better understanding of our own nature has bridged the gulf that was once thought to lie between ourselves and other species." He added that "once the religious mumbo-jumbo surrounding the term 'human' has been stripped away . . . we will not regard as sacrosanct the life of each and every member of our species, no matter how limited its capacity for intelligent or even conscious life may be." In his view, "if we compare a severely defective human infant with a nonhuman animal, a dog or a pig, for example, we will often find the nonhuman to have superior capacities, both actual and potential, for rationality, self-consciousness, communication, and anything else that can plausibly be considered morally significant."[96] While nonchalant about killing defective human infants, Singer is vigorously defends the rights of animals, although he objects to using the term "animals" to distinguish other living creatures from human beings.

"You shouldn't say animals to distinguish between humans and non-humans. We are all animals," he emphasized to an interviewer in 2004. In the same interview, Singer made clear that Darwinism lay at the foundation of his view: "All we are doing is catching up with Darwin. He showed in the nineteenth century that we are simply animals. Humans had imagined we were a separate part of Creation, that there was some magical line between Us and Them. Darwin's theory undermined the foundations of that entire Western way of thinking about the place of our species in the universe."[97]

If killing human infants is permissible, what about harvesting their organs for transplants? Anencephalic babies are born with their brain stems intact but with their higher brains missing.[98] If not stillborn, they typically survive hours or days, although on occasion they can live longer. By the 1980s and '90s there was widespread interest in using these babies as organ donors. The problem was that most major organs in anencephalic babies were unusable by the time the baby met the traditional criteria for brain death, which required cessation of all brain activity, including brain-stem activity.[99] During the hours or days that it took an anencephalic baby to die a natural death, his bodily organs seriously degraded. In order to harvest usable organs, the organs would have to be removed from anencephalic babies while they were still alive, which was clearly illegal.

In the late 1980s, Dr. Joyce Peabody of Loma Linda University tried to find a way around the problem without running afoul of the law and ethics. Loma Linda researchers came up with a way to prolong the lives of anencephalic babies long enough so that when brain death finally occurred the major organs would still be in usable condition. The process involved keeping the babies on ventilators until their own ability to breathe had irreversibly ended. Periodically the ventilators would be turned off to see if the babies could start breathing again on their own. If they did not start breathing within a certain number of minutes, the babies would be declared brain dead and their organs removed. The ethics of keeping babies alive solely to function as organ donors attracted criticism, and Loma Linda dropped the effort after initial experiments supplied lackluster results.[100] Most babies treated in this manner still produced unusable organs by the time of brain death.

But other proponents of harvesting organs from anencephalic babies proposed a cynical solution: redefine anencephalic babies as dead. In California, a state senator actually proposed such a bill in 1986. It stated that "an individual born with the condition of anencephaly is dead."[101] In 1992, the parents of an anencephalic baby in Florida named Theresa went to court to have her declared legally dead so her organs could be removed. Florida courts refused, however, and by the time baby Theresa died her organs were unusable. Medical personnel were incredulous. Leslie Olson, director of organ procurement for the University of Miami, complained to journalists that baby Theresa "better fit the category of benign tumor, rather than human being. She was a ball of tissue. The question is whether she existed at all."[102]

Dr. Peabody of Loma Linda offered a different view. "If you were to declare anencephalic infants dead for purposes of organ donation, it would mean that you would be removing hearts from babies that breathe, suck, kick and cry," she said in an interview a couple of months later. "I would need to have the individuals who passed that law feel that if it were not for organ donation they would be equally comfortable in burying a baby who was breathing, sucking, kicking and crying." Asked whether she could feel that way, Peabody replied: "Absolutely not."[103]

Ignoring the qualms expressed by many doctors and ethicists, the American Medical Association Council on Ethical and Judiciary Affairs in 1994 endorsed organ harvesting from living anencephalic infants, although it eventually reversed course after encountering sharp criticism.[104] Arguments about the humanity of fetuses and infants had typically centered on the question of when life begins, but the debate over anencephalic babies brought to the forefront an equally vexing question: When does life end? As we shall see in the next chapter, that question has just as far-reaching implications for adults as it does for babies.

15

Redefining Death

The plight of Terri Schiavo attracted the attention of America during the fall of 2003. Severely disabled after collapsing in her Florida home in 1990, Terri was unable to care for herself or verbalize her wishes, and she received food and hydration through a tube.[1] At the request of her husband Michael, a judge ordered the removal of Terri's feeding tube, and she began the slow process of death by dehydration. Michael claimed that Terri had previously expressed the wish not to be kept alive artificially, although other family members disagreed.

Terri's parents, Bob and Mary Schindler, went to court to stop the removal of the feeding tube, but their efforts were rebuffed. While Terri hovered near death, the Florida legislature convened in an emergency session and enacted a law designed to compel the courts to restart Terri's food and hydration.[2] After the law was enacted and signed by Governor Jeb Bush, Terri's tube was reinserted, but legal and political wrangling over her fate continued. In 2005 the tube was removed again, and Terri died shortly thereafter.

Terri's husband and the doctors he hired insisted that Terri had been in a "persistent vegetative state," or PVS. A "vegetative state" is typically defined as "a clinical condition of complete unawareness of the self and the environment, accompanied by sleep-wake cycles," and a "persistent vegetative state" applies if the vegetative state has lasted for at least a month.[3] Much of the news media uncritically accepted the claim that Terri was in a persistent vegetative state and reported it as fact, even though more than a dozen other doctors marshalled by Terri's parents (including six neurologists) claimed that Terri was not in a PVS.[4]

At one point, Bob and Mary Schindler held a press conference with some of their doctors to complain that journalists were spiking the views of their medical experts.[5] To be in a persistent vegetative state under Florida law, a person must show "the absence of voluntary action or cognitive behavior of any kind."[6] The Schindlers insisted that Terri was able to track them with her eyes, respond to stories and music, and even try to give simple verbal responses. A speech therapist also claimed that Terri had tried to verbalize simple words.[7] Videos filmed during visits with Terri appeared to show limited responsiveness on her part, which would contradict the claim she was in a PVS, but Michael Schiavo's experts dismissed the possibility.[8]

"It looks like she's looking at you, but really she's not," insisted Minnesota neurologist Ronald Cranford, since deceased. "It looks like she's grinning at you, but she's really not. You can believe what you want and see what you want, but it's just not there."[9] After Terri's death, it was widely—but inaccurately—reported that her autopsy "proved" that she had been in a PVS and that she had been completely blind as well.[10]

The battle over Terri Schiavo is one of a flood of cases over the past two decades dealing with the removal of food and hydration from disabled patients. Many of these cases have focused on patients like Terri who were supposed to be in a persistent vegetative state. But just as in the Schiavo court battles, there often was conflicting medical testimony in these other cases. In litigation over Nancy Ellen Jobes of New Jersey, for instance, two nationally recognized neurologists testified that Jobes was in a persistent vegetative state, while two other nationally recognized neurologists disagreed.[11]

The judge in the case dismissed the testimony of those who believed that Jobes demonstrated cognitive functioning because they morally opposed removing food and hydration from patients, which the judge thought tainted the credibility of their testimony. After a bedside visit to view Jobes for himself, the judge found "that all of the movements of Nancy Ellen Jobes described by the various witnesses are startle responses and reflex reactions to external stimuli." While he did not doubt the integrity of the medical experts who disagreed with his assessment, he concluded that "their sincere opposition to the withholding of nutrition from any patient . . . has caused them to see signs of intelligence where no such intelligence exists."[12]

Perhaps the best-known case involving the withdrawal of food and fluids from a PVS patient prior to Terri Schiavo involved Nancy Cruzan of Missouri, which ended up before the U.S. Supreme Court.[13] Nancy sustained severe head injuries in a car accident in 1983. After three weeks in a coma, she improved sufficiently so that she could chew and swallow pureed and soft foods. Medical records indicated that she drank juice and ate mashed potatoes and bananas, poached eggs, and even link sausage, but a gastrotomy tube was inserted to aid her feeding.[14] Subsequent efforts to rehabilitate her failed, and in 1987 Nancy's parents sought to stop the food and hydration provided through the tube, arguing that their daughter was in a "persistent vegetative state," manifesting no awareness of herself or her environment.

Yet whether Nancy was really in a PVS was disputed. The trial court heard testimony from nurses who testified that Nancy tracked with her eyes, smiled after being told stories, and cried after family visits.[15] James Dexter, head of the department of neurology at the University of Missouri–Columbia School of Medicine, testified that "Nancy is not in a persistent vegetative state at the present time, on the basis of fact that she is sensitive to environment and does in fact respond, though limitedly . . . to her environment."[16] Dr. Dexter originally believed Nancy to be in a PVS until he examined her himself and heard testimony from her nurses.[17] During his examination of Nancy, he said he found she had the ability to focus her eyes on him, a behavior that required the higher brain.[18] He also found evidence that she reacted to pain.[19]

By contrast, Minnesota neurologist Ronald Cranford was absolutely certain that Nancy was in a PVS and could not experience pain of any kind. He insisted that "in order to make the diagnosis" of PVS "there can be no interaction with the environment, no voluntary interaction with the environment, no consciousness, no thinking, no feeling. If there is any small amount of thinking or feeling or awareness or consciousness, then the patient is not in a persistent vegetative state. This is a complete state of lack of consciousness."[20] Yet as he testified it became apparent that this unqualified claim was misleading.

Cranford later indicated that it is typical in a PVS case to have numerous notations in the medical records from nurses about apparent patient interactions with the environment.[21] In Nancy's case, for example, a nurse had noted that Nancy smiled after being shown a Valentine card. Cranford stated that during his forty-five-minute examination of Nancy "she looked right at [her father] Joe and it was for three, four or five seconds."[22] He then acknowledged that "if you see that by itself . . . you have got to say that the person interacts with the environment."[23] But he discounted the response because he did not think it could be shown that such fixations were consistent enough. At the same time, he conceded that it was normal for a person in a PVS to respond to sound. "A patient in a persistent vegetative state can attend [to] a sound. In other words, you walk into a room, you call out her name, the person can look towards that sound." But he said such a response is "perfectly compatible with the diagnosis of vegetative state because that is clearly a brain stem mediated response."[24] The state's neurologist, Dr. Dexter, disagreed, stating that "that kind of alert response . . . happens to involve the temporal lobe and that's cortex."[25]

Dr. Cranford admitted that Nancy did respond to pain, including grimacing when he pinched her skin "very hard."[26] "Nancy grimaced at that point and she looked like she appeared to have pain on her face and as I continued the painful stimulation, her head drew off the bed . . . I did this several times, because I knew she couldn't experience pain."[27] Dr. Cranford repeatedly asserted as an unquestioned fact that Nancy could not experience pain.

But how did he know this? He admitted that "just looking at the grimacing, you can't say in and of itself that she doesn't experience pain and suffering."[28] Then how did he know? "You have to know that most patients in a persistent vegetative state will manifest this grimacing response."[29] But that response simply begged the original question. The fact that most PVS patients display the grimace response in no way proves that these patients cannot experience pain. Indeed, the fact that PVS patients typically react negatively to painful stimuli may be evidence for the exact opposite conclusion.

A second reason Dr. Cranford supplied for assuming that Nancy could not experience pain was her CT scan, which showed significant brain damage. Cranford's logic seemed to go something like this: By definition, a PVS patient cannot have a functioning neocortex. If the neocortex does not function, then "by definition the patient cannot experience pain or suffering."[30] Since the CT scan confirms that Nancy is in a PVS, then by definition she cannot experience pain. QED. The only problem with this argument was that the CT scan did *not* establish that Nancy was in a PVS or that her neocortex was completely nonfunctional.

Dr. Dexter testified that CT scans do not tell the doctor "anything at all about . . . the physiology of" the patient's brain. "I can show you children who had arrested hydrocephalus who are walking around and bowling . . . and they have a brain that looks just like that [Nancy's CT scan] and they are in school."[31] Even Dr. Cranford conceded that if he looked at Nancy's CT scan "by itself without knowing the clinical condition" he "could not diagnose the vegetative state by looking at this."[32] In other words, the CT scan did not provide independent evidence that Nancy was in a PVS or that her neocortex was completely nonfunctional. As a consequence, the CT scan did not provide independent evidence that Nancy could not experience pain.

Nancy's feeding tube was removed in December 1990. She died twelve days later on the day after Christmas.[33] Six years later her father, Joe Cruzan, committed suicide by hanging himself. Dr. Ronald Cranford commented that Joe "ran out of energy after" the death of his daughter and called his death "a rational suicide" since "he was never going to get better."[34]

The gap between Dr. Cranford's confident assertions at the Cruzan trial and the medical evidence he cited to back up his assertions is disconcerting. Although Cranford had defined PVS as a condition in which the patient demonstrates absolutely "no interaction with the environment," his testimony acknowledged that it was typical for PVS patients to respond to sounds, to be able to focus their eyes, to react with grimaces to painful stimuli, and for nurses to observe various interactions by the patient. Cranford simply dismissed the evidence as nurses and others seeing what they wanted to see. But that same criticism might be leveled at him. He thought that Nancy's life was not worth living and that it should be ended. Was he seeing only what he wanted to see?

A court case from New York that occurred during the same time as the Cruzan litigation dramatized the possibility that some doctors were too hasty when advocating the removal of food and hydration. Eighty-six-year-old stroke victim Carrie Coons was declared to be in a PVS by her doctors, who insisted that her condition was irreversible. Based on this medical testimony, the judge in the case ordered Coons's feeding tube removed. But before the order could be carried out, Carrie's roommate in the hospital pled with her to do something, "Come on, Carrie, you need to eat, or you're going to die." Nurses also gave Carrie another chance, and soon she was eating pudding, drinking juice—and talking. "I didn't realize anyone cared," she commented. The judge rescinded his earlier order, but was left wondering how medical experts who had been so sure of themselves could have been so completely wrong.[35]

Other food-and-hydration cases have expanded the circle well beyond PVS patients to include those who are clearly conscious. In 1987, the daughter of ninety-two-year-old Anna Hirth of California sought to have her mother's feeding tube removed over the objections of her mother's doctor. An Alzheimer's patient, the elderly Mrs. Hirth had suffered additional brain damage after a choking incident. Yet her doctor insisted that Mrs. Hirth was conscious, felt pain, tracked people with her eyes, and could even take in small amounts of fluid by mouth.[36] Mrs. Hirth was removed from the care of her doctor and transferred to another nursing facility, where she died just a few days later.

During the same year, Michigan resident Michael Martin received an incapacitating brain injury in a car-train accident. Although mentally disabled, he was not in a persistent vegetative state. He could "carry out some voluntary motor commands . . . recognize faces, and . . . communicate with other people through head nods." He was "conscious, alert, appears happy, plays card games, and loves country-western music." His postaccident IQ was apparently in "the range for someone mildly retarded." As in the Schiavo case, it was Michael's spouse who sought to end his life by removing his feeding tube, but his mother and sister fought the effort in court. Unlike in the Schiavo case, however, Michael's spouse, Mary, openly acknowledged that her husband was "conscious, awake" and not in a PVS. But that did not seem to matter. She still regarded him as less than human. "He does nothing but smile," she complained at a conference in 1998. She added that patients like Michael "could be the organ donors who are so desperately needed."[37]

Another recent case involved California resident Robert Wendland, who was severely impaired after an automobile accident that occurred while he was drunk. Robert could not speak, and he required a feeding tube. His wife Rose went to court asking that the feeding tube be removed, stating that Robert had indicated he would not want to be kept alive if he were in a situation similar to Rose's comatose father.[38] Yet the comparison was inapt.

Robert Wendland was neither comatose nor in a PVS, and he was partially mobile. He could operate a manual wheelchair by himself, follow simple commands, and even answer some yes/no questions by the use of special buttons. The attorney for Florence Wendland, Robert's mother, claimed: "Robert is CONSCIOUS, alert, interactive with his environment. He responds to commands. He enjoys participating in activities in the hospital's multipurpose room, including painting pictures and bowling. He is now learning to golf."[39] Florence Wendland intervened in court to oppose the removal of Robert's feeding tube.

Three times a week or more Florence rode a bus to visit her son for several hours at a time. Florence was certain that Robert was still a human being, not a vegetable. "I'll say, Robert, let me hold your hand, and he'll hold my hand. I'll say, kiss my hand, and he'll kiss my hand . . . When I talk to him, he understand[s] me. He looks at me. When I read to him, he looks at me."[40] California courts refused to order the removal of Robert's feeding tube; he died of pneumonia in 2001.[41]

Although recent cases have focused primarily on the provision of food and hydration artificially through a tube, the legal principle being established is open to a much broader application. The ultimate issue is not whether brain-impaired patients have to be supplied with food artificially, but whether they have to be supplied with food at all. There were inklings of this as early as the case of Nancy Cruzan. Given that Nancy had previously taken in food and fluids through her mouth, there was a question about whether she could do so again. But to her parents' medical expert, Dr. Ronald Cranford, the particular method of feeding Nancy was irrelevant. He argued that in Nancy's case even spoon-feeding would be "medical treatment" that should not be provided.[42] The point was that Nancy should not continue to live, and so any feeding—even through the mouth—had to be forbidden. To do otherwise "would morally be repugnant."[43] Cranford acknowledged being involved in "many situations where we could have tried syringe feeding and could have tried spoon feeding" after removing a gastrotomy tube, but did not.[44] He conceded that it might have made "some sense medically" to try spoon- or syringe-feeding in such cases "if you wanted to exhaust all possibilities," but it "wouldn't make any moral sense."[45] Dr. Cranford thought that spoon- and syringe-feeding should be considered removable medical treatment not only for PVS patients but for other "patients who have markedly impaired consciousness who may have a swallowing reflex and may not be able to swallow very normally, such patients who are severely and profoundly demented," which included people in the end stages of Alzheimer's disease.[46]

The issue of spoon-feeding arose more centrally in the case of Terri Schiavo. Some doctors believed that Terri could be weaned from her feeding tube and retaught how to take food through her mouth. "The average human being produces one-and-a-half to two pints of saliva a day," explained neurologist William Hammesfahr. "If you can't swallow it, you drool." Since Terri does not drool, she "can swallow that amount of her own saliva, which means that she can swallow liquids."[47] And if she can swallow liquids, she may be able to relearn how to swallow solid food. But Terri's husband, Michael, consistently refused requests by Terri's parents to give their daughter therapy to see if she could learn to eat by mouth, and so did the courts. At one point Terri's brother and sister were forbidden from visiting Terri unless they agreed not to try to spoon-feed her. "I don't want anyone trying to feed that girl," the judge in the case is said to have ordered.[48]

The decision to treat brain-impaired individuals as nonentities has led in some cases to the Orwellian scenario of ordering the deaths of people who apparently want to continue living. In the case of Michael Martin, evidence was presented at trial that Martin had indicated that he wanted to continue to live, but the probate judge determined that his current wishes were irrelevant because by definition he was not legally competent to make decisions for himself.[49] Fortunately for Michael, the effort to remove his feeding tube was rejected on appeal.

Ohio businesswoman Marjorie Nighbert was not as lucky. After she had a stroke, she was conscious but required a feeding tube. Her brother, who had her power of attorney, directed that the feeding tube be removed, believing that he was following the wishes she had expressed to him before she became ill. Yet as Marjorie began the slow process of dehydration she pled with the staff of her nursing home to be fed. "She was saying things like, 'Please feed me . . . I'm hungry, I'm thirsty, and I want food,' " says an attorney who briefly represented her as a court-appointed guardian.[50] The court ruled, however, that the feeding tube should remain disconnected because Marjorie was not legally competent to override the decision that had been made on her behalf by her brother. Marjorie Nighbert died soon thereafter.

In an even more shocking case from the 1980s, eighty-nine-year-old Ella Bathhurst was denied life-sustaining care by a Minnesota hospital at the request of her daughter. "From that point on food, fluids, diuretics, and antibiotics were withheld, despite the fact that Mrs. Bathhurst begged for water for six days after this was done."[51] Mrs. Bathhurst's excruciating plight is seared into the memories of some who visited her hospital room. Tia Willin's great-grandmother was a roommate of Ella Bathhurst's, and so Tia was a regular visitor with her mother to Ella's hospital room. What she saw and heard there as a twelve-year-old still haunts her.

"It was much worse than anything you can imagine," she recalls. Mrs. Bathhurst "begged and pleaded day and night for food. I can still hear her crying. She was forced to lay in a bed and watch while my great-grandmother, herself a stroke victim, received food and water. There was a nurse who took pity on her and snuck a few ice chips to her or would wet her lips with a sponge, but other than that she just suffered."[52]

Mrs. Bathhurst was never declared legally incompetent; she originally had been hospitalized for a broken hip. "The physicians and staff who let Ella Bathhurst die . . . were merely given minor reprimands," reports bioethicist Robert Barry.[53]

The underlying justification in all of these cases—sometimes implicit, sometimes explicit—is pure scientific materialism: Human beings with significant brain impairments are regarded as no longer really human. The human soul is assumed to be reducible to brain chemistry. Once a person's physical brain has deteriorated so that he can no longer function at a normal level, that person, strictly speaking, is no longer supposed to exist. This line of thought can be seen in the very terminology employed, such as "Persistent Vegetative State," which rhetorically reduces a whole class of human beings to the same level as vegetables.[54]

However, if supporters of withdrawing basic life support from PVS patients are correct—if such patients really are no longer human beings— then why should they be considered alive at all? Why shouldn't they simply be declared legally dead, which would make it considerably easier to harvest their organs for transplantation? This proposal is under serious consideration by the medical community.

Expanding Death

The currently accepted definition of "brain death" requires the cessation of functions of the whole brain, the brain stem as well as the higher brain, which is believed to direct higher-order activities such as cognition and voluntary action.[55] But now some medical and legal experts are arguing that brain death should be redefined to require only the cessation of higher brain functions, which would open the door to declaring PVS patients and perhaps others legally dead.

Bioethicist Joseph Fletcher laid the intellectual groundwork for this proposal in the 1970s. Fletcher declared that "humans without some minimum of intelligence or mental capacity are not persons, no matter how many of their organs are active, no matter how spontaneous their living processes are. If the cerebrum is gone . . . and only the mid-brain or brain-stem is keeping autonomic functions going, they are only objects, not subjects."[56]

In an essay about the nature of "brain death," Fletcher used evolutionary biology to justify his argument. Fletcher said that human beings have three brain parts, each of which reflects their evolutionary ancestry: The "reptilian" brain is the "brain stem," which controls the "autonomic functions." The "mammalian" brain is the "mid-brain, the so-called limbic system in which emotional controls are centered." Finally, there is the genuinely "human . . . brain, which Homo sapiens has developed in the past one million years." This "human brain" he identified with the "cerebral cortex."[57] Only those beings with a functioning cerebral cortex were genuinely human.

"The heart of the matter . . . is that it is cerebral or mental function that is the key to authentically human life—and therefore the key to human death."[58] Fletcher criticized the accepted definition of brain death as being insufficient because it recognized too many people as still living whom he thought should be classified as nonpersons.

By the 1990s, there was growing support for expanding the definition of brain death along the lines discussed by Fletcher.[59] A key motivation for change was the desire to make more organs available for transplantation. As an article in the prestigious British medical journal *The Lancet* explained in 1997, "If the legal definition of death were to be changed to include comprehensive loss of higher brain function, it would be possible to take the life of a patient (or more accurately to stop the heart, since the patient would be defined as dead) by a 'lethal' injection and then remove the organs needed for transplantation."[60]

While to some these views may seem gruesome, they are in fact widely shared in the medical community. A survey of American doctors published in the *Annals of Internal Medicine* reported that 54 percent of medical directors and 44 percent of neurologists agreed that PVS patients "should be considered dead."[61] Moreover, 65 percent of those surveyed believed it was ethical to harvest organs from such patients for transplantation once the decision has been made to removal all life-support, including food and hydration.[62] Where the redefinition of death will stop is anybody's guess. If PVS patients are declared dead, why not mentally impaired individuals like Robert Wendland? Indeed, his wife Rose insisted on CNN in 2001 that "Robert's actually brain dead," and that he "die[d] seven years ago."[63]

Within the framework of scientific materialism, such an analysis is perfectly reasonable. If man is solely a physical being, what meaningful life exists apart from the physical functioning of his brain? Arguments against the humanity of mentally disabled adults are based on the same principles of scientific materialism. Of course, if man is more than mere matter, if he is a fusion of matter and eternal spirit, then brain impairment alone might not be sufficient to deny someone his status as a human being. If a man's personal identity survives death—as traditional religious teachings assert, and as most Americans still believe—then it might be possible for an immaterial soul to persist inside a body whose physical brain has seriously eroded. It might even be possible for self-awareness to persist despite the lack of the physical characteristics usually associated with consciousness in healthy individuals.

Trying to determine whether there is any self-aware or rational activity in comatose or PVS persons is certainly problematic. By definition, they cannot be asked about their feelings or thoughts while they are unconscious. But there have been studies of individuals diagnosed as unconscious who later returned to consciousness and were asked about their unconscious experience. A study of one hundred unconscious patients treated at Hartford Hospital in Connecticut revealed that 73 percent of these patients were self-aware during the period doctors and nurses believed they were unconscious.[64] Twenty-seven percent of the patients could recall specific conversations, emotional reactions, or being touched or moved while unconscious. The author of the study concluded, "The integrity of the person's personality in this state is maintained. The people all kept the same attitudes, likes, and dislikes they had when overtly conscious. They also responded emotionally to events and people in the same way they did before they became unconscious."[65] Nine percent of the patients reported inner awareness of their own thoughts or dreams.

Thirty-seven percent of the patients reported a variety of hallucinations, out-of-body experiences, or near-death experiences.

There are numerous other attested cases of people who experienced mental and emotional states while judged to be unconscious or in a PVS. Kate Adamson, who filed an affidavit in the Schiavo case, was classified by doctors as vegetative after she experienced a severe stroke. Yet from her perspective, she remained mentally aware: "I was conscious and could hear, see, and feel everything. I could feel pain, but was unable to move any part of my body."[66] At one point, her feeding tube was removed and she began to starve. It was a horrific experience. "I went without food for eight days. I thought I was going insane. Inside my body I was trying to scream out that I needed to eat but I could not communicate. The hunger pains overrode every thought I had. The pain was sheer torture."[67]

Patients who have recovered enough to communicate again with the outside world frequently express emotional pain over having been treated like a nonentity by doctors and family members. After being placed in an ICU with continual seizures tied to untreated lupus, Rus Cooper-Dowda was subjected to dehumanizing comments by the medical staff, who assumed that she could not hear their comments. Her early attempts to communicate went unnoticed or were misinterpreted as reflexive actions. "I desperately started trying to communicate. When talk of my pointless life would commence, I first tried moving to show the topic really mattered to me. In response, I got sedated for seizures." Cooper-Dowda says she "was treated like a piece of furniture. As I was not 'really there,' it was okay to complain about my care while 'having' to 'move and dust me.' I heard that last phrase more than once. I also heard allegedly funny staff arguments over what kind of vegetable I was in the intensive care garden patch."[68]

These accounts by those formerly presumed to be unconscious or in a PVS are open to different interpretations. It is possible that some of the people are recalling memories that did not in fact happen. Perhaps their brains created the memories after the fact. It is also difficult to determine whether the experiences of those who come out of an unconscious or vegetative state can be validly extrapolated to explain the experiences of those who persist in the state. Finally, it is possible that some of these people were simply the victims of sloppy diagnostic practices—perhaps they never really met the criteria for unconsciousness or PVS to begin with.

Nevertheless, the data should at least raise questions about dogmatic assertions that unconscious patients must be "vegetables" who have lost their personhood. And the possibility that some of these persons were misdiagnosed also should raise alarms. A British study published in 1996 reported that seventeen of forty supposed PVS patients referred to the Royal Hospital for Neurodisability in Britain were misdiagnosed—an error rate of nearly 43 percent![69] Although most of these patients stayed almost completely immobile, the vast majority could answer questions, perform mathematical tasks such as subtraction and division, and spell out short messages. Nearly half could compose letters to their relatives. In addition to the patients who were misdiagnosed, nearly a third of the patients in the study came out of their PVS after being given rehabilitation therapy.

Only a quarter of patients remained in the PVS after therapy.[70] The lead researcher of the British study, Keith Andrews, wrote that "it is disturbing to think that some patients who were aware had for several years been considered to be, and treated as being, vegetative." Given the high misdiagnosis rate, some observers might question the legitimacy of PVS as a diagnostic category in the first place.

Not Ronald Cranford, who expressed skepticism that the study's results were relevant to the United States. Although he praised the British study, Cranford speculated to the *Los Angeles Times* that only 2 percent of PVS diagnoses by American neurologists might be wrong, and 5 to 10 percent of the diagnoses by nonspecialists.[71] In fact, an earlier American study reported a PVS misdiagnosis rate of 38 percent.[72]

Data for the self-awareness of unconscious patients are consistent with arguments for a nonmaterial soul, but they are also consistent with a conclusion that consciousness is a more complicated physical process than previously realized, and that current diagnostic methods are insufficient to determine whether someone is self-aware while in an unconscious state.

The difficulty of determining awareness in unconscious patients is exacerbated by the fact that we still do not know the precise contributions made by various parts of the brain to consciousness. When the President's Commission for the Study of Ethical Problems in Medicine and Biomedical and Behavioral Research reaffirmed the "whole brain" formulation of brain death in the early 1980s, one of its reasons was the lack of definitive scientific knowledge about the physiological aspects of consciousness: "It is not known which portions of the brain are responsible for cognition and consciousness; what little is known points to substantial interconnections among the brainstem, subcortical structures and the neocortex. Thus, the 'higher brain' may well exist only as a metaphorical concept, not in reality."[73] In the case of babies born with only their brain stems, neurologist Alan Shewmon has argued that we should keep "an open mind" about whether they "have any subjective awareness":

> In experimental animals, brain stem structures have been shown to mediate complex behaviors, sometimes traditionally assumed to be cortical, including binocular depth perception, habituation, learning, and discriminative

conditioning. Similarly, decerebrate (anencephalic or hydranencephalic) human newborns with relatively intact brain stems can manifest a surprising repertory of complex behaviors, including distinguishing their mothers from others, consolability, conditioning, and associative learning.[74]

The fact that, according to Keith Andrews, "few [PVS] patients remain totally unresponsive after undergoing a rehabilitation programme" should make one wonder about absolutist claims that PVS patients have no capability of cognitive function, especially when in many PVS cases there is at least some clinical evidence of responsiveness.[75] The same goes for absolutist claims that PVS patients cannot experience pain. As Chris Borthwick points out, speculative assertions that PVS patients cannot have the proper brain function to experience pain

> would be convincing only if one believed that we know everything there is to be known about the neurology of consciousness and pain—only if we were certain, for example, that no part of pain is experienced purely through the brainstem. Not only are we not at that stage of knowledge, it is difficult to conjecture how one would go about testing, or even operationalising, the hypothesis.[76]

As for empirical claims that the higher brain is completely inoperative in PVS patients, the evidence marshalled thus far seems far from conclusive, and the reasoning employed can only be called sloppy. When Ronald Cranford pointed out that "the metabolic rates of the neocortex in patients in a persistent vegetative state are 50–60% below the level of patients with a normal level of consciousness," one wonders why this fact did not offer possible evidence of some functionality in the neocortex.[77] It certainly did not prove there is no function, which is the claim repeatedly made in court by experts like Cranford. Cranford's flawed logic becomes even more apparent when he admits that these "markedly decreased metabolic levels" in PVS patients "are comparable to levels obtained in deep anesthesia or deep coma."[78]

People under anesthesia are not cognitively dead, so why should we assume that PVS patients are? If anything, recent neurological research indicates that there may be more auditory, visual, and pain processing going on in the brains of PVS patients than previously thought.[79] Brain scans of one twenty-six-year-old woman ("Kate") who was diagnosed as being in a persistent vegetative state "showed evidence of [her] perceiving and processing visual stimuli by the fact that her responses to photographs of familiar faces differed from responses to scrambled images with the same colours and brightness."

Because Kate returned to full consciousness after several months, some have now suggested that she originally was in a "minimally conscious state" rather than a "persistent vegetative state." But at the time she was diagnosed she fit the standard criteria for PVS. While she "was opening her eyes spontaneously and had sleep-wake cycles," she "showed no consistent spontaneous or elicited motor responses or eye movements, nor was she able to communicate."[80] Her condition called for the denial of food and water under policies advocated by some.

Given the considerable evidence that many people diagnosed as unconscious or PVS may in fact be cognitively aware human beings, why has there been such a rush among medical professionals to categorize them as no longer human?

One reason is that healthy individuals find it hard to believe that persons with profound disabilities have lives worthy of existence. When faced with someone with a severe mental disability, a doctor or nurse may think that the disabled person would be "better off dead." Medical staff may also see a disabled patient as a strain on the patient's family or as a waste of scarce medical resources.

Such views were expressed with clarity by Ronald Cranford. When confronted with evidence from the British study that many supposed PVS patients may in fact be cognitively aware, Cranford emphasized that the misdiagnosed PVS patients in the study were still "severely disabled," and "speculate[d] that most people would find this condition far more horrifying than the vegetative state itself, and some might think it an even stronger reason for stopping treatment than complete unconsciousness."[81] When testifying on behalf of removing food and fluids from Robert Wendland, Cranford repeatedly stressed the burden on Wendland's family. "You can go on and on about the psychological harm to the family. I think the family should be able to go through the grieving process. Four years is enough . . . [T]he family should be allowed to live their lives . . . Robert should be allowed to die so the family can grieve."[82]

In other venues Cranford was more outspoken about the economic costs of continuing to treat PVS patients and those who are cognitively disabled. "The estimated annual cost of maintaining persistent vegetative state patients in the United States is $120 million to $1.2 billion," he claimed in one journal article. "Why should society spend this much money for patients who can never benefit from continued treatment in any way, when 37 million people with no medical insurance or coverage have such a reduced opportunity for even a minimal level of health care? Is this social justice?"[83]

Arguments stigmatizing cognitively disabled patients as "life unworthy of life" and as intolerable burdens on their families and society reach back to the origins of the modern euthanasia movement in America, which drew significant inspiration in its early years from the eugenics crusade. After 1915, many American eugenists supported euthanasia for at least some of the unfit, and eugenists had a formative influence on the founding of the Euthanasia Society of America (ESA) in the late 1930s.[84] "A striking 73 percent of ESA's founders were supporters of eugenics," writes Ian Dowbiggin. "By the early 1940s, the list of ESA advisory council members who had defended eugenics to one degree or another was long."[85]

That is an understatement. The ESA advisory council included not merely those "who had defended eugenics," but some of the most prominent leaders in the eugenics movement. These included Henry Goddard (the godfather of hysteria over the "feeble-minded"), Arthur Estabrook (who testified in the Carrie Buck sterilization case), Albert Wiggam (eugenics popularizer extraordinaire), and even Margaret Sanger.

Dowbiggin argues that not every eugenist joined the ESA "solely for eugenic reasons," but he acknowledges that there were clear ideological connections between the eugenics and euthanasia movements.[86] ESA member Ann Mitchell justified euthanasia as part of a "biological house cleaning," and advocated "euthanasia as a war measure, including euthanasia for the insane, feebleminded monstrosities."[87]

More generally, observes Dowbiggin, "eugenics, with its affinities to social Darwinism and scientific naturalism, convinced thinkers such as [Albert] Wiggam that, when faced with difficult decisions about birth, sex, and death, the old gods were dead and the old rules no longer applied."[88] Current justifications for ending the lives of PVS patients also echo arguments in pre–World War II Germany for ending the lives of "mentally dead" patients.[89]

The scientific materialism that motivated the early euthanasia movement operates in current controversies. Perhaps the most fundamental reason why many medical professionals dismiss the humanity of "vegetative" patients is their acceptance of the idea that human beings are wholly material organisms whose personal identities cannot survive the impairment of their brains. In a 1980s Gallup survey, only 32 percent of leading doctors and 16 percent of leading scientists believed in life after death.[90] By contrast, nearly 70 percent of the general public believe in life after death.

These are two fundamentally different views of the nature of human personhood, and they have dramatically different consequences for how to view the humanity of persons with cognitive disabilities. Believing that personal identity remains intact after physical death makes it easier to view cognitively disabled patients as retaining their fundamental identities even in the midst of severe brain impairment. On the other hand, believing that personal identity is tied completely to the proper functioning of the physical brain makes it more plausible to regard patients with brain impairments as less than human or even as already dead.

The materialistic understanding of human beings profoundly demeans those who are physically incapacitated. Although sometimes cloaked in the rhetoric of compassion, the materialistic view can easily result in writing off the humanity of entire classes of persons, whether they be welfare recipients (in the case of eugenics), certain races (in the case of the Darwinian justification of racial inferiority), or disabled infants and adults (in the case of euthanasia).

Occasionally the brutality of this view is exposed for all to see. In an article about Terri Schiavo in 2003, commentator Bryan Zepp Jamieson compared Schiavo alternately to a "slab of meat," "insects . . . born without a head," a "carcass," and "a dead parrot." "There's nobody home. The wheel is spinning, but the hamster has died," he quipped cruelly. In short, Terri Schiavo was a nonentity, already dead. Jamieson correctly predicted that the courts would ultimately "reinstate the original decision to remove the feeding tube from the carcass. If they don't then America will be just as dead as Terri is."[91]

Others might respond that to treat Terri Schiavo and similar individuals as "carcass[es]" would be the true indication that American society had died.

Conclusion: The Abolition of Man

Julian West went to sleep in Boston one night in 1887. He awoke in the same city in the year 2000. When he fell asleep, Boston was being torn apart by growing battles between rich and poor. When he awoke, Mr. West was astonished to find himself in a brave new world.

Slums and tenements had given way to fountains, tree-lined avenues, and buildings of "architectural grandeur."[1] Poverty had been abolished, capitalism had been replaced by centralized planning, and the state was now everyone's employer. Social problems had all but disappeared. Jails were a relic of the past, and the few remaining criminals were treated as evolutionary throwbacks and interned in hospitals to be treated rather than punished.[2] The continuing improvement of the race was assured by a program of positive eugenics based on Darwin's law of "sexual selection."[3] There was little need for checks and balances against corruption because environmental conditioning had produced a new race of men who were both equal and morally good.

According to Mr. Barton, a celebrated clergyman of the new era, "the ten commandments became well-nigh obsolete in a world where there was no temptation to theft, no occasion to lie either for fear or favor, no room for envy where all were equal, and little provocation to violence where men were disarmed of power to injure one another. Humanity's ancient dream of liberty, equality, fraternity . . . at last was realized."[4] Human evolution had finally made realistic the goal of creating heaven on earth. "The long and weary winter of the race is ended," declared Mr. Barton. "Its summer has begun. Humanity has burst the chrysalis. The heavens are before it."[5]

This fanciful depiction of life in the twenty-first century was the invention of Edward Bellamy in his popular utopian novel, *Looking Backward* (1887). Bellamy's novel perfectly embodied the optimistic vision offered by scientific materialism at the end of the nineteenth century. During an era when science seemed to be uncovering the material basis of all human problems, it was widely believed that science could lead to the transformation of society, bringing about greater human freedom, dignity, and happiness in the process.

A transformation of society did occur, but not of the sort anticipated by the early boosters of scientific materialism. Human nature was not reformed, crime did not disappear, and scientific materialism did not usher in a new age of "liberty, equality, fraternity."

What went wrong?

An idea's consequences may not be fully anticipated by its proponents. Nathaniel Hawthorne wryly observed that "no human effort, on a grand scale, has ever yet resulted according to the purpose of its projectors . . . We miss the good we sought, and do the good we little cared for."[6] Scientific materialism was supposed to be a great engine of human progress in politics and culture. It was not. And its failures continue to influence American public policy.

Technocracy

One consequence of scientific materialism for politics was the elevation of technocracy—rule by scientific experts—over democracy. Since science was supposed to be the true source of objective information about the world, proponents of scientific materialism logically concluded that scientists—not the general public, or their elected representatives—should be the ultimate arbiters of public policy.

At its core, this message was profoundly anti-egalitarian and anti-democratic. Speaking before the Second International Congress of Eugenics in 1921, Alleyne Ireland declared that current conditions had rendered America's original form of government established by the Constitution and the Declaration of Independence "utterly unsuitable." America's Founders believed that "governments derive their just powers from the consent of the governed," and they set up arrangements "designed with a view to making abuse of power difficult." But in an age when government must increasingly provide a wide range of social services, society could no longer afford to rely on government by nonexperts. Ireland stated that it was "imperative . . . that the omnipresent activity of government should be guided by the light of scientific knowledge and conducted through the instrumentality of a scientific method."[7]

The claim that society should place its faith in scientific experts rather than ordinary citizens or elected officials was a common refrain in public-policy debates colored by scientific materialism. To be sure, few were as blunt as Ireland in directly attacking the Constitution or demanding a governing role for scientists. Yet in controversy after controversy, the message was unmistakable. Whether the issue was education or welfare or crime, members of the public were urged to place their trust in the findings of scientific experts rather than their own core beliefs or the views of political and religious leaders. Science with a capital "S" dictated the replacement of punishment with treatment in the criminal-justice system, the enactment of forced sterilization in the welfare system, and the substitution of "value-free" information from sex researchers for traditional moral teachings about family life in public schools. In each of these areas, the claim was made at least implicitly that scientific expertise should trump other sources of knowledge, including ethics, philosophy, tradition, religion, and common sense.

Much could be said in favor of the authority of scientific expertise in modern life. In an increasingly complex and technologically driven world, the need for scientific input on public policy would seem obvious. Since many policy questions today arise in such science-based fields as medicine, transportation, and ecology, why shouldn't politicians and voters simply defer to the authority of scientific experts in these areas?

Although this line of reasoning exhibits a surface persuasiveness, it ignores the natural limits of scientific expertise. Scientific knowledge may be necessary for good public policy in certain areas. But it is not sufficient. Political problems are preeminently moral problems, and scientists are ill-equipped to function as moralists. C. S. Lewis warned about this drawback of technocracy in the 1950s. "I dread specialists in power, because they are specialists speaking outside their special subjects," Lewis wrote. "Let scientists tell us about sciences. But government involves questions about the good for man, and justice, and what things are worth having at what price; and on these a scientific training gives a man's opinion no added value."[8]

To cite a concrete example: Wildlife biologists may be able to provide policymakers with information about which species are in danger of extinction and perhaps predict some of the costs of their extinction to biodiversity. But they have no more authority than anyone else in determining whether a particular endangered species is more valuable than the jobs that may be lost trying to save that species from extinction. Politics is largely about ranking and reconciling competing goods. But the ranking of goods involves questions of justice and morality, and as Lewis pointed out, "a scientific training gives a man's opinion no added value" on such questions.

Technocracy poses a further difficulty: Experts can be wrong, sometimes egregiously. If the history of scientific materialism in politics shows anything, it is that scientific experts are as fallible as anyone else. They are capable of being blinded by their own prejudices and going beyond the evidence in order to promote the policies they favor. Alfred Kinsey's empirical claims about the sexual behavior of the general American public were junk science, given his deeply flawed sample population; yet that did not stop him from boldly making his claims and vigorously defending them as sound science.

What is true of individual scientists can be true of the scientific community as a whole. For decades, eugenics was embraced as legitimate by America's leading scientists and scientific organizations such as the American Association for the Advancement of Science. Critics of eugenics, meanwhile, were stigmatized as antiscience and religious zealots. Yet the critics turned out to be right. Similarly, the lobotomy was uncritically embraced for years by the medical community as a miracle cure, and the scientist who pioneered the operation in human beings won a Nobel Prize for his efforts. Only after tens of thousands of individuals had been lobotomized did healthy skepticism prevail.

To cite a more recent example, various scientists and medical professors into the 1990s continued to invoke Haeckel's discredited theory of embryonic recapitulation to supply a scientific justification for abortion. And in 2003, hundreds of scientists in Texas defended inaccurate biology textbooks they likely had never read because they were more interested in safeguarding the public image of Darwin's theory of evolution than they were in presenting students with accurate facts.

Any suggestion that policymakers should simply rubber-stamp the advice of the current majority of scientists is profoundly subversive of the fundamental principles of representative democracy. As equal citizens before the law, scientists have every right to inform policymakers of the scientific implications of their actions. But they have no special right to demand that policymakers listen to them alone.

Unfortunately, a growing chorus urges that public policy be dictated by the majority of scientific experts without input from anyone else. This bold assertion is made not just with regard to evolution, but concerning a host of other controversial issues such as sex education, euthanasia, embryonic stem-cell research, cloning, and global warming. Any dissent from the orthodoxy of "experts" on these issues allegedly represents a "war on science."[9] But that's just not the case.

Utopianism

A second consequence of scientific materialism for public policy was the cultivation of a vigorous form of utopianism. Believing they possessed the key to understanding and ultimately controlling human behavior, defenders of scientific materialism were confident that science could usher in heaven on earth—if only they tried hard enough.

Their heady optimism is not difficult to understand. By the late nineteenth century, science had produced marvelous advances in medicine, agriculture, sanitation, and transportation. Why couldn't the triumphs of the scientific method over the natural world be extended to the social sphere? If science could prevent the spread of physical diseases like smallpox, why couldn't it also prevent outbreaks of social diseases like crime and poverty? If science could breed better strains of cattle and corn, why couldn't it breed better kinds of people?

Addressing the American Breeders Association in 1913, U.S. Secretary of Agriculture James Wilson acknowledged that the wholesale replacement of "inferior" human stocks with "the best part of the human race . . . at first seems like an Utopian vision," but he then quickly added: "Why should it not come? Must science stop in its beneficence with the plant and the animal? Is not man, after all, the architect of his own racial destiny?"[10] Wilson's rosy rhetoric revealed the startling naïveté at the heart of the scientific-materialist agenda.

Scientists and policymakers who were readily skeptical of claims made by religion or tradition turned out to be supremely credulous when it came to claims made in the name of science. They accepted at face value the purported benefits of such procedures as lobotomies, psychosurgery, and forced sterilization. They made grand promises about how science could solve intractable social problems such as crime and poverty. They showed little appreciation for the fact that science, like all human endeavors, could be misused, especially when allied with political power. Eugenist Herbert Walter sanguinely predicted that nothing like "the Spanish Inquisition or . . . the Salem witchcraft persecution" would take place in an age of modern science. Only two decades before the Nazis ascended to power in Germany, Walter predicted that "it is unlikely that the world will ever see another great religious inquisition, or that in applying to man the newly found laws of heredity there will ever be undertaken an equally deplorable eugenic inquisition."[11] Harry Laughlin asserted with confidence that no one—not even one person—had been wrongly sterilized in America.[12] AAAS president Charles Eliot at least acknowledged the prospect that physical and chemical science could be enlisted "as means of destruction and death." But even he thought the application of biology to society held no danger: "Biological science has great advantage in this respect over physical and chemical [science]. It can not so frequently or easily be applied to evil ends."[13] Eliot wrote those words in 1915 as the eugenics movement was well on its way to compelling the sterilization of thousands of people across America.

Prior to the rise of scientific materialism, a strong anti-utopian sentiment in American political culture counterbalanced the zealousness of reformers. America's Founders, in addition to their idealism, displayed a keen realism about the imperfections of human nature. "If men were angels, no government would be necessary," wrote James Madison in *The Federalist*.[14] "The best Institutions may be abused by human depravity . . . they may even . . . be made subservient to the vilest of purposes," echoed George Washington.[15] Nathaniel Hawthorne satirized the overblown hopes of contemporary reformers in his short story "Earth's Holocaust" (1844). There he described how militant do-gooders planned to cleanse the earth of imperfection by creating a giant bonfire out on the western prairies on which they could throw every conceivable cause of social evil.[16] The great conflagration burned for days and consumed everything thrown into it, but the fire still did not produce the perfect society. Hawthorne's punch line was that the reformers failed because they could not reach the ultimate cause of human misery, the human heart. Social conditions might wax and wane, but sinful human nature was unchangeable this side of heaven.

Scientific materialism tried to refute this kind of political realism. Human nature, said the scientific materialists, was not fixed; it could be remade through the methods of modern science. Men may not be angels now, but under the right biological and environmental conditioning they might become angelic. Scientific breeding and medical treatment could usher in a new age only dreamt of by previous reformers.

One would like to believe that Americans have learned from the excesses of scientific utopianism, but current political controversies inspire no confidence in this regard. The miracle cures may be different today, but the utopian rhetoric is remarkably similar. Seventy years ago, eugenics promised to cure America's social problems through better breeding. Today, mental-health crusaders promise to eliminate behavioral problems among America's children by screening every schoolchild for mental illness and then putting millions of them on psychoactive drugs. Like the eugenics crusade, the current push to dramatically increase the number of children on psychoactive drugs reduces all behavioral problems to a material basis. And like the eugenics crusade, it is accompanied by grandiose claims that go far beyond the actual science. Like the eugenics crusade, it is justified in humanitarian terms even while it raises serious issues about civil liberties and human dignity. How many children will be hurt before this latest crusade runs out of steam?

Dehumanization

A third consequence of scientific materialism for public policy was dehumanization. Although its boosters saw scientific materialism as a way to solve social problems and advance human dignity, the historical record shows that it often denigrated entire classes of humanity. The belief that men and women could be reduced to their physical capacities plus their material inputs could be profoundly dehumanizing.

In criminal justice, the belief that a person was "no more 'responsible' for becoming wilful and committing a crime than the flower for becoming red and fragrant"[17] may have led to more humane treatment in some cases, but it also robbed the criminal offender of the dignity of being treated as a rational being whose choices matter. At the same time, it opened the door to horrific forms of "scientific" rehabilitation that never would have been allowed if they had been imposed as punishments.

In sex education, the depiction of human sexuality as little more than mammalian behavior reduced human beings to the level of animals and drained human relationships of the moral and spiritual context that gave them their deepest meaning.

In the corporate world, scientific materialism fed eugenic employment policies and the use of advertising to scientifically manipulate consumers into purchasing products.

In the welfare system, the quest to identify the biological roots of poverty paved the way for forced sterilization, anti-immigrant hysteria, and the demonization of anyone who was regarded as physically or mentally imperfect.

The impact of scientific materialism on welfare policy is especially worth noting, because it directly challenged the guiding principles of the existing social-welfare system. Traditional charity was premised on the idea that all human beings were created in the image of God and therefore worthy of assistance, mercy, and redemption. Eugenic welfare reformers denounced such humanitarian views as false and dangerous. Edward East attacked as unscientific the idea that "man is created in the image of God"[18] and suggested that the claim that all human beings have equal worth is ludicrous. Margaret Sanger warned of the "dangers inherent in the very idea of humanitarianism and altruism, dangers which have today produced their full harvest of human waste, of inequality and inefficiency."[19]

America's experience with the dehumanizing effects of scientific materialism was far from exceptional. The three regimes of the twentieth century best known for being founded explicitly on the principles of scientific materialism—Soviet Russia, Nazi Germany, and Communist China—are all remembered for their horrific brutality rather than any advancement of human dignity. In Germany, the connection between scientific materialism and Nazi crimes against humanity is unmistakable, as historian Richard Weikart has ably demonstrated in his recent book on the influence of Darwinian ethics in Germany.[20]

The dehumanizing effects of scientific materialism remain a live issue for public policy today, especially in so-called right-to-die cases. Efforts to redefine mentally and physically disabled infants and adults as already dead, the widespread careless diagnosis of the "persistent vegetative state," and the demeaning rhetoric of bioethicists such as Peter Singer, raise, chillingly, the ghosts of evils past.

Relativism

A fourth consequence of scientific materialism for public policy was relativism. Darwinian theory in particular supplied a powerful justification for evolving standards in politics and morality. Part of the justification was by way of analogy: If evolution was the normal state of the natural world, why shouldn't it be regarded as the normal state of politics?

The preeminent achievement of applying the evolutionary paradigm to politics was the doctrine of the evolving Constitution championed by Woodrow Wilson and other progressives. No longer would American government be hamstrung by a static understanding of human nature or human rights. It must adapt and evolve to meet the challenges of new conditions. In the words of Wilson, "living political constitutions must be Darwinian in structure and in practice. Society is a living organism and must obey the laws of Life . . . all that progressives ask or desire is permission . . . to interpret the Constitution according to the Darwinian principle."[21]

But the link between Darwinian theory and relativism was not merely analogical. In *The Descent of Man*, Darwin depicted morality as the evolving product of natural selection. Rather than reflecting timeless standards of truth sanctioned by God or nature, moral codes evolved by natural selection to promote survival. As the conditions for survival changed, so did what was moral for any species. In one situation, maternal love might be moral; in another situation, infanticide. In one situation, kindness might be moral; in another situation, cruelty.

While Darwin surely hoped that traditional virtues were biologically beneficial in nineteenth-century Britain, if circumstances changed and those virtues no longer promoted survival, he would have to grant that they would no longer be virtues. To recall a startling passage by Darwin quoted earlier in this book: "If, for instance . . . men were reared under precisely the same conditions as hive-bees, there can hardly be a doubt that our unmarried females would, like the worker-bees, think it a sacred duty to kill their brothers, and mothers would strive to kill their fertile daughters; and no one would think of interfering."

Whatever his personal moral preferences, Darwin's reductionistic account of the development of morality left little room for objectively preferring one society's morality over another. Each society's moral code developed to promote the survival of that society, and so each society's moral code could be considered equally "natural." Darwin's evolutionary explanation of the origin of the family was just as relativistic. It was clear from his account that there could be no superior form of marriage or family life for every time and place. In Darwin's framework, everything that regularly occurred in nature must be regarded as normal almost by definition. While for the most part Darwin did not press his relativistic analysis of morality to its logical conclusion, he laid the groundwork for others who came after him. The ultimate result of Darwinian moral relativism can be seen in the sex research of zoologist Alfred Kinsey and the moral pluralism embraced by the sex-education reformers at SIECUS and similar organizations. Their efforts to convince the public that all variations of sexual behavior are "normal"—including adult-child sex and even incest—were a logical culmination of the approach Darwin pursued in *The Descent of Man*.

Stifling Free Speech

A final consequence of scientific materialism was the stifling of free speech and debate over the public-policy implications of science. This is surely one of the most striking ironies of the effort to enlist scientific materialism to reform society. In their own minds, proponents of scientific materialism were the defenders of enlightenment against superstition and rational debate against unreasoning dogmatism.

But the rhetoric they employed against their opponents was often far from conducive to open discussion. The repeated insistence that scientists know best and thus politicians and the public should blindly accept the policy views of scientists did not encourage critical scrutiny of scientific claims made in politics. Even more stifling of genuine debate was the frequent playing of the religion card in policy disputes involving science. With the help of sympathetic journalists, proponents of scientific materialism tried to turn every policy dispute into a battle pitting the enlightened forces of science against bigoted religious extremists. Promoters of eugenics heaped scorn on Catholic and fundamentalist critics of forced sterilization. Advocates of Kinsey-style sex education demonized parents who raised objections as Bible-thumpers who were conspiring against democracy. Today, defenders of a Darwin-only biology curriculum accuse their opponents of trying to insert Genesis into science classes, no matter the facts.

Instead of addressing the policy arguments raised by critics of sex education or Darwin-only science education, defenders of scientific materialism try to make the religious beliefs of their opponents the central issue, arguing that their real or perceived religious motivations somehow disqualify them from being active participants in the public square.

America is a deeply religious country, and no doubt many critics of the agenda of scientific materialism are motivated in part by their religious beliefs. So what? Many opponents of slavery were motivated by their Christian beliefs, and many leaders of the civil-rights movement were even members of the clergy. All of them had an equal right with other citizens to raise their voices in public debates. So long as religious persons in politics offer secular justifications for their policy proposals, they have every right to demand that their ideas be heard on the merits regardless of their private religious views.

Although evolutionists portray themselves as the victims of fundamentalist intolerance, in most places today it is the critics of Darwin's theory who are being intimidated or silenced. Some universities are even adopting the equivalent of evolution "speech codes" to muzzle free speech by science professors who may be skeptical of Darwin's theory. At the University of Idaho, for example, President Timothy White issued a letter in 2005 forbidding faculty from "teaching . . . views that differ from evolution . . . in our life, earth, and physical science courses or curricula."[22] The directive targeted tenured microbiology professor Scott Minnich, a proponent of intelligent design.[23]

"The University of Idaho's statement does not simply ban discussions of evolution that are unrelated to the subjects of courses being taught," noted Gonzaga University law professor David DeWolf. "Nor does it merely forbid religious-based views of evolution from being taught in science classes. The statement offers a blanket prohibition on any 'views that differ from evolution,' no matter how scientific, and no matter how related to the courses under study." DeWolf concluded: "this is viewpoint discrimination in its most naked form."[24]

These politically correct efforts to silence the academic critics of Darwinism are fueled by increasingly toxic rhetoric on the part of evolutionists. Rather than defend the scientific merits of evolution, Darwinists have become obsessed with denouncing their opponents as dangerous zealots hell-bent on imposing theocracy. They routinely apply the label of "Taliban" to anyone who supports teaching students about scientific criticisms of Darwinian theory. Biology professor P. Z. Myers at the University of Minnesota, Morris, meanwhile, has demanded "the public firing and humiliation of some teachers" who express doubts about Darwin.[25] He says that evolutionists should "screw the polite words and careful rhetoric. It's time for scientists to break out the steel-toed boots and brass knuckles, and get out there and hammer on the lunatics and idiots."[26]

Defenders of evolution who claim to fear blind zealotry might want to look in the mirror. The new "Darwinian fundamentalists" have become just as intolerant as the religious fundamentalists they despise.

Such intolerance should raise concerns for thoughtful citizens from across the political spectrum. True liberals—those who favor free and open debate—should be appalled by the growing campaign of intimidation against academic critics of Darwinism. Whatever one's personal view of Darwinism, the current atmosphere is unhealthy for science, and it is unhealthy for a free society.

A couple of caveats are in order. First, what does it mean to identify the above outcomes as "consequences" of scientific materialism? Is it being asserted that technocracy, utopianism, dehumanization, moral relativism, and the stifling of free speech were somehow *necessary* outcomes of scientific materialism for public policy?

No. Few actions are the necessary result of any particular set of ideas. People often act inconsistently with their own beliefs. Sometimes they act better than their beliefs, as with the bigot who treats an individual member of a minority group fairly. Sometimes they act worse than their beliefs, as with the professed pacifist who takes revenge on an enemy. People also interpret and apply their beliefs in different ways. Adherents to the same religion may disagree about the morality of the death penalty or the use of military force. Hence, no claim is being made that the above consequences were somehow mandated by scientific materialism.

That does not mean there was no rational connection between scientific materialism and the outcomes described. Undoubtedly there was. If one believes that science supplies the most authoritative information about the world, technocracy would seem to be a reasonable option. If one believes that human behavior can be reduced to biochemistry, the idea that science can reform human nature no longer may seem utopian. If one believes that humanity reached its current stage of development due to a Darwinian process of natural selection acting on random variations, the case for negative eugenics may appear well-nigh compelling. Again, these consequences may not have been necessitated by scientific materialism, but they were certainly natural and logical conclusions of the materialist worldview.

A second caveat relates to what these consequences mean for the truth of scientific materialism. The consequences identified here are likely to be regarded as pernicious by most people. The fact that scientific materialism may have bad effects, however, does not mean that scientific materialism is wrong. Most people may be revolted by claims that they do not have a free will or that their love of their spouse can be reduced to a chemical reaction. But revulsion is not the same as refutation. This book therefore does not argue that the bad effects of scientific materialism disprove the materialist hypothesis.

Yet that is not the end of the discussion. Bad effects may not disprove an idea, but they may raise enough questions to spur a critical reevaluation of the evidence on which the idea rests. After all, it is possible that an idea like scientific materialism has pernicious consequences because it was false to begin with.

A growing number of scientists and other scholars are presenting evidence that would appear to confirm that conclusion. They say that recent developments in biology, chemistry, physics and related sciences undermine the scientific-materialist worldview. If they are correct, their ideas could have a significant impact on the public-policy debates of the future.

Life after Materialism?

"There is now little doubt that living things owe their origin entirely to certain physical and chemical properties of the ancient earth," announced biologist Paul Weisz triumphally in his biology textbook in 1959.[27] "Nothing supernatural was involved." Weisz's assertion was a classic expression of the scientific-materialist creed. Yet more than four decades later, Weisz's statement stands out for its startling hubris.

Stephen Meyer is a Cambridge-trained philosopher of science whose research focuses in large part on origin-of-life biology and chemistry. As he tells it, for much of the twentieth century origin-of-life researchers had high hopes of explaining the origin of life in terms of basic biochemistry.

In the late nineteenth century, Darwinists Ernst Haeckel and T. H. Huxley proposed simple chemical processes that they thought could generate the first living cell out of nonlife. "Just as salt could be produced spontaneously by adding sodium to chloride, so, thought Haeckel and Huxley, could a living cell be produced by adding several chemical constituents together and then allowing spontaneous chemical reactions to produce the simple protoplasmic substance that they assumed to be the essence of life." During the 1920s and '30s, Russian scientist Alexander Oparin similarly "envisioned a series of chemical reactions that he thought would enable a complex cell to assemble itself gradually and naturalistically from simple chemical precursors."[28] In the 1950s, University of Chicago researchers Harold Urey and Stanley Miller conducted a celebrated experiment that seemed to corroborate Oparin's theory, leading many scientists to believe that the mystery of the origin of life was well on its way to a materialist solution.

But in the ensuing decades, explains Meyer, the high hopes of early researchers were dashed. Not only did the Miller-Urey experiment fail under updated simulations of the earth's early atmosphere,[29] the entire supposition that the origin of life could be explained by a series of brute chemical reactions dictated by chance or necessity was thrown into question by discoveries at the very foundations of life.

According to the tenets of scientific materialism, the more scientists looked into the biochemical foundations of life, the more they were supposed to discover that every living thing could be reduced simply to matter in motion. Instead, says Meyer, the more scientists investigated the foundations of life, the more they were confronted with what disconcertingly looked like evidence of mind.

This was sharply apparent in the case of proteins, key building blocks of cellular life. According to the materialist worldview, the development of proteins should be completely reducible to some kind of mechanical process of chemical attraction along with chance. But with advances in molecular biology, it became apparent that proteins could not be explained adequately by either chance or mechanical laws. In fact, the development of proteins was dependent on highly detailed instructions encoded in each cell's DNA. Meyer argues that these biological instructions constitute a form of information, and that the DNA that encode them function as a language:

> Just as the letters in the alphabet of a written language may convey a particular message depending on their arrangement, so too do the sequences of nucleotide bases . . . inscribed along the spine of a DNA molecule convey a precise set of instructions for building proteins within the cell. The nucleotide bases in DNA function in much the same way as symbols in a machine code or alphabetic characters in a book.

In each case, the arrangement of the characters determines the function of the sequence as a whole. As Richard Dawkins has noted, "The machine code of the genes is uncannily computer-like." Or as Bill Gates has noted, "DNA is like a computer program, but far, far more advanced than any software we've ever created."[30]

The information content of DNA poses a fundamental challenge to materialist explanations of life because materialist explanations in principle cannot account for the most important characteristic of information: its meaning.

Consider a front-page story published in the *New York Times*. A chemist could come up with a detailed chemical breakdown of the newsprint and ink used to print the story, and even offer a comprehensive physical explanation for how the ink bonded to the newsprint. Yet this rundown of the material components of the published article would not help anyone understand one iota about the message the story's author was trying to communicate. The material composition of the printed newspaper story and the information conveyed by the story are separate and distinct. Indeed, the material composition of the physical newspaper is completely irrelevant in explaining the information content of the newspaper. Put another way, one could say that the medium is definitely *not* the message. Meyer explains:

> One thing I do in classes to get this idea across to students is that I hold up two computer disks. One is loaded with software, the other one is blank. And I ask, "What is the difference in mass between these two computer disks as a result of the difference in the information content that they possess?" And of course the answer is zero, none, there is no difference as a result of the information. That is because information is a massless quantity. Information is not a material entity. Then how can any materialistic explanation explain its origin? How can any material cause explain its origin?[31]

Meyer contends that the origin of biological information is one of the most vexing problems facing modern biology, and it has far-reaching implications all the way up the line in the development of life, not just at the molecular level. If mind-numbingly complicated biological instructions are needed to create even simple functioning proteins, how much more biological information is needed for the development of physical structures such as the origin of animal body plans roughly five hundred million years ago during the event known as the "Cambrian Explosion"?

"To produce a new organism like a trilobite," explains Meyer, "you need a whole bunch of new cell types; and . . . then you need new proteins to service the different unique cell types; and to build the proteins you need genetic information in the form of DNA." "The big question that the Cambrian Explosion poses," continues Meyer, "is where does all that new information come from? Where does the new information come from needed to build those proteins, to service those new cell types, to build these fundamentally new forms of animals?" Meyer argues that supporters of neo-Darwinism and scientific materialism "are really at a loss to answer that question. It's a sudden emergence of [a] huge amount of new information and really defies the capacity of the natural selection-mutation mechanism to produce all that information." In the words of Meyer, biological information poses "really a grave difficulty" for materialistic theories of the origin and development of life. "This is not a minor anomaly."[32]

"In the nineteenth century, we thought that there were two fundamental entities of science: Matter and energy," says Meyer. "At the beginning of the twenty-first century, we now recognize that there is a third fundamental entity, and it is information. It is not reducible to matter; it is not reducible to energy."[33] The old materialism held that at the bottom of mind is matter. The new evidence suggests that at the bottom of matter is mind. If information is truly an irreducible property of the universe, this fact has profound implications for the continued viability of scientific materialism.

Other scholars are raising comparable objections to scientific materialism from the standpoint of quantum physics. According to theoretical physicist Henry Stapp at the Lawrence Berkeley National Laboratory, classical physics from Newton onward attempted to reduce everything to "purely mechanical processes that are specified without acknowledging or entailing the existence of . . . mind . . . Now, however, that material conception of nature, which was the cause of so much philosophical dispute, has been found to be fundamentally false."[34] Stapp maintains that the new physics for all practical purposes overthrows the old reductionist scientific materialism. In his view, quantum physics "describes a world built not out of bits of matter, as matter was understood in the nineteenth century, but rather out of a fundamentally different kind of stuff. According to the revised notion, physical reality behaves more like spatially encoded information that governs tendencies for experiential events to occur, than like anything resembling material substance."[35] Stapp says that

the older theory was about tiny bits of matter, and how their behaviors were governed by the effects of the neighboring bits. The new theory is about bits of information or knowledge that agents acquire by performing purposeful actions. It is about the freedom provided by the theory for human agents to choose which actions they will take—and when they will take them—and about the useful knowledge we can derive from our experiencings of Nature's response to our probing actions.[36]

Stapp claims that public policy has not yet caught up to the findings of the new physics, and as a result many political and social decisions are still based on the outdated science of the old physics:

> Although more than three quarters of a century have passed since the emergence of the new basic physics, the old idea of mechanical determinism has not yet been rooted out of educated thought. The force of that persisting idea continues to have a profound impact upon your life. It still drives the decisions of governments, schools, courts, and medical institutions, and even your own choices, to the extent that you are influenced by what you read and are told by pundits who expound as current the conclusions of an out-of-date science.[37]

Whether or not quantum physics requires the antimaterialistic view championed by Stapp is a matter of keen dispute. But at a minimum, it does seem to open the door to such a view. Contrary to widespread claims that science necessitates a completely materialistic account of mind, the new physics allows for other possibilities. "It is widely held that science tells us that the microphysical realm is causally closed, so that there is no room for mental states to have any effects," acknowledges philosopher of mind David Chalmers.[38] But Chalmers believes that quantum physics opens this stock response to serious challenge.

Neuropsychiatrist Jeffrey Schwartz at the University of California, Los Angeles believes that brain research is also raising questions about materialistic accounts of mind. According to Schwartz, while traditional brain psychiatry assumes "that all aspects of emotional response are passively determined by neurobiological mechanisms," recent research has lent support to the idea that the relationship between brain and mind is a two-way street.[39]

Not only do changes in a person's brain influence one's mental states, but a person's conscious thoughts can be shown to produce physical changes in one's brain circuitry and metabolism. In other words, there is growing scientific evidence that mind influences brain, just as brain influences mind. Schwartz is offering a dualistic account of the human mind that refuses to reduce all mental activity to a product of neurobiology. Human choices and desires are genuine entities that produce real effects.

"Is this a good argument for mind being fundamentally distinct from brain?" asks mathematician and philosopher William Dembski. "It depends what you are looking for. If you want a knock-down argument against materialism and materialist accounts of mind, this won't do it. But if you are looking for consilience, in which multiple lines of independent evidence converge on the same target, then Schwartz's argument is a good one to have in your arsenal."[40] Dembski's point is that the new challenges to materialism are not dependent on any single scholar or discipline. They are being advanced by different scholars in multiple fields, including biochemistry, genetics, cosmology, physics, philosophy of mind, and information theory. Consilience is probably a good word to apply to these intellectual developments, because it suggests the hope of eventually fashioning a unified alternative to scientific materialism.

Of course, consilience is supposed to be one of the great advantages of scientific materialism. Sociobiologist E. O. Wilson even wrote a book a few years ago titled *Consilience*, in which he touted materialism as the ultimate way to unify human knowledge across disciplines. "We are approaching a new age of synthesis, when the testing of consilience is the greatest of all intellectual challenges," declared Wilson.[41]

Wilson's book predicted that every discipline—from economics and politics to literature and ethics—would eventually be explicable in material terms. Moral sentiments, for example, would become explainable by analyzing their "underlying neural and endocrine responses," identifying their "prescribing genes," and exploring "their contributions to survival and reproductive success during the long periods of prehistoric time in which they genetically evolved."[42] Consilience, on Wilson's terms, meant reducing everything to the products of mindless matter.

More than sixty years ago, C. S. Lewis anticipated Wilson's version of consilience in *The Abolition of Man* (1944). Reducing men to matter, said Lewis, was tantamount to abolishing man altogether. Lewis was not against the quest for unified knowledge, nor was he antiscience. But he yearned for a new kind of natural science that "when it explained it would not explain away. When it spoke of the parts it would remember the whole. While studying the It it would not lose what Martin Buber calls the Thou-situation."[43] Lewis wondered whether such a "regenerate science" was even possible, almost despairing that it may not be.

Today, a growing number of scholars are arguing that such a science may well be possible, even as defenders of the old order insist that everything can be reduced to matter in motion. The debate has been joined, and the long-term consequences may be profound. The view that prevails will likely have as dramatic an impact on the politics and culture of the future as scientific materialism has had on the politics and culture of the past.

Afterword to the Paperback Edition: Scientism in the Age of Obama— and Beyond

Darwin Day in America was first published in the fall of 2007. A year later, the United States elected Barack Obama as president. Obama's election may have proved to be a watershed in the ongoing debate over science, scientism, and American society.

Obama began invoking the authority of science right from the start, promising during his first inaugural address to "restore science to its rightful place."[1] He left unstated precisely what the rightful place of science was, but a few months later he elaborated in a speech to the National Academy of Sciences.

"Under my administration, the days of science taking a back seat to ideology are over," he announced to applause. ". . . To undermine scientific integrity is to undermine our democracy." Obama pledged to the scientists "a new effort to ensure that federal policies are based on the best and most unbiased scientific information." Declaring that "science forces us to reckon with the truth as best as we can ascertain it," he noted that some of the truths revealed by science "fill us with awe," while "others force us to question long-held views." Near the end of his address, as if it were an afterthought, Obama acknowledged that "science can't answer every question. . . . Science cannot supplant our ethics or our values, our principles or our faith." Nevertheless, "science can inform those things and help put those values . . . to work—to feed a child, or to heal the sick, to be good stewards of this Earth."[2]

In the years that followed, the Obama administration claimed for itself the mantle of scientific authority like no other presidency in American history. Administration officials regularly sought to communicate the imprimatur of science. In 2009 the White House hosted 150 medical doctors for a carefully staged photo-op as the president advocated his proposed overhaul of health insurance. Doctors were told to wear white lab coats for the event, presumably to lend an aura of scientific credibility for the attending media. Extra lab coats were even handed out to those who ignored the dress code.[3]

Notwithstanding President Obama's pledge to make "scientific decisions based on facts, not ideology," his administration's actions often sent a different message.

Prophet of the Apocalypse

Even before taking the oath of office, Obama announced his selection of a scientist named John Holdren to lead the White House Office of Science and Technology Policy. It was a strange choice if the administration truly aimed to keep ideology from interfering with science.

Without question, Holdren had a distinguished scientific résumé. A physicist as well as a professor of environmental policy at Harvard, he was a member of the National Academy of Sciences and past president of the American Association for the Advancement of Science. At the same time, he had a lengthy history of ideological activism in the name of science. In particular, he was a past collaborator with population-control zealot Paul Ehrlich, author of *The Population Bomb* (1968), which one critic has called (not without some justification) "the most spectacularly wrong book ever written."[4] Insisting that "the battle to feed all of humanity is over," Ehrlich had predicted a rising world death rate and mass starvation of "hundreds of millions of people" by the 1970s.[5] He was wrong on both counts (and many others).[6]

Ehrlich was a classic doomsayer, continually using science to predict that the Earth's destruction was right around the corner. Holdren had been one of his comrades-in-arms, coauthoring articles and books that predicted the demise of civilization if humanity did not make radical changes. In a book of readings Holdren and Ehrlich edited together titled *Global Ecology* (1971), they warned of catastrophic climate change. But the climate change they worried about most in 1971 was not global warming; it was global cooling. Conceding that warming would eventually become a problem, they worried that pollution could instigate a new global ice age well before then. Even worse, they announced, this new ice age, by adding weight to the Antarctic ice cap, "could generate a tidal wave of proportions unprecedented in recorded history."[7]

A decade later, Holdren was still predicting climate catastrophe, but now it was because "carbon-dioxide climate-induced famines could kill as many as a billion people before the year 2020."[8] By the turn of the new century, Holdren was supplying help to former vice president Al Gore in the creation of the now widely discredited documentary *An Inconvenient Truth* (which Holdren later insisted was "scientifically solid").[9]

Along with his regular prophesies of doom, Holdren promoted various authoritarian proposals in the name of science. In 1977 he coauthored the book *Ecoscience* with Ehrlich and Ehrlich's wife. Warning of a coming population catastrophe, the book seemed to recommend compulsory population control as the solution. The authors emphasized that "counting on either a spontaneous demographic transition or on voluntary family planning programs . . . to reduce population growth and thereby ensure successful development would . . . be a serious mistake."[10] They also praised the success of China's harsh population-control policies (while denying that China's policies were actually coercive).[11] Criticizing the "pronatalist bias" of U.S. income tax laws that "implicitly encouraged marriage and childbearing," Holdren and the Ehrlichs proposed discouraging marriages by imposing "high marriage fees" and discouraging new children by raising "taxes on luxury baby goods and toys." They also declared that "social pressures on both men and women to marry and have children must be removed." Holdren and his coauthors had little patience for those who believed that family size should be none of the government's business: "The number of children in a family is a matter of profound public concern. The law regulates other highly personal matters. For example, no one may lawfully have more than one spouse at a time. Why should the law not be able to prevent a person from having more than two children?"[12]

Holdren and the Ehrlichs proposed transferring political power from America's democratically elected leaders to what they called a "Planetary Regime," an international technocracy that would "control the development, administration, conservation, and distribution of all natural resources, renewable or nonrenewable, at least insofar as international implications exist. . . . The Regime might also be a logical central agency for regulating all international trade . . . including all food on the international market." The new world government could "be given responsibility for determining the optimum population for the world and for each region and for arbitrating various countries' shares within their regional limits." The authors didn't spell out how the "Planetary Regime" would enforce population limits, but in another part of their book they suggested the creation of "an armed international organization, a global analogue of a police force."[13]

Holdren's history of end-of-the-world predictions and coercive utopianism provided a foretaste of his approach as the Obama administration's top science adviser.

Critics accused Holdren of employing scare tactics and going beyond the scientific evidence to promote climate-change initiatives as head of the White House Office of Science and Technology Policy. He became embroiled in controversy because he highlighted recent extreme cold weather, droughts, and wildfires as evidence of the growing impact of global warming. As even some global warming advocates conceded, those claims were scientifically suspect. During the winter of 2014, for example, Holdren released a short video hyping the purported link between warming and cold weather.[14] A few weeks later, five climatologists responded in the journal *Science*: "As climate scientists, we share the prevailing view in our community that human-induced global warming is happening. . . . But we consider it unlikely that those consequences will include more frigid winters."[15]

Later in 2014, environmental studies professor Roger Pielke at the University of Colorado at Boulder tweeted citations to studies that contradicted Holdren's claims about links between global warming and droughts.[16] A political progressive who accepted the idea of human-caused global warming, Pielke was nevertheless a stickler for following what the evidence said. His contrarianism earned him abuse from Holdren and his defenders,[17] but Pielke was unrepentant, ultimately responding in the liberal magazine the *New Republic*: "Exaggerations by advocates of climate action, like those of science advisor Holdren, undermine that trust when they go beyond what the science is telling us. Efforts to quash mainstream, legitimate voices will further undermine that trust."[18]

In August of that year, Holdren tried to capitalize on wildfires blazing throughout the western United States by producing a YouTube video that connected wildfires to climate change. Titled "It Only Takes Three Minutes to See Why We Must Act on Climate Change," the video, according to the White House, was intended to show "how climate change is making America's wildfires more dangerous and why we must act now."[19]

Unfortunately for Holdren, three new studies published during the same year by different researchers argued that wildfires were not more severe today than they were in the past. "If we use the historical baseline as a point in time for comparison, then we have not seen a measurable increase in the size or the severity of fires," one of the researchers reported. "In fact, what we have seen is actually a deficit in forest fires compared to what early settlers were dealing with when they came through this area."[20] Not only that, but the 2014 wildfire season turned out to be significantly less damaging than the average season, with almost 50 percent fewer acres burned than the yearly average from 2004 to 2013.[21]

Scientism Unchained

John Holdren was not the only member of the Obama administration to make questionable scientific claims to further political goals or to justify various forms of coercive utopianism. Government coercion extended all the way into school lunchrooms. First Lady Michelle Obama became the administration's point person for a dramatic revamp of school lunch menus and food aid to the poor, all done under the banner of "sound science."

Stories soon began to circulate of students and local school district officials protesting the unintended consequences of the new regulations.[22] Photos of unappetizing meals prepared under the new standards spread across the Internet, including a student-produced YouTube video, "We Are Hungry," that ended up being viewed more than 1.4 million times.[23] The administration responded by impugning those who disagreed with its efforts as the enemies of science.

Writing in the *New York Times*, Mrs. Obama argued that "the initiatives we undertake are evidence-based, and we rely on the most current science. Research indicated that kids needed less sugar, salt and fat in their diets, so we revamped school lunch menus accordingly." She also attacked Congress for attempting to "override science" by considering a bill that would restore white potatoes to the list of foods poor women were allowed to purchase with funds from the Women, Infants, and Children (WIC) program.[24]

The science behind Mrs. Obama's claims about potatoes was disputed,[25] but even if the new dietary standards were based on "sound science," the Obama administration's apparent belief that science alone should determine what everyone in the nation could eat was classic scientism. Public policy is largely about reconciling competing goods, and attaining an ideal calorie count dictated by a government scientist is surely not the only human good. Other goods might include enjoying appetizing food, freely determining one's own diet according to one's own wishes, exercising local control of school menus, and maintaining the flexibility to tailor menus to a diversity of students, including athletes, teenage mothers, and others who need more calories.[26] The administration's if-we-impose-it-they-will-eat-it approach wasn't exactly a rousing success. According to government auditors, almost 1.1 million fewer students ate school lunches during the 2012–13 school year compared with the previous year.[27]

The Environmental Protection Agency (EPA) displayed even more hubris and insularity when it refused demands from Congress to disclose the scientific data the agency had used to establish sweeping new air pollution standards. Congress wanted the data released so that independent experts could evaluate whether the EPA's standards were justified. Yet for more than a year, EPA officials stonewalled.[28] When Congress continued to press the issue, EPA head Gina McCarthy unleashed a furious attack in a speech before the National Academy of Sciences. Lauding science as "our professor and our protector" as well as "our North Star," McCarthy denounced those seeking more transparency. "Those critics are playing a dangerous game by discrediting the sound science our families and our businesses depend on every day," McCarthy warned. Complaining that "our science seems to be under constant assault by a small—but vocal—group of critics," McCarthy claimed that the controversy over the EPA's secrecy was really "about challenging the credibility of world-renowned scientists and institutions like Harvard University and the American Cancer Society."[29] In other words, the government's scientists were so superior that their conclusions ought to be accepted without questioning.

McCarthy justified the EPA's secrecy as an effort to "protect confidential personal health data from those who are not qualified to analyze it—and won't agree to protect it." The privacy issue was a red herring. Congress already had made clear that it did not object to personal information being removed or protected.[30] One suspects that the real sticking point for McCarthy was her condescending attitude that outside experts who might disagree with the EPA "are not qualified to analyze" the scientific data. McCarthy ended petulantly: "If EPA is being accused of 'secret science' because we rely on real scientists to conduct research, and independent scientists to peer review it, and scientists who've spent a lifetime studying the science to reproduce it—then so be it!"[31] McCarthy's sweeping assertion that government scientists and experts should be above the norms of democratic accountability was breathtaking.

Congress eventually issued a subpoena to get the EPA to release the data, but even then, more months of stonewalling ensued. Finally the EPA supplied some but not all of the data. The agency conceded that the data supplied were "not sufficient" to allow independent researchers "to replicate the analyses in the epidemiological studies" the EPA had relied on.[32]

Even more disturbing than the administration's lack of transparency was its recurring use of "science" as a trump card to override both ethical concerns and religious liberty. This could be seen early on when it overturned the Bush administration's ban on federal funding for some kinds of embryonic stem-cell research.[33] Embryonic stem-cell research (as opposed to adult stem-cell research) can involve the destruction of human embryos, making the practice ethically problematic for a significant number of people.[34] Compelling taxpayers to fund the objectionable research intensified the ethical objections. But the Obama administration seemed oblivious to the larger ramifications of its new policy. In its view, advancing the cause of science was more important than accommodating the ethical objections of a significant portion of the nation's citizenry.

An especially egregious case of the administration's use of science to trump ethics involved its mandate on employers to cover contraceptives and potentially abortion-inducing drugs in their health-care plans.[35] According to the administration, the mandate was required by good science. "Scientists have abundant evidence that birth control has significant health benefits for women," declared Secretary of Health and Human Services Kathleen Sebelius.[36] Science supports contraception; thus employers (including many religious employers) ought to be compelled to provide it, including drugs that may induce abortions. QED.

Just how far some administration officials were willing to take the idea that science should override ethical concerns became apparent with the disclosure of a multiyear experiment funded by the National Institutes of Health (NIH) involving more than 1,300 premature infants. As part of the experiment, premature infants were randomly assigned to receive higher or lower levels of oxygen. Those receiving lower levels of oxygen were more likely to die, while those receiving higher levels of oxygen suffered serious eye damage that could lead to blindness. Parents were not informed of the possible increased risk of death for infants enrolled in the study. Nor were most of them informed that researchers recalibrated oxygen equipment to generate false readings, thus preventing medical staff from adjusting oxygen levels based on the individual needs of the infants in their care.[37]

Medical ethicists were appalled. "The word 'unethical' doesn't even begin to describe the egregious and shocking deficiencies in the informed-consent process for this study," said Michael Carome, MD, the director of the Health Research Group at the nonprofit (and politically liberal) group Public Citizen. "Parents of the infants who were enrolled in this study were misled about its purpose. . . . They were misled to believe everything being done was in the 'standard of care' and therefore posed no predictable risk to the babies."[38] Carome, who previously served in the Office for Human Research Protections in the U.S. Department of Health and Human Services, helped lead the effort to expose the misconduct of researchers and to ensure that the abuses did not recur.

The premature-infant study began during the administration of George W. Bush, but it was Obama administration officials who had to respond to the ethical objections raised. They had a choice: acknowledge there was a problem and fix it, or deny any wrongdoing. They chose the latter option.

Early in 2013 it became clear that the NIH's study was in trouble. The Office for Human Research Protections (OHRP) of the U.S. Department of Health and Human Services issued an enforcement letter against the

University of Alabama at Birmingham (UAB) because the researchers on the premature-infant study had failed to obtain adequate informed consent from participants. The OHRP required researchers to submit a plan to fix the problem. Yet only a few months later, the OHRP sent a follow-up letter placing its previous enforcement action on hold.

What had happened in the interim? According to Public Citizen, documents released under the Freedom of Information Act "strongly suggest that NIH launched an aggressive campaign to undermine OHRP's regulatory authority."[39] Although OHRP was supposed to act as an independent watchdog, NIH officials were allowed to review and rewrite the OHRP's second compliance letter. A coinvestigator of the study was also allowed to review the draft compliance letter. The full extent of the NIH's changes to the draft letter could not be ascertained because the Obama administration almost completely redacted the draft versions of the compliance letter it released under the Freedom of Information Act.

"NIH interference in the conduct of an ongoing compliance oversight investigation appears to be unprecedented in the history of OHRP," wrote Public Citizen. "This interference has seriously compromised the integrity and independence of OHRP's compliance oversight investigation."[40]

Public Citizen compared NIH's efforts to "a pharmaceutical company's being permitted by . . . the FDA Commissioner's office to review and edit a warning letter drafted by [the] FDA Office of Scientific Investigations about violations of the FDA's human subjects protection regulations involving a clinical trial sponsored by that company." Public Citizen noted that such an occurrence "obviously would be viewed as grossly unacceptable and, presumably, would never be permitted."[41]

Chief among the defenders of the premature-infant study was NIH head Francis Collins. One of Obama's key science appointees, Collins was known for his work as head of the Human Genome Project as well as for being an outspoken evangelical Christian. Unlike most evangelicals, however, Collins had supported Obama for president in 2008, and many of his views were out of sync with those of other evangelicals.[42] He was among the NIH officials permitted to review the OHRP's second compliance letter, and according to Public Citizen, he led a public relations campaign to undermine the OHRP's initial findings. Citing e-mail messages, Public Citizen accused Collins of seeking to have the second OHRP compliance letter issued the day before an article coauthored by Collins was to be published in the *New England Journal of Medicine* defending the premature-infant study.[43] Public Citizen found it "disturbing" that Collins and his coauthors "essentially leaked" to journal editors "the fact that OHRP soon would be issuing a compliance oversight letter to UAB putting on hold all compliance actions related to the investigation."[44]

In their public defense of the NIH-funded study, Collins and his coauthors insisted that "investigators had no reason to foresee that infants in one study group would have a higher risk of death than would those in the other group."[45] Public Citizen later called that claim "disingenuous," providing documentation showing that key researchers *were* aware of and discussed the possibility of a differential death rate from lower oxygen levels.[46] Indeed, one of the purposes of the study was to find out whether there was a differential death rate. In their article, Collins and his coauthors also neglected to disclose that researchers had recalibrated the oxygen equipment to prevent individualized care or that most parents had never been informed of this crucial fact. Science trumped ethics yet again.

The Obama administration's embrace of scientism was not limited to public policy. In 2014 President Obama ventured into the broader culture wars over science by taping a video introduction to the *Cosmos* television series hosted by astrophysicist Neil deGrasse Tyson. The creators of the series revealed that they had not asked for Obama's involvement; the White House had sought them out.[47] *Cosmos* was a reboot of an earlier series by the same name hosted by agnostic physicist Carl Sagan. Sagan had been criticized for trying to use science to promote metaphysical materialism, and in that sense Tyson's new series was a worthy heir to Sagan's original production.[48] Tyson had previously dismissed God as "an ever-receding pocket of scientific ignorance,"[49] and the producers of the new *Cosmos* were known for believing that "religion sucks"[50] and for warning students: "Stay away from the church. In the battle over science vs. religion, science offers credible evidence for all the serious claims it makes. The church says, 'Oh, it's right here in this book, see? The one written by people who thought the sun was magic?' "[51] Given such views, it wasn't surprising that the new *Cosmos* portrayed religion as the enemy of science, claimed that science shows how life originated through unguided processes, and even compared climate-change skeptics to Nazis.[52] Immediately after Obama's videotaped introduction, the 2014 series replayed a classic clip from the original series in which Carl Sagan professes his allegiance to materialism: "The cosmos is all that is, or ever was, or ever will be."

The Rise of Totalitarian Science

In many ways, the Obama administration's scientism reflected the trends documented in the rest of this book, trends that span both political parties and have become ever more pronounced during the past several years. Our culture is witnessing the rise of what could be called totalitarian science—science so totalistic in its outlook that its defenders claim the right to remake every sphere of human life, from public policy and education to ethics and religion. The evidence for the rise of this kind of scientific authoritarianism is not just anecdotal. A study published in 2010 confirmed that in recent years there has been a dramatic increase in what some have called the "authoritarian tone" of science, exemplified by the growing use in science journalism of phrases such as "science requires," "science dictates," and "science tells us we should."[53]

The area of ecology has seen increasing calls for coercive measures to control human population in the name of saving nonhuman life. Zoologist Eric Pianka at the University of Texas urges the reduction of the Earth's human population by up to 90 percent and calls on the government to confiscate all the earnings of any couple who has more than two children. "You should have to pay more when you have your first kid— you pay more taxes," he insists. "When you have your second kid you pay a lot more taxes, and when you have your third kid you don't get anything back, they take it all."[54] Shades of John Holdren and Paul Ehrlich in the 1970s.

In his recent book *The War on Humans*, lawyer and bioethicist Wesley J. Smith has documented how the growing coercive utopianism of many environmentalists is grounded in a visceral hatred of humans and the denial that human beings are special or unique.[55] In the words of zoologist Pianka, "Humans are no better than bacteria,"[56] and "Other things on this earth have been here longer than us . . . and they have a right to this planet too—that includes wasps that sting you, ants that bite you, scorpions and rattlesnakes."[57] Pianka goes on to criticize humans for "sucking everything we can out of mother Earth and turning it into fat human bio-mass."[58]

This disregard for humans reflects a reductionist form of Darwinian theory. Christopher Manes, one of the early leaders of the environmentalist group Earth First!, explains:

> Taken seriously, evolution means there is no basis for seeing humans as more advanced or developed than any other species. Homo sapiens is not the goal of evolution, for as near as we can tell evolution has no telos— it simply unfolds, life-form after life-form. Elephants are no more developed than toadstools, fish are no less advanced than birds, cabbages have as much ecological status as kings. Darwin invited humanity to face the fact that the observation of nature has revealed not one scrap of evidence that humankind is superior or special, or even particularly more interesting than, say, lichen.[59]

A similar Darwinian worldview inspired ecoterrorist James Lee, who in 2010 took staff of the Discovery Channel hostage. Lee called on the Discovery Channel to "talk about Evolution. Talk about Malthus and Darwin until it sinks into the stupid people's brains until they get it!" Lee's stated goal was to save "what's left of the non-human Wildlife by decreasing the Human population. That means stopping the human race from breeding any more disgusting human babies!"[60]

The Darwinian denial of human exceptionalism can be found on both sides of the political spectrum. On the Left, there is Princeton bioethicist Peter Singer, author of *A Darwinian Left*.[61] Singer's view that "the life of a newborn baby is of less value than the life of a pig, a dog, or a chimpanzee" was discussed in chapter 14, as was his insistence that "Darwin's theory undermined the foundations" of our traditional "Western way of thinking about the place of our species in the universe." On the Right, there is John Derbyshire, who was a longtime writer for *National Review* until being dismissed in 2012 after authoring an article for another publication arguing that blacks are more antisocial and less intelligent than whites.[62] Derbyshire argues that racial differences are the products of evolution.[63] He has also written that "the broad outlook on human nature implied by Darwinian ideas contradicts the notion of human exceptionalism, without which the Abrahamic religions lose their point." He added that "modern biologists, informed by Darwin," know "we are merely another branch on Nature's tree."[64]

Scientism has expanded in the area of medicine and bioethics, where the old idea of eugenics is being revived in the name of good science. In 2012 Nancy Snyderman, chief medical editor for NBC News, publicly defended eliminating handicapped babies through abortion: "I am pro-science, so I believe that this is a great way to prevent diseases."[65] Of course, if it is "pro-science" to support eradicating babies with genetic flaws, it must be "anti-science" to oppose it. In a 2014 *Huffington Post* article titled "Let's (Cautiously) Celebrate the 'New Eugenics,'" Jon Entine of the Genetic Literacy Project went further, assuring readers that they had nothing to fear from efforts to improve humans through genetic engineering.[66]

Still on the horizon are the even more radical "transhumanists," who argue that humanity's goal should be bioengineering a new race of supermen. According to transhumanist Nick Bostrom, "human nature" is "a work-in-progress, a half-baked beginning that we can learn to remold in desirable ways." "Current humanity need not be the endpoint of evolution," proclaims Bostrom. "Transhumanists hope that by responsible use of science, technology, and other rational means we shall eventually manage to become posthuman, beings with vastly greater capacities than present human beings have."[67] Bostrom is a professor of philosophy at Oxford University, an indication of how ideas that used to be on the fringe have seeped into the mainstream.

In the areas of sexuality and family life, the major development related to scientism since the original publication of this book has been the revolution in the definition of marriage—often imposed by judges, but in some states adopted by voters. The arguments over same-sex marriage are many and complicated, but scientism certainly has played a key role. In 2011 U.S. Attorney General Eric Holder justified his refusal to defend the Defense of Marriage Act by declaring that "a growing scientific consensus accepts that sexual orientation is a characteristic that is immutable." Indeed, "social science regarding sexual orientation . . . make[s] clear that sexual orientation is not a characteristic that generally bears on legitimate policy objectives."[68] Never mind that Holder's scientific claims were highly debatable[69]; his operating assumption, like that of many American judges, lawyers, and policymakers, appeared to be that "science" should be the arbiter of whether the traditional definition of marriage should be allowed to survive. Such an approach dismissed philosophy, theology, reflections on the human condition, and moral reasoning as outdated or irrelevant. Unless male-female marriage could be justified by the state by "scientific" evidence, it must fail.

The same-sex marriage controversy is a good example of how scientism often ends up producing greater coercion for many, not freedom of choice. As florists, bakers, justices of the peace, and even members of the clergy are now discovering, once same-sex marriage is mandated, freedom of choice for those who do not want to participate in or bless same-sex marriages largely disintegrates.[70] Even the freedom to disagree publicly with the idea of same-sex marriage comes into question, as a number of people have discovered after losing their jobs because of their support for traditional marriage.[71]

The effort to curtail freedom of choice in the name of science now extends to those struggling with their own sexual feelings. Once "science" has normalized all sexual orientations and gender identities, those experiencing conflict between their feelings and their moral or religious beliefs may be legally prevented from receiving the help they want. In 2012 and 2013 California and New Jersey enacted bans on so-called sexual orientation change efforts by licensed counselors for young people under eighteen.[72] The bans were marketed to the public with claims that young people were being forced to undergo electric shock, ice baths, and similar barbaric treatments to change their sexual orientation. These lurid stories often turned out to have no basis in fact or to involve nonlicensed counselors. For example, New Jersey legislators heard from an individual who claimed to have been sent to a summer camp in Ohio where the person was subjected to horrific electric-shock therapy. It turned out that the camp did not exist, and the entire story appears to have been fabricated, inspired by a fictional Hollywood film from 1999.[73]

While media organizations seemed eager to publicize the most unsubstantiated claims about sexual orientation change efforts, most failed to note that the extreme therapies being discussed had already been rejected by the main professional organization of licensed therapists involved with helping those with unwanted sexual feelings. That group stated categorically: "We do not advocate the use of shock therapy, aversion therapy, holding therapies or any other intervention that has demonstrated potential for harm."[74]

Even if treatments such as electric shock to change sexual orientation were actually occurring, the laws adopted by California and New Jersey were a classic bait and switch: they did not specifically target such extreme treatments, and they defined sexual orientation change efforts so broadly that they banned even talk therapy "to change behaviors . . . or gender expressions."[75] Under that expansive definition, a boy in distress because he wanted to dress like a girl could no longer receive professional counseling to help him change his unwanted behavior.

It did not take long for the cruelty of the new restrictions to become evident. A Catholic teenager in New Jersey was so distraught at his unwanted sexual feelings and gender confusion that he had attempted suicide. His parents sought help from an unlicensed counselor, but that counselor only made matters worse. Then they found a licensed professional who helped their son deal with his unwanted feelings in a way consistent with his religious and ethical beliefs. As a result, their son experienced significant improvement and change. Yet New Jersey's new law prevented this young man from receiving the help he urgently desired from a licensed therapist.[76]

A few voices in the mental health community protested this assault on patients' rights. Nicholas Cummings formerly served as president of the American Psychological Association and as chief of mental health for Kaiser Permanente, the largest managed-health-care organization in the United States. A strong supporter of gay rights, Cummings sponsored the American Psychological Association's resolution declaring that homosexuality is not a mental disorder. Nevertheless, based on his own professional expertise and clinical experience with thousands of patients, Cummings strongly opposed efforts to ban sexual orientation change therapy for clients who desired it. "Attempting to characterize all sexual reorientation therapy as 'unethical' violates patient choice and gives an outside party a veto over patients' goals for their own treatment," he wrote. "A political agenda shouldn't prevent gays and lesbians who desire to change from making their own decisions."[77] Cummings's objections fell on deaf ears. Patient choice apparently did not extend to those who ran afoul of the new "scientific" consensus on sexuality.

Laws banning sexual orientation change efforts communicate a message that a young person's current sexual feelings and behaviors are absolutely unchangeable—a message that is not, in fact, based on good science.

Although sexual orientation may not be changeable for some young people (and it is probably difficult to change for many others), recent research has shown remarkable fluidity in sexual attraction, behavior, and identity among young people with same-sex attractions.

Survey data from more than fourteen thousand teens tracked by the National Longitudinal Study of Adolescent Health show that "approximately 80% of adolescent boys and half of adolescent girls who expressed either partial or exclusive same-sex romantic attraction" eventually became heterosexual as young adults.[78] Refusing to accept this evidence, some researchers have speculated that the results must have been due to prank answers.[79] But other surveys have demonstrated similar fluidity in sexual identity among those with same-sex attractions.[80] In fact, the evidence for changeability is becoming so pervasive that prominent sex researcher Lisa Diamond at the University of Utah, a self-identified lesbian, has urged the LGBT community to jettison their talking point that sexual orientation is immutable: "The queers have to stop saying 'Please help us, we're born this way and we can't change.' " She said that this argument "is going to bite us in the ass, because now we know there is enough data out there" showing it's not true.[81]

Even more alarming is the revival of efforts to promote the legitimacy of child-adult sex. As discussed in chapter 13, the effort to legitimize such interactions was a major item on the agenda of sexual reformers up through the 1970s. This effort provoked a backlash and was pushed to the fringes. But now the issue is returning to the mainstream. A conference held at Cambridge University in 2013 featured apologists for pedophilia such as sexuality researcher Philip Tromovitch, who insisted that "paedophilic interest is natural and normal for human males."[82] In the United States, an American Psychiatric Association working group expressed concern that the association's diagnostic criteria for pedophilia technically applied only to adults who desire sexual interactions with children ten and under. The group proposed amending the criteria to make clear that they applied to adults who desire to have sex with children between the ages of eleven and fourteen as well. But in 2012 the American Psychiatric Association rejected the proposal, leading many to wonder whether the association now considered sexual contacts between adults and children in early puberty as normal.[83]

Evangelizing Atheists

The growing militancy of scientism can be seen in renewed claims that faith and science are incompatible.

In the fall of 2014, University of Washington evolutionary psychologist David Barash took to the pages of the *New York Times* to explain "The Talk" he gives to students at the beginning of each academic year. The Talk, he wrote, "isn't, as you might expect, about sex, but about evolution and religion, and how they get along. More to the point, how they don't." Barash's take-home point for his students? "The more we know of evolution, the more unavoidable is the conclusion that living things, including human beings, are produced by a natural, totally amoral process, with no indication of a benevolent, controlling creator."[84]

Barash is a scientist who resides in politically and socially progressive Seattle. But even in the nation's heartland, and even in humanities classes, the science-refutes-religion message can be a prominent theme at colleges and universities. At Ball State University in Indiana, for example, an English professor has taught an honors course on "Dangerous Ideas." The sole textbook for the course is an anthology edited by a prominent atheist in which the authors assert that "Science Must Destroy Religion," that "There is no God; no Intelligent Designer; no higher purpose to our lives," and even that scientists should function as our society's "high priests."[85]

Chapter 11 discussed previous efforts to misuse science, especially Darwinian biology, to promote a metaphysical agenda. In recent years the efforts to promote scientific atheism have grown more pronounced, with troubling effects on the rising generation.

Among the growing number of Americans who describe themselves as unaffiliated with any particular faith, nearly a third (32 percent) believe that "modern science proves religion is superstition," while nearly a quarter (23–24 percent) identify this belief as an "important reason" they became unaffiliated. The percentages are even higher for those raised as mainline Protestants: 39 percent of those currently unaffiliated who grew up as mainline Protestants now believe that "modern science proves religion is superstition," and 31 percent identify this belief as an "important reason" they became unaffiliated.[86]

Behind these statistics are people like Kyle Simpson. Raised as a devout Christian, the twenty-something Kyle now no longer believes in God, although he says he wants to believe.[87] Scientific claims played a key role in his change of mind. Interviewed by National Public Radio as part of a larger story on young people who leave their faith, Kyle explained: "I don't [believe in God] but I really want to. That's the problem with questions like these is you don't have anything that clearly states, 'Yes, this is fact,' so I'm constantly struggling. But looking right at the facts—evolution and science—they're saying, no there is none."

Kyle's story is not surprising given that young people grow up in a culture where 55 percent of American adults now believe that "science and religion [are] often in conflict."[88] This same attitude can be found among young adults originally raised as Christians. Nearly a third of young adults with a Christian background believe that "churches are out of step with the scientific world we live in," and a quarter of them think that "Christianity is anti-science."[89]

Although many colleges and universities are undoubtedly fertile ground for promoting the idea that science and faith are in conflict, recent research suggests that a significant percentage of students already embrace such a view by the time they enter college. In a survey of more than 100,000 incoming college freshmen, nearly a third of first-year male students (30.6 percent) and nearly a fourth of first-year female students (22 percent) agreed with the statement "The universe arose by chance." Moreover, almost a third (30.8 percent) of all freshmen viewed the relationship between science and faith as one of "conflict." Another third of all freshmen (31.9 percent) viewed the relationship between science and faith as one of "independence"—suggesting a compartmentalized view of religion as irrelevant to questions raised by science.[90]

One can expect the gospel of scientific materialism to spread to still younger students in future years. In a foretaste of things to come, Boston University psychologist Deborah Kelemen has conducted research showing that children may be "intuitive theists" who are predisposed to see nature as the product of intelligent design. In her view, rather than something to be celebrated, this finding is a cause for alarm because it means "children's science failures may, in part, result from inherent conflicts between intuitive ideas and the basic tenets of contemporary scientific thought."[91] Given children's "teleological biases," which lead them to see "goal-directed design" in nature, Kelemen believes science educators need to be more aggressive in reaching students at younger ages, and she even recommends staging "storybook interventions" with elementary school children to convince them of the logic of unguided natural selection.[92]

Burning the Heretics

The proponents of scientism are becoming even more extreme in their efforts to shut down anyone who disagrees with the purported scientific "consensus" on any given topic.

Testifying before Congress in 2011, Obama administration science adviser John Holdren sought to explain to members of Congress why they should disregard the views of scientists skeptical of various claims about global warming. "There are always heretics," said Holdren dismissively.[93]

It was a revealing comment. Once scientism turns science into a religion, disagreement with the dogmas of the orthodox becomes tantamount to heresy.

The very issue Holdren was testifying about—climate change—provides a disturbing example of the growing effort to treat scientific dissent as heresy. One of America's leading daily newspapers, the *Los Angeles Times*, announced in 2013 it would no longer publish letters to the editor that expressed skepticism about the human role in climate change.[94] Since one of the original purposes of printing letters to the editor was to air community viewpoints that might differ from a newspaper's official position, the *Times*'s decision represented a dramatic departure from historic journalistic standards.

Others go much further, calling for the criminal prosecution of global warming skeptics. In 2014 Professor Lawrence Torcello at the Rochester Institute of Technology published an essay urging that global warming skeptics who receive funding for their work be charged with criminal negligence. Torcello labeled any criticisms of his proposal on grounds of free speech as "misguided."[95] Journalist Adam Weinstein published a follow-up essay in which he declared that "denialists should face jail. They should face fines. They should face lawsuits from the classes of people whose lives and livelihoods are most threatened by denialist tactics."[96]

Such lawsuits are already occurring. As I write this chapter, the conservative journal *National Review* faces potentially ruinous litigation from Penn State University climatologist Michael Mann, one of the leading global warming activists in academia, because *National Review* writers vigorously criticized his research.[97]

Other than climate change, the science topic on which intimidation and censorship is probably the most rampant right now is the debate between Darwinian biology and intelligent design. Chapter 12 documented the efforts to silence both those who dissent from Darwin and those who think there is evidence of intelligent design in nature. Since the original publication of this book, the intolerance of dissent in this area has continued to worsen.

Martin Gaskell is by all accounts an outstanding astronomer. He was a top applicant to become the head of an observatory at the University of Kentucky. In the words of one university faculty member there, "his qualifications . . . stand far above those of any other applicant."[98] But Gaskell was ultimately rejected for the job after the biology faculty waged an internal war against his choice. Why did they want to prevent him from getting the job? First, Gaskell was perceived by other faculty to be "potentially evangelical."[99] Worse, although he identified himself as a supporter of evolution, in online notes for a talk on science and faith Gaskell respectfully discussed the views of intelligent-design proponents and acknowledged that modern evolutionary theory had unresolved problems, just like any scientific theory. That was apparently too much for the biologists at the University of Kentucky.

Eric Hedin is an associate professor of physics at Ball State University. Like Gaskell, he is an outstanding scientist with a long list of professional peer-reviewed publications.[100] For many years he taught an interdisciplinary honors class at his university called "The Boundaries of Science" that explored the limits of science, and during one small part of the course he discussed the debate over intelligent design in physics and cosmology, on issues such as the fine-tuning of the universe for life.[101] The course received positive student reviews.[102] But in 2013 atheist evolutionary biologist Jerry Coyne at the University of Chicago and the Freedom from Religion Foundation lodged complaints against Hedin.[103] Ball State then violated its own procedures and appointed an ad hoc committee stacked with avowed critics of intelligent design, including two who were speakers at a previous Darwin Day conference organized by the Ball State Freethought Alliance,[104] a group whose "original goal," according to its president, was "belittling religion."[105] Ball State eventually canceled Hedin's class. In addition, the university president issued a campus speech code banning professors not only from covering intelligent design in science classes but also from expressing support for the concept in social science and humanities classes.[106] Recall that Ball State is the same university that allowed an English professor to teach a course emphasizing that "Science Must Destroy Religion." When it comes to academic freedom, some professors are freer than others.

David Coppedge was a senior computer systems administrator for the Cassini Mission to Saturn at NASA's Jet Propulsion Lab in California. He faced demotion and discharge after he offended his supervisor by occasionally offering to loan colleagues DVDs about intelligent design.[107] No one had ever complained to Coppedge about his offers of DVDs, but when the supervisor found out, Coppedge faced a punitive investigation. His employment evaluations, which had been outstanding, suddenly took a nosedive. He ultimately lost his job, supposedly as part of budget reductions.

Efforts to suppress the discussion of intelligent design now extend to trying to shut down private events. In 2009 the California Science Center in Los Angeles, a branch of the state government, rented its theater to a private group to screen a pro–intelligent design film titled *Darwin's Dilemma*. After local pro-Darwin scientists complained, the center broke the contract and canceled the screening.[108] During the same year, activists mounted a massive denial-of-service attack designed to shut down online registration for a conference sponsored by another pro–intelligent design group.[109]

Some activists are now engaging in what are essentially mob actions to muzzle intelligent-design supporters. Stanley Wilson is a biologist who was scheduled to teach a course on the debate over evolution and intelligent design at Amarillo College in Texas. The seminar was to be offered as a noncredit personal enrichment course for adult education. But then members of a local "freethinkers" group threatened the college, saying that if the course went ahead they would physically disrupt the class. So the college canceled it.[110]

In addition to outright suppression, proponents of scientism try to squelch public debate by demonizing those who dare to speak out. Fortunately, some scientists and scholars refuse to be silenced.

Doubting Thomas

If someone prior to 2012 had predicted that Oxford University Press would publish a book with the title *Mind and Cosmos: Why the Materialist Neo-Darwinian Conception of Nature Is Almost Certainly False*, one might have wondered about his sanity, or at least about how familiar he was with current discourse in elite academia. But Oxford did in fact publish the book, and the intellectual aftershocks have yet to subside.[111]

The book's author, philosopher Thomas Nagel, is a professor of long standing at New York University and the recipient of numerous awards and honors, including an honorary doctorate from Oxford University, fellowships from the National Science Foundation and the National Endowment for the Humanities, and election to such august bodies as the American Academy of Arts and Sciences and the American Philosophical Society.[112] It is a testament to Professor Nagel's stature that his dissent from Darwinian theory was allowed to be published at all. But his stature has not prevented a flood of abuse and even occasional suggestions of creeping senility.

It's not often that a book by a professional philosopher attracts the notice—let alone the ire—of the cultural powers-that-be. *Mind and Cosmos* has been denounced in *The Nation* and the *Huffington Post*,[113] attacked by prominent academics including evolutionary psychologist Steven Pinker at Harvard and biologist Jerry Coyne at the University of Chicago,[114] dubbed the "most despised science book of 2012" by the London *Guardian*,[115] defended in the *New Republic* (where Nagel's critics were blasted as "Darwinist dittoheads" and a "mob of materialists"),[116] reported on in a feature story for the *New York Times*,[117] and put on the cover of the *Weekly Standard*, which depicted poor Professor Nagel being burned alive while surrounded by a cabal of demonic-looking men in hoods.[118]

Nagel attracted special displeasure for praising Darwin skeptics like mathematician David Berlinski and intelligent-design proponents like biochemist Michael Behe and philosopher of science Stephen Meyer.[119] As the *New York Times* explained, many of Nagel's fellow academics view him unfavorably "not just for the specifics of his arguments but also for what they see as a dangerous sympathy for intelligent design."[120] Now there is a revealing comment: academics, typically blasé about everything from justifications of infanticide to the pooh-poohing of pedophilia, have concluded that it is "dangerous" to give a hearing to scholars who think nature displays evidence of intelligent design.

Unfortunately for Nagel, he is a serial offender when it comes to listening to the purveyors of such disreputable ideas. In 2009 he selected *Signature in the Cell: DNA and the Evidence for Intelligent Design* as a book of the year for the *Times Literary Supplement*.[121] Written by my Discovery Institute colleague Stephen Meyer (whose ideas are discussed in the original conclusion to this book), *Signature in the Cell* made the case for purpose in nature from the existence of the digital information embedded in DNA. After being denounced by one scientist for praising Meyer's book, Nagel dryly recommended that the scientist should "hold his nose and have a look at the book" before dismissing it.[122]

Apparently unconcerned about being accused of consorting with the enemy, Nagel insisted in *Mind and Cosmos* that "the defenders of intelligent design deserve our gratitude for challenging a scientific world view that owes some of the passion displayed by its adherents precisely to the fact that it is thought to liberate us from religion." Nagel added that he thinks this antireligious materialist worldview "is ripe for displacement"—an intriguing comment considering that he himself remains an unrepentant atheist.[123]

Nagel ultimately offered a simple but profound objection to Darwinism: "Evolutionary naturalism provides an account of our capacities that undermines their reliability, and in doing so undermines itself."[124] In other words, if our mind and morals are simply the accidental products of a blind material process like natural selection acting on random genetic mistakes, what confidence can we have in them as routes to truth?

The basic philosophical critique of Darwinian reductionism offered by Nagel had been made before, perhaps most notably by Sir Arthur Balfour, C. S. Lewis, and Alvin Plantinga.[125] But around the same time as the publication of Nagel's book came new *scientific* discoveries that undermined Darwinian materialism as well. In the fall of 2012, the Encyclopedia of DNA Elements (ENCODE) project released results showing that much of so-called junk DNA actually performs biological functions.[126] The ENCODE results overturned long-repeated claims by leading Darwinian biologists that most of the human genome is genetic garbage produced by a blind evolutionary process.[127] At the same time, the results confirmed predictions made during the previous decade by scholars who think nature displays evidence of intelligent design.[128]

New scientific challenges to orthodox Darwinian theory have continued to proliferate. In 2013 Stephen Meyer published *Darwin's Doubt: The Explosive Origin of Animal Life and the Case for Intelligent Design*, which threw down the gauntlet on the question of the origin of biological information required to build animal body plans in the history of life. The intriguing thing about Meyer's book was not the criticism it unleashed from the usual suspects but the praise it attracted from impartial scientists. Harvard geneticist George Church lauded it as "an opportunity for bridge-building rather than dismissive polarization—bridges across cultural divides in great need of professional, respectful dialogue."[129] Paleontologist Mark McMenamin, coauthor of a major book from Columbia University Press on animal origins, called it "a game changer for the study of evolution" that "points us in the right direction as we seek a new theory for the origin of animals."[130]

Even critics of *Darwin's Doubt* found themselves at a loss to come up with a convincing answer to Meyer's query about biological information. University of California at Berkeley biologist Charles Marshall, one of the world's leading paleontologists, attempted to answer Meyer in the pages of the journal *Science* and in an extended debate on British radio. But as Meyer and others pointed out, Marshall tried to explain the needed information by simply presupposing the prior existence of even more unaccounted-for genetic information. "That is not solving the problem," said Meyer. "That's just begging the question."[131]

C. S. Lewis perceptively observed in his final book that "nature gives most of her evidence in answer to the questions we ask her."[132] Lewis's point was that old paradigms often persist because they blind us from asking certain questions. They begin to disintegrate once we start asking the *right* questions. Scientific materialism continues to surge, but perhaps the right questions are finally beginning to be asked.

It remains to be seen whether as a society we will be content to let those questions be begged or whether we will embrace the injunction of Socrates to "follow the argument . . . wherever it may lead."[133] The answer to *that* question may determine our culture's future.

Notes

Introduction

1. James Kalat, *Introduction to Psychology*, 4th ed. (Pacific Grove: Brooks/Cole Publ.,1996): 8.
2. Interview with John Lofton, "Rape, for Good or Ill?" [March 25, 2000], http://www.worldnet-daily.com/bluesky lofton/20000325 xclof rape good.shtml (accessed March 26, 2000). The book cited is Randy Thornhill and Craig T. Palmer, *A Natural History of Rape: Biological Bases of Sexual Coercion* (Cambridge, MA: MIT Press, 2000).
3. "The Evolution of Biological Social Systems," Press Release, Office of Legislative and Public Affairs, National Science Foundation, September 2003, http://www.nsf.gov/od/lpa/news/03/pr03106_slimemold.htm (accessed March 7, 2005). Although hopeful, the press release for the project did caution that the "mechanisms for social behavior in slime molds may not correspond closely to those in other organisms."
4. Hugh Elliot, *Modern Science and Materialism* (London: Longmans, Green, and Co., 1927), 138.
5. For a good discussion of scientism, see Michael D. Aeschliman, *The Restitution of Man: C. S. Lewis and the Case Against Scientism* (Grand Rapids, MI: 1998).
6. Positivism originated with French philosopher Auguste Comte (1798–1857), although its roots go back further, and there have been many variations after Comte. For discussions of Comte's system of positivism, see Guido de Ruggiero, "Positivism," in Edwin R. A. Seligman, editor, *Encyclopedia of the Social Sciences* (New York: The Macmillan Co., 1937), 11:260–66; and Robert Lowry Clinton, "Democracy, Nineteenth Century Social Science, and the Devil's Pitchfork," paper presented at the Southwestern Political Science Association Annual Meeting, March 2005, New Orleans. For more general discussions of positivism and the ways in which it may be distinguished from strict philosophical materialism, see Neal C. Gillespie, *Charles Darwin and the Problem of Creation* (Chicago: University of Chicago Press, 1979), especially 8–11, 153, and Maurice Mandelbaum, *History, Man, and Reason: A Study in Nineteenth-Century Thought* (Baltimore: The Johns Hopkins Press, 1971), 10–37.

7. Gillespie, *Charles Darwin*, 8.
8. For a good summary of E. O. Wilson's views, see Edward O. Wilson, *Consilience: The Unity of Knowledge* (New York: Alfred A. Knopf, 1998).
9. John O. McGinnis, "The Origin of Conservatism," *National Review* (December 22, 1997), 31.
10. Larry Arnhart, *Darwinian Conservatism* (Charlottesville, VA: Imprint Academic, 2005), 1. See also James Q. Wilson, *The Moral Sense* (New York: The Free Press, 1993) and Larry Arnhart, *Darwinian Natural Right: The Biological Ethics of Human Nature* (New York: State University of New York Press, 1998). For a critique of Arnhart's views, see John G. West, *Darwin's Conservatives: The Misguided Quest* (Seattle: Discovery Institute Press, 2006).
11. Peter Singer, "Darwin for the Left," *Prospect* (June 1998); also see Peter Singer, *A Darwinian Left: Politics, evolution and cooperation* (New Haven: Yale University Press, 2000).
12. J. McKeen Cattell, "Homo Scientificus Americanus," *Science*, April 10, 1903, 569.
13. Charles W. Eliot, "The Fruits, Prospects, and Lessons of Recent Biological Science," *Science*, December 31, 1915, 926.
14. Quoted in Frederick Albert Lange, *History of Materialism*, trans. Ernest Chester Thomas (London: Kegan Paul, Trench, Trübner, & Co. Ltd., 1892), 2:312.
15. C. S. Lewis, *The Abolition of Man* (New York: Macmillan Publishing, 1947), 84.

Chapter 1: Nothing Buttery from Atomism to the Enlightenment

1. Quoted in Nicholas Wade, "Animal's Genetic Program Decoded, in a Science First," *New York Times*, December 11, 1998.
2. Quoted in ibid.
3. Quoted in Maggie Fox, "Fly Gene Map May Have Many Uses, Scientists Say," Reuters, March 23, 2000, http://dailynews.yahoo.com/h/nm/20000323/sc/fly_uses_2.html.
4. Patricia Reaney, "Are You Man or Mouse? Check Your Genes . . .," Reuters, Dec. 4, 2002, http://story.news.yahoo.com/news?tmpl=story2&cid=570&u=/nm/20021204/.
5. Derek E. Wilman, Monica Uddin, Guozhen Liu, Lawrence Grossman, and Morris Goodman, "Implications of Natural Selection in Shaping 99.4% Nonsynonymnous DNA Identity between Humans and Chimpanzees: Enlarging Genus Homo," *Proceedings of the National Academy of Sciences* 100, no. 12 (June 10, 2003): 7181. Goodman is identified as the contributor of this article to *Proceedings*.
6. Randolph Schmid, "Team Sees Primate as Chimp off the Old Block," *Seattle Times*, May 20, 2003.
7. Morris Goodman, "Current Research," http://www.med.wayne.edu/anatomy/department/goodman.htm.
8. Richard Lewontin, "Billions and Billions of Demons," *New York Review of Books* electronic edition, January 9, 1997, http://www.nybooks.com/articles/1297.
9. Leo Strauss, *Liberalism Ancient and Modern* (Ithaca, NY: Cornell University Press, 1968), 207. In the original context, Strauss in this statement was describing the impact of "the new political science, or the universal science of which the new political science is a part." The "universal science" described by Strauss is described in this book as "scientific materialism."
10. Donald MacKay, *Human Science and Human Dignity* (Downers Grove, IL: InterVarsity Press, 1979), 27.
11. Quoted in Fox, "Fly Gene."
12. Michael Ruse and E. O. Wilson, "The Evolution of Ethics" in *Religion and the Natural Sciences: The Range of Engagement*, by James Huchingson (New York: Harcourt Brace Jovanovich, 1993), 210.
13. Francis Crick, *The Astonishing Hypothesis: The Scientific Search for the Soul* (New York: Touchstone, 1995), 3.
14. Stephen Jay Gould, *Ever Since Darwin* (New York: W. W. Norton, 1977), 13.
15. Introduction to Plato, *Laws*, trans. R. G. Bury (Cambridge, MA: Harvard University Press, 1917), 1:vii.

16. Plato, *Laws*, 889B–C, in *The Dialogues of Plato*, trans. Benjamin Jowett (New York: Random House, 1937), 631.
17. Ibid.
18. Ibid., 890A, 631–32.
19. Ibid., 890B, 632.
20. Ibid, 896C-D, 638.
21. Introduction to Plato, *Laws*, trans. R. G. Bury, 1:vii; David Ritchie, *Plato* (New York: Charles Scribner's Sons, 1902), 175.
22. W. K. C. Guthrie criticizes the use of the term "materialist" for sixth-century Milesian philosophers. "That is a word currently applied to those who deliberately deny any place to the spiritual among first principles. It denotes all those who, well aware that a distinction has been drawn between material and non-material, are prepared to maintain in argument that nothing in fact exists which has not its origin in material phenomena. To suppose the sixth-century philosophers to have been capable of thinking like this is a serious anachronism." Guthrie, *A History of Greek Philosophy* (Cambridge: Cambridge University Press, 1965), 1:64.
23. For a discussion of what little is actually known about the biographies of Leucippus and Democritus, consult W. K. C. Guthrie, *A History of Greek Philosophy*, 2:382–89; and G. S. Kirk, J. E. Raven, and M. Schofield, *The Presocratic Philosophers*, second edition (New York: Cambridge University Press, 1983), 402–6. For good discussions of the thought of the early atomists, see *History of Greek Philosophy*, 2: 382–507; Kirk, *Presocratic Philosophers*, 402–33; Jonathan Barnes, *The Presocratic Philosophers* (New York: Routledge, 1982), 342–77; Frederick Albert Lange, *History of Materialism*, trans. Ernest Chester Thomas, 4th ed. (London: Kegan Paul, Trench, Trübner, & Co., 1892), 1: 3–32.
24. Quoted in *History of Greek Philosophy*, 2: 451.
25. Ibid., 441–46.
26. Ibid., 454. The atomists combined their mechanical theory of thought with skepticism about human knowledge. Just how far the atomists pressed their skepticism is a matter of debate, but according to the ancient authority Sextus, Democritus sometimes denied the reality of "what appears to the senses," arguing that the only true reality was "atoms and void" (Kirk, *Presocratic Philosophers*, 410). However, Sextus went on to report that Democritus did believe genuine knowledge could be apprehended through the "intellect" even if it could not be produced by "sight, hearing, smell, taste, [and] touch" (ibid., 412). Given the fragmentary nature of the evidence, a definitive assessment about the extent of the atomists' skepticism is probably impossible.
27. Ibid., 420.
28. Ibid., 419.
29. Guthrie, *History of Greek Philosophy* 2: 501.
30. Epicurus to Herodotus in *The Stoic and Epicurean Philosophers: The Complete Extant Writings of Epicurus, Epictetus, Lucretius, Marcus Aurelius*, ed. Whitney J. Oates (New York: Random House, 1940), 4–5.
31. Ibid., 11.
32. Epicurus to Menoeceus in *Stoic and Epicurean Philosophers*, 31. Somewhat inconsistently, Epicurus defended the existence of immortal gods, but he may only have been trying to shield himself from persecution (ibid., 30). His beliefs about the gods certainly were far from traditional. He disparaged "the beliefs of the many" when it came to religion, going so far as to say that "the impious man is not he who denies the gods of the many, but he who attaches to the gods the beliefs of the many" (ibid.). According to Epicurus, the gods had no effect on the physical world, and human beings should neither fear nor expect divine interference in human affairs (Epicurus to Herodotus, *Stoic and Epicurean Philosophers*, 13). Epicurus implied that men ought to be thankful for this fact, because "if God listened to the prayers of men, all men would quickly have perished: for they are for ever praying for evil against one another" (Fragment 58, *Stoic and Epicurean Philosophers*, 50; the source of this fragment is uncertain).
33. Epicurus to Menoeceus, *Stoic and Epicurean Philosophers*, 33.
34. Epicurus to Herodotus, *Stoic and Epicurean Philosophers*, 5.
35. Lange, *History of Materialism* 1:142n68. Lange in this passage is referring most directly to Lucretius's belief in swerving atoms, but he notes that the same position was held by Epicurus.
36. Epicurus to Menoeceus, *Stoic and Epicurean Philosophers*, 32.

37. George Depue Hadzsits, Lucretius and His Influence (New York: Cooper Square Publishers, 1963), 5.
38. Lucretius, "On the Nature of Things," trans. William Ellery Leonard, http://classics.mit.edu/Carus/nature_things.html. One change has been made to the above translation. The word "design" has been substituted for the word "counsel" in conformity with at least two other translations. See Lucretius, De Rerum Natura, I.1025, transz W. H. D. Rouse (Cambridge, MA: Harvard University Press, 1959), 75; Lucretius, "On the Nature of Things," in Stoic and Epicurean Philosophers, trans. H. A. J. Munro, 88.
39. Introduction, Stoic and Epicurean Philosophers, xviii.
40. Hadzsits, Lucretius, 30–35.
41. Aristotle, Physics, 194b.17–195a.5, trans. Hippocrates Apostle (Grinnell, Ia.: The Peripatetic Press), 29–30.
42. Ibid., 194b.24–26, 29.
43. Ibid., 194b.30–33, 30.
44. Ibid., 194b.27–30, 29.
45. Harry V. Jaffa, The Reichstag Is Still Burning: The Failure of Higher Education and the Decline of the West (Montclair, CA: Claremont Institute and Pi Sigma Alpha, Delta Epsilon Chapter, 1989), 37.
46. Ibid., 37–38.
47. See Tertullian, A Treatise on the Soul, in The Ante-Nicene Fathers, vol. 3, Master Christian Library [CD-Rom], version 6, trans. Peter Holmes (Albany, Ore.: Ages Software, 1998). While Tertullian argued that the human soul was corporeal, he also regarded it as immortal and intrinsically rational. Contradicting the atomists' view that the soul and its rational faculties were produced by the purposeless collisions of atoms, Tertullian wrote that "the soul is rather the offspring of God than of matter" and its rationality was "impressed upon it from its very first creation by its Author, who is Himself essentially rational." [Treatise on the Soul, chap. 22, 364, and chap. 16, 348–49].
48. However, one begins to see glimmers of materialism's revival by the 1300s. In 1348 in Paris, Nicolaus de Autricuria was forced to recant several doctrines, including a belief that "in the processes of nature there is nothing to be found but the motion of the combination and separation of atoms." Lange, History of Materialism 1: 225.
49. Edward Jesup, The Life of the Celebrated Monsieur Pascal (London: Printed by W. Burton for J. Hooke, 1723), 7–9; Donald Adamson, Blaise Pascal, Mathematician, Physicist and Thinker about God (New York: St. Martin's Press, 1995), 1–2, 21–25.
50. Jesup, The Life of Pascal, 15–16.
51. Quoted in Bertram Vivian Bowen, Faster Than Thought: A Symposium on Digital Computing Machines (New York: Pitman, 1953), 333. Bowen does not give the original source of this quote, and I have not been able to locate it. Stanley Jaki was also unable to discover the original source, and so its authenticity may be questioned. See Stanley Jaki, Brain, Mind and Computers (Washington, D.C.: Regnery Gateway, 1989), 24n35.
52. Regarding Hobbes's timidity, see Leslie Stephen, Hobbes (London: Macmillan & Co., 1904), 3, 156; D. G. James, The Life of Reason: Hobbes, Locke, Bolingbroke (London: Longmans, Green, 1949), 7.
53. Thomas Hobbes, Leviathan, ed. C. B. Macpherson (Harmondsworth, UK: Penguin Books, 1986), 186.
54. Ibid., 81.
55. Ibid., 85.
56. Ibid., 85–99.
57. Hobbes, Leviathan, 263.
58. Hobbes, Of Liberty and Necessity, 262.
59. Hobbes, Leviathan, 171.
60. Hobbes, An Answer to a Book Published by Dr. Bramhall, in The English Works of Thomas Hobbes, 4:350–54.
61. Hobbes, Leviathan, 489.
62. Ibid., 522.
63. Stanley Jaki, Brain, Mind and Computers, 24.
64. Bacon, The Advancement of Learning, ed. G. W. Kitchin (London: J. M. Dent and Sons, 1973), XI.I, 118.

65. Bacon, *The New Organon* in *The New Organon and Related Writings,* ed. Fulton Anderson (Indianapolis: Bobbs-Merrill Educational Publishing, 1960), bk. 2, aphorism XVI, 151. When discussing the human soul in *The Advancement of Learning,* Bacon acknowledged the existence of questions about "how far it is exempted from laws of matter, and of the immortality thereof, and many other points," but he held "that in the end" answers to such questions "must be bounded by religion . . . for as the substance of the soul in the creation was not extracted out of the mass of heaven and earth by the benediction of a producat but was immediately inspired from God: so it is not possible that it should be (otherwise than by accident) subject to the laws of heaven and earth, which are the subject of philosophy" (Bacon, *The Advancement of Learning,* ed. G. W. Kitchin [London: J. M. Dent and Sons, 1973], XI.I, 118). In *The New Organon,* meanwhile, Bacon called the mind "a kind of divine fire" (in *The New Organon and Related Writings,* ed. Fulton Anderson [Indianapolis: Bobbs-Merrill Educational Publishing, 1960], bk. 2, aphorism XVI, 151), and he even suggested that the operation of magnets "may be an instance of divorce between corporeal nature and natural action," which supplied proof from merely human philosophy of the existence of essences and substances separate from matter and incorporeal. For allow that natural virtue and action, emanating from a body, can exist for a certain time and in a certain space altogether without a body, and you are not far from allowing that it can also emanate originally from an incorporeal substance (ibid., bk. 2, aphorism XXXVII, 204).

66. For the origins of the view that Descartes was either a closet materialist or that his ideas naturally led to materialism, see Stephen Gaukroger, *Descartes: An Intellectual Biography* (Oxford: Clarendon Press, 1997), 1–2. Elizabeth Haldane provides a nice summary of the two divergent streams following from Descartes' conception: "[Descartes'] system has produced strange antitheses: on the one hand, it has developed into materialism of the kind which has characterised the physiology founded on his preliminary researches; on the other, it has led to the idealism of Berkeley and of Kant." [Elizabeth S. Haldane, *Descartes, His Life and Times* (London: John Murray, 1905), 177]

67. Descartes, *Principles of Philosophy,* in *The Philosophical Writings of Descartes,* 315.

68. Descartes, *Treatise on Man,* in *The Philosophical Writings of Descartes,* 99.

69. Descartes, *Discourse on the Method,* in *The Philosophical Writings of Descartes,* 127.

70. Descartes, *Principles of Philosophy,* in *The Philosophical Writings of Descartes,* 195.

71. For the early controversy sparked by Locke's statements, see John W. Yolton, *Thinking Matter: Materialism in Eighteenth-Century Britain* (Minneapolis: University of Minnesota Press, 1984); and John W. Yolton, *Locke and French Materialism* (Oxford: Clarendon Press, 1991).

72. Locke, *An Essay Concerning Human Understanding,* Bk. IV, chap. 3, §6 in *Great Books of the Western World,* ed. Robert Maynard Hutchins (Chicago: Encyclopedia Britannica, 1952), XXV, 313–15.

73. Ibid., 315.

74. "It is no less than a contradiction to suppose matter (which is evidently in its own nature void of sense and thought) should be that Eternal first-thinking Being." Ibid., 314.

75. Ibid., Bk. IV, chap. 10, §5, 350.

76. Ibid., Bk. IV, chap. 10, §5–6.

77. Quoted in Michael Ayers, *Locke* (London: Routledge, 1991), 2: 179.

78. George Gilder, "The Materialist Superstition," *The American Enterprise* 9 (September–October 1998): 38. For a provocative treatment of Newton as a key stepping-stone to modern materialism, see Benjamin Wiker, *Moral Darwinists: How We Became Hedonists* (Downers Grove, IL: InterVarsity Press, 2002), 124–42.

79. Quoted in H. S. Thayer, ed., *Newton's Philosophy of Nature: Selections from His Writings* (New York: Hafner Pub. Co., 1953), 10.

80. According to Newton, "it is not to be conceived that mere mechanical causes could give birth to so many regular motions [in celestial bodies], since the comets range over all parts of the heavens in very eccentric orbits . . . This most beautiful system of the sun, planets, and comets could only proceed from the counsel and dominion of an intelligent and powerful Being." [Quoted in Thayer, *Newton's Philosophy of Nature,* 42.] For a good summary of Newton's metaphysical views, see Robert H. Hurlbutt III, *Hume, Newton and the Design Argument* (Lincoln, NE: University of Nebraska Press, 1965), especially 9, 79–80.

81. Mary Wollstonecraft Shelley, *Frankenstein: Or, the Modern Prometheus* (New York: Collier Books, 1961), 48–49.

82. David Ketterer, *Frankenstein's Creation: The Book, The Monster, and Human Reality* (Victoria: University of Victoria English Monograph Series, 1979), 78–79.

83. Giovanni Aldini, *An Account of the Late Improvements in Galvanism* (London: Printed for Cuthell and Martin, 1803), 2–6, 87.

84. Ibid., 189.

85. Ibid., 63.

86. Ibid., 8.

87. Ibid., 67.

88. Ibid., 68.

89. Ibid., 73.

90. Ibid., 189.

91. Ibid., 194.

92. See Anne K. Mellor, *Mary Shelley: Her Life, Her Fiction, Her Monsters* (New York: Methuen, 1988), 105–106. By 1819, Aldini himself seems to have tired of his experiments on human cadavers, stating that they should not be repeated without reason. See Giovanni Aldini, *General views on the application of galvanism to medical purposes* (London: J. Callow, 1819), 27.

93. Julien Offray De La Mettrie, *Man a Machine* (La Salle, IL: Open Court, 1912), 93.

94. Ibid., 97. The full sentence from La Mettire actually states that "the diverse states of the soul are always correlative with those of the body." [*Man a Machine*, 97] I have substituted "mind" for "soul" in my paraphrase in order to bring out more clearly La Mettrie's real meaning. According to La Mettrie, "soul is . . . but an empty word" in its traditional usage, and so "an enlightened man should use [it] only to signify the part in us that thinks." That is how La Mettrie used the term.

95. Ibid., 95.

96. Ibid., 92–97.

97. Ibid., 95.

98. Ibid.

99. Ibid., 119.

100. Ibid., 118.

101. Ibid., 119.

102. Ibid., 120.

103. Ibid., 103.

104. Ibid., 117.

105. Ibid., 100.

106. Ibid.

107. "Such is the likeness of the structure and functions of the ape to ours that I have very little doubt that if this animal were properly trained he might at last be taught to pronounce, and consequently to know, a language." Ibid., 103; also see 100–103.

108. Ibid., 103–4.

109. Ibid., 115.

110. Ibid., 113.

111. "What animal would die of hunger in the midst of a river of milk? Man alone . . . He knows neither the foods suitable for him, nor the water that can drown him, nor the fire that can reduce him to ashes. Light a wax candle for the first time under a child's eyes, and he will mechanically put his fingers in the flame . . . Or, put the child with an animal on a precipice, the child alone falls off; he drowns where the animal would save itself by swimming . . . For one more example, let us observe a dog and a child who have lost their master on a highway: the child cries and does not know to what saint to pray, while the dog, better helped by his sense of smell than the child by his reason, soon finds his master." Ibid., 113–14.

112. Ibid., 114.

113. Ibid., 128.

114. Ibid., 122–26.

115. Ibid., 146–47.

116. Paul Henri Thiery, Baron d'Holbach, *The System of Nature* (New York: Garland Publishing, 1984), 3:13.

117. Ibid., 78–79.

118. Ibid., 209.

119. Ibid., 2.

120. Ibid., 1.

121. Ibid., 340.

122. Ibid., 220.

123. Ibid., 247.

124. D'Holbach offered a lengthy passage where he allowed those lacking in virtue to offer excuses in their own voices: "The voluptuary argues,—You pretend that I should resist my desires; but was I the maker of my own temperament, which unceasingly invites me to pleasure? . . . The choleric man vociferates,—You advise me to put a curb on my passions; to resist the desire of avenging myself: but can I conquer my nature? The zealous enthusiast exclaims,—You recommend to me mildness, you advise me to be tolerant, to be indulgent to the opinions of my fellow-men; but is not my temperament violent? . . . In short, the actions of man are never free." Ibid., 226–27.

125. Ibid.

126. Ibid., 256.

127. Ibid., 262.

128. Ibid., 271.

129. Pierre-Simon Laplace, *Philosophical Essay on Probabilities*, translated by Andrew I. Dale, (New York: Springer-Verlag, 1995), 2.

130. The social and academic ostracism over materialism continued well into the nineteenth century. See discussion in Howard Gruber, *Darwin on Man, A Psychological Study of Scientific Creativity*, second edition (Chicago: University of Chicago Press, 1981), 34–45, 203–206.

131. Elie Luzac, *Man More than a Machine* (L'homme plus que machine, 1748).

132. *Temple of Nature*, Canto I, lines 235–48, quoted in Maureen McNeil, *Under the Banner of Science: Erasmus Darwin and His Age* (Manchester, UK: Manchester University Press, 1987), 115–16.

133. For more information about Erasmus Darwin and his views, see McNeil, *Under the Banner.*

Chapter 2: Darwin's Revolution

1. Alfred Russel Wallace, *An Anthology of His Shorter Writings*, ed. by Charles H. Smith (New York: Oxford University Press, 1991), 31.

2. Ibid., 31–32.

3. Ibid., 32.

4. Ibid., 33.

5. See Francis Darwin, ed., *The Autobiography of Charles Darwin and Selected Letters* (New York: Dover, 1958), 43–44, 196–200; John Bowlby, *Charles Darwin: a New Life* (New York: W. W. Norton, 1990), 331–332; Adrian Desmond and James Moore, *Darwin: The Life of a Tormented Evolutionist* (New York: Warner Books, 1991), 467–70.

6. Darwin to Wallace, April 14, 1869, in *The Life and Letters of Charles Darwin*, ed. Francis Darwin (New York: D. Appleton and Co., 1905), 297.

7. Quoted in Bowlby, *Charles Darwin*, 394.

8. Quote praising the labors of Darwin and Wallace is from George Mivart, *On the Genesis of Species*, 2nd ed. (New York: Macmillan, 1871), 2. More generally, for information about the relationship between Darwin and Mivart, see Gertrude Himmelfarb, *Darwin and the Darwinian Revolution* (Chicago: Elephant Paperbacks, 1996), 359–62; Desmond and Moore, *Darwin*, 570–71, 577–78, 582–86, 589–91, 610–13; Bowlby, *Charles Darwin*, 403–4, 421.

9. Desmond and Moore, *Darwin*, 613.

10. Quoted in interview clip during "Evolution Under the Microscope," story by CBN News for *The 700 Club* (October 6, 1999).

11. Kenneth R. Miller and Joseph Levine, *Biology*, 5th ed., teacher's edition (Upper Saddle River, NJ: Prentice Hall, 2000), 270.

12. Desmond and Moore, *Darwin*, 45–97; Darwin, *Autobiography*, 17–19; James Moore, *The Darwin Legend* (Grand Rapids, MI: Baker Books, 1994), 30.

13. Darwin, *Autobiography*, 61.

14. Charles Darwin, *The Origin of Species by Means of Natural Selection*, 6th ed. (New York: D. Appleton and Co., 1896), 2:304.

15. Ibid., 1:228; 2:294, 304.

16. Ibid., 2:294.

17. Ibid., 306.

18. Darwin, *Autobiography*, 44.

19. The chapters referred to here are from the first edition. In a later edition of *Descent*, the original chapter 4 was inserted after chapter 1, resulting in different numbering for the other early chapters.

20. Although most attention is paid to Darwin's first book, there are a number of scholars who have explored in various ways Darwin's discussion of mind and morality in *The Descent of Man*. These include Robert J. Richards, *Darwin and the Emergence of Evolutionary Theories of Mind and Behavior* (Chicago: University of Chicago Press, 1987), especially 185–242; Paul Lawrence Farber, *The Temptations of Evolutionary Ethics* (Berkeley, CA: University of California Press, 1994), especially 10–21; James Rachels, *Created from Animals: The Moral Implications of Darwinism* (New York: Oxford University Press, 1998); David Loye, *Darwin's Lost Theory of Love: A Healing Vision for the New Century* (Lincoln, NE: toExcel Press, 2000); Benjamin Wiker, *Moral Darwinism: How We Became Hedonists* (Downers Grove, IL: Intervarsity Press, 2002), 215–54.

21. Charles Darwin, *The Descent of Man, and Selection in Relation to Sex* (Princeton, NJ: Princeton University Press, 1981), 1:35. This is a reprint of the first edition, and hereafter cited as Darwin, *Descent* (1871).

22. Ibid., 1:36, 39–40.

23. Ibid., 40.

24. Ibid., 41–42.

25. Ibid., 42.

26. Charles Darwin, *The Descent of Man, and Selection in Relation to Sex*, rev. ed. (New York: D. Appleton and Co., 1896), 1:71. Hereafter cited as Darwin, *Descent* (1896).

27. *Descent* (1871), 1:44.

28. Ibid., 46.

29. *Descent* (1896), 1:75.

30. Ibid. I have used Darwin's wording here from a later edition because in the original *Descent* Darwin's wording was slightly more awkward. Instead of "only a few persons now dispute," he wrote, "Few persons any longer dispute." *Descent* (1871), 1:46.

31. *Descent* (1871), 1:49.

32. Ibid., 49. In future editions, Darwin watered down this section, admitting that man was "capable of incomparably greater and more rapid improvement" than other animals, even while denying that this difference in degree was a difference in kind. See *Descent* (1896), 1:79.

33. *Descent* (1871), 1:50.

34. Ibid., 53.

35. Ibid., 56.

36. Ibid., 57.

37. Ibid., 57.

38. This sort of explanation had been popularized by French evolutionist Jean Lamarck (1744–1829). While today Lamarckian evolution is usually contrasted with Darwinian evolution, the differences between the theories of Darwin and Lamarck are not nearly so great as is often supposed. It was not until later that Darwinism became allied with modern genetics, which made the idea of the inherited effects of use insupportable.

39. *Descent* (1871), 1:62.

40. Ibid.

41. Ibid., 65.

42. Ibid., 65–67.

43. Ibid., 65.

44. Ibid., 68.

45. Ibid., 395.

46. See Larry Arnhart, *Darwinian Natural Right: The Biological Ethics of Human Nature* (New York: State University of New York Press, 1998). For a critique of Arnhart's effort to square Darwinian theory with traditional ethics, see John G. West, *Darwin's Conservatives: The Misguided Quest* (Seattle: Discovery Institute Press, 2006), 19–32; and J. Budziszewski, "Accept No Imitations:

The Rivalry of Naturalism and Natural Law," in *Uncommon Dissent: Intellectuals Who Find Darwinism Unconvincing,* ed. William Dembski (Wilmington, DE: ISI Books, 2004), 99–113.

47. *Descent of Man* (1871), 1:106.
48. Ibid., 82.
49. Ibid., 2:394.
50. Ibid., 1:83–84; 2:363–65.
51. Ibid., 1:76–77.
52. Ibid., 103.
53. Ibid., 104.
54. Ibid.,73.
55. For an account of Darwin's kindly and sympathetic nature in private life, see Darwin, *Autobiography,* especially 303–4.
56. *Descent* (1871), 1:89.
57. Ibid., 91.
58. Paul Barrett, et. al., *Charles Darwin's Notebooks, 1836–1844* (New York: Cornell University Press, 1987), 608.
59. Darwin himself credited Greg, Wallace, and Galton in the text of *Descent* (1871), 167–68. See, in particular, William R. Greg, "On the Failure of 'Natural Selection' in the Case of Man," *Fraser's Magazine* 78 (1868): 353–62; and Francis Galton, "Hereditary Talent and Character," second paper, *Macmillan's Magazine,* August 1865, 318–27. For a discussion of the views of Greg, Wallace, and Galton, see Richards, *Darwin and the Emergence of Evolutionary Theories of Mind,* 161–76. For additional information on the background sources of this section of *Descent,* see John C. Greene, "Darwin as a Social Evolutionist," in *Science, Ideology, and World View: Essays in the History of Evolutionary Ideas* (Berkeley, CA: University of California Press, 1981), 95–127.
60. *Descent* (1871), 1:168.
61. Ibid., 168–69.
62. Ibid., 169.
63. Ibid., 170.
64. Ibid., 172–73.
65. Ibid., 174.
66. Ibid., 177–78.
67. *Descent* (1896), 1:143.
68. Wallace, "Human Selection," in Wallace, *An Anthology,* 51. This essay by Wallace was published in 1890, after Darwin's death.
69. *Descent* (1871), 1:180.
70. Ibid., 2:403.
71. On Darwin's opposition to contraception, see Desmond and Moore, *Darwin,* 627–28.
72. *Descent* (1871), 2:403.
73. Ibid., 1:256–57.
74. Ibid., 2:361.
75. Ibid., 358–59.
76. Ibid., 361–62.
77. Ibid., 360, 362, 367.
78. Ibid., 367–68.
79. *Descent* (1896), 1:83.
80. G. K. Chesterton, *The Everlasting Man* (San Francisco: Ignatius, 1993), 37–38.
81. Ibid., 26.
82. See La Mettrie, *Man a Machine,* 100–105, 115–19.
83. Darwin, *Origin,* 306.
84. Darwin, *Autobiography,* 272. For background see Browne, *Charles Darwin,* 213–14.
85. Desmond and Moore, *Darwin,* 573; Moore, *Darwin Legend,* 45. But see the contrary view expressed by Janet Browne, *Charles Darwin: The Power of Place* (New York: Alfred Knopf, 2002), 347.
86. Nora Barlow, ed., *The Autobiography of Charles Darwin, 1809–1882, with Original Omissions Restored* (New York: Norton, 1969), 86–87.
87. Ibid., 92.
88. Ibid., 87.

89. Ibid., 90.
90. Ibid., 91.
91. Ibid., 94.
92. Darwin, *Autobiography*, 249. For a similar statement later in life, see *Descent of Man* (1871), 2:396: "The birth both of the species and of the individual are equally parts of that grand sequence of events, which our minds refuse to accept as the result of blind chance."
93. Darwin, *Autobiography*, 249.
94. Ibid., 68.
95. Ibid., 66–67.
96. Quoted in Desmond and Moore, *Darwin*, 645.
97. Ibid., 655.
98. Ibid., 591, 643.
99. Barrett, *Notebooks*, 165.
100. Ibid., 614.
101. Ibid., 613.
102. Ibid., 536.
103. Ibid.
104. Ibid., 533–34.
105. Stephen Jay Gould, *Ever Since Darwin: Reflections in Natural History* (New York: Norton, 1977), 13. For a somewhat different view, see the discussion of Darwin's views in his notebooks and elsewhere in Neal Gillespie, *Charles Darwin and the Problem of Creation*, 138–45, 153. Gillespie acknowledges Darwin's affirmative statements about materialism but tries to give them a narrow reading. He eventually concludes that "Darwin's materialism, it would seem, was nothing more than positivism. It committed him, not to a metaphysics of matter in motion as the ultimate reality, but only to a system of naturalistic and lawful science." Ibid., 140. Even if one accepts Gillespie's reading of the relevant texts, one is still left with the conclusion that Darwin's view of human behavior and beliefs was thoroughly materialistic, making Gillespie's finely calibrated distinctions seem like distinctions without a difference, at least insofar as Darwin's view of human culture is concerned. Whether or not Darwin ultimately embraced a complete philosophical materialism, his understanding of human society was thoroughly based on materialistic reductionism.
106. Asa Gray, "Review of *What Is Darwinism?*" reprinted in Charles Hodge, *What Is Darwinism? and Other Writings on Science and Religion*, ed. Mark A. Noll and David N. Livingstone (Grand Rapids, MI: Baker Books, 1994), 162, 165.
107. Ibid., 163.
108. Ronald L. Numbers, *Darwinism Comes to America* (Cambridge, MA: Harvard University Press, 1998), 40–43; also see discussion in Edward Larson, *Summer for the Gods: The Scopes Trial and America's Continuing Debate over Science and Religion* (New York: Basic Books, 1997), 18–20.
109. Larson, *Summer for the Gods*, 41.
110. Elliot, *Modern Science and Materialism*, 138.

Chapter 3: Criminal Science

1. Unless otherwise noted, the factual details about the Leopold-Loeb case are taken from Hal Higdon, *Leopold and Loeb: The Crime of the Century* (Chicago: University of Illinois Press, 1999).
2. "In Search of History: Born Killers, Leopold and Loeb" [television documentary on the History Channel] (A&E Television Networks, 1998).
3. Higdon, *Leopold and Loeb*, 241.
4. Alvin Sellers, *The Loeb-Leopold Case, with Excerpts from the Evidence of the Alienists and Including the Arguments to the Court by the Counsel for the People and the Defense* (Brunswick, GA: Classic Publishing Company, 1926), 156.
5. Ibid., 134.
6. Ibid., 133.
7. Ibid., 155, 156.

8. Ibid., 169.

9. Ibid., 155.

10. Ibid., 182.

11. Ibid., 180–81.

12. Ibid., 151–52.

13. Ibid., 315.

14. Orville Dwyer, "Great Trial Ends," Chicago Tribune, Aug. 29, 1924, 1.

15. Sellers, *The Loeb-Leopold Case*, 312.

16. Ibid., 313.

17. Ibid., 314.

18. Ibid., 314, 315.

19. Ibid., 315–16.

20. Ibid., 319–21.

21. Immanuel Kant, *The Metaphysical Elements of Justice*, trans. John Ladd (New York: Library of Liberal Arts, Macmillan, 1965), 100.

22. Enrico Ferri, "The Positive School of Criminology," in *Criminology: A Book of Readings*, ed. Clyde Vedder, Samuel Koenig, and Robert Clark (New York: The Dryden Press, 1953), 134–35. The importance of moral blame to the traditional justification of punishment can be seen early in the history of civilization. As far back as the celebrated Code of Hammurabi (circa 1700–1800 BC), lawgivers attempted to assign punishments based on the moral culpability of the lawbreakers. For a good treatment of ideas about moral responsibility and crime from the ancient world to the development of the English common law, see Eugene R. Milhizer, "Justification and Excuse: What They Were, What They Are, and What They Ought to Be," *St. John's Law Review* 78:725–93. For an interesting discussion of criminal culpability in the works of Cicero and Roman law, see Richard A. Bauman, *Crime and Punishment in Ancient Rome* (New York: Routledge, 1996), 39–40.

23. See Elio Monachesi, "Cesare Beccaria," and Gilbert Geis, "Jeremy Bentham," in *Pioneers of Criminology*, ed. Hermann Mannheim (London: Stevens and Sons, 1960), 36–50, 51–67.

24. Ludwig Büchner, *Force and Matter, or Principles of the Natural Order of the Universe* (New York: Peter Eckler, 1891); Desmond and Moore, *Darwin*, 656–57.

25. Büchner, *Force and Matter*, 375.

26. Ibid., 369.

27. Ibid., 375.

28. Ibid., 377.

29. Ibid., 376.

30. Ibid., 378.

31. Maurice Parmelee, "Introduction," in Cesare Lombroso, *Crime: Its Causes and Remedies*, trans. Henry Horton (Montclair, N.J.: Patterson Smith, 1968), xiv–xv; William Noyes, "The Criminal Type," *Journal of Social Science* 24 (April 1888): 32–34. For a discussion of Lombroso and Social Darwinism, see Mike Hawkins, *Social Darwinism in European and American Thought, 1860–1945* (Cambridge: Cambridge University Press, 1997), 74–80.

32. Noyes, "Criminal Type," 34.

33. quoted by Parmelee in Lombroso, *Crime*, xviii.

34. Lombroso, *Crime*, 376.

35. Ferri, "The Positive School," 137–38.

36. Lombroso, *Crime*, 432.

37. Ibid., 424.

38. Ibid., 428.

39. Ibid., 342.

40. Ibid., 353–60.

41. C. S. Lewis, "The Humanitarian Theory of Punishment," in *God in the Dock: Essays on Theology and Ethics*, ed. Walter Hooper (Grand Rapids, MI: Eerdmans, 1970), 288.

42. James Weir Jr., "Criminal Anthropology," *Medical Record* 45 (January 1894): 43.

43. Ibid., 44.

44. Ibid., 45.

45. Ibid.

46. Ibid.

47. William Goring, *The English Convict: A Statistical Study* (London: Darling and Son, 1913); James Q. Wilson and Richard Herrnstein, *Crime and Human Nature* (New York: Simon and Schuster, 1985), 75–77.

48. This movement, known as "eugenics," will be covered in detail in chapters 5, 7, and 8.

49. Sellers, *The Loeb-Leopold Case*, 26–27, 31–32.

50. Louis Berman, "Crime and the Endocrine Glands," *American Journal of Psychiatry* (September 1932), 223.

51. Ibid., 234.

52. Edward Podolsky, "The Chemical Brew of Criminal Behavior," *Journal of Criminal Law, Criminology, and Police Science* 45 (March–April 1955): 676, 677.

53. Ibid, 678.

54. Wilson and Herrnstein, *Crime and Human Nature*, 79. For a general discussion of the shift from biological explanations to cultural/social explanations in the social sciences (not tied to crime), see Carl N. Degler, *In Search of Human Nature: The Decline and Revival of Darwinism in American Social Thought* (New York: Oxford University Press, 1991), 59–211.

55. Sigmund Freud, *A General Introduction to Psychoanalysis* (New York: Washington Square Press, 1960), 26–27, 296, 363–66. For more about the influence of Darwin on Freud, see Frank J. Sulloway, *Freud, Biologist of the Mind: Beyond the Psychoanalytic Legend* (New York: Basic Books, 1979), 13–17, 199–201, 238–76; and Lucille B. Ritvo, *Darwin's Influence on Freud: A Tale of Two Sciences* (New Haven, CT: Yale University Press, 1990). Ritvo concludes that "the effect of Darwin's work on Freud's appears to have been massive and primarily of a positive nature" (197). Ritvo's book contains a helpful appendix listing more than two dozen references to Darwin in the writings of Freud (199–201).

56. Freud, *A General Introduction to Psychoanalysis*, 209.

57. Ibid., 379–80.

58. Jack London, *Before Adam* (Oakland, CA: Star Rover House, 1982), 13–14.

59. David Abrahamsen, *Who Are the Guilty? A Study of Education and Crime* (Westport, CT: Greenwood Press, 1972), 36.

60. Ibid., 21.

61. Ibid., 22–23.

62. B. F. Skinner, *Beyond Freedom and Dignity* (New York: Bantam Books, 1972).

63. For a discussion of recent developments in psychology more favorable to a traditional conception of human beings, see Paul Vitz, "Back to Human Dignity: From Modern to Postmodern Psychology," *Intercollegiate Review* 31 (Spring 1996): 15–23.

64. John Staddon, "On Responsibility and Punishment," *The Atlantic Monthly* 275 (February 1995): 89–90.

65. Emory Bogardus, *Sociology*, 3rd ed. (New York: Macmillan, 1949), 436.

66. Ibid., 437.

67. Bernard Diamond, "From M'Naghten to Currens, and Beyond," *California Law Review* 50 (May 1962): 198.

68. Ibid.

69. Vernon Mark and Frank Ervin, *Violence and the Brain* (New York: Harper and Row, 1970).

70. Wilson and Herrnstein, 100–102.

71. For a comprehensive overview of the research into the early 1990s, see Albert J. Reiss Jr., Klaus A. Miczek, and Jeffrey A. Roth, editors, *Understanding and Preventing Violence, Volume 2: Biobehavioral Influences* (Washington, D.C.: National Academy Press, 1994). For a general discussion of the return of biological explanations in the social sciences, see Degler, *In Search of Human Nature*, 215–349.

72. J. P. Shalloo, "The Emergence of Criminology," in Vedder, et. al., *Criminology*, 2–3.

73. Nathaniel F. Cantor, *Crime, Criminals and Criminal Justice* (New York: Henry Holt and Company, 1932), 265.

74. Ibid., 266.

75. John F. Cuber, *Sociology: A Synopsis of Principles* (New York: D. Appleton Century Company, 1947), 248–49.

76. Cuber, *Sociology*, 3rd ed. (New York: Appleton-Century-Crofts, 1955), 236 (emphasis added).

77. "General Introduction to the Modern Criminal Science Series," in Lombroso, *Crime*, vii.

Chapter 4: Crime as Mental Illness

1. Guiteau, quoted by Charles Rosenberg, *The Trial of the Assassin Guiteau: Psychiatry and Law in the Gilded Age* (Chicago: University of Chicago Press, 1968), 141.

2. Quoted in *Cases and Materials on Criminal Law and Procedure*, ed. Rollin Perkins and Ronald Boyce, 6th ed. (Mineola, NY: The Foundation Press, 1984), 597. Although M'Naghten is the most famous formulation of the "right-wrong" test, the test itself predates M'Naghten. See *Durham v. United States*, 214 F.2d 862 (1954), 869.

3. It should be noted that by the time of the Leopold-Loeb case, Illinois had abandoned a strict right-wrong test for insanity and had adopted something akin to the irresistible impulse test discussed below. See Henry Weihofen, *Insanity as a Defense in Criminal Law* (New York: Commonwealth Fund, 1933), 118–19. Even under this expanded standard, Clarence Darrow did not think he could prevail on the issue of legal guilt.

4. See *State v. Jones*, 50 N.H. 369 (1871) and *State v. Pike*, 49 N.H. 399 (1870).

5. Cases applying versions of the "irresistible impulse" test include *Roberts v. State*, 3 Ga. 310 (Georgia, 1847); *State v. Ferrer*, 9 West.L.J. 513 (Ohio, 1852); *Fouts v. State*, 4 Greene 500 (Iowa, 1854); *Com. v. Shurlock*, 14 Leg.Int. 33 (Pennsylvania, 1857); *Com. v. Winnemore*, 1 Brewst. 356 (Pennsylvania, 1867); *Stevens v. State*, 31 Ind. 485 (Indiana, 1869); *Parsons v. State*, 2 So. 854 (Alabama, 1887); *Plake v. State*, 23 N.E. 273 (Indiana, 1890); *Morgan v. State*, 130 N.E. 528 (Indiana, 1921); *Cline v. Commonwealth*, 59 S.W.2d 577 (Kentucky, 1933); *Com. v. Chester*, 150 N.E.2d 914 (Massachusetts, 1958); *Early v. People*, 352 P.2d 112 (Colorado, 1960); *State v. Noble*, 384 P.2d 504 (Montana, 1963); *State v. Arthur*, 160 N.W.2d 470 (Iowa, 1968); *State v. Holt*, 449 P.2d 119 (Utah, 1969); *People v. Russell*, 173 N.W.2d 816 (Michigan, 1969); *State v. Vennard*, 270 A.2d 837 (Connecticut, 1970); *Com. v. Walzack*, 360 A.2d 914 (Pennsylvania, 1976); *Smith v. State*, 397 N.E.2d 959 (Indiana, 1979); *People v. Horn*, 205 Cal. Rptr. 119 (California, 1984); *People v. Duckett*, 209 Cal. Rptr. 96 (California, 1984); *Lawrence v. State*, 454 S.E.2d 446 (Georgia, 1995); *Herbin v. Com.*, 503 S.E.2d 226 (Virginia, 1998).

6. Cases rejecting or severely qualifying the "irresistible impulse" test include *State v. Felter*, 25 Iowa 67 (Iowa, 1868); *State v. Stickley*, 41 Iowa 232 (Iowa, 1875); *Wright v. People*, 4 Neb. 407 (Nebraska, 1876); *Cunningham v. State*, 56 Miss. 269 (Mississippi, 1879); *Sanders v. State*, 94 Ind. 147 (Indiana, 1884); *State v. Pagels*, 4 S.W. 931 (Missouri, 1887); *State v. Alexander*, 8 S.E. 440 (South Carolina, 1889); *State v. Scott*, 43 N.W. 62 (Minnesota, 1889); *State v. Harrison*, 15 S.E. 982 (West Virginia, 1892); *State v. Miller*, 20 S.W. 243 (Missouri, 1892); *Wilcox v. State*, 28 S.W. 312 (Tennessee, 1894); *Mackin v. State*, 36 A. 1040 (New Jersey, 1897); *People v. Hubert*, 51 P. 329 (California, 1897); *Davis v. State*, 32 So. 822 (Florida, 1902); *State v. Lyons*, 37 So. 890 (Louisiana, 1904); *Turner v. Territory*, 82 P. 650 (Oklahoma, 1905); *State v. Hassing*, 118 P. 195 (Oregon, 1911); *Wade v. State*, 92 So. 97 (Alabama, 1921); *State v. White*, 209 P. 660 (Kansas, 1922); *State v. Simenson*, 262 N.W. 638 (Minnesota, 1935); *State v. Moore*, 76 P.2d 19 (New Mexico, 1938); *State v. Cumberworth*, 43 N.E.2d 491 (Ohio, 1942); *State v. Wallace*, 131 P.2d 222 (Oregon, 1942); *State v. Marsh*, 185 P.2d 486 (Washington, 1947); *Thompson v. State*, 68 N.W.2d 267 (Nebraska, 1955); *Chase v. State*, 369 P.2d 997 (Alaska, 1962); *People v. Wolff*, 394 P.2d 959 (California, 1964); *State v. Sikora*, 210 A.2d 193 (New Jersey, 1965); *Price v. State*, 347 S.E.2d 608 (Georgia, 1986); *State v. Smith*, 592 N.W.2d 143 (Nebraska, 1999).

7. Weihofen, *Insanity as a Defense in Criminal Law*, 100–103.

8. Ibid., 151–58.

9. Ibid., 158–60; *Davis v. United States*, 160 U.S. 469 (1895).

10. Brief for Appellee, *Powell v. Texas*, 392 U.S. 514 (1968), 14–15.

11. *Durham v. United States*, 874–875.

12. Ibid., 875.

13. Lawrence Friedman, *Crime and Punishment in American History* (New York: Basic Books, 1993), 403; Diamond, "From M'Naghten to Currens," 192.

14. *Campbell v. United States*, 307 F.2d 597 (District of Columbia, 1962).

15. Ibid., 261.

16. Ibid., 262.

17. Ibid., footnote 8, 265.

18. Ibid., 269.

19. Ibid., footnote 8, 269.

20. Ibid., 269.

21. Ibid., 272.

22. Ibid., 266.

23. Ralph Slovenko, "The Mental Disability Requirement in the Insanity Defense," *Behavioral Sciences and the Law* 17 (1999): 173.

24. *Campbell v. United States*, 275.

25. Quoted in Perkins and Boyce, *Cases and Materials on Criminal Law*, 624. For examples of state courts adopting the ALI standard, see *Terry v. Com.*, 371 S.W.2d 862 (Kentucky, 1963); *Com. v. McHoul*, 226 N.E.2d 556 (Massachusetts, 1967); *People v. Drew*, 149 Cal. Rptr. 275 (California, 1978); *State v. Johnson*, 399 A.2d 469 (Rhode Island, 1979); *Blake v. United States*, 407 F.2d 908 (5th Cir. 1969); *United States v. Brawner*, 471 F.2d 969 (U.S.App.D.C., 1972). By the 1990s, twenty-one states plus the District of Columbia had adopted the ALI standard, and two more states adopted a combination of the ALI test and M'Naghten. See Rita Buitendorp, "A Statutory Lesson from 'Big Sky Country' on Abolishing the Insanity Defense," *Valparaiso University Law Review* 30 (Summer 1996): 975n64.

26. Quoted in Perkins and Boyce, *Cases and Materials on Criminal Law*, 624.

27. Unless otherwise noted, the following facts of the case are taken from the transcript of the trial filed with the appeal to the U.S. Supreme Court. See "Record from Municipal Court of Los Angeles Judicial District, County of Los Angeles, State of California, in the case of *People of the State of California, Plaintiff, v. Lawrence Robinson, Defendant*," reprinted as part of the appellate file for *Lawrence Robinson v. California*, Supreme Court of the United States, October Term, 1961, No. 554, 1–116.

28. California Health and Safety Code, §11721.

29. Appellant's Opening Brief, *Robinson v. California* (1962), especially 23–31.

30. Ibid., 19.

31. Ibid., 10–14.

32. Ibid., 20.

33. Ibid., 22.

34. See Leo Alexander, "Medical Science Under Dictatorship," *The New England Journal of Medicine* 241, no.2 (July 14, 1949): 39–41.

35. The brief for the State of California outlined some of the consequences of the view being argued for by McMorris: "Of course it is generally conceded that a narcotic addict, particularly one addicted to the use of heroin, is in a state of mental and physical illness. So is an alcoholic. So is the arsonist who sets fires for the thrill of the flame the victim of a mental illness, pyromania. The uncontrollable urge of the kleptomaniac to steal makes a thief of an often otherwise respectable person. Yet certainly these crimes must subject their perpetrators to punishment in spite of the fact that they are associated with an illness." Brief of Appellee, *Robinson v. California* (1962), 15.

36. Statement of Samuel Carter McMorris, Oral Argument before the U.S. Supreme Court in the case of *Robinson v. California* (April 17, 1962), audio file at the United States Supreme Court Multimedia Database, http://www.oyez.org/oyez/audio/332/argument-ra.smil (accessed July 16, 2005).

37. Ibid.

38. In fact, as soon as civil libertarians were successful in getting the courts to strike down criminal penalties against narcotic addiction and public drunkenness, they also started to question the use of civil commitment as an alternative. See discussion in Brief for the Appellee, *Powell v. Texas*, 392 U.S. 514 (1968), 30–31.

39. At the time, only ten states punished addiction per se. Appellant's Opening Brief, *Robinson v. California*, 15.

40. See *Mapp v. Ohio*, 367 U.S. 643 (1961); *Abington v. Schempp*, 374 U.S. 203 (1963); *Miranda v. Arizona*, 384 U.S. 436 (1966); *Roe v. Wade*, 410 U.S. 113 (1973); *Engel v. Vitale*, 370 U.S. 421 (1962).

41. *Robinson v. California*, 370 US 666, p. 667.

42. Ibid., 666.

43. Ibid., 664–65.

44. Ibid., 686.

45. Ibid., 688. While White added that he could not believe that the Court "would forbid the application of the criminal laws to the use of narcotics," he thought that the majority opinion in

Robinson v. California cast "serious doubt" about state power in this area, and states as well as the federal government would be forced wait until the Court chose to clear up the ambiguity.

46. Ibid., 689.

47. "Drug Case Victory Won After Death of Addict," *Los Angeles Times* (June 26, 1962), pt. 1, 10.

48. Statement of Samuel Carter McMorris, Oral Argument before the U.S. Supreme Court in the case of *Robinson v. California*.

49. "Drug Case Victory Won," pt. 1, 10.

50. *Robinson v. California*, 371 US 905 (Mem.) (1962).

51. Michael Asimow, Note, "Constitutional Law: Punishment for Narcotic Addiction Held Cruel and Unusual.—*Robinson v. California* (U.S. 1962)," *California Law Review* 51 (1963): 225; John Bagalay, Jr., Note, "Constitutional Law—Criminal Law—Penal Sanctions Applied to Narcotics are Unconstitutional as Cruel and Unusual Punishment. Robinson v. California, 370 U.S. 660 (1962)," *Texas Law Review* 41 (February 1963): 446; Calvin Robinson, Note, "Constitutional Law—Statute Making the Status of Being a Drug Addict a Crime Held Unconstitutional—Robinson v. California (Sup. Ct. 1962)," *Nebraska Law Review* 42 (April 1963): 691; Recent Cases, [Anonymous], "Constitutional Law—Imprisonment for the 'Crime' of Narcotics Addiction Held Unconstitutional as Cruel and Unusual Punishment," *University of Pennsylvania Law Review* 111 (November 1962): 126–27; Recent Cases, [Anonymous], "Due Process: Criminal Penalty for Narcotic Addiction is Cruel and Unusual Punishment," *Minnesota Law Review* 47 (February 1963): 492–93; Case Comment, "Criminal Law—Narcotics—Criminal Prosecution for Addiction is a Cruel and Unusual Punishment Violating Eighth and Fourteenth Amendments," *Vanderbilt Law Review* 16 (December 1962): 218–19. But see the view in Recent Decisions, [Anonymous], "Constitutional Law—Narcotics—Criminal conviction for the status of drug addiction is a cruel and unusual punishment," *Duquesne University Law Review* 1 (Spring 1963): 264.

52. Bagalay, "Constitutional Law," 448; [Anonymous] "Constitutional Law," 127–28; Robinson, "Constitutional Law," 691, 693.

53. Bagalay, 448.

54. Hugh R. Manes, "*Robinson v. California*: A Farewell to Rationalism?" *Law in Transition* 22 (Winter 1963): 244; Robinson, "Constitutional Law," 691–92.

55. Manes, "Robinson," 244; Robinson, "Constitutional Law," 691–92; Bagalay, "Constitutional Law," 448; [Anonymous], "Due Process," 493.

56. Robinson, "Constitutional Law," 696. Diane Yockey similarly argued that "acceptance of the conclusion that the narcotics addict cannot be held responsible for crimes he commits pursuant to his illness would eventually require that most laws effecting [sic] him be reappraised." Diane Yockey, "Constitutional Law—Criminal Sanctions Against the Narcotics Addict," *Tulane Law Review* 37 (December 1962), 124.

57. *Blouin v. Walker*, 154 So.2d 368 (1963); *State v. Margo*, 191 A.2d 43 (1963); *Browne v. State*, 129 N.W.2d 175 (1964); *Lloyd v. United States*, 343 F.2d 243 (1964); In re Carlson, 410 P.2d 379 (1966); *People v. Nettles*, 213 N.E.2d 536 (1966).

58. *People v. O'Neil*, P.2d 928 (1965).

59. *Driver v. Hinnant*, 356 F.2d 761 (4th Cir. 1966).

60. Ibid., 764–65.

61. *Easter v. District of Columbia*, 361 F.2d 50 (D.C. Cir. 1966).

62. The reactions of law reviews are summarized in the Amicus Brief of the American Civil Liberties Union, et. al., *Powell v. Texas*, 392 U.S. 514 (1968), Appendix H, 1-5.

63. See discussion of cases in Robert Keller, *Recent Cases*, "Constitutional Law—Cruel and Unusual Punishment—Conviction of a Chronic Alcoholic for Public Intoxication not Violative of Eighth Amendment Proscription of Cruel and Unusual Punishment," *Buffalo Law Review* 18 (1969): 345–347.

64. Ibid., 345–46. Oregon's Supreme Court took the opposite view. Ibid., 347.

65. *State v. Freiberg*, 151 N.W.2d 1 (1967).

66. The case in which the court declined review was *Budd v. California*, cert. denied, 385 U.S. 909 (1966).

67. After the decision in *Powell*, Gerald Stern noted in the *American Bar Association Journal* that the court's decision was "contrary to the predictions of most commentators." Gerald Stern, "Handling Public Drunkenness: Reforms Despite *Powell*," *American Bar Association Journal* 55 (July 1969): 656.

68. Gary V. Dubin, "The Ballad of Leroy Powell," *UCLA Law Review* 16, no. 1 (November 1968): 139.
69. Brief for Appellee, *Powell v. Texas*, 2.
70. Amicus Brief of the American Civil Liberties Union, *Powell v. Texas*, 8.
71. Brief for Appellee, *Powell v. Texas*, 2–3. Under cross-examination, defense had shown that Mr. Powell also had no more money available to buy drinks that morning, indicating another reason for his apparent self-control.
72. Ibid., 16.
73. Ibid.
74. Ibid., 4.
75. Ibid., 5–6.
76. *Powell v. Texas*, 392 U.S. 514 (1968), 535.
77. Ibid., 535–36.
78. Ibid., 536.
79. Ibid., 547.
80. Ibid., 538.
81. Ibid., 527.
82. Ibid., 529.
83. Ibid., 532.
84. Ibid., 542.
85. Ibid., 567.
86. Ibid., 569.
87. Ibid., 559n2.
88. Ibid., 534. Justice Black drove home this same point to devastating effect in his concurrence: "The rule of constitutional law urged upon us by appellant would have a revolutionary impact on the criminal law, and any possible limits proposed for the rule would be wholly illusory. If the original boundaries of Robinson are to be discarded, any new limits too would soon fall by the wayside . . . But even if we were to limit any holding in this field to 'compulsions' that are 'symptomatic' of a 'disease' . . . the sweep of that holding would still be startling. Such a ruling would make it clear beyond any doubt that a narcotics addict could not be punished for 'being' in possession of drugs or, for that matter, for 'being' guilty of using them. A wide variety of sex offenders would be immune from punishment if they could show that their conduct was not voluntary, but part of the pattern of a disease. More generally speaking, a form of the insanity defense would be made a constitutional requirement throughout the Nation, should the Court now hold it cruel and unusual to punish a person afflicted with any mental disease whenever his conduct was part of the pattern of his disease and occasioned by a compulsion symptomatic of the disease." Ibid., 544–45.
89. Ibid., 548–49.
90. Ibid., 552n4.
91. Ibid., 549–51, 553–54.
92. See Keller, "Constitutional Law—Cruel and Unusual Punishment," 348–55; John Gibbons, *Recent Decisions*, "Criminal Law—Chronic Alcoholism as a defense to a charge of public drunkenness," *Journal of Urban Law* 46 (1968): 117–18; Mike Stevenson, "Chronic Alcoholism and Criminal Responsibility," *Gonzaga Law Review* 4 (Spring 1969): 339–40, 342–43; Louis Henkin, "The Supreme Court, 1967 Term," *Harvard Law Review* 82 (November 1968): 103–11; Gerard Robert Lear, Note, "Constitutional Law—Chronic Alcohol Addiction and Criminal Responsibility—Statute which Attaches Criminality to Chronic Alcoholics who Appear in Public While Intoxicated is not Cruel and Unusual Punishment," *The American University Law Review* 18 (June 1969): 574–75; Lynd Mische, *Recent Cases*, "Constitutional Law—Chronic Alcoholism as a Defense to a Charge of Public Intoxication," *Missouri Law Review* 34 (Fall 1969): 603–4.
93. See Keller, "Constitutional Law—Cruel and Unusual Punishment," 353–354; Henkin, "The Supreme Court, 1967 Term," 110–11; Lear, "Constitutional Law—Chronic Alcohol Addiction," 574–75.
94. *Powell v. Texas*, 551.
95. Carolyn Alexander, "Oregon's Psychiatric Security Review Board: Trouble in Paradise," *Law and Psychology Review* 22 (Spring 1998): 5.
96. James Q. Wilson, *Moral Judgment: Does the Abuse Excuse Threaten Our Legal System?* (New York: Basic Books, 1997), 44–45.

97. Buitendorp, "A Statutory Lesson from 'Big Sky Country,'" 971–72; Michael Perlin, "Unpacking the Myths: The Symbolism Mythology of Insanity Defense Jurisprudence," *Case Western Reserve Law Review* 40 (1990): 648–49.

98. Eric Silver, Carmen Cirincione, and Henry Steadman, "Demythologizing Inaccurate Perceptions of the Insanity Defense," *Law and Human Behavior* 18 (1994): 67.

99. Jocelyn Lymburner and Ronald Roesch, "The Insanity Defense: Five Years of Research," *International Journal of Law and Psychiatry* 22 (1999): 216; Eric Silver, "Punishment or Treatment? Comparing the Length of Confinement of Successful and Unsuccessful Insanity Defendants," *Law and Human Behavior* 19 (1995): 380.

100. Carmen Cirincione and Charles Jacobs, "Identifying Insanity Acquittals: Is it Any Easier?" *Law and Human Behavior* 23 (1999): 493–494. In 1981, there were 250 insanity acquittals in California, 100 in Hawaii, 115 in New York, ninety-eight in Oregon, and seventy-three in Washington. Ibid., 493.

101. Silver, "Demythologizing," 67–68.

102. Ibid., 67.

103. Alexander, "Oregon's Psychiatric Security Review Board," 5.

104. Ibid., 6; California Penal Code §25 (b) (1982); Perlin, "Unpacking the Myths," 638.

105. Perlin, "Unpacking the Myths," 603–4.

106. See Michael Davidson, "Post-Traumatic Stress Disorder: A Controversial Defense for Veterans of a Controversial War," *William and Mary Law Review* 29 (Winter 1988): 415–40; Karl Kirkland, "Post-Traumatic Stress Disorder vs. Pseudo Post-Traumatic Stress Disorder," *Alabama Lawyer* 56 (March 1995): 90–93; Falk, "Novel Theories of Criminal Defense Based Upon the Toxicity of the Social Environment," *North Carolina Law Review* 74 (March 1996): 731–811; Christopher Slobogin, "Psychiatric Evidence in Criminal Trials: To Junk or Not to Junk?" *William and Mary Law Review* 40 (October 1998): 1–56; Jennifer Grossman, "Postpartum Psychosis—A Defense to Criminal Responsibility or Just Another Gimmick?" *University of Detroit Law Review* 67 (1990): 311–44; Beth Bookwalter, "Throwing the Bathwater out with the Baby: Wrongful Exclusion of Expert Testimony on Neonaticide Syndrome," *Boston University Law Review* 78 (October 1998): 1185–1210; Gardner, "Postpartum Depression Defense: Are Mothers Getting Away with Murder?" *New England Law Review* 24 (Spring 1990): 953–89; Amy Nelson, "Postpartum Psychosis: a New Defense?" *Dickinson Law Review* 95 (Spring 1991) 625–50; Larua Reece, "Mothers Who Kill: Postpartum Disorders and Criminal Infanticide," *UCLA Law Review* 38 (February 1991) 699–757.

107. Brenda Barton, "When Murdering Hands Rock the Cradle: An Overview of America's Incoherent Treatment of Infanticidal Mothers," *Southern Methodist University Law Review* 51 (March-April 1998): 604–5.

108. *Commonwealth v. Conaghan*, 720 N.E.2d 48 (Massachusetts, 2000).

109. *State v. Young*, 999 P.2d 230 (Hawaii, 2000).

110. *Reyna v. State*, 11 S.W.3d 401 (Texas, 2000).

111. *State v. Greene*, 984 P.2d 1024 (Washington, 1999).

112. *Chatman v. Commonwealth*, 518 S.E.2d 847 (Virginia, 1999).

113. Cirincione and Jacobs, "Identifying Insanity Acquittals," 493. In 1995, for example, there were 117 insanity acquittals in California, ninety-six in Florida, and 102 in New Jersey.

114. George Blau and Hugh McGinley, "Use of the Insanity Defense: A Survey of Attorneys in Wyoming," *Behavioral Sciences and the Law* 13 (1995): 517–528. "The insanity defense is more often employed as a defense or as a defense strategy in the resolution of cases than is generally suggested." Ibid.,. 527.

115. Wilson, 46.

Chapter 5: Turning Punishment into Treatment

1. Ronald Fitten, "Former Teacher Won't Go to Prison for Rape of Boy," *Seattle Times*, Nov. 14, 1997, http://archives.seattletimes.nwsource.com/cgi-bin/texis/web/vortex/search/+mZ6eOXhK.

2. Fitten, ibid.; Carol Ostrom, "Bipolar Disorder: Valid Excuse for Letourneau?" *Seattle Times*,

February 15, 1998, http://archives.seattletimes.nwsource.com/cgi-bin/texis/web/vortex/disp lay?slug=mary&date=19980215&query=Letourneau.

3. See Francis Allen, *The Decline of the Rehabilitative Ideal: Penal Policy and Social Purpose* (New Haven, CT: Yale University Press, 1981).

4. Allen, 4–16; Adam Jay Hirsch, *The Rise of the Penitentiary: Prisons and Punishment in Early America* (New Haven, CT: Yale University Press, 1992), 3–68.

5. See the discussion of punishment in Puritan Massachusetts in Hirsch, *The Rise of the Penitentiary*, 3–6, 33–35.

6. See ibid., 13–68.

7. Ibid., 66; Allen, *The Decline of the Rehabilitative Ideal*, 15–16.

8. A. L. Jacoby, "The New Approach to the Problem of Delinquency: Punishment versus Treatment," in *Proceedings of the National Conference of Social Work, Fifty-Third Annual Session, May 26–June 2, 1926* (Chicago: University of Chicago Press), 180.

9. "Report of the Advisory Committee on Penal Institutions, Probation and Parole," 181.

10. Ibid., 213.

11. Ibid., 183.

12. Francis T. Cullen and Paul Gendreau, "Assessing Correctional Rehabilitation: Policy, Practice, and Prospects," in *Policies, Processes, and Decisions of the Criminal Justice System*, vol. 3 of *Criminal Justice 2000* (Washington, DC: National Institute of Justice, 2000), 122. Also see Allen, *The Decline of the Rehabilitative Ideal*, especially 9–10.

13. G. I. Giardini, "Social Dynamics in Probation and Parole," in *Criminology: A Book of Readings*, 613.

14. Reed Clegg, *Probation and Parole: Principles and Practices* (Springfield, IL: Charles Thomas, 1964), 65.

15. Ibid., 66.

16. Ibid., 101, 104.

17. Charles Lionel Chute and Marjorie Bell, *Crime, Courts, and Probation* (New York: Macmillan, 1956), 164.

18. Clegg, 104.

19. Albert Morris, "As the Criminal Sees Himself," in *Criminology: A Book of Readings*, 72.

20. "Gee, Officer Krupke," lyrics, http://www.westsidestory.com/site/level2/lyrics/krupke. html.

21. "Treatment, Not Time: LeTourneau Sentenced for Student's Rape," *Seattle Times*, November 15, 1997, http://archives.seattletimes.nwsource.com/cgi-bin/texis/web/vortex/display?slug=mary&date=19971115&query=Letourneau.

22. Barton, "When Murdering Hands Rock the Cradle," 605–6, 615–16.

23. *State v. Heartfield*, 998 P.2d 1080 (Arizona, 2000).

24. Eric Silver, "Punishment or Treatment? Comparing the Lengths of Confinement of Successful and Unsuccessful Insanity Defendants," *Law and Human Behavior* 19 (1995): 382.

25. "Montana's reform only eliminated the insanity defense from its statutes and not from its courtrooms. The evidence once offered to support an insanity defense has been redirected through the incompetency process, proof of *mens rea*, and sentencing. As a result, insanity pleas and insanity acquittals still exist in Montana." Buitendorp, "A Statutory Lesson from 'Big Sky Country,'" 970.

26. Table 15, Reasons Given for Departures Below the Guidelines, Pennsylvania Commission on Sentencing 2000 Annual Report, http://pcs.la.psu.edu/2000%20Report--Tables-Only.pdf. This data from Pennsylvania should be used cautiously, because in a substantial number of cases judges did not provide a reason for a below-standard sentence. So these statistics only tell us about the cases in which the judges did provide reasons.

27. In fiscal year 2004, "diminished capacity" was cited 225 times by judges as a reason for downward departures and "mental and emotional conditions" was cited eighty-six times. Judges often provide multiple reasons for downward departures, so there is likely some overlap in these two categories. "Diminished capacity" consistently ranked as the fifth or sixth most cited reason for nongovernment initiated downward sentencing departures between fiscal years 1998 and 2004, while "mental and emotional conditions" was usually the tenth most cited reason. For the data see Table 25, "Reasons Given by Sentencing Courts for Other Departures Below the Guideline Range, Fiscal Year 2004, Pre-Blakely (October 1, 2003, through June

24, 2004)" and Table 25, "Reasons Given by Sentencing Courts for Other Departures Below the Guideline Range, Fiscal Year 2004, Post-Blakely (June 25, 2004, through September 30, 2004)" *2004 Sourcebook of Federal Sentencing Statistics* (Washington, D.C. U.S. Sentencing Commission), http://www.ussc.gov/ANNRPT/2004/SBTOC04.htm; Table 25A, "Reasons Given by Sentencing Courts for Other Departures Below the Guideline Range, Fiscal Year 2003," *2003 Sourcebook of Federal Sentencing Statistics* (Washington, D.C. U.S. Sentencing Commission), http://www.ussc.gov/ANNRPT/2003/SBTOC03.htm; Table 25, "Reasons Given by Sentencing Courts for Downward Departures, Fiscal Year 2002," *2002 Sourcebook of Federal Sentencing Statistics* (Washington, D.C. U.S. Sentencing Commission), http://www.ussc.gov/ANNRPT/2002/SB-TOC02.htm; Table 25, "Reasons Given by Sentencing Courts for Downward Departures, Fiscal Year 2001," *2001 Sourcebook of Federal Sentencing Statistics* (Washington, D.C. U.S. Sentencing Commission), http://www.ussc.gov/ANNRPT/2001/SBTOC01.htm; Table 25, "Reasons Given by Sentencing Courts for Downward Departures, Fiscal Year 2000," *2000 Sourcebook of Federal Sentencing Statistics* (Washington, D.C. U.S. Sentencing Commission), http://www.ussc.gov/ANNRPT/2000/SBTOC00.htm; Table 25, "Reasons Given by Sentencing Courts for Downward Departures, Fiscal Year 1999," *1999 Sourcebook of Federal Sentencing Statistics* (Washington, D.C. U.S. Sentencing Commission), http://www.ussc.gov/ANNRPT/1999/Sbtoc99.htm; Table 25, "Reasons Given by Sentencing Courts for Downward Departures, Fiscal Year 1998," *1998 Sourcebook of Federal Sentencing Statistics* (Washington, D.C. U.S. Sentencing Commission), http://www.ussc.gov/ANNRPT/1998/Sbtoc98.htm.

28. See Alan Ellis, "Answering the 'Why' Question: The Powerful Departure Grounds of Diminished Capacity, Aberrant Behavior, and Post-Offense Rehabilitation," *Federal Sentencing Reporter* (May/June 1999), 11: 322; also see Carlos Pelayo, " 'Give Me a Break! I Couldn't Help Myself!'?: Rejecting Volitional Impairment as a Basis for Departure Under Federal Sentencing Guidelines Section 5K2.13," *University of Pennsylvania Law Review* 147 (January 1999): 729–69.

29. Ibid. In 2003, the Sentencing Commission acted to curtail some downward sentencing departures, including those based on a diminished capacity defense of "gambling addiction." See "Sentencing Commission Acts to Reduce Number of Sentences below the Guidelines," News Release, October 8, 2003 (Washington, D.C.: United States Sentencing Commission), http://www.ussc.gov/PRESS/rel100803a.htm (accessed June 18, 2005).

30. Jesse Ewell, "A Plea for Castration to Prevent Criminal Assault" (1907), quoted in Haller, *Eugenics: Hereditarian Attitudes in American Thought* (New Brunswick, NJ: Rutgers University Press, 1963), 209.

31. Haller, *Eugenics*, 46.

32. G. Frank Lydston, "Asexualization in the Prevention of Crime," *The Medical News* 68 (May 23, 1896): 575.

33. Hunter McGuire and G. Frank Lydston, "Sexual Crimes among Southern Negroes" (1893), quoted in Haller, 46.

34. Lydston, "Asexualization in the Prevention of Crime," 573.

35. Haller, *Eugenics*, 48–49.

36. Martin Barr, *Mental Defectives: Their History, Treatment, and Training* (Philadelphia: P. Blakiston's Son, 1910), 191.

37. Ibid., 195.

38. Ibid., 194.

39. Horatio Hackett Newman, *Evolution, Genetics, and Eugenics*, third edition. (Chicago: University of Chicago Press, 1932), 441.

40. For Darwin's endorsement of voluntary marriage restrictions, see *Descent* (1871), 2:403. In both *Descent* and in his correspondence with Francis Galton, Darwin did worry about whether the eugenics crusade might be "utopian," but he nevertheless thought it was a worthy effort. See Bowlby, *Charles Darwin*, 415–16. On the eugenic efforts of sons George and Leonard, see Kevles, *In the Name of Eugenics: Genetics and the Uses of Human Heredity* (Cambridge: Harvard University Press, 1995), 60 and Desmond and Moore, *Darwin*, 610–611.

41. For more information on Francis Galton, see Haller, *Eugenics*, 8–20, and Kevles, *In the Name of Eugenics*, 3–19.

42. Kevles, *In the Name of Eugenics*, xiii.

43. See Kevles, *In the Name of Eugenics*, 4; Haller, *Eugenics*, 17–18.

44. Haller, *Eugenics*, 11.

45. Ibid.
46. "Eugenic Legislation, A Review," *The Journal of Heredity* 6 (March 1915): 142.
47. Haller, *Eugenics*, 50.
48. Quoted in ibid.
49. Ibid., 137.
50. Ibid., 134–35.
51. "Heredity as a Factor in the Improvement of Social Conditions," *The American Breeders Magazine* 2 (Fourth Quarter, 1911): 249, 252.
52. Margaret Sanger to the Human Betterment Foundation (February 1951), in Margaret Sanger Papers, Series 3, Subseries 4, Articles and Speeches, #306308.
53. Haller, *Eugenics*, 141.
54. Quoted in "The Jukes in 1915," *The Journal of Heredity* 7 (October 1916): 474.
55. "Feeblemindedness," *The Journal of Heredity* 6 (January 1915): 33. This same idea can be seen in the comments of an immigration commissioner in New York who argued that "the feeble-minded contribute largely to the criminal class." Cited in "First Report of the Committee on Immigration of the Eugenics Section," *American Breeders Magazine* 3 (Fourth Quarter, 1912): 252.
56. Haller, *Eugenics*, 7.
57. *Skinner v. State of Oklahoma Ex Rel. Williamson*, 316 U.S. 535 (1942).
58. Stephan Chorover, "Psychosurgery: A Neuropsychological Perspective," in *Psychosurgery: A Multidisciplinary Symposium* (Lexington, MA: Lexington Books, 1974), 16–17; David Shutts, *Lobotomy: Resort to the Knife* (New York: Van Nostrand Reinhold Co., 1982), 40–41; Elliot Valenstein, "The Practice of Psychosurgery: A Survey of the Literature (1971–1976)," in *National Commission for the Protection of Human Subjects of Biomedical and Behavioral Research*, Appendix: Psychosurgery (Washington D.C.: U.S. Department of Health, Education and Welfare, 1977), I-15.
59. Chorover, "Psychosurgery," 17–18; Shutts, *Lobotomy*, 35–40.
60. Shutts, *Lobotomy*, 47, 53, 56.
61. Valenstein, "The Practice of Psychosurgery," I-5 and I-6; Shutts, *Lobotomy*, 59–218.
62. Shutts, *Lobotomy*, 194, 250; Chorover, "Psychosurgery," 19.
63. Chorover, 18.
64. Freeman, quoted in ibid.
65. For more information on the tragic fate of Rosemary Kennedy, see Doris Kearns Goodwin, *The Fitzgeralds and the Kennedys: An American Saga* (New York: St. Martin's Griffin, 1991), 641–43; Laurence Leamer, *The Kennedy Men: 1901–1963* (New York: Harper Perennial, 2002), 153–55.
66. Shutts, *Lobotomy*, 154.
67. Ibid., 154–56; George Annas and Leonard Glantz, "Psychosurgery: The Law's Response," in *Psychosurgery: A Multidisciplinary Symposium*, 34–35.
68. Quoted in Shutts, *Lobotomy*, 156.
69. Ibid., 156.
70. Ibid., 134–35, 156–57, 252.
71. Quoted in ibid., 157.
72. Chorover, "Psychosurgery," 19; Valenstein, "The Practice of Psychosurgery," I-7.
73. Vernon Mark, William Sweet, and Frank Ervin, "Role of Brain Disease in Riots and Urban Violence," *Journal of the American Medical Association* 201 (September 11, 1967): 895. Emphasis in the original.
74. Mark and Ervin, *Violence and the Brain*, 156–58.
75. Ibid., 158.
76. Ibid., 160.
77. R. K. Procunier, director of corrections for the state of California, quoted in Samuel Chavkin, *The Mind Stealers: Psychosurgery and Mind Control* (Boston: Houghton Mifflin Co., 1978), 60.
78. Chavkin, *The Mind Stealers*, 95–104.
79. Ibid., 95–96.
80. Ibid., 98.
81. Ibid., 63–88. Also see *Quality of Health Care – Human Experimentation*, 1973. Hearings before the Subcommittee on Health of the Committee on Labor and Public Welfare, United States Senate, 93rd Congress, First Session, on S. 974, S. 878, S.J. 71, March 7 and 8, 1973, Part 3 (Washington, DC: U.S. Government Printing Office, 1973), 836–43, 1026–33, 1125, 1133.

82. Testimony of Bernard L. Diamond, M.D., School of Criminology, University of California, in *Quality of Health Care*, 842.

83. Jean Heller, "Human Experimentation," AP news story (December 18, 1972), in Ibid., 1125.

84. Chavkin, *The Mind Stealers*, 72.

85. Report on Penal Institutions, 172.

86. Peter R. Breggin and Ginger Ross Breggin, *The War Against Children* (New York: St. Martin's Press, 1994), 116.

87. Mark and Ervin, *Violence and the Brain*, 93.

88. Ibid., 97.

89. Breggin and Breggin, *The War Against Children*, 120.

90. Ibid., 121.

91. Ibid., 123.

92. See U.S. Senate, Judiciary Committee, Subcommittee on Constitutional Rights, *Individual Rights and the Federal Role in Behavior Modification* (Washington, D.C.: Government Printing Office, November 1974); U.S. Senate Committee on Labor and Public Welfare, Subcommittee on Health, *Quality of Health Care – Human Experimentation, Part 2* (Washington, D.C.: Government Printing Office, February 23, 1973). Also see Public Law 93-348 (1973), which charged the National Commission for the Protection of Human Subjects of Biomedical and Behavioral Research to study the use of psychosurgery.

93. Mark and Ervin, *Violence and the Brain*, 139.

94. Breggin and Breggin, *The War Against Children*, 118–19.

95. Quoted in Chavkin, *The Mind Stealers*, 177. More generally, for a summary of Delgado's views about the value of brain control, see 145–48, 173–78.

96. Ibid, 177.

97. Quoted in Samuel Shuman, *Psychosurgery and the Medical Control of Violence: Autonomy and Deviance* (Detroit: Wayne State University Press, 1977), 97.

98. National Commission for the Protection of Human Subjects of Biomedical and Behavioral Research, Reports and Recommendations: Psychosurgery (Washington, D.C.: Department of Health and Human Services, 1977).

99. Chavkin, *The Mind Stealers*, 80.

100. Breggin and Breggin, *The War Against Children*, 128–30; Shuman, 11–23. Shuman's book reprints the decision of the Michigan court in this case as well as many court documents. Ibid., 221–324.

101. Robert Martinson, "What Works? Questions and Answers About Prison Reform," *The Public Interest* (Spring 1974), 22–54.

102. Francis T. Cullen and Paul Gendreau, "Assessing Correctional Rehabilitation: Policy, Practice, and Prospects," in *Policies, Processes, and Decisions of the Criminal Justice System*, vol. 3 or *Criminal Justice 2000* (Washington, D.C.: National Institute of Justice, 2000), 119–24.

103. See Joan Mullen, et. al., *Summary Findings and Policy Implications of a National Survey*, vol. 1 of *American Prisons and Jails* (Washington, D.C.: National Institute of Justice, 1980), 2–11.

104. J. S. Bainbridge Jr., "The Return of Retribution," *American Bar Association Journal* 71 (May 1985): 61–63.

105. Cullen and Gendreau, "Assessing Correctional Rehabilitation," 124–62.

106. Ibid., 137–38. Cullen and Gendreau claim that this figure may underestimate the effect of rehabilitation programs because the "control groups" used in the studies cited may themselves have received some form of treatment. In that case, the re-offense rates of both the "treatment" groups and the "control" groups would reflect the impact of rehabilitation. But this conclusion seems dubious. If it is in fact true that studies of rehabilitation programs do not have real control groups, then there is no way to know whether the 10 percent figure underestimates or overestimates the effect of rehabilitation on recidivism. Without a real control group, very little can be said about the effectiveness of the programs. Cullen and Gendreau also suggest that the 10 percent figure may underestimate the effect of rehabilitation because reoffense rates are difficult to determine accurately. Again, if this is really the case, it may invalidate the 10 percent figure, but it does not necessarily tell one whether the figure should be higher or lower. See ibid., 138.

107. See Richard Gebelein, "The Rebirth of Rehabilitation: Promise and Perils of Drug Courts," *Sentencing and Corrections: Issues for the Twenty-First Century*, no. 6 (U.S. Department of Justice,

May 2000); *Defining Drug Courts: The Key Components* (Washington, DC: U.S. Department of Justice, Office of Justice Programs, Bureau of Justice Assistance: January 1997, reprinted October 2004).

108. For overviews of recent research findings about the effectiveness and problems of drug courts, see *Drug Courts: The Second Decade* (Washington, D.C.: U. S. Department of Justice, Office of Justice Programs, National Institute of Justice, June 2006); *Adult Drug Courts: Evidence Indicates Recidivism Reductions and Mixed Results for Other Outcomes* (Washington, DC: General Accounting Office, February 2005); Kevin Whiteacre, "The Jury's Still Out on Drug Courts," (August 18, 2004), http://www.jointogether.org/news/yourturn/commentary/2004/the-jurys-still-out-on-drug.html; Douglas B. Marlowe, "Drug Court Efficacy vs. Effectiveness," (September 24, 2004), http://www.jointogether.org/news/yourturn/commentary/2004/drug-court-efficacy-vs.html. For an early critique of drug courts in a popular journal, see Eric Cohen, "The Drug Court Revolution," *The Weekly Standard* 15 (December 27, 1999): 20–23.

109. "Georgia Accountability Courts," http://www.georgiacourts.org/courts/accountability/overview.php.

110. Marc Levin, "Drug Courts: The Right Prescription for Texas," *Policy Perspective* (Texas Public Policy Foundation, February 2006), 1 and 3.

111. Cullen and Gendreau, "Assessing Correctional Rehabilitation," 146–47.

112. Jeremy Travis, "But They All Come Back: Rethinking Prisoner Reentry," *Sentencing and Corrections: Issues for the Twenty-First Century*, no. 7 (U.S. Department of Justice, May 2000), 7. For more information about the "restorative justice" movement, see Restorative Justice Online, www.restorativejustice.org. A number of states are incorporating restorative justice into their correctional programs. For examples, see "Restorative Justice" (Montana Department of Justice website), http://www.doj.mt.gov/victims/restorativejustice.asp; "DPS Hosts Conference Highlighting Restorative Justice Best Practice," (Missouri Department of Public Safety press release, September 13, 2006), http://www.dps.mo.gov/dps/NEWS/Releases06/September/RestorativeJusticeConference.htm; "Neighborhood Restorative Justice," Florida Statutes 2006, Chapter 985.155, http://www.leg.state.fl.us/statutes/index.cfm?App_mode=Display_Statute&URL=Ch0985/ch0985.htm; "Restorative Justice" (Vermont Department of Corrections), http://www.doc.state.vt.us/justice/restorative-justice/; and "More Information About Restorative Justice," (Alabama Center for Dispute Resolution), http://alabamaadr.org/flash-Site/restorativejustice/info.cfm.

113. See discussion of these kinds of punishments in Stephen Garvey, "Can Shaming Punishments Educate?" *University of Chicago Law Review* 65 (Summer 1998): 733–94.

114. Dora Schriro, "Correcting Corrections: Missouri's Parallel Universe," *Sentencing and Corrections: Issues for the Twenty-First Century*, no. 8 (U.S. Department of Justice, May 2000).

115. "Letourneau Sent to Prison for More than Seven Years," *Seattle Times* (February 6, 1998), http://archives.seattletimes.nwsource.com/cgibin/texis/web/vortex/display?slug=mary&date=199 80206&query=Letourneau. After Letourneau was finally released she married the former student she had been convicted of raping. "Letourneau Marries Fualaau amid Media Circus," *Seattle Post-Intelligencer*, May 21, 2005, http://seattlepi.nwsource.com/local/225292_wedding21.html?searchpagefrom=1&searchdiff=27.

116. Albert J. Reiss Jr. and Jeffrey A. Roth, eds., *Understanding and Preventing Violence* 1 (Washington, D.C.: National Academy Press, 1993):24.

117. For a fuller treatment of the controversy over the anti-violence initiative, see Peter R. Breggin and Ginger Ross Breggin, *The War Against Children of Color: Psychiatry Targets Inner City Youth* (Monroe, ME: Common Courage Press, 1998), especially 1–20.

118. Tony Fabelo, " 'Technocorrections': The Promises, the Uncertain Threats," Sentencing and Corrections: Issues for the 21st Century, no. 5 (May 2000): 3.

119. Ibid., 2–3.

120. Ibid., 1, 6.

121. Remarks by Harold Koplewicz, reprinted in "Remarks by the First Lady at White House Conference on Mental Health" (Washington, D.C.: The White House, Office of the Press Secretary, June 7, 1999), http://clinton4.nara.gov/WH/EOP/First_Lady/html/generalspeeches/1999/19990607.html.

122. Even proponents of the biological model of mental illness sometimes acknowledge the lack of evidence for their view. In *Mental Health: A Report of the Surgeon General*, the integration of men-

tal health with neuroscience is praised because it "offers a way to circumvent the antiquated split between the mind and the body that historically has hampered mental health research." However, later in the report, the authors acknowledge that "few lesions or physiologic abnormalities define the mental disorders, and for the most part their causes remain unknown." See section titled "Chapter Conclusions" in chap. 1 of *Mental Health: A Report of the Surgeon General* (1999), http://www.surgeongeneral.gov/library/mentalhealth/chapter1/sec4.html.

123. Harold Koplewicz, quoted in Rob Waters, "Johnny Get Your Pills: Are We Overmedicating Our Kids?" Salon.com, June 17, 1999, http://archive.salon.com/health/feature/1999/06/17/antidepressants/print.html.

124. Peter Breggin, *Reclaiming Our Children: A Healing Plan for a Nation in Crisis* (Cambridge, MA: Perseus Books, 2000), 24.

125. Lawrence H. Diller, M.D., "Kids on Drugs," Salon.com (March 9, 2000), http://dir.salon.com/health/feature/2000/03/09/kid_drugs/index.html; Testimony of Terence Woodworth, Deputy Director, Office of Diversion Control, Drug Enforcement Administration, before the Committee on Education and the Workforce, Subcommittee on Early Childhood, Youth and Families (Washington, D.C.: U.S. House of Representatives, May 16, 2000), http://www.house.gov/ed_workforce/hearings/106th/ecyf/ritalin51600/woodworth.htm.

126. See table 3-3, chap. 3, *Mental Health: A Report of the Surgeon General.*

127. "There is wide variation among types of practitioners with respect to frequency of diagnosis of ADHD. Data indicate that family practitioners diagnose more quickly and prescribe medication more frequently than psychiatrists or pediatricians." ["Diagnosis and Treatment of Attention Deficit Hyperactivity Disorder," National Institutes of Health Consensus Development Conference Statement (Nov. 16–18, 1998), available online at http://consensus.nih.gov/cons/110/110_statement.htm.] Similar variation exists in the percentage of ADHD patients who are treated with psychostimulants. "Substantial evidence exists of wide variations in the use of psychostimulants across communities and physicians, suggesting no consensus among practitioners regarding which ADHD patients should be treated with psychostimulants." [Ibid.]

128. The "Consensus Statement" issued under the auspices of the National Institutes of Health in 1998 conceded that "after years of clinical research and experience with ADHD, our knowledge about the cause or causes of ADHD remains speculative. Consequently, we have no strategies for the prevention of ADHD." ["Diagnosis and Treatment of Attention Deficit Hyperactivity Disorder."]

129. See *Testimony of Lawrence Diller before the Committee on Education and the Workforce, Subcommittee on Early Childhood, Youth and Families* (Washington, D.C.: U.S. House of Representatives, May 16, 2000), http://www.house.gov/ed_workforce/hearings/106th/ecyf/ritalin51600/diller.htm; Diller, "Kids on Drugs."

130. Testimony of Terence Woodworth. The 500 percent figure is supplied by the Drug Enforcement Administration. Others have claimed that the figure is actually 700 percent. See *Testimony of Lawrence Diller.* A study published in 1996 concluded that there was only a "2.5-fold" increase in methylphenidate treatment of youths between 1990 and 1995, but this data is for only half of the decade, and the study only looked at youths between five and eighteen. See Daniel J. Safer, Julie M. Zito, Eric M. Fine, "Increased Methylphenidate Usage for Attention Deficit Disorder in the 1990s," *Pediatrics* 98 (December 1996): 1084–88.

131. Ibid.; Diller; Breggin, *Reclaiming Our Children*, 144.

132. Breggin, *Reclaiming Our Children*, 144.

133. Testimony of Terence Woodworth.

134. For documentation of the various side effects of psychostimulants such as Ritalin, see Peter R. Breggin, "Psychostimulants in the Treatment of Children Diagnosed with ADHD: Risks and Mechanism of Action," *International Journal of Risk & Safety in Medicine* 12 (1999): 3–35; Esther Cherland, Renee Fitzpatrick, "Psychotic Side Effects of Psychostimulants: A 5-Year Review," *Canadian Journal of Psychiatry* 44 (October 1999): 811–13.

135. *Testimony of Peter Breggin before the Subcommittee on Education and Investigations*, Committee on Education and the Workforce (Washington, DC: House of Representatives, Sept. 29, 2000), http://www.house.gov/ed_workforce/hearings/106th/oi/ritalin92900/breggin.htm.

136. The MTA Cooperative Group, "A 14-month Randomized Clinical Trial of Treatment Strategies for Attention-Deficit/Hyperactivity Disorder," *Archives of General Psychiatry* 56, no. 12

(December 1999): 1073–1086; The MTA Cooperative Group, "Moderators and Mediators of Treatment Response for Children with Attention-deficit/hyperactivity disorder: The Multimodal Treatment Study of Children with Attention-Deficit/Hyperactivity Disorder," *Archives of General Psychiatry* 56, no. 12 (December 1999): 1088–1096.

137. See critique offered by Peter Breggin, "A Critical Analysis of the NIMH Multimodal Treatment Study for Attention-Deficit/Hyperactivity Disorcer (The MTA Study)," available online at http://www.breggin.com/mta.html.

138. For example, in a recent widely reported study of the treatment of adolescent depression with the drug fluoxetine, the positive effect of the drug in nearly six out of every ten patients could be accounted for by the placebo effect. While 60.6 percent of those taking the drug responded positively, 34.8 percent of children taking placebos had a similar response as did 43.2 percent of children receiving behavioral therapy alone. This suggests that the clear majority of children on this drug could receive just as much benefit from taking a placebo or receiving behavioral therapy. See Treatment for Adolescents With Depression Study (TADS) Team, "Fluoxetine, Cognitive-Behavioral Therapy, and Their Combination for Adolescents With Depression," *Journal of the American Medical Association* 292: 807–20.

139. Breggin, "A Critical Analysis"; MTA Cooperative Group, 1082–83.

140. "The MTA Cooperative Group (1999a, 1999b) presented analyses of outcome data measured after 14 months of treatment. Dependent measures evaluated in this initial report included parent and teacher ratings, classroom observations, and academic achievement. At the 14-month point, behavioral treatment had been faded, that is, parent training sessions were either terminated or in a maintenance phase (i.e., held monthly or less-frequently), contact with the children's schools had ended except for parent–teacher contact and infrequent therapist–teacher phone calls, and the intensive part of the child treatment (the STP and a part-time classroom aide) had ended 6–9 months before the posttreatment measures were taken. In contrast, at 14-month endpoint, medication was continued in its most intensive phase (monthly contacts, ongoing dosage adjustments, and highest dosages). The results of these analyses indicated, not surprisingly, that active medication (38 mg MPH or equivalent per day) produced greater effects than did faded behavioral treatment on most measures." [William E. Pelham, Jr., Elizabeth M. Gnagy, Andrew R. Greiner, Betsy Hoza, Stephen P. Hinshaw, James M. Swanson, Steve Simpson, Cheri Shapiro, Oscar Bukstein, Carrie Baron-Myak, and Keith McBurnett, "Behavioral versus Behavioral and Pharmacological Treatment in ADHD Children Attending a Summer Treatment Program," *Journal of Abnormal Child Psychology* 28, no. 6 (2000): 509.] Also see Breggin, "A Critical Analysis."

141. See Pelham, et. al., "Behavioral versus Behavioral and Pharmacological," 507–525. For a recent journal article describing the effectiveness of one particular type of non-drug intervention for preschoolers with ADHD, see Edmund J. S. Sonuga-Barke, Margaret Thompson, Howard Abikoff, Rachel Klein, and Laurie Miller Brotman, "Nonpharmacological Interventions for Preschoolers with ADHD: The Case for Specialized Parent Training," *Infants & Young Children* 19, no. 2 (April-June 2006): 142–53.

142. "Only 13/61 (22%) or subjects randomized to best-dose MPH and 7/53 (13%) of those randomized to placebo met criteria for excellent response on the SNAP composite score, a difference that did not reach statistical significance." Laurence Greenhill, Scott Kollins, Howard Abikoff, James McDracken, et. al., "Efficacy and Safety of Immediate-Release Methylphenidate Treatment for Preschoolers with ADHD," Journal of the American Academy of Child and Adolescent Psychiatry 45, no. 11 (November 2006): 1290.

143. Tim Wigal, Laurence Greenhill, Shirley Chuang, James McGough, et. al., "Safety and Tolerability of Methylphenidate in Preschool Chidren with ADHD," Journal of the American Academy of Child and Adolescent Psychiatry 45, no. 11 (November 2006): 1302.

144. See New Freedom Commission on Mental Health, Achieving the Promise: Transforming Mental Health Care in America. Final Report. DHHS Pub. No. SMA-03-3832. Rockville, MD: 2003. Available online at http://www.mentalhealthcommission.gov/reports/FinalReport/toc.html (accessed June 17, 2005). According to the Commission, "in a transformed mental health system, the early detection of mental health problems in children and adults—through routine and comprehensive testing and screening—will be an expected and typical occurrence. At the first sign of difficulties, preventive interventions will be started to keep problems from escalating." This "screening and early intervention will occur in both readily accessible, low-

stigma settings, such as primary health care facilities and schools, and in settings in which a high level of risk exists for mental health problems, such as criminal justice, juvenile justice, and child welfare systems. Both children and adults will be screened for mental illnesses during their routine physical exams." ["Executive Summary," Achieving the Promise, http://www.mentalhealthcommission.gov/reports/FinalReport/FullReport.htm (accessed on June 17, 2005).] Regarding the role of schools in screening for mental illness, the Commission argues that "schools are in a key position to identify mental health problems early and to provide a link to appropriate services. Every day more than 52 million students attend over 114,000 schools in the U.S. When combined with the six million adults working at those schools, almost one-fifth of the population passes through the Nation's schools on any given weekday." ["Goal 4: Early Mental Health Screening, Assessment, and Referral to Services Are Common Practice," Achieving the Promise, http://www.mentalhealthcommission.gov/reports/Final-Report/FullReport-05.htm (accessed June 17, 2005).]

145. Some teachers are obviously uncomfortable with the new expectations being placed on them. In a statement submitted as part of a congressional hearing about the use of medications in schools, Utah teacher Ann Boley complained that she and other teachers were expected to do mental health assessments and referrals of their students after a "one hour orientation." "I, along with others of the teaching staff, was appalled with the assigned responsibilities. Our backgrounds are as teachers in various of the secondary education academic disciplines . . . We are not human behavior specialists any more than we are astronauts." Boley further noted that teachers engaged in mental health assessments and referrals may be subjected to legal liability if things go wrong. See "Written Statement of Ann Boley," in *Protecting Children: The Use of Medication in Our Nation's Schools and H.R. 1170, Child Medication Safety Act of 2003*, Hearing before the Subcommittee on Education Reform of the Committee on Education and the Workforce, United States House of Representatives, 108th Congress, First Session, Serial No. 108-14 (Washington: U.S. Government Printing Office, 2003), 141–42.

146. Rick Karlin, "Ritalin use splits parents, school," *Albany Times Union* (May 7, 2000), A1.

147. Ibid.

148. Testimony of Patti Johnson before the Subcommittee on Education and Investigations, Committee on Education and the Workforce (Washington, DC: House of Representatives, Sept. 29, 2000), http://www.house.gov/ed_workforce/hearings/106th/oi/ritalin92900/johnson.htm (accessed June 17, 2005). Also see the statement from a Utah parent in Protecting Children, 139–140.

149. For the record of the hearing, see *Protecting Children*.

Chapter 6: Law of the Jungle

1. James Roark, Michael Johnson, et. al., *The American Promise: A History of the United States* (Boston: Bedford Books, 1998), 710.

2. Joseph Conlin, *The American Past, Part Two: A Survey of American History Since 1865*, 3rd ed. (San Diego: Harcourt Brace Jovanovich, 1990), 498.

3. John Mack Faragher, Mari Jo Buhle, et. al., *Out of Many: A History of the American People*, 2nd ed. (Upper Saddle River, New Jersey: Prentice Hall, 1997), 2:654.

4. Carl N. Degler, Thomas C. Cochran, et. al., *The Democratic Experience*, 5th ed. (Glenview, Illinois: Scott, Foresman and Co., 1981), 53.

5. Faragher, *Out of Many*, 591.

6. Mary Beth Norton, David M. Katzman, et. al., *A People and a Nation: A History of the United States*, 2nd edition (Boston: Houghton Mifflin Company, 1986), 490.

7. Alan Brinkley, *American History: A Survey* (New York: McGraw-Hill, Inc., 1995), 488. Some more recent textbooks have moderated their discussions of Social Darwinism. "It is not clear how seriously the business community took social Darwinism." Irwin Unger, *These United States: The Questions of Our Past, Volume II: since 1865*, 5th ed. (Englewood Cliffs, NJ: Prentice-Hall, 1992), 474; "Few men of practical affairs were directly influenced by Darwin's ideas." John A. Garraty, Robert A. McCaughey, *The American Nation: A History of the United States Since 1865*,

Volume Two, 6th ed. (New York: Harper and Row, 1983), 490; "Despite their seeming lack of regard for common people, many of the robber barons who had embraced Social Darwinism also supported the spirit of charity." Gary B. Nash, *American Odyssey: The United States in the Twentieth Century* (New York: Glencoe/McGraw-Hill, 1997), 210; "But few American millionaires were true Social Darwinists. The very ruthlessness of the theory made it unpalatable to rich families who, in their personal lives, were committed to traditional religious values. Moreover, businessmen are rarely intellectuals, and Spencer's philosophy and writing style were as thick and murky as Rockefeller's crude oil." Conlin, *The American Past*, 498; "The influence of Social Darwinism on American thinking has been exaggerated, but in the powerful hands of Sumner and others it did influence some journalists, ministers, and policymakers." Robert A. Divine, T. H. Breen, et. al., *America, Past and Present, Volume Two from 1865*, 2nd ed. (Glenview, IL: Scott Foresman and Company, 1987), 568.

8. Richard Hoftstadter, *Social Darwinism in American Thought*, rev. ed. (Boston: The Beacon Press, 1955), 5.

9. See Darwin, *Origin of Species*, 74.

10. Quoted in Hoftstadter, *Social Darwinism in American Thought*, 41.

11. Hofstadter, *Social Darwinism in American Thought*, 51.

12. Quoted in ibid., 45.

13. Quoted in ibid.

14. Quoted in ibid., 45.

15. Irvin G. Wyllie, "Social Darwinism and the Businessman," *Proceedings of the American Philosophical Society* 103, no. 5 (October, 1959): 633.

16. Ibid., 632.

17. Ibid. Ironically, Rockefeller's talk did not even promote the view that was attributed to it. "The title of his talk had been 'Christianity in Business,' and . . . he had entered a plea for more Christian virtue in the transaction of business." Ibid., 635.

18. See, for example, Adam Smith, *An Inquiry into the Nature and Causes of the Wealth of Nations*, ed. Edwin Cannan (London: Methuen and Co., Ltd., 1904), especially bk. 1, chap. 7–11; bk. 2, chap. 2, 5; bk. 4, chap. 7; bk. 5, chap. 1. Available online from http://www.econlib.org/library/Smith/smWN13.html (accessed August 22, 2004); Frederic Bastiat, *Economic Harmonies*, trans. W. Hayden Boyers, ed. George B. de Huszar (Irvington-on-Hudson, NY: The Foundation for Economic Education, Inc., 1996), chap. 10. Available online from http://www.econlib.org/library/Bastiat/basHar10.html (accessed August 22, 2004).

19. Wyllie, "Social Darwinism," 634.

20. Ibid.

21. Thomas Robert Malthus, *First Essay on Population 1798* (London: Macmillan & Co. Ltd., 1926).

22. See Darwin, *Autobiography*, 42–43; Darwin, *Origin of Species*, 75. For an able discussion of the different views advanced by historians on the role of Malthus in the development of Darwin's theory, see David R. Oldroyd, "How Did Darwin Arrive at His Theory? The Secondary Literature to 1982," *History of Science* 12 (1984): 325–74.

23. Malthus, *First Essay*, 27.

24. Ibid.

25. Ibid., 346.

26. Darwin, *Autobiography*, 42–43.

27. Malthus, *First Essay*, 18–38.

28. Ibid., 131.

29. Francis Bowen, "Malthusianism, Darwinism, and Pessimism," *North American Review* 129 (1879): 452.

30. Ibid., 454.

31. Ibid., 455, 456.

32. Edward Atkinson, *The Industrial Progress of the Nation* (New York: G. P. Putnam's Sons, 1889; reprint edition by Arno Press [New York], 1973), 156, 159.

33. Edward Atkinson, "The Unlearned Professions," *The Atlantic Monthly* 45, no. 272 (June 1880): 747.

34. Atkinson, *Industrial Progress*, 60.

35. Edward Atkinson, "Commercial Development," *Harper's* 51, no. 302 (July 1875): 267.

36. Atkinson, "The Unlearned Professions," 745.

37. Atkinson, "Commercial Development," 260. Also see discussion in *The Industrial Progress of the Nation*, 383–388.
38. Wyllie, "Social Darwinism," 634.
39. Recent revisionist scholars have downplayed the influence of natural selection and evolution in Spencer and Sumner. For example, see Robert Bannister, *Social Darwinism: Science and Myth in Anglo-American Social Thought*, with a new preface (Philadelphia: Temple University, 1988), 34–56, 97–113. However, see the able response by Mike Hawkins, *Social Darwinism in European and American Thought*, 82–103, 108–18. Although Sumner drew on Darwinism to justify his economic views, even he did not fit the left-wing stereotype of a Social Darwinist. Sumner acknowledged that society had an obligation to the physically incapacitated, and he supported laws regulating working conditions for women and children. In his own mind, he was a defender of the working poor. See William Graham Sumner, "The Forgotten Man" in *Social Darwinism: Selected Essays of William Graham Sumner* (Englewood Cliffs, NJ: Prentice Hall, 1963), 118, 123–24, 134.
40. For a good discussion of German biologists who used Darwinism to critique socialism and uphold laissez faire, see Richard Weikart, *Socialist Darwinism: Evolution in German Social Thought from Marx to Bernstein* (San Francisco: International Scholars Publications, 1999), 109–22. For discussions of Darwin's relation to "Social Darwinism," see the following note.
41. See Richard Weikart, "Laissez-Faire Social Darwinism and Individualist Competition in Darwin and Huxley," *The European Legacy* 3, no. 1 (1998): 17–30, especially 19–25; Richard Weikart, "A Recently Discovered Darwin Letter on Social Darwinism," *Isis* 86 (1995): 609–11. For a scholar who tries to distance Darwin from the social applications of his theory, see James Allen Rogers, "Darwinism and Social Darwinism," *Journal of the History of Ideas* 33, no. 2 (April–June, 1972), 265–80. Although Rogers states that "[t]he term, Social Darwinism, was extremely unfortunate because it linked Darwin's theory of natural selection with various theories of human social evolution for which Darwin was in no way responsible" (280), his article provides evidence that Darwin was more connected to social applications of his theory than sometimes thought.
42. Weikart, "Laissez-Faire Social Darwinism," 25–27.
43. "Despite small forways into the province of social thought, Darwin was discreet and never placed heavy emphasis on political or economic applications of his theory." Weikart, *Socialist Darwinism*, 108.
44. Charles Darwin to H. Thiel, February 25, 1869 in *The Life and Letters of Charles Darwin*, ed. Francis Darwin, (New York: D. Appleton and Co., 1905), 2 294.
45. Bannister, *Social Darwinism*, 114.
46. Upton Sinclair, *The Jungle* (1906), available online at http://www.gutenberg.net/dirs/etext94/jungl10.txt (accessed March 8, 2005).
47. Ibid., chap. 2.
48. Ibid., chap. 7.
49. Ibid., chap. 11.
50. Ibid., chap. 29.
51. "A Declaration of Principles," adopted by the Nationalist Club in 1889, quoted in Arthur E. Morgan, *Edward Bellamy* (New York: Columbia University Press, 1944), 262. Also see Bannister, *Social Darwinism*, 124–25.
52. Henry Demarest Lloyd, *Wealth against Commonwealth* (New York: Harper & Brothers, 1898), 495.
53. Jack London, "The Class Struggle," in Jack London, *War of the Classes* (Oakland: Star Rover House, 1982), 18.
54. Bannister, *Social Darwinism*, 126.
55. Wyllie, "Social Darwinism," 635.
56. See discussion in Tom McKay, "Social Darwinism and Business Thought in the 1920's" (MA Thesis, Western Illinois University, 1973), 47–69.
57. Donald Hanson, financial editor of the Boston Evening Transcript, quoted in McKay, "Social Darwinism," 60.
58. W. W. Galbreath, president of Youngstown Pressed Steel Company, quoted in McKay, "Social Darwinism," 61.
59. G. R. Martin, vice president of the Great Northern Railway, quoted in McKay, "Social Darwinism," 63.

60. Ludwig von Mises, *Socialism: An Economic and Sociological Analysis*, trans. J. Kahane (Indianapolis: Liberty Fund, 1981), 281.

61. Ibid., 285.

62. Ibid., 285–86.

63. F. A. Hayek, "Freedom, Reason and Tradition," *Ethics* 68, no. 4 (July, 1958): 233.

64. Milton Friedman, *Essays in Positive Economics* (Chicago: University of Chicago Press, 1953), 21, 22. Emphasis in original. For other examples of trying to apply Darwinian theory to economics see Armen Alchian, "Uncertainty, Evolution and Economic Theory," *Journal of Political Economy* 58 (June 1950): 211–221; Richard Nelson and Sidney Winter, *An Evolutionary Theory of Economic Change* (Cambridge: Belknap Oress, 1982).

65. Edith Tilton Penrose, "Biological Analogies in the Theory of the Firm," *The American Economic Review* 42, no. 5 (December, 1952): 804.

66. Alexander Rosenberg, *Darwinism in Philosophy, Social Science and Policy* (New York: Cambridge University Press, 2000), 172.

67. Hayek, "Freedom," 232. For a more detailed discussion of the contribution of political economists to the development of Darwin's theory, see Silvan S. Schweber, "Darwin and the Political Economists: Divergence or Character," *Journal of the History of Biology* 13, no. 2 (Fall 1980): 195–289; Oldroyd, "How Did Darwin Arrive at His Theory?"

68. Arnhart, *Darwinian Conservatism*, 18. Also see 16, where Arnhart writes: "A spontaneous order is an unintended order. It is a complex order that arises not as the intended outcome of any intelligent design of any mind or group of minds, but as the unintended outcome of many individual actions to satisfy short-term needs. Spontaneous order is design without a designer."

69. See, for example, Robert Hagstrom, *Latticework: The New Investing* (New York: Texere, 2000), especially 55–96; Clay Carr, *Choice, Chance and Organizational Change: Practical Insights from Evolution for Business Leaders and Thinkers* (New York: Amacom, 1996).

70. William Tucker, "The New Science of Spontaneous Order," *Reason*, January 1996, 37.

71. Ibid., 38.

72. Ibid.

73. Arnhart, *Darwinian Conservatism*, 16.

74. Ibid., 18, emphasis added.

75. For a more detailed discussion of Arnhart's claims about "spontaneous order" see West, *Darwin's Conservatives*, 44–47.

76. Molly Ivins, quoted in Deirdre Donahue, "What's the Story on 'Story'? Great Reading," *USA Today* (October 6, 1989), 4D; Kevin Phillips, "Bold Ideas Reduced to Movie Talk, Budget Debate Demands More," *Los Angeles Times* (February 9, 1986), 5-1.

77. Ernest Conine, "Milk of Human Kindness Hasn't Borne a GOP Label," *Los Angeles Times* (October 27, 1987), 2-7.

78. Mario Cuomo, quoted in William Schneider, "The Democrats in '88," *Atlantic Monthly* (April 1987), online edition, http://www.theatlantic.com/politics/policamp/demo88.htm (accessed June 22, 2005); Walter Mondale, quoted in Steven Tipton, "Religion and the Moral Rhetoric of Presidential Politics," http://www.religion-online.org/showarticle.asp?title=1429 (accessed June 22, 2005).

79. Robert Scheer, "Taking Food From Mouths of Babes, Gingrich's Tax Cut Proposals Would Just Help Government Better Serve the Rich," *Los Angeles Times* (May 16, 1995), B-7.

80. Roger Boesche, "Homeless? Hungry? It's All Your Fault; The Gingrich Era Means Class-Based Politics, 'Us' vs. 'Them' in a War on the Poor," *Los Angeles Times* (December 1, 1994), B-7.

81. "Decision '92, Special Voters' Guide to State and Local Elections: The Third Parties," *Los AngelesTimes* (October 25, 1992), T-5.

82. John Vernon, "Libertarians" (Letter to the Editor), *Los Angeles Times* (October 31, 1992), B-7.

83. Robert Reich, "The Political Center, Straight Up; It's Bogus: Real Leaders Don't Take People Where They Already Are," *Washington Post* (June 17, 2001), B-1.

84. Kevin Phillips, "The Cycles of Financial Scandal," *New York Times* (July 17, 2002), 19.

85. For the influence of Gilder's book on the incoming Reagan administration, see George Gilder, *Wealth and Poverty: A New Edition of the Classic* (San Francisco: ICS Press, 1993), xv.

86. George Gilder, *Wealth and Poverty* (New York: Basic Books, 1981), 21, 27. Gilder further expanded his argument about the connection between altruism and capitalism in the revised paperback edition of his book, for which he added six thousand words to the chapter "The Returns of Giving." See *Wealth and Poverty: A New Edition*, xxii, 21–38.

87. Gilder, *Wealth and Poverty* (1981), 259–269.

88. For examples of Gingrich's approach toward the poor, see his articles "How kids can learn in the inner city" (Washington Post, Sept. 25, 1995), available at http://www.newt.org/index.php?src=news&submenu=press&prid=406&category=Opinion (accessed June 23, 2005), and "What Good Is Government and Can We Make It Better?" (*Newsweek*, April 10, 1995), available at http://www.newt.org/index.php?src=news&submenu=press&prid=401&category=Opinion n (accessed June 23, 2005).

89. Marvin Olasky, *The Tragedy of American Compassion* (Washington, D.C.: Regnery Publishing, 1992). For Olasky's discussion of Social Darwinism, see 60–98.

90. Quoted in Mike Allen, "President Urges War on Poverty; At Notre Dame, Bush Touts Faith Plan," *Washington Post* (May 21, 2001), A1.

91. Jack London, "Wanted: A New Law of Development," in London, *War of the Classes*, 243–44.

92. *Frederick Engels, Ludwig Feuerbach and the End of Classical German Philosophy* (originally published in 1886; Progress Publishers edition, 1946), pt. 4; available online at http://www.marxists.org/archive/marx/works/1886/ludwig-feuerbach/index.htm (accessed June 23, 2005).

93. Karl Marx to Ferdinand Lassalle, Jan. 16, 1861, Karl Marx, *Frederick Engels: Collected Works* (New York: International Publishers, 1975–2005), 41:245.

94. Engels went on to add: "Never before has so grandiose an attempt been made to demonstrate historical evolution in Nature, and certainly never to such good effect. One does, of course, have to put up with the crude English method." Frederick Engels to Karl Marx, December 11 or 12, 1859, *Collected Works*, 40:550. The views of Marx and Engels on Darwin's death-blow to teleology were echoed by other German socialists. "German socialists . . . were elated with Darwin's elimination of teleology from nature, which they regularly summoned in defense of their materialist world view." Weikart, *Socialist Darwinism*, 221. The same view was adopted by later Soviet thinkers. "In Soviet opinion this theory of Darwin's deals a devastating blow against the teleological point of view." Gustav A. Wetter, *Dialectical Materialism: A Historical and Systematic Survey of Philosophy in the Soviet Union*, trans. Peter Heath (New York: Frederick A. Praeger, Publishers, 1958), 379–380.

95. Nikolai Krementsov, *Stalinist Science* (Princeton, NJ: Princeton University Press, 1997), 73.

96. Quoted in Wetter, *Dialectical Materialism*, 461.

97. T. D. Lysenko, *Heredity and Its Variability*, trans. Theodosius Dobzhansky (Morninside Heights, NY: King's Crown Press, 1946), 24.

98. Krementsov, *Stalinist Science*, 74, 73.

99. Karl Marx to Frederick Engels, December 19, 1860, *Collected Works*, 41:232.

100. Karl Marx to Ferdinand Lassalle, January 16, 1861, *Collected Works*, 41:245.

101. Weikart, *Socialist Darwinism*, 35. Despite his praise of Darwin, Marx was ambivalent about the application of Darwinism to human society, and he could be very negative about Darwin's theory in private. Ibid., 15–51. For an interesting summary of the contacts between Darwin and Marx, see Ralph Colp, Jr., "The Contacts between Karl Marx and Charles Darwin," *Journal of the History of Ideas* 35, no. 2 (April-June 1974): 329–38. For many years it was believed that Marx offered to dedicate an edition of *Das Kapital* to Darwin, but this idea was based on a misidentification of the recipient of one of Darwin's letters. See Margaret A. Fay, "Did Marx Offer to Dedicate Capital to Darwin?: A Reassessment of the Evidence," *Journal of the History of Ideas* 39, no. 1 (January–March, 1978): 133–46.

102. Bannister, *Social Darwinism*, 133–35; Weikart, *Socialist Darwinism*, 83–101.

103. London, "Wanted: A New Law of Development," 244, 245, 218–20.

104. Frederick Engels to Pyotr Lavrov, Nov. 12–17, 1875, *Collected Works*, 45:106.

105. Upton Sinclair, *The Jungle*, chap. 30.

106. See Weikart, *Socialist Darwinism*, 4–5, 15–82, 221–22.

107. Himmelfarb, *Darwin and the Darwinian Revolution*, 422.

108. As Marx put it, "philosophers have hitherto only interpreted the world . . . the point is to change it." Karl Marx, "Theses On Feuerbach" (1845), trans. Cyril Smith, http://www.marxists.org/archive/marx/works/1845/theses/ (accessed June 24, 2005).

109. For a discussion of the impact of Germanic political science on America during the Progressive era, see Dennis John Mahoney, "A New Political Science for a World Made Wholly New: The Doctrine of Progress and the Emergence of American Political Science," (Ph.D. Dissertation, Claremont Graduate School, 1984), especially 25–45. Mahoney discusses the evolutionary character of Hegelian political science on 33–35.

110. See Mahoney, "A New Political Science," 103–42.

111. Woodrow Wilson, *Constitutional Government* in *The Papers of Woodrow Wilson*, ed. Arthur S. Link, (Princeton, NJ: Princeton University Press, 1974), 18:105.

112. Woodrow Wilson, *The New Freedom*, with an introduction and notes by William Leuchtenburg (Englewood Cliffs, NJ: Prentice-Hall, Inc., 1961), 41–42.

Chapter 7: Breeding Our Way out of Poverty

1. Henry Herbert Goddard, *The Kallikak Family: A Study in the Heredity of Feeble-Mindedness* (New York: The Macmillan Company, 1925), 7.

2. Michael W. Perry, ed., *The Pivot of Civilization in Historical Perspective by Margaret Sanger* (Seattle: Inkling Books, 2001), 80.

3. "Dr. Henry Goddard, Psychologist, Dies," *New York Times* (June 22, 1957), 15.

4. J. David Smith, *Minds Made Feeble: The Myth and Legacy of the Kallikaks* (Rockville, MD: Aspen Systems, 1985), 5.

5. Goddard, *The Kallikak Family*, 16.

6. Ibid., 18.

7. Ibid., 104.

8. Ibid., 7.

9. Ibid., 11–12.

10. Ibid., 29, 71.

11. Quoted in "Would Check Birth of All Defectives," *New York Times* (September 21, 1912), 7.

12. Quoted in "Hope of Better Brains for All," *New York Times* (September 27, 1912), 9.

13. Quoted in "Bishops Approve Plan to Apply Eugenics to Marriage," *New York Times* (March 31, 1912), SM-8.

14. Christine Rosen, *Preaching Eugenics: Religious Leaders and the American Eugenics Movement* (New York: Oxford University Press, 2004).

15. "Eugenics Organizations," Image Archive on the American Eugenics Movement, http://www.eugenicsarchive.org/eugenics/topics_fs.pl?theme=19&search=&matches= (accessed February 15, 2007); also see "Description of the American Breeders Association," (American Breeders Association, 1909), available as document 410 at the Image Archive on the American Eugenics Movement, http://www.eugenicsarchive.org (accessed August 15, 2003).

16. James Wilson, "Presidential Address," *American Breeders Magazine* 4, no. 1 (First Quarter, 1913), 53.

17. "Eugenics Seeks to Improve the Natural, the Physical, Mental and Temperamental Qualities of the Human Family," (Cold Spring Harbor, NY: Eugenics Record Office, 1927), 1, available as document 248 at the Image Archive on the American Eugenics Movement, http://www.eugenicsarchive.org (accessed August 15, 2003).

18. Wilson, "Presidential Address," 55.

19. "First Eugenics Congress," *New York Times* (July 25, 1912), 5.

20. "Our Work in Eugenics," *New York Times* (July 26, 1912), 4.

21. See Marvin Olasky, *The Tragedy of American Compassion.*

22. Bleecker Van Wagenen, "The Eugenic Problem," in *Proceedings of the National Conference of Charities and Correction at the Thirty-Ninth Annual Session, Held in Cleveland, Ohio, June 12–19, 1912*, ed. Alexander Johnson (Fort Wayne, IN: Fort Wayne Printing Company, 1912), 277.

23. Robert M. Yerkes, "Eugenics: Its Scientific Basis and Its Program," (Abstract) in *Proceedings of the National Conference of Charities and Correction* (1912), 279–280.

24. Edward Larson, *Sex, Race, and Science: Eugenics in the Deep South* (Baltimore: The Johns Hopkins University Press, 1995), 23.

25. C. B. Davenport, "Eugenics and Charity," *Proceedings of the National Conference of Charities and Correction* (1912), 281.

26. Miss Adams, remarks during discussion, in *Proceedings of the National Conference of Charities and Correction* (1912), 282.

27. Olasky, *Tragedy of American Compassion*, 60–98.

28. Quoted in "Nature or Nurture?" *Journal of Heredity* 6, no. 5 (May 1915): 227.
29. Washington Gladden, *Present Day Theology*, 3rd ed. (Columbus, Ohio: McClelland and Co., 1913), 36–37.
30. Edwin Black, *War Against the Weak: Eugenics and America's Campaign to Create a Master Race* (New York: Four Walls Eight Eindows, 2003), 12.
31. "Understanding Evolution," http://evolution.berkeley.edu/evosite/misconceps/IIIBmight.shtml (accessed August 26, 2006).
32. See John G. West, "Museum Exhibit Supresses Darwin's Real Views on Eugenics, Race, and Capitalism," December 1, 2005, http://www.evolutionnews.org/2005/12/suppressing_the_truth_about_da.html (accessed February 15, 2007). The exhibit later went on tour. See John G. West, "American Museum of Natural History Darwin Exhibit Goes on the Road," http://www.evolutionnews.org/2006/10/american_museum_of_natural_his.html (accessed February 15, 2007).
33. David Micklos, "None Without Hope: Buck vs. Bell at 75," http://www.dnalc.org/resources/buckvbell.html (accessed August 15, 2003). Micklos is director of the DNA Learning Center at the Cold Spring Harbor Laboratory, Long Island, New York.
34. Darwin, *The Descent of Man* (1871), 1:168–84. Also see discussion of this passage in chapter 2.
35. Frank Lowden, "Social Work in Government," *Proceedings of the National Conference of Social Work, May 16–23, 1923* (Chicago: University of Chicago Press), 150–51.
36. Edward M. East, *Heredity and Human Affairs* (New York: Charles Scribner's Sons, 1927), 311.
37. Edwin Conklin, "Value of Negative Eugenics," *Journal of Heredity* 6, no. 12 (December 1915): 539–540.
38. Paul Popenoe and Roswell Johnson, "Eugenics and Euthenics" in Horatio Hackett Newman, *Evolution, Genetics and Eugenics*, 3rd ed. (Chicago: University of Chicago Press, 1932), 517–518.
39. H. E. Jordan, "Heredity as a Factor in the Improvement of Social Conditions," *American Breeders Magazine* 2, no. 4 (Fourth Quarter, 1911): 253.
40. East, *Heredity and Human Affairs*, 237.
41. Information about this committee and its work can be found in a letter from Charles Davenport to Claxton, Bureau of Education, Dept. of the Interior (May 20, 1922) in the Charles Davenport Papers, Manuscript Collection B D27, American Philosophical Society, Philadelphia.
42. "Eugenic Legislation: A Review," *Journal of Heredity* 6, no. 3 (March 1915): 144.
43. Lena K. Sadler, "Is the Abnormal to Become Normal?" in *A Decade of Progress in Eugenics*, Scientific Papers of the Third International Congress of Eugenics (Baltimore: Williams and Wilkins Co., 1934), 198.
44. Charles Davenport, *Heredity in Relation to Eugenics* (NewYork: Henry Holt and Co., 1911), 1.
45. Jordan, "Heredity as a Factor," 249.
46. Conklin, "Value of Negative Eugenics," 538.
47. Alexander Graham Bell, "How to Improve the Race," *Journal of Heredity* 5, no. 1 (January 1914): 2.
48. Albert Edward Wiggam, "Does Heredity or Environment Make Men?" in Newman, *Evolution, Genetics and Eugenics*, 501.
49. Ibid., 502.
50. Caleb Williams Saleeby, "The Promise of Race Culture," in Newman, *Evolution, Genetics and Eugenics*, 536–37. Saleeby was British, but this excerpt from one of his books appeared in an American textbook.
51. Charles Davenport, "The Duty of Society: Its Members Must Be Eugenically Responsive or It Will Die" [letter], *New York Times* (July 8, 1913), 6.
52. "Nature or Nurture?" 228.
53. "Feeblemindedness: A Review by the Editor" *Journal of Heredity* 6, no. 1. (January 1915): 33.
54. Ibid.
55. Popenoe and Johnson, "Eugenics and Euthenics," 511.
56. C. B. Davenport, "Report of Committee on Eugenics," in *American Breeders Association Report of the Meeting held at Omaha, Neb., December 8, 9, and 10, 1909, and for the Year ending December 31, 1909* 6 (Washington D.C.: American Breeders Association, 1911): 94.
57. Quoted in Saleeby, "The Promise of Race Culture," 543.
58. Wiggam, "Does Heredity . . . Make Men?" 505.
59. Jordan, "Heredity as a Factor," 253.

60. Ibid., 254.
61. Maynard Metcalf, "Evolution and Man," *Journal of Heredity* 7, no. 8 (August 1916): 360.
62. Ibid., 364.
63. Quoted in Perry, *Pivot of Civilization*, 129.
64. Wiggam, "Does Heredity . . . Make Men?" 506.
65. Conklin, "Value of Negative Eugenics," 539.
66. Herbert Walter, "Human Conservation", in Newman, *Evolution, Genetics and Eugenics*, 531.
67. Wilson, "Presidential Address," 56.
68. Quoted in "To Improve the Race," *New York Times* (March 24, 1912), C5.
69. Larson, *Sex, Race, and Science*, 22.
70. Ibid.
71. W. C. Rucker, "More 'Eugenic Laws,' " *Journal of Heredity* (May 1915), 225.
72. Rosen, *Preaching Eugenics*, 53–83.
73. Ibid., 223–26.
74. Walter, "Human Conservation," 524.
75. "Statement of Harry H. Laughlin," *Biological Aspects of Immigration: Hearings before the Committee on Immigration and Naturalization, House of Representatives, Sixty-Sixth Congress, Second Session, April 16–17, 1920* (Washington, D.C.: Government Printing Office, 1921), available as document 1113 at the Image Archive on the American Eugenics Movement, http://www.eugenicsarchive.org (accessed August 15, 2003).
76. Kevles, *In the Name of Eugenics*, 103.
77. See Roswell Johnson, "Eugenics and Immigration" *New York Times* (February 16, 1924), 12; "1890 Census Urged as Immigrant Base," *New York Times* (January 7, 1924), 8.
78. "Like-Minded or Well-Born" *New York Times* (February 10, 1924), E6.
79. Joseph DeJarnette, "Mendel's Law Poem," (1921), available as document 1235 at the Image Archive on the American Eugenics Movement, http://www.eugenicsarchive.org (accessed August 15, 2003).
80. Unless otherwise noted, the facts about Carrie Buck's life are taken from Paul Lombardo, "Three Generations, No Imbeciles: New Light on *Buck v. Bell*," *New York University Law Review* 60, 30.
81. Quoted in Harry Laughlin, *The Legal Status of Eugenical Sterilization, Supplement to the Annual Report of the Municipal Court of Chicago for the Year 1929* (Chicago: Municipal Court of Chicago, 1930), 25.
82. Figures derived from tables in Harry Laughlin, *Eugenical Sterilization: 1926* (New Haven, CT: American Eugenics Society, 1926), 10–18.
83. Kevles, *In the Name of Eugenics*, 110.
84. Laughlin, *Eugenical Sterilization: 1926*, 14–15.
85. For the text of the model statute, see Harry Laughlin, *Eugenical Sterilization in the United States* (Chicago: Psychopathic Laboratory of the Municipal Court of Chicago, 1922), 446–51; also see Laughlin, *Eugenical Sterilization: 1926*, 64–75.
86. Lombardo, "Three Generations," 54–58.
87. Excerpt from Laughlin's deposition to the court, quoted in Laughlin, *Legal Status of Eugenical Sterilization*, 17.
88. Ibid., 19.
89. Ibid., 22.
90. Kevles, 110.
91. Excerpt from trial testimony of A. H. Estabrook in Laughlin, *Legal Status of Eugenical Sterilization*, 25.
92. Excerpt from trial testimony of A. S. Priddy in ibid.
93. Quoted in Albert W. Alschuler, *Law without Values: The Life, Work, and Legacy of Justice Holmes* (Chicago: University of Chicago Press, 2000), 27–28. For a thorough discussion and analysis of Holmes's view of the world, see pages 14–83 of Alschuler's book.
94. *Buck v. Bell*, 274 U.S. 200 (1927), 207.
95. Holmes to Harold Laski (May 12, 1927) in *Holmes-Laski Letters: The Correspondence of Mr. Justice Holmes and Harold J. Laski, 1916–1935*, ed. Mark DeWolfe Howe (Cambridge, MA: Harvard University Press, 1953), 942.
96. Larson, *Sex, Race, and Science*, 107.

97. Haller, *Eugenics*, 141; Kevles, *In the Name of Eugenics*, 116.
98. Sanger, "The Function of Sterilization," *Birth Control Review* (October 1926), 299.
99. Perry, *The Pivot of Civilization*, 215.
100. Davenport, quoted in Herbert Walter, "Human Conservation", in Newman, *Evolution, Genetics and Eugenics*, 523.
101. East, *Heredity and Human Affairs*, 307, 239.
102. Sanger in Perry, *Pivot of Civilization*, 204, 212. In an unpublished article, "We Must Breed a Race of Thoroughbreds," Sanger wrote that scientific birth control would solve the problem of "the dependent and the delinquent" who were "breeding like weeds." Perry, *Pivot of Civilization*, 212.
103. Burbank, quoted in Margaret Sanger, "Is Race Suicide Probable?" (1925) in Perry, *Pivot of Civilization*, 176.
104. Sadler, "Is the Abnormal to Become Normal?" 196.
105. Sanger in Perry, *Pivot of Civilization*, 214.
106. Sadler, "Is the Abnormal to Become Normal?" 193.
107. East, *Heredity and Human Affairs*, 29.
108. Edward M. East, *Mankind at the Crossroads* (New York: Charles Scribner's Sons, 1924), 232.
109. Sadler, "Is the Abnormal to Become Normal?" 196–197.
110. Quoted in Kevles, 116.
111. Quoted in Theodore Russell Robie, "Selective Sterilization for Race Culture," in *A Decade of Progress in Eugenics*, 203.
112. See Nancy L. Gallagher, *Breeding Better Vermonters: The Eugenics Project in the Green Mountain State* (Hanover, NH: University Press of New England, 1999).
113. Ibid., 7.
114. Valerie Parker, "Breeding Better Citizens: A Hidden Chapter of American History," March 22, 2000, *ABCNews.com*, http://abcnews.go.com/onair/2020/2020_000322_eugenics_feature.html (accessed March 25, 2000).
115. Robie, "Selective Sterilization," 208.
116. "Table: of sterilizations done in state institutions under state laws up to and including the year 1940," from the Harry H. Laughlin Archives, Truman State University, available as document 1191 at the Image Archive on the American Eugenics Movement, http://www.eugenicsarchive.org (accessed August 15, 2003).
117. Laughlin, *Legal Status of Eugenical Sterilization*, 79.
118. Kevles, 345n18.
119. Ibid.
120. Ibid., 345–46.
121. Smith, *Minds Made Feeble*, 111.
122. Ibid., 21–33.
123. Ibid., 33.
124. Ibid., 27. In her "last years she was offered the alternative of leaving the institution to live in the community from which she had been segregated for almost all of her life. She declined the opportunity . . . the institution was the only community she understood and trusted." Ibid., 33.
125. Quoted in Stephen Jay Gould, *The Mismeasure of Man* (New York: W. W. Norton, 1981), 174. J. David Smith, however, argues that Goddard's recantation "was far from total or permanent." See discussion in Smith, *Minds Made Feeble*, 130–134.
126. Lombardo, "Three Generations," 61; for an example of a letter written by Carrie Buck, see Carlos Santos, "Historic Test Case: Wrong Done to Carrie Buck Remembered," *Charlottesville Times-Dispatch* (February 17, 2002).
127. Quoted in Santos, "Historic Test Case."
128. Ibid.
129. J. David Smith, *The Eugenic Assault on America* (Fairfax, VA: George Mason University Press, 1993), 5–6.
130. Ben Franklin, "Teen-Ager's Sterilization An Issue Decades Later," *New York Times* (March 7, 1980), A16.
131. Quoted in Gould, *Mismeasure of Man*, 336.
132. See Jack Lessenberry, "Genocide, Michigan Style," (August 5, 1998), metrotimes, www.metrotimes.com/news/columns/18/jl/45.html (accessed March 5, 2005); Jack Lessenberry, "The

unkindest cut of all," (January 12, 2000), metrotimes, www.metrotimes.com/20/15/Columns/jackless.html (accessed March 5, 2005); Christa Piotrowski, "U.S. Court Battle Over Forced Sterilization: A Dark Chapter of American History," (July 20, 2000), NZZ Online, www.nzz.ch/english; Transcript, "Breeding Better Citizens," ABC News 20/20 (airdate: March 22, 2000), http://www.abcnews.go.com (accessed March 25, 2000).

133. Quoted in Piotrowski, "U.S. Court Battle Over Forced Sterilization."
134. Quoted in Lessenberry, "Genocide, Michigan Style."
135. Quoted in Lessenberry, "The Unkindest Cut of All."
136. Quoted in "Breeding Better Citizens."
137. Ibid.
138. Ibid.; Lessenberry, "Genocide, Michigan Style."
139. Ibid.
140. Lessenberry, "The Unkindest Cut of All"; Lessenberry, "Genocide, Michigan Style."
141. Ellsworth Huntington, *The Character of Races* (New York: Charles Scribner's Sobns, 1924), 59.
142. Charles Davenport, "Race Crossing in Man," typescript in Charles Davenport Papers, Collection B D27, American Philosophical Society, Philadelphia, 4.
143. Charles Davenport, "Scientific Cooperation with Nature: Eugenics," typescript in Charles Davenport Papers, Collection B D27, American Philosophical Society, Philadelphia, 2.
144. Bannister, Social Darwinism, 181. For Darwin's opposition to slavery, see Desmond and Moore, *Darwin*, 120–22, 328–29, 521, 541.
145. "It is . . . highly probable that with mankind the intellectual faculties have been gradually perfected through natural selection." Darwin, *Descent* (1871), 1:160. "The variability or diversity of the mental faculties in men of the same race, not to mention the greater differences between the men of distinct races, is so notorious that not a word need here be said." Ibid., 1:109–10. "There is . . . no doubt that the various races, when carefully compared and measured, differ much from each other, —as in . . . the form and capacity of the skull, and even in the convolutions of the brain . . . Their mental characteristics are likewise very distinct; chiefly as it would appear in their emotional, but partly in their intellectual, faculties." Ibid., 1:216. For an example of an American who cited Darwin on this topic, see Jerome Dowd, *The Negro in American Life* (New York: The Century Co., 1926), 372.
146. Darwin, *Descent* (1871), 1:201. Other disparaging comments toward blacks include: "Even the most distinct races of man, with the exception of certain negro tribes, are much more like each other in form than would at first be supposed." Emphasis added, ibid., 1:215–16; "the Negro differs more . . . from the other races of man than do the mammals of the same continents from those of the other provinces." Ibid., 1:219; "The inferior vitality of mulattoes in spoken of in a trustworthy work as a well-known phenomenon." Ibid., 1:221. Seemingly contrary to these remarks, Darwin at one point emphasized "the numerous points of mental similarity between the most distinct races of man. The American aborigines, Negroes and Europeans differ as much from each other in mind as any three races that can be named; yet I was incessantly struck, whilst living with the Fuegians on board the 'Beagle,' with the many little traits of character, shewing how similar their minds were to ours; and so it was with a full-blooded negro with whom I happened once to be intimate." Ibid., 1:232. This passage might be read as an endorsement of the essential mental equality of the races. But that is probably a misreading. In context, Darwin is trying to argue that despite major differences between the races in the area of mental abilities there are small "unimportant" similarities, and these "unimportant" similarities are proof that the races share a common ancestor, because it "is extremely improbable that they should have been independently acquired by aboriginally distinct species or races." Ibid. Darwin makes the same point at the very end of the book. Ibid., 2:388. On a related issue, although Darwin believed in racial inferiority, he did not maintain that a race need be permanently saddled with specific inferiorities. He noted that "the distinctive characters of every race of man were highly variable," and "[i]t may be doubted whether any character can be named which is distinctive of a race and is constant." Ibid., 1:225. He also noted that there are numerous differences between individuals in each race. "Savages even within the limits of the same tribe, are not nearly so uniform in character, as has often been said." Ibid. These comments would seem to hold out the hope for racial progress. However, one must counterbalance such comments with the passage cited in the next note, wherein Darwin predicts the inevitable extermination of "the savage races" by the "civilised races."

147. Darwin, *Descent* (1871), 1:201.

148. Bannister, *Social Darwinism*, 183.

149. Eric D. Anderson, "Black Responses to Darwinism, 1859–1915," in *Disseminating Darwinism: The Role of Place, Race, Religion, and Gender*, ed. Ronald L. Numbers and John Stenhouse, (New York: Cambridge University Press, 1999), 261. Despite my disagreement with Anderson's assessment of Darwin, the essay provides a fascinating look at the muted and ambivalent response of American blacks to Darwinism.

150. Bannister, *Social Darwinism*, 180–88; Anderson, "Black Responses," 261. For a good discussion of the racist views of a leading critic of Darwin, see Gould's treatment of American naturalist Louis Agassiz. Gould, *Mismeasure of Man*, 42–50.

151. Dowd, *Negro in American Life*, 357–58, 395–96, 583–84. It should be noted that Dowd's professed sympathy for blacks was rather limited. Although he urged that efforts be undertaken to stop lynchings in the South and to provide better public services to blacks, he opposed widespread negro suffrage. See 583–586. And while he criticized Nordic supremacy, his criticism was embedded within a statement acknowledging the inequality of the races: "While I do not believe in the equality of the human races, any more than in the equality of dogs or of tobacco, I would not wish to be understood as having the slightest sympathy with the idea that the Nordic race is the paramount race of the world" (396).

152. Ibid., 394.

153. Ibid., 395.

154. Charles Davenport, "How Early in Ontogeny Do Human Racial Characters Show Themselves?" (December 21, 1933) in Charles Davenport Papers, Collection B D27, American Philosophical Society, Philadelphia, 4.

155. Madison Grant, *The Passing of the Great Race*, 4th rev. ed. (New York: Charles Scribner's Sons, 1921), 18. But see the criticism of this view by eugenist Edward East in *Heredity and Human Affairs*, 183.

156. Ibid., 263.

157. Ibid., 262.

158. East, *Mankind at the Crossroads*, 137–38. In a later book, however, East acknowledged that "[i]t is of no importance whether the negro or the white man is more closely related to the apes." East, *Heredity and Human Affairs*, 189.

159. Charles Davenport, "A Biologist's View of the Negro Problem," handwritten mss. in Charles Davenport Papers, Collection B D27, American Philosophical Society, Philadelphia, 3.

160. Charles Davenport, "Protection of the National Germ Plasm" (Dec 31, 1920) in Charles Davenport Papers, Collection B D27, American Philosophical Society, Philadelphia, 1.

161. Davenport, "Scientific Cooperation with Nature," 2; East, *Heredity and Human Affairs*, 194–95; East, Mankind at the Crossroads, 134–36.

162. Davenport, "Do Races Differ in Mental Capacity?" in Charles Davenport Papers, Collection B D27, American Philosophical Society, Philadelphia, [hand pagination], 10, also 3–4.

163. "[I]t does not follow that because a particular cross is generally not favorable that all race crossing is bad." Davenport, "Race Crossing," in Charles Davenport Papers, Collection B D27, American Philosophical Society, Philadelphia, numbered as p. 9 but actually fifth sheet of typescript. "Intercrossing between two or more such stocks, if their genetic constitutions are different, increases variability. This is an advantage, provided each ethnical element entering the mixture has something valuable to contributed and later selection is not dysgenic . . . In view of these facts, it is sheer nonsense to attribute every defect of a mongrel stock to hybridisation." East, *Heredity and Human Affairs*, 182.

164. East, *Mankind at the Crossroads*, 132.

165. This point is implied in the following passage by East: "The question put to the geneticist was this: 'Given two stocks of the human race widely differentiated, but with equivalent hereditary endowments, do you advise intermarriage?' He replied: 'Do not risk it; it is dangerous ground.' If this be true when the races are relatively equal in innate qualities, how much more emphasis can be put into the reply if it can be shown that one is far superior to the other." Ibid., 128.

166. Ibid., 133,

167. Ibid., 141, emphasis in the original.

168. Davenport, "Race Crossing," 7.

169. Charles Davenport, "Is the Crossing of Races Useful?" (November 12, 1929) in Charles Davenport Papers, Collection B D27, American Philosophical Society, Philadelphia, 2.

170. "An Act to Preserve Racial Integrity," in *Virginia Health Bulletin* 16 (March 1924), Virginia Department of Health, 4.

171. For detailed accounts of Plecker's views and tactics, see Black, *War Against the Weak*, 161–82; Smith, *Eugenic Assault on America*, 59–82.

172. W. A. Plecker, "Virginia's Effort to Preserve Racial Integrity," in *A Decade of Progress in Eugenics*, 105–12. Plecker also supplied a display on anti-miscegenation laws for the companion exhibition on eugenics held at the American Museum of Natural History. See *A Decade of Progress in Eugenics*, Plate 14.

173. Plecker, "Virginia's Effort to Preserve Racial Integrity," 105.

174. For more information on Davenport's views of miscegenation restrictions, see discussion in Dowd, *Negro in American Life*, 382 [Get ERO Bulletin No. 9].

175. Davenport, "A Biologist's View of the Negro Problem."

176. Grant, *Passing of the Great Race*, 228.

177. Ibid., title page.

178. "Race Genetics Problems," *American Breeders Magazine* 2, no. 3 (Third Quarter, 1911): 232.

179. Quoted in Larson, *Sex, Race, and Science*, 109.

180. Kevles, *In the Name of Eugenics*, 118.

181. Quoted in Black, *War Against the Weak*, 276.

182. Dr. G. Gyssling (German Consul) to Dr. K. Burchardi, March 24, 1936. The author found a copy of a typescript of this letter in the E. S. Gosney Papers, California Institute of Technology Archives, Folder 8.13.

183. Paul Popenoe, "The German Sterilization Law," *Journal of Heredity* 25, no. 7 (July 1934): 260.

184. Ibid., 257.

185. Kevles, 116.

186. See "Eugenics in Virginia," http://www.healthsystem.virginia.edu/internet/library/historical/eugenics/4-influence.cfm (accessed February 15, 2007).

187. Laughlin, *Legal Status of Eugenical Sterilization*, 79, 80.

188. John H. Morehead, "Veto Message" (April 14, 1913) in Harry Laughlin, *Eugenical Sterilization*, 48.

189. "Religious leaders pursued eugenics precisely when they moved away from traditional religious tenets. The liberals and modernists in their respective faiths—those who challenged their churches to conform to modern circumstances—became the eugenics movement's most enthusiastic supporters." Rosen, *Preaching Eugenics*, 184.

190. Bryan, "Mr. Bryan's Last Speech," reprinted in William Jennings Bryan and Mary Baird Bryan, *The Memoirs of William Jennings Bryan* (reprint of 1925 edition; New York: Haskell House Publishers, 1971), 548; Larson, *Summer for the Gods*, 27–28.

191. Pope Pius XI, "Casti Connubii," (December 31, 1930), http://www.vatican.va/holy_father/pius_xi/encyclicals/documents/hf_p-xi_enc_31121930_casti-connubii_en.html (accessed March 5, 2005), § 68; more generally, see §§68–71. For a discussion of Catholic opposition to eugenics, see Kevles, 118–21. Supporters of eugenics frequently identified Catholics as the main obstacle to sterilization: "Apart from mere ignorance, almost the only opposition to the conservative use of selective sterilization comes from Roman Catholics." E. S. Gosney, "Sterilization, Catholicism, and Politics," Typescript, June 27, 1935, in the E. S. Gosney Papers, Folder 3.8; "The only explanation I can give [for the absence of sterilizations in Idaho] is this: Idaho is very largely Catholic and as you know the Catholic Church religiously opposes sterilization." Lewis Williams, Director, Department of Charitable Institutions, State of Idaho to Human Betterment Foundation, January 13, 1942, in the E. S. Gosney Papers, Folder 7.11.

192. Larson, *Sex, Race, and Science*, 107–15.

193. The facts of the controversy are taken from Joseph Andrieux, "To the citizens of Bottineau County [I]," *The Bottineau Courant and Bottineau County Herald* (July 7, 1937), 1; "To the citizens of Bottineau County [II]," *The Bottineau Courant and Bottineau County Herald* (July 28, 1937), 1; Minutes of the Special Meeting of the Bottineau County Welfare Board, July 6, 1937 (on file at Bottineau County Social Services, Bottineau, North Dakota).

194. Minutes of the Special Meeting of the Bottineau County Welfare Board.

195. Minutes of the Special Meeting of the Bottineau County Welfare Board.

196. "Resolution of the Bottineau County Welfare Board," (July 6, 1937).
197. Andrieux, "To the citizens of Bottineau County [I]."
198. Margaret Sanger to Sam Ter Haas, County Welfare Board (July 19, 1937), Sophia Smith Collection, Margaret Sanger Papers, A9492, Reel 513, Frame 0342-3.
199. Andrieux, "To the citizens of Bottineau County [II]."
200. Andrieux, "To the citizens of Bottineau County [II]"; Minutes of the Meeting of the Bottineau County Welfare Board (July 17, 1937).
201. Andrieux, "To the citizens of Bottineau County [II]." The article to which Andrieux responded was Ernest Schoonover, "To the citizens of Bottinuea County," *The Bottineau Courant and Bottineau County Herald* (July 14, 1937), 1.
202. Newsclipping from *Duluth Minnesota Herald* (January 9, 1915), American Philosophical Society, MS Coll 77, Eugenics Record Office, Series VI, Box 1, Folder: Eugenics in Law, 1913–1934.
203. For a brief biography of his achievements see "Thomas H. Morgan—Biography," http://nobelprize.org/medicine/laureates/1933/morgan-bio.html (accessed June 25, 2005).
204. Thomas Hunt Morgan, *Evolution and Genetics*, 2nd ed. (Princeton, NJ: Princeton University Press, 1925), 205.
205. Ibid., 206.
206. In 1942, E. S. Gosney, president of the Human Betterment Foundation, sent a letter to Morgan notifying him that he had been unanimously voted in as a member of the Foundation. The letter references a previous conversation with Morgan, implying that Morgan had expressed a willingness to become a member. However, there is no letter from Morgan accepting the honor. E. S. Gosney to Thomas H. Morgan, March 24, 1942, E.S. Gosney Papers, Folder 1.4.
207. Kevles, *In the Name of Eugenics*, 122–23, 142–47.
208. Edwin Grant Conklin, *Heredity and Environment in the Development of Men*, 5th ed. rev. (Princeton, NJ: Princeton University Press, 1923), 295; also, more generally, 292–315. For similar thoughts in other editions, see Conklin, *Heredity and Environment in the Development of Men* (Princeton, NJ: Princeton University Press, 1915), 404–40; Conklin, *Heredity and Environment in the Development of Men*, 2nd ed. (Princeton, NJ: Princeton University Press, 1916), 416–53; Conklin, *Heredity and Environment in the Development of Men*, 3rd ed. (Princeton, NJ: Princeton University Press, 1919), 275–97.
209. Conklin was also supportive of the idea of positive eugenics. "No eugenical reform can fail to take account of the fact that the decreasing birth rate among intelligent people is a constant menace to the race. We need not 'fewer and better children' but more children of the better sort and fewer of the worse variety." Conklin, *Heredity and Environment* (1923), 314. He went on to criticize birth control reformers for believing that "the race is to be regenerated through sterilization or birth control. But unfortunately this reform begins among those who because of good hereditary traits should not be infertile." Ibid.
210. Jennings, "Eugenics," in *Encyclopedia of Social Sciences* (New York: Macmillan, 1937), 621 [volume was originally published in 1931]. Jennings's support for negative eugenics, even as he criticized parts of the eugenics movement, was repeated again and again in different publications: "[T]here are a few social practices that recognizably and directly work toward racial degeneration; these should obviously be stopped. Such are the freedom and encouragement of reproduction among the feeble-minded, the criminal, the insane." H. S. Jennings, *Prometheus, or Biology and the Advancement of Man* (New York: E. Dutton, 1925), 78. "To stop the propagation of the feebleminded, by thoroughly effective measures, is a procedure for the welfare of future generations that should be supported by all enlightened persons. Even though it may get rid of but a small proportion of the defective genes, every case saved is a gain, is worth while in itself." H. S. Jennings, *The Biological Basis of Human Nature* (New York: W. W. Norton, 1930), 238. "[Selective elimination in fruit flies] encourages us to hope that if measures are indeed taken to prevent the multiplication of individuals with defective genes, the human race need not deteriorate." Ibid., 357. "The fuller, more adequate life—the higher life—is to be given preference over the lower and simpler. Since some representatives of life must give away to others, it is the lower that must give way to the higher." H. S. Jennings, *The Universe and Life* (New Haven, CT: Yale University Press, 1933), 82, also see 90–91. This comment appears as part of a discussion about applying the principle of selection to living things, including human beings. After granting that "the lower that must give way to the higher," Jennings cautions that "nothing in the biological principle that life is to be promoted calls for any unnecessary sacrifice of the

slightly lower to the slightly higher. There is nothing in that principle to justify the destruction of millions of men to satisfy the ambitions of a Napoleon." Ibid., 85. Of course, no other mainstream eugenist would have disagreed with this comment. They were not calling for the senseless slaughter of millions of men, but for the prevention of new "defectives" who they regarded as a threat to the "normal" population. Kevles also makes much of how Jennings acknowledged the role played by environment as well as heredity in the formation of human beings, and his disagreement with the notion that programs to improve social conditions are useless. Again, this is true, but the role Jennings assigned to environment should not be overstated. In his 1935 textbook on genetics, for example, Jennings presented research from a human twin study to demonstrate the purported role of heredity in producing criminals. The conclusion he drew from the study shows how far even he connected at least some problems to heredity: "Thus it is clear that everything that helps to determine whether a man shall be a criminal or a good citizen is deeply affected by the materials of which the individual is made, by his 'heredity.' Mentality, morality, conduct, career, fate, are affected by the hereditary materials with which the individual starts life." H. S. Jennings, *Genetics* (New York: W. W. Norton, 1935), 203; for his discussion of the twins study, see 202–4.

211. Jennings, *The Universe and Life*, 90–91.
212. Haller, 7; Smith, *Minds Made Feeble*, 169.
213. Smith, *Minds Made Feeble*, 169–90.
214. Sanger, "Sterilization: A Modern Medical Program for Human Health and Welfare," (June 5, 1951), Sophia Smith Collection, Records of Planned Parenthood Federation of America, 2.
215. Sanger, "Address," Supplement to The Malthusian (January 1951) in Sophia Smith Collection, Margaret Sanger Papers.
216. Ibid. For other articles and speeches of this period where Sanger continued to express concerns about the "feeble-minded," see Sanger, "Sterilization: A Modern Medical Program," 3–4; Sanger, "Sterilization" [Letter to the Human Betterment Federation of Des Moines] (February 1951), Sophia Smith Collection, Margaret Sanger Papers.
217. For a summary of Osborn's biography, see "Background Note," Frederick Henry Osborn Papers, American Philosophical Society, http://www.amphilsoc.org/library/mole/o/osborn. htm (accessed February 20, 2007).
218. For examples, see Kevles, *In the Name of Eugenics*, 170–75; Rosen, *Preaching Eugenics*, 166–68.
219. Frederick Osborn, "Summary of Proceedings" of the Conference on Eugenics in Relation to Nursing, held on February 24, 1937, typescript, American Eugenics Society Records, Manuscript Collection 575.06 Am3, American Philosophical Society, 2.
220. Ibid. Regarding the German sterilization program, Osborn noted that "it is generally doubted whether equal or better results might not have been obtained by a voluntary instead of a compulsory system," but he still insisted that "the German sterilization program is apparently an excellent one." Ibid., 1.
221. Frederick Osborn, "Galton and Mid-Century Eugenics," *Eugenics Review* 48, no. 1 (April 1956): 21.
222. Ibid., 17, emphasis added.
223. "Freedom of Choice for Parenthood: A Program of Positive Eugenics," *Eugenical News* 38, no. 2 (June 1953): 227. Although voluntary family planning came to the forefront of eugenics after World War II, there had been a close connection between eugenists and proponents of birth control for some time. Eugenists had supported efforts to control the births of the unfit. However, they also had worried about the use of birth control to depress the birth rates of the fit. For Osborn's earlier ambivalence toward birth control, see Frederick Osborn, "Memorandum of Remarks at Birth Control Meeting at Mrs. Roland Redmond's house, February 1937," dated March 1, 1937, American Eugenics Society Records, Manuscript Collection 575.06 Am3, American Philosophical Society.
224. Frederick Osborn, "Preparing for Parenthood," undated pamphlet in the American Eugenics Society Records, Manuscript Collection 575.06 Am3, American Philosophical Society, 11. This pamphlet was adapted from Chapter 7 of Osborn's book *Preface to Eugenics* (New York: Harper and Brothers, 1951).
225. See "About the Population Council: Highlights in Council History," http://www.popcouncil. org/about/timeline.html (accessed Feb. 21, 2007).
226. Osborn, *Preparing for Parenthood*, 1.

227. Letter from Frederick Osborn to Charles Lindbergh, December 27, 1966, in the American Eugenics Society Records, Manuscript Collection 575.06 Am3, American Philosophical Society. Also see earlier letter from Lindbergh to Osborn, March 22, 1965, in the same collection.

228. Frederick Osborn, "The Eugenic Hypothesis, Part II: Negative Eugenics," *Eugenical News* 37, no. 1 (March 1952): 8

229. Osborn, "Preparing for Parenthood," 9–10.

230. Osborn, "The Eugenic Hypothesis, Part II," 9.

231. See the discussion in the text associated with Albert Priddy's comments cited in note 82.

232. For examples, see letters from Frederick Osborn to Philip Boffey, October 26, 1964; to Chaplain Sheldon E. Hermanson, April 12, 1965; and to Shibdas Burman, Sept. 8, 1965; all in American Eugenics Society Records, Manuscript Collection 575.06 Am3, American Philosophical Society.

233. Frederick Osborn, "New Trends in Human Evolution," typescript dated Nov. 12, 1952, American Eugenics Society Records, Manuscript Collection 575.06 Am3, American Philosophical Society, 2.

234. "Eugenics Credo" typescript, undated but circa July August 1954, American Eugenics Society Records, Manuscript Collection 575.06 Am3, American Philosophical Society, 3. It is not entirely clear who initiated or wrote the Credo, but Osborn was encouraged to publish it by Wickliffe Draper of the Pioneer Fund, a major donor to the AES. In addition to the version of the Credo cited here, there is a shorter typed version dated July 27, 1954 filed in the same folder, as well as a copy of the July 27 version with extensive additions that appear to be in the handwriting of Osborn. These additions (which include the language opposing miscegenation) were incorporated into the July August typescript of the Credo. In a letter to one of his staff, Osborn indicated that "I would like to publish it, and I judge you think well of it." At the same time, he was "afraid of it because it raises the ghosts of the old racial and social class bias' for which the eugenic society was damned in the past, and also because I am not sure it is good genetics." Letter from Frederick Osborn to Mrs. Helen Hammons, Aug. 15, 1954, American Eugenics Society Records, Manuscript Collection 575.06 Am3, American Philosophical Society. Also see discussion in William H. Tucker, *The Funding of Scientific Racism: Wickliffe Draper and the Pioneer Fund* (Urbana, IL: University of Illinois Press, 2002), 57–58.

235. Tucker, *The Funding of Scientific Racism*, 58.

236. Frederick Osborn, "Eugenics: The Human Future," *Science News Letter*, July 25, 1964, 55.

237. Ibid., 62.

238. Letter from Frederick Osborn to P.S. Barrows, Aug. 25, 1965, American Eugenics Society Records, Manuscript Collection 575.06 Am3, American Philosophical Society.

239. Letter from Frederick Osborn to P.S. Barrows, April 8, 1965, American Eugenics Society Records, Manuscript Collection 575.06 Am3, American Philosophical Society.

240. George Neumayr, "The New Eugenics," *The American Specator*, July 13, 2005, http://theamericanprowler.com/dsp_article.asp?art_id=8418 (accessed Feb. 24, 2007).

241. Melinda Tankard Reist, *Defiant Birth: Women Who Resist Medical Eugenics* (North Melbourne, Australia: Spinifewx Press, 2006), 1–2.

242. John J. Donahue and Steven D. Levitt,, "The Impact of Legalized Abortion on Crime" (2000), http://ssrn.com/abstract=174508 (accessed Feb. 24, 2007), Abstract.

243. Ibid., 13–14.

244. For critiques of the Donahue and Levitt paper, see John R. Lott and John E. Whitley, "Abortion and Crime: Unwanted Children and Out-of-Wedlock Births" (April 30, 2001), Yale Law & Economics Research Paper No. 254, http://ssrn.com/abstract=270126 (accessed Feb. 24, 2007); Christopher L. Foote and Christopher F. Goetz, "Testing Economic Hypotheses with State-Level Data: A Comment on Donohue and Levitt (2001)," Federal Reserve Bank of Boston, Working Papers, No. 05-15, Nov. 22, 2005, http://www.bos.frb.org/economic/wp/wp2005/wp0515.pdf (accessed Feb. 24, 2007).

245. Alexander Sanger, *Beyond Choice: Reproductive Freedom and the 21st Century* (New York: Public Affairs, 2004), 292.

246. Ibid., 59.

247. Ibid., 302.

248. Quoted in David Rogers, "Sanger: Pro-choice stance stale," March 20, 2004, *Palm Beach Daily News*, http://www.palmbeachdailynews.com/news/newsfd/auto/feed/news/2004/03/20/10 79760835.26609.3982.6211.html (accessed March 20, 2004).

249. Davenport, "Scientific Cooperation with Nature: Eugenics," typescript, Charles Davenort Papers, Manuscript Collection B D27, American Philosophical Society.
250. East, *Heredity and Human Affairs*, 307, 306.
251. For example, consider the academic positions of the following leading American eugenists: Zoologist Edwin Conklin was a professor at Princeton. Biologist Edward East was a professor at Harvard. Paleontologist Henry Fairfield Osborn was a professor at Columbia University. Zoologist David Starr Jordan was president of Stanford University, and zoologist Vernon Kellogg was a professor there. [See "Conklin, Edwin Grant," "East, Edward Murray," "Jordan, David Starr," "Kellogg, Vernon Lyman," "Osborn, Henry Fairfield," *The New Columbia Encyclopedia*, fifth edition (New York: Columbia University Press, 1993), 628, 822, 1427–28, 1461, 2030.
252. For a general summary of the coverage of eugenics in secondary school biology texts, see Gerald Skoog, "Topic of Evolution in Secondary School Biology Textbooks: 1900–1977," *Science Education* 63, no. 5 (1979): 628. For a more detailed discussion of this issue, see the discussion in Chapter 11 and accompanying footnotes.
253. Kevles, 121.
254. Samuel Pennypecker, "Veto Message," (March 30, 1905), in Laughlin, *Eugenical Sterilization*, 36.

Chapter 8: The Science of Business

1. Katherine M. H. Blackford and Arthur Newcomb, *Analyzing Character: The New Science of Judging Men; Misfits in Business, the Home and Social Life* (New York: Henry Alden, 1917), 310.
2. Ibid.
3. Katherine M. H. Blackford and Arthur Newcomb, *The Job, the Man, the Boss* (New York: Doubleday, Page & Company, 1914), 106.
4. Ibid., 117.
5. Ibid., 113.
6. See Katherine M. H. Blackford, *Character Analysis by the Observational Method*, 5th ed., ed. Arthur Newcomb (New York: Independent Corporation, 1920).
7. Katherine M. H. Blackford and Arthur Newcomb, *The Right Job: How to Choose, Prepare for, and Succeed in It.* 2 vols. (New York: The Review of Reviews Corporation, 1924).
8. "Many Women in Campaign," *New York Times*, October 19, 1924, 28; "Women's Art Show Will Open Tonight," *New York Times*, September 21, 1925, 12.
9. See "Schwab, Charles Michael," in *Encyclopedia Americana* (1948 ed.), 24:409.
10. Blackford and Newcomb, *The Job*, 123.
11. Ibid., 124.
12. Ibid., 137–38.
13. Ibid., 138.
14. Major Woodruff, quoted in ibid., 136.
15. Ibid., 124–25.
16. Ibid., 139–40.
17. Blackford and Newcomb, *Analyzing Character*, 436.
18. Major Woodruff, quoted in Blackford and Newcomb, *The Job*, 136.
19. Ibid., 140–41.
20. Blackford and Newcomb, *Analyzing Character*, 433.
21. Ibid., 438.
22. Ibid., 353–54.
23. Cattell, "HomoScientific Americanus," 569.
24. For more information about the scientific management movement and its critics, see Frederick Winslow Taylor, *The Principles of Scientific Management* (New York: Harper and Brothers, 1919); H. S. Person, ed., *Scientific Management in American Industry* (New York: Harper and Brothers, 1929); Clarence Bertrand Thompson, ed., *Scientific Management: A Collection of the More Significant Articles Describing the Taylor System of Management* (Cambridge, MA: Harvard University Press, 1914); Horace Bookwalter Drury, *Scientific Management: A History and Criticism* (New

York: Columbia University, 1915); Milton Nadworny, *Scientific Management and the Unions, 1900–1932: A Historical Analysis* (Cambridge, MA: Harvard University Press, 1955); Judith Merkle, *Management and Ideology: The Legacy of the International Scientific Management Movement* (Berkeley, CA: University of California Press, 1980).

25. Frank Parsons, *Choosing a Vocation* (Boston: Houghton Mifflin Co., 1909), 3.
26. Ibid., 14.
27. Ibid., 20.
28. Ibid., 21.
29. Ibid., 22.
30. J. Adams Puffer, *Vocational Guidance: The Teacher as a Counselor* (Chicago: Rand McNally and Co., 1914), 57.
31. Ibid., 57–58.
32. Ibid., 58.
33. Ibid., 60.
34. Ibid., 83.
35. Ibid., 84.
36. Ibid., 59.
37. Ibid., 84.
38. Gustave Blumenthal, "Vocational Analysis," in *Some Aspects of Vocational Guidance* (New York: The Central Committee on Vocational Guidance, 1912), 14.
39. Ibid., 15.
40. Ibid., 16.
41. Ibid., 17.
42. Franklin Giddings, "The Child as a Member of Society," *Teachers College Record* 16 (1915): 443.
43. Ibid., 438.
44. Dexter Kimball, *Plant Management* (New York: Alexander Hamilton Institute, 1919), 285.
45. Ibid., 285–86.
46. John Brewer, *The Vocational-Guidance Movement: Its Problems and Possibilities* (New York: Macmillan, 1919), 148.
47. Ibid., 149.
48. Ibid., 167.
49. Quoted in H. L. Hollingworth, *Vocational Psychology and Character Analysis* (New York: D. Appleton and Co., 1931), 41.
50. Hugo Münsterberg, *Business Psychology* (Chicago: LaSalle Extension University, 1922), 255–56.
51. Blackford and Newcomb, *Analyzing Character* (1917), 346.
52. Herman Schneider, "Selecting Young Men for Particular Jobs," in *Readings in Vocational Guidance*, ed. Meyer Bloomfield (Boston: Ginn and Co., 1915), 374.
53. Ibid., 378.
54. Charles Josiah Galpin, *Rural Life* (New York: The Century Co., 1918), 32–36.
55. Ibid., 32.
56. Ibid., 33.
57. Ibid., 35.
58. Ibid., 41.
59. Ibid., 42–46.
60. Ibid., 40–41.
61. Henry C. Taylor, *An Introduction to the Study of Agricultural Economics* (New York: Macmillan, 1905), 36.
62. Carl Kelsey, quoted in Taylor, *An Introduction*, 37–38. The book cited by Taylor was Carl Kelsey, *The Negro Farmer* (Chicago: Jennings and Pye, 1903). Although Kelsey claimed to believe in the ability of blacks to improve their condition (8), he rejected the idea that their problems were due to unfair social conditions or mistreatment by whites. He further seemed to believe that evolution had made blacks inferior to whites. "The trouble is that we at the North are unable to disabuse ourselves of the idea that the Negro is a dark skinned Yankee and we think, therefore, that if all is not as it should be that something is wrong, that somebody or some social condition is holding him back. We accuse slavery, attribute it to the hostility of the Southern white. Something is holding him back, but it is his inheritance of thousands of years in Africa, not slavery nor the Southern whites. It is my observation that the white of the black belt deal

with the Negro more patiently and endure far more of shiftless methods than the average Northerner would tolerate for a day." (67, emphasis in the original) Interestingly, one reason Kelsey thought that blacks would be able to improve was because of the operation of natural selection in weeding out the most deficient blacks. "The transition from slavery to freedom set in operation the forces of natural selection, which are sure and steadily working among the people and are weeding out those who for any reason can not adapt themselves to the new environment." (66)

63. Henry C. Taylor, *Agricultural Economics* (New York: Macmillan, 1922), 131. Taylor added that those workers weeded out by evolution should be cared for humanely, but in a "way which will not encourage their reproduction." Ibid.

64. Henry C. Taylor, *Outlines of Agricultural Economics* (New York: Macmillan, 1931), 561.

65. Harry C. McDean, " 'Reform' Social Darwinists and Measuring Levels of Living on American Farms, 1920–1926," *Journal of Economic History* (March 1983), 43:79.

66. Ibid., 80.

67. Ibid., 83.

68. Harry C. McDean, "Social Scientists and Farm Poverty on the North American Plains, 1933–1940," *Great Plains Quarterly* (Winter 1983), 3:20–21.

69. Quoted in Harry C. McDean, "Professionalism, Policy, and Farm Economists in the Early Bureau of Agricultural Economics," *Agricultural History* (1983), 74.

70. Ibid., 82.

71. Taylor, *Outlines*, 579.

72. Ibid., 562.

73. Gallagher, *Breeding Better Vermonters*, 92.

74. *Rural Vermont: A Program for the Future by Two Hundred Vermonters* (Vermont Commission on Country Life, 1931). This document is available online at the University of Vermont, http://chipmunk.uvm.edu:6336/dynaweb/eugenics/ruralvt/ (accessed June 29, 2005).

75. Fred R. Yoder, *Introduction to Agricultural Economics* (New York: Thomas Y. Crowell Co., 1929), 29–30.

76. Ibid., 45.

77. Ibid., 46. Worries about the number of mental defectives in farming communities was a significant concern among rural sociologists of the period. See the articles on this topic in John Phelan, *Readings in Rural Sociology* (New York: Macmillan, 1920), 203–25.

78. McDean, " 'Reform' Social Darwinists," 84–85.

79. McDean, "Social Scientists," 20–24. Canadian farm policy during this same period represents a sharp contrast. "While American social scientists worked during the Great Depression to transform chronically poor plains farmers into factory workers, their Canadian counterparts planned to keep all of their poor plains farmers in the agricultural economy. Their research in plains farm poverty during the Great Depression led them to view all poor plains farmers as a group that was impoverished by unusual conditions of depression and of drought. They never sought to identify in this group a class of unprogressive or chronically poor farmers." Ibid., 24.

80. Paterson and Ludgate, "Blonde and Brunette Traits," *Journal of Personnel Research* (1922), 1:122–27. Also see Cleeton and Knight, "Validity of Character Judgments Based on External Criteria," *Journal of Applied Psychology* 8 (1924): 215–31.

81. Münsterberg, *Business Psychology*, 262–89.

82. Ibid., 287, 288.

83. The biographical and business information about Rapaille cited here was taken from "Discovering What Makes Brands and Products Sell," (Boca Raton, Florida: Archetype Discoveries Worldwide), 11, 13–15, available on Rapaille's website, http://www.archetypediscoveriesworldwide.com (accessed June 29, 2005).

84. Liz Doup, "Primal urges define the wheel you: 'Reptilian brain' in driver's seat?" *Seattle Times*, April 30, 2004, F-1.

85. Herbert N. Casson, *Ads and Sales* (Chicago: A. C. McLurg & Co., 1911), 5, 6–7.

86. Walter Dill Scott, *The Theory of Advertising* (Boston: Small, Maybard & Co., 1907), 2. This book was originally published in 1903.

87. Quoted in Scott, *Theory*, 4–5.

88. Melvin S. Hattwick, *How to Use Psychology for Better Advertising* (New York: Prentice-Hall, 1950), 13.

89. Münsterberg, *Business Psychology*, 10–11.
90. Ibid., 35.
91. Albert T. Poffenberger, *Psychology in Advertising* (Chicago: A. W. Shaw Company, 1925), 6.
92. Scott, *Theory*, 53–54.
93. Ibid., 55.
94. Ibid., 57.
95. Münsterberg, *Business Psychology*, 154–55.
96. Ibid., 156.
97. Ibid., 160.
98. Ibid., 161.
99. Scott, *Theory*, 59.
100. Walter Dill Scott, *The Psychology of Advertising*, 5th ed. (Boston: Small, Maynard, & Co., 1913), 79.
101. Scott, *Psychology*, 52.
102. S. Roland Hill, *The Advertising Handbook* (New York: McGraw-Hill Book Co., 1921), 100.
103. Scott, *Psychology*, 56, 57, 58, 60, 64.
104. Ibid., 60.
105. Ibid., 64–65.
106. Ibid., 65–66.
107. Edward K. Strong, *The Psychology of Selling and Advertising* (New York: McGraw-Hill Book Co., 1925), 144.
108. Poffenberger, *Psychology*, 7.
109. Ibid., 26.
110. Hattwick, *How to Use Psychology*, 23.
111. Ibid., 6.
112. Henry C. Link, *The New Psychology of Selling and Advertising* (New York: The Macmillan Co., 1932), 83–84.
113. John B. Watson, *Behaviorism*, rev. ed. (New York: W. W. Norton, 1930), 2–3. Watson at certain points seemed to assert that thinking is little more than the vibration of the larynx, although he later backpedaled on some of those statements, claiming that "I have never believed that the laryngeal movement . . . as such played the predominating role in thought," although he acknowledged that he had previously "expressed [himself] . . . in ways which can be so interpreted." (Ibid., 238) Regardless of the role of larynx, Watson continued to assert that his "theory does hold that the muscular habits learned in overt speech are responsible for implicit or internal speech (thought)." (Ibid., 239)
114. Ibid., 3.
115. Ibid., p. ix.
116. Ibid.
117. Ibid., 149–50.
118. Ibid., 152–53.
119. Ibid., 152.
120. Ibid., 154.
121. Ibid., 155.
122. For information about Watson's extramarital affair with graduate student Rosalie Rayner and his subsequent divorce and academic dismissal, see Kerry W. Buckley, "Misbehaviorism: The Case of John B. Watson's Dismissal from Johns Hopkins University," in James T. Todd and Edward K. Morris, eds., *Modern Perspectives on John B. Watson and Classical Behaviorism* (Westport, CT: Greenwood Press, 1994), 19–36.
123. See Peggy Kreshel, "John B. Watson at J. Walter Thompson: The Legitimation of 'Science' in Advertising," *Journal of Advertising* 19, no. 2 (1990): 49–59; Deborah Coon, " 'Not a Creature of Reason': The Alleged Impact of Watsonian Behaviorism on Advertising in the 1920s," in Todd and Morris, *Modern Perspectives*, 37–63. Both authors are somewhat critical of what they see as exaggerated claims of Watson's impact on the advertising industry.
124. Watson in Link, *New Psychology*, viii
125. Hattwick, *How to Use Psychology*, 33–34.
126. Hattwick, *How to Use Psychology*, 35.
127. Watson, *Behaviorism*, 104.

128. Kreshel, "John B. Watson," 53.
129. Ibid.
130. Hattwick, *How to Use Psychology*, 308.
131. Quoted in Vance Packard, *The Hidden Persuaders* (New York: Pocket Books, 1957), 20–21.
132. For information about Dichter and his significance, see Barbara B. Stern, "The Importance of Being Ernest: A Tribute to Dichter," *Journal of Advertising Research*, July August 2002, 19–21; "Ernest Dichter: Motive Interpreter" [an interview with Ernest Dichter], *Journal of Advertising Research*, February–March 1986, 15–20; Eric Clark, *The Want Makers* (New York: Viking, 1989), 70–75; Packard, *Hidden Persuaders*, 20–25.
133. "Ernest Dichter: Motive Interpreter," 15.
134. Ibid., 16.
135. Louis Cheskin, "What Data Does Unconscious-Level testing Reveal?" *Advertising Agency Magazine*, July 19, 1957, 31.
136. Packard, *Hidden Persuaders*, 29.
137. Quoted in ibid.
138. "Importance of Being Ernest," 19.
139. Clark, *The Want Makers*, 94–95.
140. See, for example, Sidney Weinstein, Ronald Drozdenko, Curt Weinstein, "Advertising Evaluation Using Brain-Wave Measures: A Response to the Question of Validity," *Journal of Advertising Research* 24, no. 2 (April–May 1984): 67–71; Michael L. Rothschild, Yong J. Hyun, "Predicating Memory for Components of TV Commercials from EEG," *Journal of Consumer Research* 16, no. 4 (March 1990): 472–78; John R. Rossiter, Richard B. Silberstein, "Brain-Imaging Detection of Visual Scene Encoding in Long-term Memory for TV Commercials," *Journal of Advertising Research*, March–April 2001, 13–21. But note the caveats in Stephen L. Crites, Jr., Shelley N. Aikman-Eckenrode, "Making Inferences Concerning Physiological Responses: A Reply to Rossiter, Silberstein, Harris, and Nield," *Journal of Advertising Research*, March–April 2001, 23–25.
141. See Stuart Rogers, "How a publicity blitz created the myth of subliminal advertising," *Public Relations Quarterly* 37 (Winter 1992–1993): 12–17. A good online summary of the hoax can be found at http://www.snopes.com/business/hidden/popcorn.asp (accessed June 29, 2005).
142. See, for example, Joel Saegert, "Why Marketing Should Quit Giving Subliminal Advertising the Benefit of the Doubt," *Psychology & Marketing* 4, no. 2 (Summer 1987): 107–20; Philip M. Merikle, "Subliminal Auditory Messages: An Evaluation," *Psychology & Marketing* 5, no. 4 (Winter 1988): 355–72; Anthony R. Pratkanis, Anthony G. Greenwald, "Recent Perspectives on Unconscious Processing: Still No Marketing Applications," *Psychology & Marketing* 5, no. 4 (Winter 1988): 337–53; Timothy E. Moore, "The Case Against Subliminal Manipulation," *Psychology & Marketing* 5, no. 4 (Winter 1988): 297–316; Kirk H. Smith, Martha Rogers, "Effectiveness of Subliminal Messages in Television Commercials: Two Experiments," *Journal of Applied Psychology* 79, no. 6 (December 1994): 866–74.
143. Martin Evans, Agnes Nairn, Alice Maltby, "The Hidden Sex Life of the Male (and Female) Shot," *International Journal of Advertising* 19, no. 1 (2000): 43.
144. Tom Reichert, Susan E. Heckler, Sally Jackson, "The Effects of Sexual Social Marketing Appeals on Cognitive Processing and Persuasion," *Journal of Advertising* 30, no. 1 (Spring 2001): 15.
145. Ibid., 23.
146. See Richard Morin, "Citrus Smell Makes Cents, Study Finds," *Seattle Times*, March 1, 2005, http://archives.seattletimes.nwsource.com/cgi-bin/texis.cgi/web/vortex/display?slug=smell 01&date=20050301 (accessed July 9, 2005); Richard Michon, Jean-Charles Chebat, L.W. Turley, "Mall Atmospherics: The Interaction Effects of the Mall Environment on Shopping Behavior," *Journal of Business Research* 58 (2005): 576–83; Ronald E. Milliman, "Using Background Music to Affect the Behavior of Supermarket Shoppers," *Journal of Marketing* 46 (Summer 1982): 86–91; L. W. Turley and Ronald E. Milliman, "Atmospheric Effects on Shopping Behavior: A Review of the Experimental Evidence," *Journal of Business Research* 49 (2000): 193–211; Barry J. Babin, David M. Hardesty, Tracy A. Suter, "Color and Shopping Intentions: The Intervening Effect of Price Fairness and Perceived Affect," *Journal of Business Research* 56 (2003): 541–51; Joseph A. Bellizzi and Robert E. Hite, "Environmental Color, Consumer Feelings, and Purchase Likelihood," *Psychology and Marketing* 9 (September/October 1992): 347–63; Ronald E. Milliman, "The Influence of Background Music on the Behavior of Restaurant Patrons," *Journal of Consumer Research* 13 (September 1986): 286–89.

147. Mark S. Cary, "Ad Strategy and the Stone Age Brain," *Journal of Advertising Research*, January–April 2000, 103.
148. Ibid., 105.
149. Ibid., 104.
150. Alan Mitchell, "Getting Under the Skin with Effective Advertising," *Marketing Week*, November 8, 2001, 34.
151. Quoted in *The Want Makers*, 67.
152. Aristotle, *Rhetoric*, trans. W. Rhys Roberts (New York: The Modern Library, 1954), bk. 2, chap. 1–12, 90–123.
153. For a good discussion of the rational basis of rhetoric, see Aristotle, *Rhetoric*, especially bk. 1, chap. 1–3, 19–34. Regarding the improper use of emotion, Aristotle notes that "it is not right to pervert the judge by moving him to anger or envy or pity—one might as well warp a carpenter's rule before using it." Ibid., bk. 1, chap. 1, line 1354a.24, 20.

Chapter 9: Building the World of Tomorrow

1. F. C. Yohn, "Plinth to Pinnacle: A Pictorial Span of a Century and a Half," *Chicago Daily Tribune*, February 22, 1931, C1.
2. Eleanor Jewett, "Building for the World's Fair," *Chicago Daily Tribune*, March 22, 1931, H4.
3. Loren H. B. Knox, "Voice of the People: World's Fair Architecture," *Chicago Daily Tribune*, February 17, 1931, 14.
4. F. W. Fitzpatrick, "World's Fair Architecture," *Chicago Daily Tribune*, March 15, 1931, 14.
5. George W. Gray, "Bizarre Patterns for Chicago's Fair," *New York Times Magazine*, March 6, 1932, 10.
6. President Roosevelt's comment is cited in S. J. Duncan-Clark, "Preview of Fair Thrills Chicago," *New York Times*, May 21, 1933, E6.
7. Elizabeth G. Trow, "Architecture," *New York Times*, July 2, 1933, E5.
8. J. Wendell Clark, "The Architects of the Fair," *Chicago Daily Tribune*, December 4, 1933, 16.
9. Quoted in Gray, "Bizarre Patterns," 10.
10. For detailed information and archival materials about the first Chicago world's fair, consult "The World's Columbian Exposition of 1893," Paul V. Galvin Library Digital History Collection, Illinois Institute of Technology, available online at http://columbus.gl.iit.edu/ (accessed July 15, 2005).
11. Quoted in Mary P. Abbott, "Their Text the Fair," *Chicago Daily Tribune*, May 7, 1893, 38.
12. Thomas Tallmadge, *The Story of Architecture in America*, rev. ed. (New York: W. W. Norton, 1936), 207.
13. Knox, "Voice of the People: World's Fair Architecture," 14.
14. Louis Sullivan, *The Autobiography of an Idea* (New York: Press of the American Institute of Architects, Inc., 1924), 325.
15. Louis Sullivan, *Kindergarten Chats (revised 1918) and other writings* (New York: Wittenborn Art Books, 1947), 208.
16. Tallmadge, *Story of Architecture*, 217.
17. Sullivan, *Kindergarten Chats*, 219–20.
18. Ibid., 48.
19. Sullivan, *Autobiography of an Idea*, 254.
20. Maurice English, ed., *The Testament of Stone: Themes of Idealism and Indignation from the Writings of Louis Sullivan* (Evanston, IL: Northwestern University Press, 1963), 113.
21. Sullivan, *Kindergarten Chats*, 234.
22. Sullivan, *Autobiography of an Idea*, 254–55.
23. Sullivan, *Kindergarten Chats*, 45.
24. "All is growth, all is decadence. Functions are born of functions, and in turn, give birth or death to others. Forms emerge from forms, and other arise or descend from these. All are related, interwoven, intermeshed, interconnected, interblended. They exomose and endomose. They sway and swirl and mix and drift interminably. They shape, they reform, they dissipate. They

respond, correspond, attract, repel, coalesce, disappear, reappear, merge and emerge: slowly or swiftly, gently or with cataclysmic force—from chaos into chaos, from death into life, from life into death, from rest into motion, from motion into rest, from darkness into light, from light into darkness." Ibid.

25. Robert Twombly, *Louis Sullivan: The Public Papers* (Chicago: University of Chicago Press, 1988), 202.
26. Sullivan, *Kindergarten Chats*, 47.
27. Twombly, *Public Papers*, 202.
28. Sullivan, *Kindergarten Chats*, 45.
29. English, *Testament of Stone*, 121.
30. Twombly, *Public Papers*, 34.
31. Ibid., 204.
32. Sullivan, *Kindergarten Chats*, 97, 99, 139–40; English, *Testament of Stone*, 189–92.
33. English, *Testament of Stone*, 189.
34. Ibid., 107.
35. Ibid., 137.
36. Sullivan, *Kindergarten Chats*, 45.
37. English, *Testament of Stone*, 189.
38. Twombly, *Public Papers*, 96.
39. English, *Testament of Stone*, 190.
40. For the influence of Whitman on Sullivan, see Donald Drew Egbert, "The Idea of Organic Expression and American Architecture," in *Evolutionary Thought in America*, ed. Stow Persons (New Haven, CT: Yale University Press, 1950), 364–65. Sullivan also read Nietzsche. See Twombly, *Public Papers*, xviii.
41. Frank Lloyd Wright, *The Future of Architecture* (New York: Horizon Press, 1953), 64.
42. Ibid., 145. Paragraph breaks have been removed from the original.
43. Ibid., 145.
44. Bruno Taut, "Down with seriousism!" (1920) in Ulrich Conrads, ed., *Programs and Manifestoes on 20th-Century Architecture*, trans. Michael Bullock (Cambridge, MA: MIT Press, 1970), 57.
45. Ibid.
46. Adolf Loos, "Ornament and Crime," (1908) in Conrads, *Programs*, 19.
47. Ibid.
48. Ibid., 20, emphasis in original.
49. Antonio Sant'Elia and Filippo Tommaso Marinetti, "Futurist Architecture" (1914) in Conrads, *Programs*, 34.
50. Ibid., 35.
51. Ibid., 38.
52. Ibid., 36.
53. Ibid., 35.
54. Naum Gao and Antoine Pevsner, "Basic Principles of Constructivism" (1920) in Conrads, *Programs*, 56.
55. Le Corbusier, "Towards a New Architecture: Guiding Principles" (1920) in Conrads, *Programs*, 62.
56. Le Corbusier, quoted in Sheldon Cheney, *The New World Architecture* (New York: Tudor Publishing, 1935), 90.
57. Barr, quoted in Hans M. Wingler, *The Bauhaus* (Cambridge, MA: MIT Press, 1969), 569.
58. Walter Gropius, "Recommendations for the Founding of an Educational Institute as an Artistic Counseling Service for Industry, the Trades, and the Crafts" (1916) in Wingler, *The Bauhaus*, 24.
59. Walter Gropius, "The Viability of the Bauhaus Idea" (1922) in Wingler, *The Bauhaus*, 52.
60. Walter Gropius, "Principles of Bauhaus production" (1926) in Conrads, *Programs*, 95.
61. Walter Gropius, "Systematic Preparation for Rationalized Housing Construction" (1927) in Wingler, *The Bauhaus*, 126.
62. Walter Gropius, *Scope of Total Architecture* (New York: Harper and Brothers, 1955), 95.
63. Hannes Meyer, "My Ejection from the Bauhaus" (1930) in *The Bauhaus: Masters and Students by Themselves*, ed. Frank Whitford , (Woodstock, NY: Overlook Press, 1992), 283.
64. Hannes Meyer, "Building" (1928) in Conrads, *Programs*, 117.

65. Ibid., 119.
66. Geoffrey Scott, *The Architecture of Humanism: A Study in the History of Taste*, 2nd ed. (London: Constable and Company, 1924), 100.
67. Ibid., 93.
68. Ibid., 169.
69. Ibid., 170.
70. Tallmadge, *Story of Architecture*, 307.
71. Henry-Russell Hitchcock and Philip Johnson, *The International Style* (New York: W. W. Norton, 1966), 44.
72. Ibid., 43.
73. Ibid., 43.
74. Wingler, *The Bauhaus*, 569.
75. Sheldon Cheney, *The New World Architecture* (New York: Tudor Publishing, 1935), 8.
76. Ibid., 62–63.
77. Ibid., 9–10.
78. Ibid., 10.
79. See Meyer, "My Ejection from the Bauhaus" (1930) in Whitford, *The Bauhaus*, 283.
80. For more information about Mies van der Rohe, see Philip Johnson, *Mies van der Rohe* (New York: The Museum of Modern Art, 1947), and Franz Schulze, *Mies van der Rohe: A Critical Biography* (Chicago: University of Chicago Press, 1985). Regarding Mies van der Rohe's relationship to politics, including his efforts to gain Nazi patronage, Schulze writes: "There is nothing to gain by picturing Mies . . . as a staunch opponent of Nazism. This was, after all, a man who within eight years' time had designed a monument to a pair of Communist martyrs, a throne for a Spanish king, a pavilion for a moderate socialist government, and another for a militantly right-wing totalitarian state! Politically, Mies was a passive soul; his active moral energies were turned toward his art and away from practically all else." Schulze, *Mies van der Rohe*, 201; more generally, see 188–202.
81. Alex Scobie, *Hitler's State Architecture: The Impact of Classical Antiquity* (London: College Art Association, 1990), especially 2–3, 13–68, 133–37; Robert Taylor, *The Word in Stone: The Role of Architecture in the National Socialist Ideology* (Berkeley, CA: University of California Press, 1974), 38–40, 71–75, 150.
82. On the Nazis' divided views toward modern architecture, see Schulze, *Mies van der Rohe*, 187, 198–199. Regarding the more favorable reception of modern architecture by Italy's Fascists, it is interesting to note that this fact was earnestly recognized by those in the Bauhaus. In 1933, one Bauhaus student wrote in a letter to a friend that "[n]aturally it hurts us 'modern artists' a lot that the new government misunderstands the new art. It's completely different in Italy! The new architecture and the new art there (Italian Futurism) are recognized as Fascist art." Wassily Kandinsky to Werner Drewes, April 10, 1933, in Whitford, *The Bauhaus*, 301.
83. See "Hannes Meyer," Whitford, *The Bauhaus*, 313.
84. See "Walter Gropius," "Ludwig Mies van der Rohe," and "László Moholy Nagy," in Whitford, *The Bauhaus*, 312–13.
85. László Moholy Nagy to Lothar Schreyer, quoted in Whitford, *The Bauhaus*, 166.
86. Gropius, *Scope*, 75.
87. Ibid., 33.
88. Ibid., 102–3. Mies van der Rohe began to move away from some of the most extreme justifications for modern architecture by the late 1920s, taking issue, for example, with Hannes Meyer's thoroughgoing materialism and extreme functionalism. According to Walter Gropius, Mies van der Rohe responded to Meyer's "assertion [that] 'life is oxygen plus sugar plus starch plus protein'" with the retort: "Try stirring all that together—it stinks." After taking over the directorship of the Bauhaus, Mies van der Rohe also assured a student wearied by Meyer's extreme functionalism that "one might still strive for beauty in architecture." Whitford, *The Bauhaus*, 286–87. Despite his changing understanding of the nature of architecture, Mies van der Rohe's architectural creations remained exemplars of modernism.
89. *Official Guide Book of the New York World's Fair, 1939* (New York: Exposition Publications, 1939), 31.
90. Ibid., 32.
91. Ibid., 42–45.

92. Wright, *Future of Architecture*, 144.
93. Tom Wolfe, *From Bauhaus to Our House* (New York: Farrar, Straus, Giroux, 1981); James Howard Kunstler, *The Geography of Nowhere* (New York: Simon and Schuster, 1993), especially 59–84.
94. Catesby Leigh, "The Last Modernist: Robert Venturi's architectural ambivalence," *The Weekly Standard* (August 13, 2001), 37.
95. Egbert, "Idea of Organic Expression," 386.
96. Hundertwasser, "Mould Manifesto Against Rationalism in Architecture," (1958) in Conrads, *Programs*, 157.
97. Reinhard Gieselmann and Oswald Mathias Ungers, "Towards a New Architecture," (1960) in Conrads, *Programs*, 165.
98. For a brief description of postmodern architecture, see Kunstler, *Geography*, 81–84; for descriptions of the new urbanism and traditional neighborhood developments, see ibid., 253–260, and Peter Katz, *The New Urbanism: Toward an Architecture of Community* (New York: McGraw-Hill, 1994).
99. Examples include Seaside, Florida, http://www.seasidefl.com/ (accessed July 15, 2005);Vermillion, North Carolina, http://www.newvermillion.com/home.htm (accessed July 15, 2005); and Northwest Landing, Washington, http://www.nwlanding.com/ (accessed July 15, 2005).
100. See, for example, Donovan Rypkema, "Historic Preservation is Smart Growth," text of a speech delivered on March 3, 1999, http://hmturnerfoundation.org/html/artsmartgrow.html (accessed Feb. 22, 2007); Donovan Rypkema, "Culture, Historic Preservation and Economic Development in the 21st Century," September 1999, http://www.columbia.edu/cu/china/DRPAP.html (accessed Feb. 22, 2007); *Urban Infill Housing: Myth and Fact* (Washington, D.C.: Urban Land Institute, 2001).
101. Quotes and information about Johnson's new building design come from David M. Levitt, "The Multiplier Effect: Architect, 95, Plans 6 Copies of Next Building," Bloomberg News story reprinted in *Seattle Times* (July 13, 2001), C1 and C6. For an obituary of Johnson, see Richard Nilsen, "Philip Johnson's Legacy: Architect Was a Champion of Postmodernist Style," *Arizona Republic*, January 30, 2005, http://www.azcentral.com/ent/arts/articles/0130johnson30.html (accessed January 30, 2005).
102. See discussion in Ron Chernow, *Titan: The Life of John D. Rockefeller, Sr.* (New York: Random House, 1998), 645–46, 669–70.

Chapter 10: Darwin Day in America

1. Information in this paragraph was obtained from "Darwin Day 2002 Events Calendar," http://www.darwinday.org/events/calendar.html (accessed June 6, 2002); and "Events for 2007," http://www.darwinday.org/englishL/home/2007.php (accessed Feb. 22, 2007). Note: The listing of events from 2002 posted on the website as of 2007 no longer contains much of the information cited from the calendar posted in 2002. It is unclear why this information was removed.
2. "Charles Darwin as a Symbol for the Celebration of Science and Humanity," http://www.darwinday.org/ (accessed Feb. 22, 2007).
3. "Does Darwin Day worship Charles Darwin?," http://www.darwinday.org/about/faq.html (accessed June 13, 2002).
4. Amanda Chesworth, "Charles Robert Darwin," [essay explaining the rationale behind Darwin Day], http://www.darwinday.org/about/intro.html (accessed June 13, 2002).
5. See "Is Darwin Day a Non-Religious or Anti-Religious Celebration," http://www.darwinday.org/about/faq.html (accessed June 13, 2002); "The Origins and History of DDC," http://www.darwinday.org/englishL/aboutddc/history.html (accessed Feb. 22, 2007).
6. Chesworth, "Charles Robert Darwin."
7. Richard Dawkins, *The God Delusion* (Boston: Houghton Mifflin, 2006). Richard Dawkins's picture along with the title "Honorary President" used to be featured prominently on the darwinday.org home page, http://www.darwinday.org (accessed June 12, 2002).
8. The first quote comes from Richard Dawkins, "Is Science a Religion?" http://www.thehumanist.org/humanist/articles/dawkins.html (accessed July 16, 2005); the second quote is from

Richard Dawkins, *The Blind Watchmaker: Why the Evidence of Evolution Reveals a Universe Without Design* (New York: W. W. Norton and Co., 1996), 6.

9. "Advisory Board," http://www.darwinday.org/englishL/aboutddc/advisory.html (accessed July 16, 2005); "Humanism and Its Aspirations: Humanist Manifesto III," (Washington, D.C.: American Humanist Association), http://www.americanhumanist.org/3/HumandItsAspirations.htm (accessed July 16, 2005); Daniel C. Dennett, *Darwin's Dangerous Idea: Evolution and the Meaning of Life* (New York: Simon and Schuster, 1995), 63; Michael Shermer, "Science Is My Savior," http://www.science-spirit.org/article_detail.php?article_id=520 (accessed Feb. 23, 2007).

10. "Darwin Day 2002 Events Calendar."

11. "Darwin Day 2002 Events Calendar"; "Events for 2007"; and "Events: Full Listing for 2004," http://www.darwinday.org/englishL/home/2004.html (accessed Feb. 23, 2007).

12. Information about this group can be found at http://www.scientificgospel.com/ (accessed Feb. 22, 2007); lyrics of the songs cited can be downloaded at http://www.scientificgospel.com/drbaird/JudgmentDayLyrics.pdf (accessed Feb. 22, 2007).

13. See Anika Smith, "Beware of Darwin Day," Feb. 12, 2007, http://www.discovery.org/scripts/viewDB/index.php?command=view&id=3906 (accessed Feb. 22, 2007).

14. See "The Clergy Letter Project Presents Evolution Sunday 2006," http://www.butler.edu/clergyproject/rel_evol_sun_orig.htm (accessed Feb. 22, 2007).

15. For a superb account of the mythology promoted by many popular accounts of the Scopes trial, see Larson, *Summer for the Gods*.

16. George William Hunter, *A Civic Biology* (New York: American Book Company, 1914), 196.

17. Ibid., 261.

18. Ibid., 261–65.

19. Ibid., 263.

20. Ibid.

21. Ibid., 261.

22. George William Hunter, *Laboratory Problems in Civic Biology* (New York: American Book Company, 1916), 182–83.

23. George William Hunter, *New Civic Biology* (New York: American Book Company, 1926), 251, 400.

24. It should be pointed out that the changes found in *New Civic Biology* paralleled the content of Hunter's *New Essentials of Biology* (New York: American Book Company, 1923), a streamlined text that also failed to mention evolution and refrained from referring to Caucasians as the "highest" race. Interestingly, *New Essentials of Biology* also contained nothing about eugenics.

25. Hunter, *New Civic Biology*, 398–400.

26. Ibid., 403.

27. Ibid., 401.

28. Ibid., 401.

29. Ibid., 403.

30. George William Hunter and Leopoldo Uichanco, *Laboratory Manual for New Civic Biology* (New York: American Book Company, 1932), 270.

31. Ibid., 271.

32. George William Hunter, *Teachers' Manual To Accompany New Civic Biology* (New York: American Book Company, 1927), 88–89.

33. For similar treatments of eugenics in secondary school biology textbooks of the period, see Clifton Hodge and Jean Dawson, *Civic Biology* (Boston: Ginn and Company, 1918), 344–45; William Smallwood, Ida Reveley, and Guy Bailey, *New Biology* (Boston: Allyn and Bacon, 1924), 660–662. Some textbooks during this period did not promote eugenics, or only did so indirectly. Alfred Kinsey's *An Introduction to Biology* (Philadelphia: J. B. Lippincott, 1926) had no discussion of eugenics per se, but it did recommend books advocating eugenics in a list of reference books supplied on 539–40. By the 1940s, discussions of eugenics were significantly diluted and focused more on environment than heredity. For an example of a later treatment of the topic, see Edwin Sanders, *Practical Biology* (New York: D. Van Nostrand. Co, 1947), 431–37.

34. Gerald Skoog, "Topic of Evolution in Secondary School Biology Textbooks: 1900–1977," *Science Education* 63, no. 5 (1979): 628.

35. Ibid., 629.

36. For example, see George William Hunter, Herbert Eugene Walter, and George William Hunter, III, *Biology: The Study of Living Things* (New York: American Book Company, 1937), 638–42;

Arthur Haupt, *Fundamentals of Biology* (New York: McGraw-Hill Book Company, 1928), 216–21; Leslie Kenoyer and Henry Goddard, *General Biology* (New York, Harper and Brothers, 1937), 533–539; William Martin Smallwood, *A Text-Book of Biology for Students in Medical, Technical and General Courses* (Philadelphia: Lea and Febiger, 1913), 257–59. Not every college textbook during this period jumped on the eugenics bandwagon. Waldo Shumway's *Textbook of General Biology* (New York: John Wiley and Sons, 1931) told students that "it would appear judicious to walk slowly in enacting [eugenics] legislation until our knowledge of biological laws is more precise," 317.

37. Haupt, *Fundamentals of Biology*, 216.

38. Ibid., 219.

39. Ibid., 220.

40. Ibid., 343. Haupt is citing Richard Swann Lull, author of *Organic Evolution* (New York: The Macmillan company, 1925).

41. Ibid., 221.

42. These schools, and many others, indicated that they were teaching courses in eugenics in response to a 1920 survey by eugenists. The records of the survey are preserved in Eugenics Record Office MS Coll No 77, Series X, Eugenics Instruction Questionnaire. (Philadelphia: American Philosophical Society).

43. Each year into the 1940s, the Human Betterment Foundation of California sent out free propaganda materials promoting California's eugenic sterilization program to college professors around the country "for classroom use and distribution to students." See letter from Frank C. Reid to Mr. D. P. Monteith, Feb. 10, 1941, in the E. R. Gosney Papers and Records of the Human Betterment Foundation, California Institute of Technology Archives, Folder 11.2.

44. Ron Good, Mark Hafner, Patsye Peebles, "Scientific Understanding of Sexual Orientation: Implications for Science Education," *The American Biology Teacher* 62, no. 5 (May 2000): 329.

45. Ibid., 328

46. Ibid.

47. Ron Good, "Human Behavioral Genetics/Sexual Orientation," *The American Biology Teacher* 62, no. 5 (May 2000): 322–25.

48. For discussions of the partial and contradictory nature of some of the scientific evidence on this question, see Jeffrey Satinover, "The Biology of Homosexuality: Science or Politics?"; George Rekers, "The Development of a Homosexual Orientation"; and Richard Fitzgibbons, "The Origins and Therapy of Same-Sex Attraction Disorder," in *Homosexuality and American Public Life*, ed. Christopher Wolfe (Dallas: Spence Publishing Company, 1999), 3–97.

49. Haupt, *Fundamentals of Biology*, 268.

50. Stephen Jay Gould, *Rock of Ages: Science and Religion in the Fullness of Life* (New York: Library of Contemporary Thought, 1999). Gould's claims in this book that faith and science represent "Non-Overlapping Magisteria," (5–6) and that Darwinism doesn't have to be incompatible with religion are rather surprising coming from the same person who in the 1970s wrote that "Darwin applied a consistent philosophy of materialism to his interpretation of nature. Matter is the ground of all existence; mind, spirit, and God as well, are just words that express the wondrous results of neuronal complexity." Gould, *Ever Since Darwin*, 13.

51. For an example of this view, see William A. Dembski and Stephen C. Meyer, "Fruitful Interchange or Polite Chitchat? The Dialogue between Science and Theology," in Michael J.Behe, William A. Dembski, and Stephen C. Meyer, *Science and Evidence for Design in the Universe* (San Francisco: Ignatius Press, 2000), 213–34.

52. Haupt, *Fundamentals of Biology*, 269.

53. Hunter, *Biology: The Story of Living Things*, 495.

54. Ibid., 494.

55. Ibid., 407.

56. " 'Mother Nature' is not a directive personality substituted by 'ungodly scientists' for the supernatural Creator of all things. There is no more personality in natural selection than there is in the wind, which 'selects' the grain from the chaff. Nor is there necessarily any more design, any more purpose or moral being to natural selection than there is in the action of the law of gravity, or in the shaping of water-worn rocks by the surf at the seashore." Ibid., 527–28.

57. Ibid., 520.

58. Paul Weisz, *The Science of Biology* (New York: McGraw-Hill Book Company, 1959), 12.

59. Ibid., 9.

60. Weisz, *The Science of Biology* (1959), 11. Also see Weisz, *The Science of Biology*, 2nd ed. (New York: McGraw-Hill Book Company, 1963), 14; Weisz, *The Science of Biology*, 3rd ed. (New York: Mc-Graw-Hill Book Company, 1967), 13; Weisz, *The Science of Biology*, 4th ed. (New York: McGraw-Hill Book Company, 1971), 11–12.

61. Weisz, *The Science of Biology* (1959), 21;

62. Weisz, *The Science of Biology* (1967), 229.; also see Weisz, *The Science of Biology* (1963), 39; Weisz, *The Science of Biology* (1971), 137.

63. Weisz, *The Science of Biology* (1963), 12. Also see Weisz, *The Science of Biology* (1967), 11; Weisz, *The Science of Biology* (1971), 10.

64. Weisz, *The Science of Biology* (1959), 11. Also see Weisz, *The Science of Biology* (1963), 15; Weisz, *The Science of Biology* (1967), 14; Weisz, *The Science of Biology* (1971), 12.

65. Weisz, *The Science of Biology* (1963), 701. Also see Weisz, *The Science of Biology* (1967), 769; Weisz, *The Science of Biology* (1971), 538.

66. For Darwin's use of the term "survival of the fittest," which he eventually praised as "more accurate" than his own term of "natural selection," See Darwin, *Origin*, 74. For Darwin's application of his theory to society, see the discussion of *The Descent of Man* in chapter 2. Also see Weikart, "Laissez-Faire Social Darwinism," 17–30, especially 19–25; Weikart, "A Recently Discovered Darwin Letter," 609–11.

67. P. D. F. Murray, *Biology: An Introduction to Medical and Other Studies* (London: Macmillan, 1956), 2. This book was published in the United States by St. Martin's Press.

68. Salvador Luria, Stephen Jay Gould, Sam Singer, *A View of Life* (Menlo Park, CA: Benjamin/Cummings Publishing Company, 1981), 586–87. More generally, see the book's subsection titled "The Radical Philosophy of Darwin's Theory," 584–87.

69. Joseph S. Levine and Kenneth R. Miller, *Biology: Discovering Life*, 2nd ed. (Lexington, MA: D. C. Heath and Company, 1994), 161; Joseph S. Levine and Kenneth R. Miller, *Biology: Discovering Life* (Lexington, MA: D. C. Heath and Company, 1992), 1:152. Emphasis in the original.

70. Douglas Futuyma, *Evolutionary Biology*, 3rd ed. (Sunderland, MA: Sinauer Associates, 1998), 5.

71. Ibid., 8, emphasis in the original.

72. William K. Purves, David Sadava, Gordon H. Orians, and H. Craig Heller, *Life: The Science of Biology*, 6th ed. (Sunderland, MA: Sinauer Associates, 2001; Salt Lake City: W. H. Freeman and Co., 2001), 3. For nearly the same wording, see William K. Purves, Gordon H. Orians, H. Craig Heller, and David Sadava, *Life: The Science of Biology*, 5th ed. (Sunderland, MA: Sinauer Associates, 1998; Salt Lake City: W. H. Freeman and Co., 1998), 10. Despite its description of evolution as inherently purposeless, the book later tries to soften the blow for religion by offering up the standard two-realms approach discussed earlier. (Purves, *Life: Science of Biology* [2001], 14; Purves, *Life: Science of Biology* [1998], 13.) According to the authors, since religion and science are two different ways of looking at the world, they really are not in conflict. Yet this claim is utterly inconsistent with the book's previous claim that evolution is not directed toward any goals. Biblical monotheism clearly asserts that the development of life is a goal-directed process, with the goals being established by God. Any claim that the development of life is a purposeless process is directly incompatible with this foundational claim made by Western monotheism. In addition, when the book goes on to state on the same page that "[r]eligious beliefs are based on faith—not on falsifiable hypotheses, as science is," it is essentially making a theological distinction, not a scientific one, and it is a theological distinction that many theologians throughout history would reject. The book's implication that faith is something radically separate from reason, and that religious truth claims cannot be established by evidence, is highly debatable.

73. See, for example, Miller and Levine, *Biology: Discovering Life* (1992), 152; Miller and Levine, *Biology: Discovering Life* (1994), 161; Futuyma, *Evolutionary Biology*, 9; and the discussion of the Purves text in note 66.

74. "NABT Unveils New Statement on Teaching Evolution," *The American Biology Teacher* (January 1996), 61.

75. Cited in "Eugenie Scott's Reply to the Open Letter," http://fp.bio.utk.edu/darwin/Open%20letter/scott%20reply.htm (accessed June 17, 2002).

76. See Larry Witham, "Evolutionists Split over Biology Lessons" (February 24, 1998), A2; "Evolution Statement Altered," *The Christian Century* (Nov. 12, 1997), 1029.

77. "Open Letter to the National Association of Biology Teachers, to the National Center for Science Education, and to the American Association for the Advancement of the Sciences," http://fp.bio.utk.edu/darwin/Open%20letter/openletter.html (accessed June 17, 2002).

78. Ibid.

79. "Eugenie Scott's Reply to the Open Letter."

80. Eugenie Scott, "NABT Statement on Evolution Evolves," http://www.ncseweb.org/resources/articles/8954_nabt_statement_on_evolution_ev_5_21_1998.asp (accessed July 16, 2005).

81. In 2004, the NABT statement evolved yet again. Apparently the "no specific direction or goal" language continued to pose a public relations problem, and so it was toned down. According to the 2004 version, "natural selection has no discernible direction or goal." "NABT's Statement on Teaching Evolution, Supporting Material," http://www.nabt.org/sub/position_statements/evolution_supportingmaterial.asp (accessed July 16, 2005), emphasis added.

82. "Eugenie Scott's Reply to the Open Letter."

83. Massimo Pigliucci, "The NABT statement, evolution, and naturalism: a reply to Scott," http://fp.bio.utk.edu/darwin/Open%20letter/pigliucci%20to%20scott.htm (accessed June 17, 2002).

84. Massimo Pigliucci, "The NABT statement, evolution, and naturalism: a reply to Scott," http://fp.bio.utk.edu/darwin/Open%20letter/pigliucci%20to%20scott.htm (accessed June 17, 2002).

85. "Humanism and Its Aspirations: Humanist Manifesto III."

86. Dorothy Matthews, "Effect of a Curriculum Containing Creation Stories on Attitudes about Evolution," *The American Biology Teacher* 63, no. 6 (August 2001): 404–9.

87. Ibid., 407.

88. Ibid., 408.

89. Ibid., 406.

90. Ron Good, "Evolution and Creationism: One Long Argument," *The American Biology Teacher* 65, no. 7 (September 2003): 515.

91. Ibid., 515–16.

92. Ira Mellman and Graham Warren, "The Road Taken: Past and Future Foundations of Membrane Traffic," *Cell* 100, no. 1 (January 7, 2000): 99.

93. See Larry Witham, *Where Darwin Meets the Bible: Creationists and Evolutionists in America* (New York: Oxford University Press, 2002), 271–73.

94. Gregory W. Graffin and William B. Provine, "Evolution, Religion and Free Will," *American Scientist* 95 (July–August 2007), 294–97; results of Cornell Evolution Project survey, http://www.polypterus.com/results.pdf (accessed July 9, 2007), 7, 9.

95. The proportion of Americans who believe in the existence of God is 80–94 percent, depending on how the question is asked. In a 2004 Gallup Poll, 89.9 percent of respondents said they believed in God. When offered a choice between God or "a universal spirit or higher power," 80.9 percent said they believed in God, while another 12.6 percent said they believed in a universal spirit or higher power, for a combined rate of 93.5 percent. This is virtually the same result as a Gallup Poll in 1988, which found that 94.6 percent of respondents said they believed "in God or a universal spirit." (Gallup Poll, 5/02/2004-5/04/2004, 12/21/1988-12/22/1988, http://brain.gallup.com/ [accessed July 16, 2005].) The proportion of Americans who believe in life after death is 68–77 percent, depending on how the question is asked. In a 1988 Gallup Poll, 68.1 percent of respondents said they believed in "life after death," but 76.8 percent of respondents said that "there is a Heaven where people who had led good lives are eternally rewarded." (Gallup Poll, 12/21/1988-12/22/1988, http://brain.gallup.com/ [accessed July 16, 2005].)

96. Kenneth R. Miller, *Finding Darwin's God: A Scientist's Search for Common Ground Between God and Evolution* (New York: HarperCollins, 1999), 244.

97. Ibid., 250.

98. Ibid., 272.

99. Ibid., 236.

100. Ibid., 238–39.

101. Kenneth Miller, comments during "Evolution and Intelligent Design: An Exchange," March 24, 2007, at the "Shifting Ground: Religion and Civic Life in America" conference, Bedford, New Hampshire, sponsored by New Hampshire Humanities Council.

102. For a discussion of the various attempts to reconcile Darwinian theory with religion, see West, *Darwin's Conservatives: The Misguided Quest*, 63–71.

103. "Darwin's Dangerous Idea," Episode 1 of *Evolution* (Clear Blue Sky Productions and WGBH, 2001).
104. Ibid.
105. *Getting the Facts Straight: A Viewer's Guide to PBS's* Evolution (Seattle: Discovery Institute Press, 2001), 28–29.
106. "Why is Evolution Controversial Anyway" (accessed July 16, 2005).
107. "Dealing with Controversy," *Teacher's Guide*: Evolution (WGBH and Clear Blue Sky Producations, 2001), 31.
108. United Presbyterian Church in the USA statement, quoted in "Dealing with Controversy," 30.
109. "Dealing with Controversy," 31.
110. According to a Gallup Poll of 1016 adults in 2001, only 11.8 percent of American adults believe that "Human beings have developed over millions of years from less advanced forms of life, but God had no part in this process." Thus, nearly 9 in 10 Americans reject the Darwinian theory of unguided evolution as the explanation for the development of human beings. Nor do a majority of Americans accept theistic evolution. Again, according to the 2001 Gallup Poll, only 36.98 percent of American adults believe that "Human beings have developed over millions of years from less advanced forms of life, but God guided this process." A plurality of Americans (44.69%) still believe that "God created human beings pretty much in their present form at one time within the last 10,000 years or so." A plurality of Americans also regard evolution as "[j]ust one of many theories and one that has not been well-supported by evidence." Only 35.56 percent of Americans say evolution is "[a] scientific theory that has been well-supported by evidence." Nearly a quarter of Americans (24.98 percent) indicate that they "[d]on't know enough to say." (Gallup Poll, Feb. 19–21, 2001, http://brain.gallup.com/ [accessed July 16, 2005].)
111. There will be further discussion of intelligent design in chapter 11.
112. "Understanding Evolution: An Evolution Website for Teachers," http://evolution.berkeley.edu/ (accessed July 16, 2005).
113. National Science Foundation Award #0096613 (April 1, 2001), http://www.nsf.gov/awardsearch/showAward.do?AwardNumber=0096613 (accessed July 16, 2005).
114. "Misconception: Evolution and Religion Are Incompatible," http://evolution.berkeley.edu/evosite/misconceps/IVAandreligion.shtml (accessed July 16, 2005).
115. United Church Board for Homeland Ministries (United Church of Christ), "Creationism, the Church, and the Public School," reprinted in *Statements from Religious Organizations* (Oakland: National Center for Science Education), http://www.ncseweb.org/resources/articles/5025_statements_from_religious_orga_12_19_2002.asp#home (accessed July 16, 2005).
116. Eugenie Scott, "Dealing with Antievolutionism," http://www.ucmp.berkeley.edu/fosrec/Scott2.html (accessed August 25, 2006).
117. In her tips for activists who want to support evolution before their school board, Eugenie Scott advises: "Call on the clergy. Pro-evolution clergy are essential to refuting the idea that evolution is incompatible with faith . . . If no member of the clergy is available to testify, be sure to have someone do so—the religious issue must be addressed in order to resolve the controversy successfully." Eugenie C. Scott, "12 Tips for Testifying at School Board Meetings," http://www.ncseweb.org/resources/articles/7956_12_tips_for_testifying_at_scho_3_19_2001.asp (accessed July 16, 2005).
118. *Congregational Study Guide for Evolution* (Oakland: National Center for Science Education, 2001), http://www.ncseweb.org/article.asp?category=11 (accessed July 16, 2005).
119. Phina Borgeson, "Introduction to the Congregational Study Guide for Evolution," http://www.ncseweb.org/resources/articles/8888_csg-int.pdf (accessed July 16, 2005).
120. For information about these initiatives, see http://www.butler.edu/clergyproject/clergy_project.htm (accessed August 25, 2006).
121. See notes 7 and 8 for documentation.

Chapter 11: Banned in Burlington

1. Much of the following account of the DeHart controversy is based on "Science or Myth? A Documentation History of Events," on the Burlington-Ediston Committee for Science Education website, http://www.scienceormyth.org/history.html (accessed October 18, 2001); personal e-mail communication from Roger DeHart to the author (August 30, 2001); and a transcript of an interview with Roger DeHart by ColdWater Media (Burlington, Wash.: September 3, 2001).

2. For a brief overview of the modern history of intelligent design, see Jonathan Witt, "The Origin of Intelligent Design: A Brief History of the Scientific Theory of Intelligent Design," Oct. 30, 2005, http://www.discovery.org/scripts/viewDB/index.php?command=view&id=3207 (accessed Feb. 27, 2007). Two early journalistic accounts of the intelligent design movement are Teresa Watanabe, "Enlisting Science to Find the Fingerprints of a Creator: Believers in 'Intelligent Design' Try to Redirect Evolution Disputes along Intellectual Lines," *Los Angeles Times*, March 25, 2001, front page; James Glanz, "Darwin vs. Design: Evolutionists' New Battle," *New York Times* (April 8, 2001), front page. Good detailed accounts of the history of the design movement can be found in Thomas Woodward, *Doubts About Darwin: A History of Intelligent Design* (Grand Rapids, MI: Baker Books, 2003) and *Darwin Strikes Back: Defending the Science of Intelligent Design* (Grand Rapids, MI: Baker Books, 2006); and Larry Witham, *By Design: Science and the Search for God* (San Francisco: Encounter Books, 2003). Books presenting the scientific case for intelligent design include Michael Behe, *Darwin's Black Box: The Biochemical Challenge to Evolution* (New York: Free Press, 1996); William Dembski, *No Free Lunch: Why Specified Complexity Cannot Be Purchased without Intelligence* (Lanham, MD: Rowman and Littlefield, 2002); John Angus Campbell and Stephen C. Meyer, eds., *Darwinism, Design, and Public Education* (East Lansing, MI: Michigan State University Press, 2003); William A. Dembski and Michael Ruse, eds., *Debating Design: From Darwin to DNA* (New York: Cambridge University Press, 2004); William Dembski, *The Design Inference: Eliminating Chance through Small Probabilities* (New York: Cambridge University Press, 1998); William Dembski, *Intelligent Design: The Bridge Between Science and Theology* (Downers Grove, IL: InterVarsity Press, 1999); William Dembski and James Kushiner, eds., *Signs of Intelligence: Understanding Intelligent Design* (Grand Rapids, MI: Brazos Press, 2001); Angus Menuge, *Agents under Fire: Materialism and the Rationality of Science* (Lanham, MD: Rowman and Littlefield, 2004); Guillermo Gonzales and Jay W. Richards, *The Privileged Planet: How Our Place in the Cosmos is Designed for Discovery* (Washington, DC: Regnery, 2004). Scientific articles presenting the case for intelligent design include Stephen C. Meyer, "The Origin of Biological Information in the Higher Taxonomic Categories," *Proceedings of the Biological Society of Washington* 117, no. 2 (2004):. 213–39; and Scott Minnich and Stephen C. Meyer, "Genetic Analysis of Coordinate Flagellar and Type III Regulatory Circuits," *Proceedings of the Second International Conference on Design & Nature*, ed. M. W. Collins and C. A. Brebbia (Great Britain: Wessex Institute of Technology Press, 2004). For a list of peer-reviewed and peer-edited scientific articles about intelligent design, see "Peer-Reviewed and Peer-Edited Scientific Publications Supporting the Theory of Intelligent Design (Annotated)," March 1, 2007, http://www.discovery.org/scripts/viewDB/index.php?command=view&id=2640&program=CSC%20-%20Scientific%20Research%20and%20Scholarship%20-%20Science (accessed March 1, 2007).

3. Witt, "The Origin of Intelligent Design," 5.

4. For critical assessments of intelligent design, see Campbell and Meyer, *Darwinism, Design and Public Education*; Dembski and Ruse, *Debating Design*; Robert Pennock, *Tower of Babel* (Cambridge, MA: MIT Press, 1999); Robert Pennock, ed., *Intelligent Design Creationism and Its Critics: Philosophical, Theological, and Scientific Perspectives* (Boston: MIT Press, 2001); Kenneth R. Miller, *Finding Darwin's God: A Scientist's Search for Coommon Ground Between God and Evolution* (New York: HarperCollins, 1999), especially 81–164; Kenneth R. Miller, "Design on the Defensive," http://biocrs.biomed.brown.edu/Darwin/DI/Design.html (accessed July 25, 2005). For responses to these critiques, see especially Michael J. Behe, "Self-Organization and Irreducibly Complex Systems: A Reply to Shanks and Joplin," *Philosophy of Science* 67 (March 2000): 155–62, also available online at http://www.discovery.org/scripts/viewDB/index.php?command=view&id=465 (accessed July 25, 2005); Michael J. Behe, "Reply to My Critics: A Response to Reviews of *Darwin's Black Box: The Biochemical Challenge to Evolution*," *Biology*

and Philosophy 16: 685–709; Michael Behe, "A Mousetrap Defended: Response to Critics" (July 31, 2000), http://www.discovery.org/scripts/viewDB/index.php?command=view&id=446 (accessed July 25, 2005); Michael Behe, "Irreducible Complexity and the Evolutionary Literature: Response to Critics (July 31, 2000), http://www.discovery.org/scripts/viewDB/index.php?command=view&id=443 (accessed July 25, 2005); Michael Behe, "In Defense of the Irreducibility of the Blood Clotting Cascade: Response to Russell Doolittle, Ken Miller and Keith Robison" (July 31, 2000), http://www.discovery.org/scripts/viewDB/index.php?command=view&id=442 (accessed July 25, 2005); Michael Behe, "'A True Acid Test': Response to Ken Miller" (July 31, 2000), http://www.discovery.org/scripts/viewDB/index.php?command=view&id=441 (accessed July 25, 2005); Michael Behe, "Philosophical Objections to Intelligent Design: Response to Critics" (July 31, 2000), http://www.discovery.org/scripts/viewDB/index.php?command=view&id=445 (accessed July 25, 2005); Michael Behe, "Comments on Ken Miller's Reply to My Essays" (January 8, 2001), http://www.discovery.org/scripts/viewDB/index.php?command=view&id=579 (accessed July 25, 2005).

5. David DeWolf, John West, Casey Luskin, and Jonathan Witt, *Traipsing Into Evolution: Intelligent Design and the* Kitzmiller vs. Dover *Decision* (Seattle: Discovery Institute Press, 2006), 17. For a timeline of the development of intelligent design, see "Intelligent Design Timeline," http://www.researchintelligentdesign.org/wiki/Intelligent_design_timeline (accessed July 16, 2007).

6. Schiller wrote that "it will not be possible to rule out the supposition that the process of Evolution may be guided by an intelligent design." F. C. S. Schiller, "Darwinism and Design Argument," in *Humanism: Philosophical Essays* (New York: The Macmillan Co., 1903), 141. This particular essay was first published in the *Contemporary Review* in June 1897.

7. ". . . one arrives at the conclusion that biomaterials with their amazing measure or order must be the outcome of intelligent design." Fred Hoyle, "Evolution from Space," in *Evolution from Space (The Omni Lecture) And Other Papers on the Origin of Life* (Hillside, NJ: Enslow Publishers, 1982), 28. For Thaxton's role in the development of intelligent design, see Witt, "The Origin of Intelligent Design," 3–4. The Horigan book mentioned in the text is James E. Horigan, *Chance or Design?* (New York: Philosophical Library, 1979).

8. For information about Jeffrey M. Schwartz, who is probably the least known as a proponent of ID, see Richard Monastersky, "Researcher Brings Intelligent Design to Mind," *The Chronicle of Higher Education*, May 26, 2006, chronicle.com/weekly/v52/i38/38a01801.htm (accessed March 15, 2007).

9. See "Questions about Intelligent Design," especially question number 2, "Is intelligent design theory incompatible with evolution?" http://www.discovery.org/csc/topQuestions.php#questionsAboutIntelligentDesign (accessed Feb. 27, 2007).

10. Ibid. Also see Behe, *Darwin's Black Box*, 5; Dembski, *No Free Lunch*, 314–15; William Dembski, "Reflections on Human Origins," *Progress in Complexity, Information, & Design*, 4.1 (Aug. 18, 2004), http://www.iscid.org/papers/Dembski_HumanOrigins_062204.pdf (accessed March 15, 2007).

11. For example, Larry Arnhart stresses "the purely negative character of intelligent design reasoning," and faults ID proponents of engaging in "the rhetorical technique of negative argumentation from ignorance." Boiled down, the intelligent design argument in his view claims that if we cannot currently imagine how a structure such as "the bacterial flagellum could have arisen by purely natural causes or by chance, then we must assume it was designed by some sufficiently powerful intelligent agent." Arnhart points out that this conclusion does not necessarily follow because such structures could have evolved due to "natural causes that we have not yet discovered." Arnhart, *Darwinian Conservatism*, 97.

12. See Dembski, *The Design Inference*; and Dembski, *No Free Lunch*.

13. Larry Arnhart insists, however, that "for intelligent design to have some positive content," its proponents "would have to explain exactly where, when, and how a disembodied intelligence designed flagella and attached them to bacteria." (Arnhart, *Darwinian Conservatism*, 97.) But why? Can't we know that something was produced by an intelligent cause even if we do not know the method used by the intelligent cause? If I am hiking in the desert and come across what appears to be the ruin of a giant building, can't I conclude that an intelligent cause was at work even if I know nothing about how the building was created or who had inhabited the area previously? Discovering the methods by which intelligent causes organized matter and

energy in a certain way is an interesting inquiry, but determining whether an intelligent cause was involved at all also seems to be a legitimate question. Faulting intelligent design theory for not answering a question it does not even purport to answer seems unreasonable.

14. As research chemist and intelligent design proponent Charles Thaxton explained in the late 1980s, uniformitarianism assumes that "the causes we observe today are . . . uniform over space and time. We assume that whatever cause and effect relationships we experience today held in the past as well. If we observe certain effects today, and determine their causes by experience, then whenever we see evidence that the same kind of effects took place in the past, we are justified in assuming the same causes were at work. In other words we rely on uniform sensory experience to extend the assignment of causes into areas where we don't have experience." Charles B. Thaxton, "A Word to the Teacher," in Percival Davis and Dean H. Kenyon, *Of Pandas and People: The Central Question of Biological Origins* (Dallas: Haughton Publishing Co., 1989), 156–57.

15. Michael Behe, "Molecular Machines: Experimental Support for the Design Inference" (March 1, 1998), http://www.discovery.org/scripts/viewDB/index.php?command=view&id=54 (accessed August 9, 2006).

16. See Stephen Meyer, "DNA and the Origin of Life: Information, Specification, and Explanation," in *Darwinism, Design, and Public Education*, 266–70.

17. See William Dembski, "Is Intelligent Design Testable? A Response to Eugenie Scott" (January 24, 2001), http://www.discovery.org/scripts/viewDB/index.php?command=view&id=584 (accessed August 9, 2006); Jay Richards and Jonathan Witt, "Intelligent Design is Empirically Testable and Makes Predictions" (January 5, 2006), http://www.evolutionnews.org/2006/01/intelligent_design_is_empirica.html (accessed August 9, 2006).

18. ColdWater Media Interview with Roger DeHart.

19. ColdWater Media Interview with Eugenie Scott (Oakland: September 25, 2001).

20. ColdWater Media Interview with Roger DeHart.

21. Stephen Jay Gould, "Abscheulich! Atrocious!" *Natural History* (March 2000), 42–49.

22. ColdWater Media Interview with Eugenie Scott.

23. Howard Pellett, "Religion Doesn't Belong in Science Class," *Skagit Valley Herald* (April 5, 2001), www.skagitvalleyherald.com (accessed October 18, 2001).

24. Robert Miller, "Program an Effort to Drive Wedge into the Community," *Skagit Valley Herald* (June 1, 2001), www.skagitvalleyherald.com (accessed October 18, 2001).

25. Ken Atkins, "DeHart Keeps Finding New Names to Call Creationism," *Skagit Valley Herald* (May 6, 2001), www.skagitvalleyherald.com (accessed October 18, 2001).

26. "Science or Myth?"

27. Personal communication from Bruce Chapman, president, Discovery Institute, to the author; also see See "Discovery Institute and Theocracy," http://www.discovery.org/scripts/viewDB/filesDB-download.php?command=download&id=463 (accessed Feb. 27, 2007).

28. For examples of the ideas presented in this program, see John G. West, *The Politics of Revelation and Reason: Religion and Civic Life in the New Nation* (Lawrence, KS: University Press of Kansas, 1996), and John G. West, "Finding the Permanent in the Political: C. S. Lewis as a Political Thinker," in *Permanent Things: Toward the Recovery of a More Human Scale at the End of the Twentieth Century*, Andrew A. Tadie and Michael H. Macdonald, eds. (Grand Rapids, MI: William B. Eerdmans Publishing Company, 1995), 137–48.

29. Marina Parr, "Burlington Reassigns Biology Teacher," *Skagit Valley Herald* (July 22, 2001); Marina Parr, "Official: Changes in Enrollment Prompted Decision on Reassignment" (July 22, 2001). Articles are available online at www.skagitvalleyherald.com (accessed October 18, 2001).

30. According to Jones, "The decisions we make on a day-to-day basis always raise someone's eyebrows and this issue because of the controversy because of the amount of media coverage it's had over the past several years certainly we knew that there would be people that would be concerned—people that would think that we were trying to do something to Roger. But that's not why we make decisions around here. We make them in what's best for kids and what's best for our organization." ColdWater Media Interview with Rick Jones (Burlington, WA: September 4, 2001); also see Parr, "Official: Changes in Enrollment." The *Skagit Valley Herald* wrote regarding the reassignment: "Unless the end justifies the means, there's something amiss in the way Burlington-Edison High School has addressed the thorny situation involving biology

teacher Roger DeHart." The newspaper was skeptical of the district's explanation that lower enrollments in biology classes necessitated the reassignment. "If the problem is fewer biology students, a more commonly used solution would be to have DeHart teach other classes in addition to biology." "DeHart Change Points to Issues beyond Debate," *Skagit Valley Herald* (July 29, 2001), www.skagitvalleyherald.com (accessed October 18, 2001).

31. Burlington-Edison School District confirmed the undergraduate major of the science teacher who replaced DeHart.

32. Theresa Goffredo, "Marysville on edge over new biology teacher," *HeraldNet* (August 10, 2001), http://www.heraldnet.com/Stories/01/8/10/14189379.cfm (accessed June 26, 2002); Todd Frankel, "Survival of the fittest teacher?" *HeraldNet* (Ausust 19, 2001), http://www.heraldnet.com/Stories/01/8/19/14230412.cfm (accessed June 26, 2002); Theresa Goffredo, "Science teacher's position approved," *HeraldNet* (August 21, 2001), http://www.heraldnet.com/Stories/01/8/21/14246219.cfm (accessed June 26, 2002).

33. Personal communication from Roger DeHart to the author.

34. According to Rick Jones, superintendent of the Burlington-Edison School District, "we have worked with Roger during the time that I've been in the school district and have asked him to change his supplemental materials and during that time he has followed those directions." (ColdWater Media Interview with Rick Jones.) Stan Relyea, an assistant principal at Burlington-Edison High School, concurred that DeHart followed district instructions: "I've not found Roger to be insubordinate nor subversive" and "he's very professional in the manner in which he worked with the students." More generally, Relyea indicated that DeHart received positive employee evaluations. "I know that as an administrator within the building and my evaluations of Roger, he did a fine job in the classroom. And when I would observe and talk and evaluate with Roger I was very pleased with what I saw him doing. He worked well with students. He worked well with staff. And I would never you know have anything contrary to that to say as far as his classroom performance. And what he did as far as contributions to the school both in and out of the classroom and in extracurricular activities as a baseball coach was very respected, still is, and did a fine job." (ColdWater Media Interview with Stan Relyea, Assistant Principal, Burlington-Edison High School, Burlington, WA: September 4, 2001.) An official of the Marysville school district, meanwhile, said that confidential recommendations written by DeHart's previous supervisors were "outstanding." (Goffredo, "Marysville on Edge over New Biology Teacher.")

35. Personal communication from Roger DeHart to the author.

36. Stephen Meyer, "A Scopes Trial for the 90s," *The Wall Street Journal* (December 6, 1993), available online at http://www.discovery.org/scripts/viewDB/index.php?command=view&id=93 (accessed July 25, 2005).

37. Quoted in *Rodney LeVake vs. Independent School District #656*, State of Minnesota Court of Appeals, C8-00-1613 (May 8, 2001); available online at http://www.lawlibrary.state.mn.us/archive/ctappub/0105/c8001613.htm (accessed July 25, 2005). Additional information on the LeVake case can be found in James Kilpatrick, "Case of Scientific Heresy is Doomed," *Augusta Chronicle* (December 23, 2001), A-4. The Minnesota Court of Appeals found that the school district's interest in maintaining its curriculum overrode LeVake's first-amendment interest in teaching material critical of Darwinian evolution.

38. Fred Heeren, "The Lynching of Bill Dembski: Scientists Say the Jury Is Out—So Let the Hanging Begin," *The American Spectator* 33, no. 9 (November, 2000): 44–51.

39. See Gordon Gregory, "Biology Instructor's Doctrine Draws Fire," *OregonLive.com* (February 18, 2000), http://www.oregonlive.com (accessed February 18, 2000); Gordon Gregory, "Creationist Instructor Likely Will Lose His Job," *OregonLive.com* (March 28, 2000), http://www.oregonlive.com (accessed March 28, 2000); Julie Foster, "Biology Professor Forced Out; Pointed to Flaws in Theory of Evolution, Encouraged Critical Thinking," *WorldNetDaily.com* (April 14, 2000), http://www.wnd.com/news/article.asp?ARTICLE_ID=17856 (accessed July 25, 2005).

40. Haley's former department chair Bruce McClelland, quoted in Gregory, "Biology Instructor's Doctrine Draws Fire."

41. Testimony of Nancy Bryson before the Texas State Board of Education, September, 2003, Transcript of the Public Hearing before the Texas State Board of Education, September 10, 2003, Austin, Texas (Austin, TX: Chapman Court Reporting Service, 2003), 504–5.

42. Geoff Brumfiel, "Intelligent Design: Who Has Designs on Your Students' Minds?" *Nature* 434 (April 28, 2005): 1062–1065, available online at http://www.nature.com/nature/journal/v434/ n7037/full/4341062a.html (accessed July 26, 2005).

43. See "Intelligent Design and Academic Freedom," on *All Things Considered*, National Public Radio, November 10, 2005.

44. For information about the Bryan Leonard case, see Catherine Candinsky, "Evolution Debate Re-emerges: Doctoral Student's Work Was Possibly Unethical, OSU Professors Argue," *The Columbus Dispatch*, June 9, 2005, http://www.dispatch.com/news-story.php?story=dispat ch/2005/06/09/20050609-C1-01.html&chck=t>http://www.dispatch.com/news-story.php?s tory=dispatch/2005/06/09/20050609-C1-01.html&chck=t (accessed June 9, 2005); "Attack on OSU Graduate Student Endangers Academic Freedom," http://www.discovery.org/scripts/ viewDB/index.php?command=view&id=2661 (accessed July 26, 2005); "Professors Defend Ohio Grad Student Under Attack by Darwinists," http://www.discovery.org/scripts/viewDB/ index.php?command=view&id=2715 (accessed July 26, 2005).

45. See David Klinghoffer, "The Branding of a Heretic," *The Wall Street Journal*, January 28, 2005, http://www.opinionjournal.com/taste/?id=110006220 (accessed July 26, 2005). For more information about the controversy surrounding the publication of the journal article supportive of intelligent design, see "Info on Meyer Article Published in Smithsonian Journal," http://www. discovery.org/scripts/viewDB/index.php?command=view&id=2399 (accessed July 26, 2005), and http://www.rsternberg.net/ (accessed July 26, 2005).

46. Letter to Richard Sternberg from the U.S. Office of Special Counsel, August 5, 2005, available at http://www.rsternberg.net/OSC_ltr.htm (accessed September 9, 2006). Also see Klinghoffer, "The Branding of a Heretic."

47. *Intolerance and the Politicization of Science at the Smithsonian: Smithsonian's Top Officials Permit the Demotion and Harassment of Scientist Skeptical of Darwinian Evolution, Staff Report Prepared for the Hon. Mark Souder, Chairman, Subcommittee on Criminal Justice, Drug Policy and Human Resources* (Washington, DC: United States House of Representatives, Committee on Government Reform, December 11, 2006), 3 and 20–21.

48. Ibid., 4.

49. Ibid., 5–6.

50. Ibid., 22. The congressional report further explained: "Dr. Sues hoped that the NCSE could unearth evidence that Dr. Sternberg had misrepresented himself as a Smithsonian employee, which would have been grounds for his dismissal as a Research Associate: 'As a Research Associate, Sternberg is not allowed to represent himself as an employee of the Smithsonian Institution, and, if he were to do so, he would forfeit his appointment.'" Ibid.

51. Ibid., 23. Emphasis in original.

52. Quoted in Michael Powell, "Editor Explains Reasons for 'Intelligent Design' Article," *Washington Post*, August 19, 2005, A19.

53. See Guillermo Gonzalez and Jay Richards, *The Privileged Planet: How Our Place in the Cosmos is Designed for Discovery* (Washington, DC: Regnery Publishing, 2004).

54. For more information, see John West, "Astronomer Guillermo Gonzalez Appeals Tenure Denial to Iowa Board of Regents, July 11, 2007, http://www.evolutionnews.org/2007/07/astronomer_ guillermo_gonzalez.html (accessed July 16, 2007); "Biography of Dr. Guillermo Gonzalez," May 14, 2007, http://www.discovery.org/scripts/viewDB/index.php?command=view&id=4058 (accessed July 16, 2007); John West, "Iowa State Professor Who Was Denied Tenure Exceeds Department's Tenure Standard by 350%," May 13, 2007, http://www.evolutionnews.org/2007/05/ iowa_state_professor_who_was_d.html (accessed July 16, 2007); Casey Luskin, "Guillermo Gonzalez Has Highest Normalized Citation Count among ISU Astronomers for Publications Since 2001," May 28, 2007, http://www.evolutionnews.org/2007/05/guillermo_gonzalez_has_ highest.html (accessed July 16, 2007). A list of publications by Guillermo Gonzalez can be downloaded from http://www.discovery.org/scripts/viewDB/filesDB-download.php?comm and=download&id=1362 (accessed July 16, 2007).

55. William Dillon, "Geoffroy Denies Tenure Appeal of Intelligent Design Proponent," *The Ames Tribune*, June 2, 2007, http://www.midiowanews.com/site/tab1.cfm?newsid=18417569&BRD=2 700&PAG=461&dept_id=554432&rfi=6 (accessed July 16, 2007).

56. John Hauptman, "Rights are intact: Vote turns on question, 'What is science?'" June 2, 2007, http://desmoinesregister.com/apps/pbcs.dll/article?AID=/20070602/OPINION01

/706020313/1035/OPINION (accessed July 16, 2007); John West, "Breaking News: Iowa State Department Faculty Acknowledge ID Played Role in Gonzalez's Tenure Denial," May 18, 2007, http://www.evolutionnews.org/2007/05/isu_department_faculty_acknowl.html (accessed July 16, 2007).

57. "In fact, *Mein Kampf* does not contain a single explicit command for genocide equivalent to those found in the Hebrew Bible." Hector Avalos, *Fighting Words: The Origins of Religious Violence* (Amherst, NY: Prometheus Books, 2005), 361; also see 316–19 and John West, "Iowa State Promotes Atheist Professor Who Equates Bible with Mein Kampf While Denying Tenure to ID Astronomer," May 22, 2007, http://www.evolutionnews.org/2007/05/iowa_state_promotes_atheist_pr.html (accessed July 16, 2007).

58. John West, "Scandal Brewing at Baylor University? Denial of Tenure to Francis Beckwith Raises Serious Questions about Fairness and Academic Freedom," March 28, 2006, http://www.evolutionnews.org/2006/03/scandal_at_baylor_university_d.html (accessed Feb. 27, 2007). For a list of Dr. Beckwith's publications, see http://www3.baylor.edu/~Francis_Beckwith/cv.html (accessed Feb. 27, 2007).

59. See, for example, Francis J. Beckwith, *Law, Darwinism, and Public Education: The Establishment Clause and the Challenge of Intelligent Design* (Lanham, MD: Rowman and Littlefield, 2003); "Science and Religion Twenty Years After McLean v. Arkansas: Evolution, Public Education, and the New Challenge of Intelligent Design," *Harvard Journal of Law & Public Policy* 26, no. 2 (Spring 2003): 455–99; "Public Education, Religious Establishment, and the Challenge of Intelligent Design," *Notre Dame Journal of Law, Ethics, & Public Policy* 17, no. 2 (2003): 461–519; "A Liberty Not Fully Evolved?: The Case of Rodney LeVake and the Right of Public School Teachers to Criticize Darwinism," *San Diego Law Review* 39, no. 4 (Nov/Dec 2002): 1311–25.

60. Robert Crowther, "Welcome News as Scholar Francis Beckwith is Granted Tenure at Baylor," Sept. 27, 2006, http://www.evolutionnews.org/2006/09/welcome_news_as_scholar_franci.html (accessed Feb. 27, 2007).

61. Peter Machamer, Marcello Pera, and Aristedes Baltas, *Scientific Controversies: Philosophical and Historical Perspectives* (New York: Oxford University Press, 2000), 6. It should be noted that the authors here are summarizing what they call the "historical-social view of science" that has developed among scholars since the late 1950s. Although they think this view "has done more justice to science as actually practiced than the logical positivists' rational reconstructions," they also believe that it has shortcomings of its own. See ibid., 6–7.

62. See the helpful discussion of the suppression of dissenting views in Gordon Moran, *Silencing Scientists and Scholars in Other Fields: Power, Paradigm Controls, Peer Review, and Scholarly Communication* (Greenwich, CT: Ablex Publishing Corporation, 1998).

63. E-mail from Jonathan Wells to the author (October 23, 2001).

64. ColdWater Media Interview with Jonathan Wells (Berkeley, CA: September 25, 2001).

65. Ibid.

66. Personal communication from Jonathan Wells to the author.

67. Jonathan Wells, *Charles Hodge's Critique of Darwinism: An Historical-Critical Analysis of Concepts Basic to the 19th Century Debate* (Lewiston, NY: Edwin Mellen Press, 1988).

68. Personal communication from Jonathan Wells to the author.

69. Jonathan Wells, *Icons of Evolution: Science or Myth? Why Much of What We Teach about Evolution is Wrong* (Washington, DC: Regnery Publishing, 2000).

70. Ibid., 7.

71. Ibid., 6.

72. See Larry D. Martin, "An Iconoclast for Evolution?" *The World & I* (February, 2001), 241–46; Jerry A. Coyne, "Creationism by Stealth," *Nature* 410 (April 12, 2001): 745–46; Massimo Pigliucci, "Intelligent Design Theory," *BioScience* 51, no. 5 (May, 2001): 411–14; Eugenie C. Scott, "Fatally Flawed Iconoclasm," *Science* 292 (June 22, 2001): 2257–58; Rudolf A. Raff, "The Creationist Abuse of Evo-Devo," *Evolution & Development* 3, no. 6 (November–December, 2001): 373–74; Kevin Padian and Alan Gishlick, "The Talented Mr. Wells," *The Quarterly Review of Biology* 77, no. 1 (March, 2002): 33–37.

73. Coyne, "Creationism by Stealth."

74. E-mail from anonymous scientist to *Nature* (April 17, 2001). The author was furnished with a copy of this e-mail, but the scientist who lodged the complaint prefers to stay anonymous.

75. Coyne, interview clip on "CNN Newsroom" (May 3, 2001).

76. Coyne, "Creationism by Stealth."
77. "The Talented Mr. Wells," 33–34, 37.
78. Ibid., 34.
79. In an unpublished letter sent to *The Quarterly Review of Biology*, Carolyn A. Larabell (Associate Professor at the Lawrence Berkeley Laboratory and the University of California, San Francisco) wrote that "Both of these claims are false. Dr. Wells and I performed numerous experiments together in my laboratory at Berkeley while he was a post-doc. That research resulted in two peer-reviewed papers to which we contributed as co-authors. [A] Some of our work has even appeared in a textbook on developmental biology. [B] I am surprised that *The Quarterly Review of Biology* would publish something with so little regard for truthfulness and professional decorum. The false claims of Padian and Gishlick unjustly damage not only the reputation of Dr. Wells, but also—indirectly—the reputations of those who worked with him. It seems to me that a retraction is in order." The publications cited by Larabell were [A] "Confocal Microscopy Analysis of Living Xenopus Eggs and the Mechanism of Cortical Rotation," *Development* 122 (April, 1996): 1281–89; "Microtubule-mediated organelle transport and localization of beta-catenin to the future dorsal side of Xenopus eggs," *Proceedings of the National Academy of Sciences USA* 94 (February, 1997): 1224–29. [B] Klaus Kalthoff, *Analysis of Biological Development*, 2nd ed. (Boston: McGraw Hill, 2001), 208. Unpublished letter from Carolyn Larabell to the editor of *The Quarterly Review of Biology* (May 2, 2002).
80. E-mail from editor at *Nature* to anonymous scientist (April 25, 2001). The author was furnished with a copy of this e-mail.
81. The letter from Jonathan Wells that *Nature* refused to publish is posted online as "Response to Jerry Coyne's Review of *Icons of Evolution*" (April 26, 2001), http://www.discovery.org/scripts/viewDB/index.php?command=view&id=626 (accessed July 25, 2005).
82. Letter from Albert Carlson to Carolyn Larabell (May 20, 2002).
83. Wells, *Icons of Evolution*, 144–53.
84. Ibid., 145.
85. Ibid.
86. Ibid., 149. Emphasis in the original.
87. Ibid., 151.
88. Ibid., 151, 156.
89. Oral remarks by Kenneth R. Miller at a meeting of the Ohio Board of Education (March 11, 2002). The remarks were transcribed from a videotape of the meeting.
90. Kenneth R. Miller, "Goodbye, Columbus," http://www.millerandlevine.com/km/evol/debate.html (accessed June 20, 2002).
91. Wells, *Icons of Evolution*, 148–51, 303–4. In addition, Miller sometimes misrepresents Wells as having claimed that peppered moths "never" rest on tree trunks (e-mail from Kenneth R. Miller to Joshua Littlefield [April 26, 2002]). But in *Icons of Evolution*, as well as subsequent writings, Wells does not argue that peppered moths "never" rest on tree trunks, only that they *"do not normally rest on tree trunks"* (Wells, *Icons of Evolution*, 149, emphasis in the original) or that they "rarely rest on tree trunks in the wild" (Jonathan Wells, "Moth-eaten Statistics: A Reply to Kenneth R. Miller" [April 16, 2002], http://www.discovery.org/scripts/viewDB/index.php?command=view&id=1147 [accessed on July 25, 2005]). For documentation of how Miller has misrepresented the scientific views of other scholars critical of Darwinism, see Mike Gene, "Miller and Behe on Origins: Guest response to Ken Miller's review of *Darwin's Black Box*" (July 10, 2000); Michael Behe, "A Mousetrap Defended: Response to Critics"; Michael Behe, "Irreducible Complexity and the Evolutionary Literature: Response to Critics; Michael Behe, "In Defense of the Irreducibility of the Blood Clotting Cascade: Response to Russell Doolittle, Ken Miller and Keith Robison"; " 'A True Acid Test': Response to Ken Miller"; "Philosophical Objections to Intelligent Design: Response to Critics" (July 31, 2000); "Comments on Ken Miller's Reply to My Essays." For a critique of Miller's accuracy on other topics besides science, see "Biologist Ken Miller Flunks Political Science on Santorum" (April 19, 2002) http://www.discovery.org/scripts/viewDB/index.php?command=view&id=1149 (accessed July 25, 2005).
92. Kenneth R. Miller, "Paying the Price," http://www.millerandlevine.com/km/evol/wells-april-2002.html (accessed June 20, 2002). The rebuttal by Wells that precipitated Miller's response was Jonathan Wells, "There You Go Again: A Response to Kenneth R. Miller" (April

9, 2002), http://www.discovery.org/scripts/viewDB/index.php?command=view&id=1144 (accessed July 25, 2005).

93. Kenneth Miller and Joseph Levine, *Biology* (Upper Saddle River, NJ: Prentice Hall, 2002).

94. Judith Hooper, *Of Moths and Men: An Evolutionary Tale* (New York: W. W. Norton and Co., 2002).

95. Jonathan Wells, "Critics Rave Over *Icons of Evolution*: A Response to Published Reviews" (June 12, 2002), http://www.discovery.org/scripts/viewDB/index.php?command=view&id=1180 (accessed July 25, 2005).

96. Michael Ruse, "How Evolution Became a Religion," *National Post* (May 13, 2000), B1, available online through InfoTrac OneFile, Thomson Gale, http://find.galegroup.com/itx/infomark. do?&type=retrieve&tabID=T003&prodId=ITOF&docId=A63295843&source=gale&userGroup Name=kcls_web&version=1.0 (accessed July 26, 2005).

97. Wells, "Critics Rave."

98. ColdWater Media Interview with Stephen Meyer (Spokane, WA: September 7, 2001).

99. For an annotated listing of relevant articles, see "Bibliography of Supplementary Resources For Ohio Science Instruction" (Seattle: Discovery Institute, March 11, 2002), http://www.discovery.org/scripts/viewDB/index.php?command=view&id=1127 (accessed July 26, 2005).

100. W. Daniel Hillis, as quoted in John Brockman, *The Third Culture: Beyond the Scientific Revolution* (New York: Simon & Schuster, 1995), 26. Hillis is described as a computer scientist on 379 of this book.

101. Detailed information about the No Child Left Behind Act can be found at the U.S. Department of Education website, http://www.ed.gov/nclb/landing.jhtml (accessed July 25, 2005). For a good summary and overview of the act's provisions, see *No Child Left Behind: A Parents Guide* (Washington: U.S. Department of Education, 2003), available online at http://www.ed.gov/parents/academic/involve/nclbguide/parentsguide.pdf (accessed July 25, 2005).

102. Grade 10, Life Sciences, Indicator #23 and Grades 9–10, Life Sciences, Benchmark H, Academic Content Standards, K-12 Science, Ohio Department of Education, http://www.ode.state.oh.us/academic_content_standards/ScienceContentStd/PDF/SCIENCE.pdf (accessed July 25, 2005), 37, 62, 132, 145.

103. "Critical Analysis of Evolution—Grade 10," Ohio Department of Education, draft reflecting changes at March 2004 meeting of the Ohio State Board of Education, http://www.ode.state.oh.us/academic_content_standards/sciencesboe/pdf_setA/L10-H23_Critical_Analysis_of_Evolution_Mar_SBOE_changes.pdf (accessed July 25, 2005).

104. Strand II: The Content of Science, Standard II (Life Science), 9–12 Benchmark II, Performance Standard 9, New Mexico Science Content Standards, Benchmarks, and Performance Standards, New Mexico State Department of Education, http://www.nmlites.org/downloads/science/science_standards.pdf (accessed July 25, 2005), 36.

105. Grade 9–12, Strand I: History and Nature of Science, Sub-Strand A: Scientific World View, Benchmark 2, Minnesota Academic Standards, Science K-12, Minnesota Department of Education, http://education.state.mn.us/content/078664.doc (accessed July 25, 2005), 16.

106. "Rationale of the State Board for Adopting these Science Curriculum Standards," Kansas Science Education Standards, as approved by the Kansas State Board of Education on Nov. 8, 2005, previously available at http://www.ksde.org/outcomes/sciencestd.pdf (accessed Aug. 8, 2006), ii.

107. Written testimony of Dr. Steven Schafersman submitted to the Texas State Board of Education for the public hearing on textbooks on July 9, 2003. This quote appears in the version of the testimony actually submitted to the Board by Dr. Schafersman. A revised version of Dr. Schafersman's testimony, cited below, later appeared on the website of Texas Citizens for Science.

108. Testimony of Wendee Holtcamp, Transcript of the Public Hearing before the Texas State Board of Education, September 10, 2003, Austin, Texas (Chapman Court Reporting Service, 2003), 312.

109. Testimony of Steven Weinberg, Transcript of the Public Hearing . . . September 10, 299.

110. Testimony of Robert Pennock, Transcript of the Public Hearing . . . September 10, 516.

111. J. Sjogren and R. DiSilvestro, "Fifty-two Ohio Scientists Call for Academic Freedom on Darwin's Theory" (March 20, 2002), http://www.discovery.org/scripts/viewDB/index.php?command=view&id=1163 (accessed July 25, 2005).

112. Andrew Welsh-Huggins (Associated Press), "State School Board Approves Evolution Lesson Plan," March 10, 2004, http://www.ohio.com (accessed March 10, 2004); Glen Needham, "Science Standards will Spur Critical Thinking," *The Columbus Dispatch*, March 1, 2004, A-6.

113. Needham, "Science Standards."

114. Quoted in Robert Crowther, "In Ohio Darwinist Admits Plan to Burn Evolution Critics," Jan. 10, 2006, http://www.evolutionnews.org/2006/01/in_ohio_darwinist_admits_plan.html (accessed Feb. 27, 2007).

115. See "Ranks of Scientists Doubting Darwin's Theory on the Rise," Feb. 7, 2007, http://www.discovery.org/scripts/viewDB/index.php?command=view&id=2732 (accessed Feb. 27, 2007); "80 Years After Scopes Trial New Scientific Evidence Convinces Over 400 Scientists That Darwinian Evolution is Deficient," July 18, 2005, http://www.discovery.org/scripts/viewDB/index.php?command=view&id=2732 (accessed July 25, 2005). This statement was originally issued in 2001; see "A Scientific Dissent from Darwinism," http://www.reviewevolution.com/press/DarwinAd.pdf (accessed July 25, 2005). For the most updated information about the list, see http://www.dissentfromdarwin.org (accessed Feb. 27, 2007).

116. Quoted in "100 Scientists, National Poll Challenge Darwinism" (Seattle: Discovery Institute, September 24, 2001), http://www.reviewevolution.com/press/pressRelease_100Scientists.php (accessed July 25, 2005).

117. Quoted in "80 Years After Scopes Trial."

118. ColdWater Media Interview with Scott Minnich.

119. Quoted in "40 Texas Scientists Join Growing National List of Scientists Skeptical of Darwin," September 5, 2003, http://www.discovery.org/scripts/viewDB/index.php?command=view&id=1555 (accessed July 26, 2005).

120. Philip S. Skell, "Why Do We Invoke Darwin? Evolutionary theory contributes little to experimental biology," *The Scientist*, no. 16 (August 28, 2005): 10. To make this point, Skell quotes A.S. Wilkins, editor of the journal *BioEssays*, who has written: "Evolution would appear to be the indispensable unifying idea and, at the same time, a highly superfluous one."

121. For examples of materials submitted to educational policymakers citing peer-reviewed research on these issues, see "Bibliography of Supplementary Resources"; "The Scientific Controversy over Whether Microevolution Can Account for Macroevolution," (Seattle: Discovery Institute), http://www.discovery.org/scripts/viewDB/filesDB-download.php?id=118 (accessed July 25, 2005); "The Scientific Controversy over the Cambrian Explosion," (Seattle: Discovery Institute), http://www.discovery.org/scripts/viewDB/filesDB-download.php?id=119 (accessed July 25, 2005); "An Analysis of the Treatment of Homology in Biology Textbooks," (Seattle: Discovery Institute), http://www.discovery.org/articleFiles/PDFs/homologyrpt.pdf (accessed July 25, 2005); "A Preliminary Analysis of the Treatment of Evolution in Biology Textbooks," (Seattle: Discovery Institute), http://www.discovery.org/articleFiles/PDFs/TexasPrelim.pdf (accessed July 25, 2005).

122. Revised version of the written testimony of Dr. Steven Schafersman submitted to the Texas State Board of Education for the public hearing on textbooks on July 9, 2003, available online at http://www.texscience.org/files/schafersman-july9-testimony.htm (accessed July 25, 2005). This version differs in certain respects from the written testimony Dr. Schafersman actually submitted to the Board of Education.

123. Written testimony of Dr. Steven Schafersman submitted to the Texas State Board of Education for the public hearing on textbooks on September 10, 2003, available online at http://www.texscience.org/files/icons-revealed/index.htm (accessed July 25, 2005).

124. Brian Leiter, "Biology Textbooks under Attack," August 11, 2003, The Leiter Reports, http://webapp.utexas.edu/blogs/archives/bleiter/000146.html#000146 (accessed July 25, 2005).

125. Testimony of David Hillis, Transcript of the Public Hearing . . . September 10, 135.

126. Written Testimony of Thomas M. Davis, submitted to the Texas State Board of Education for the public hearing on textbooks on July 9, 2003.

127. "Statement of the Reverend Mark Belletini," Ohio Citizens for Science website, http://ecology.cwru.edu/ohioscience/state-belletini.asp (accessed July 6, 2002). This statement was originally printed on a handout titled "Ohio Science Facts" that was distributed to the Ohio State Board of Education at its February 4, 2002, meeting.

128. See Austin Cline, "Texas Taliban Becoming a Role Model?" (March 12, 2005) http://atheism.about.com/b/a/152876.htm (accessed Jan. 22, 2007); "Intelligent Design: Designing Unintel-

ligent Schoolchildren in SC" August 27, 2005, ("The American Taliban is at it again, pressuring lawmakers to teach creationism alongside bona fide, indisputable, solid science"), http://www.pamspaulding.com/weblog/2005/08/intelligent-design-designing.html (accessed Jan. 22, 2007); Farb, "Comment 85283," December 25, 2006, ("In a week, the Kansas Science Taliban loses its majority on the State Board of Education, and life will return to something like normalcy."), http://www.badastronomy.com/bablog/2006/12/24/kansas-is-full-of-light-air/#comment-85283 (accessed Jan. 22, 2007); Grumpy, "Darwinism on Trial in Kansas," September 7, 2005 ("The American Taliban is on the march in Kansas and if the scientific community does nothing they will take over our childrens [sic] education."), http://forum.physorg.com/index.php?showtopic=2869&st=0&#entry25424 (accessed Jan. 22, 2007); Jeff, "Evolution's foes lose ground in Kansas," August 2, 2006 ("American taliban on Kansas school board pushed out in elections. Progress in Kansas?"), http://www.newscloud.com/read/71271 (accessed Jan. 22, 2007); Markus the Carcass, "Comment #46270," September 2, 2005 ("Basically what we are looking at here is the American version of the Taliban in action."), http://www.pandasthumb.org/archives/2005/08/intelligent_des_6.html#comment-46270 (accessed Jan. 22, 2007); DemFromCT, "Proverbs 11:29: 'He that troubleth his own house shall inherit the wind' " March 14, 2005 ("The know-nothings are really on the march, in so many ways…but that doesn't mean deliberately turning our schools into another arm of the American Taliban's war on reality."), http://thenexthurrah.typepad.com/the_next_hurrah/2005/03/proverbs_1129_h.html (last visited Jan. 22, 2007).

129. John W. Wenzel, "Creationist Posed in Scientist's Clothing," *The Columbus Dispatch*, March 5, 2004, A-8.

130. Paul Gross, "Intelligent Design and that Vast Right-Wing Conspiracy," *Science Insights*, Sept. 2003 (National Association of Scholars). Also see Barbara Forres and Paul R. Gross, *Creationism's Trojan Horse: The Wedge of Intelligent Design* (New York: Oxford University Press, 2004).

131. Barbara Forrest, untitled article in *Evolutionary Science*, June 2002 (newsletter published by Ohio Citizens for Science), available online at http://ecology.cwru.edu/ohioscience/newsletter.asp (accessed July 6, 2002).

132. Barbara Forrest, "Overview: The Newest Evolution of Creationism," *Natural History* (April 2002), available online at http://www.actionbioscience.org/evolution/nhmag.html#overview (accessed July 25, 2005).

133. New Orleans Secular Humanist Association home page, http://nosha.secularhumanism.net/index.html. Forrest is listed as a member of the board of directors on the "Who's Who" page of the website, http://nosha.secularhumanism.net/whoswho.html (accessed July 6, 2002).

134. Lawrence M. Krauss, Curriculum Vitae, http://genesis1.phys.cwru.edu/~krauss/cv.htm (accessed July 25, 2005).

135. Explaining why a small percentage of National Academy of Science members do not reject God, Krauss in 2006 said: "Scientists are people and are as full of delusions about every aspect of their life as everyone else. We all make up inventions so that we can rationalize our existence and why we are who we are." Quoted in William Dembski, "NAS at 85% Atheists—Let's Bump It Up to 100%," Dec. 7, 2006, http://www.uncommondescent.com/religion/nas-at-85-atheists-lets-bump-it-up-to-100/ (accessed Feb. 27, 2007). In fairness to Krauss, he did suggest that such scientists need not be pressured to become atheists. The point here is to show that Krauss's own personal beliefs are hostile to traditional theism.

136. *Yale Bulletin and Calendar*, http://www.yale.edu/opa/ybc/campusnotes.html (accessed March 15, 2007)

137. Steven D. Schafersman, "The History and Philosophy of Humanism and its Role in Unitarian Universalism," speech originally delivered in Sept. 1995 and revised in December 1998, http://www.freeinquiry.com/humanism-uu.html (accessed July 25, 2005).

138. Steven Schafersman, "The Challenge of the Fossil Record," http://www.freeinquiry.com/challenge.html (accessed July 25, 2005).

139. Steven Weinberg, quoted in "Free People from Superstition," http://ffrf.org/fttoday/2000/april2000/weinberg.html (accessed July 25, 2005).

140. See Larson, *Summer for the Gods*; Carol Iannone, "The Truth about *Inherit the Wind*," *First Things*, February 1997, http://www.firstthings.com/article.php3?id_article=3645 (accessed Feb. 27, 2007); Marvin Olasky and John Perry, *Monkey Business: The True Story of the Scopes Trial* (Nashville: Broadman and Holman, 2005).

141. "Cobb County Policy on Objective Origins Science," September 26, 2002, available online at http://www.discovery.org/scripts/viewDB/index.php?command=view&id=1396 (accessed July 25, 2005).

142. See "School Board Votes to Allow Creationism in Class," *CNN.com*, September 27, 2002, http://www.cnn.com/ (accessed September 27, 2002); and Barnini Chakraborty, "Ga. School District Votes to Allow Different Views on Life's Origins, including Creationism," Associated Press report, September 27, 2002, http://www.sfgate.com/cgi-bin/article.cgi?file=/news/archive/2002/09/27/national0504EDT0475.DTL (accessed July 26, 2005).

143. Constance Holden, "Creationism Edges Toward the Classroom," *Science* magazine online, September 28, 2002, http://www.sciencemag.org (accessed September 28, 2002). The print edition of *Science* later published a slightly more accurate account by the same author. See Constance Holden, "Georgia County Opens Door to Creationism," *Science* 298, no. 5591 (October 4, 2002): 35–36.

144. Quoted in Mia Taylor and Mary Macdonald, "Cobb Unanimously Approves Discussion of Other Theories," *Atlanta Journal-Constitution*, September 26, 2002, http://www.ajc.com/ (accessed September 26, 2002).

145. David Burch, "Scientists Reach Out to Cobb In Support of 'Disputed Views,'" *Daily Journal* (Marietta, Georgia), September 21, 2002, http://www.mdjonline.com/ (accessed September 21, 2002); Mary Macdonald, "28 Scholars Back Cobb on Evolution; Skeptical Approach to Darwinism Urged," *Atlanta Journal-Constitution*, September 21, 2002, http://www.ajc.com/ (accessed September 21, 2002).

146. Associated Press, "Ohio OKs Teaching Evolution with a Twist," *Salt Lake Tribune*, March 10, 2004, originally appeared online at http://www.sltrib.com (accessed March 11, 2004).

147. See "CNN Fabricates Texas Textbook Story," Discovery Institute, July 10, 2003, http://www.discovery.org/scripts/viewDB/index.php?command=view&id=1513 (accessed July 25, 2005).

148. For detailed information about the CNN report on Missouri, see "CNN Airs False Report Claiming Teachers will be Fired," Discovery Institute, April 6, 2004, http://www.discovery.org/scripts/viewDB/index.php?command=view&id=1985 (accessed July 25, 2005); " 'Bogus' Story Aired by CNN: TV Journalist Does Report on Evolution Bill, Mischaracterizes What Legislation Would Do," *WorldNetDaily.com*, April 7, 2004, http://www.worldnetdaily.com/news/article.asp?ARTICLE_ID=37914 (accessed July 25, 2005); Jim Brown and Jody Brown, "CNN Accused of Fabricating Controversy in Missouri Evolution Story," Agape-Press news service, April 12, 2004, http://headlines.agapepress.org/archive/4/122004a.asp (accessed July 25, 2005); "Missouri Legislator Asks CNN to Stop Misrepresenting His Bill," Discovery Institute, April 13, 2004, http://www.discovery.org/scripts/viewDB/index.php?command=view&id=2003 (accessed July 25, 2005); Transcript, *CNN Sunday Morning*, April 4, 2004, http://www.cnn.com/TRANSCRIPTS/0404/04/sm.04.html (accessed July 25, 2005); "CNN Defends False Report with Spin and Half-Truths," Discovery Institute, April 13, 2004, http://www.discovery.org/scripts/viewDB/index.php?command=view&id=2002 (accessed July 25, 2005).

149. For additional examples of misreporting by the national newsmedia, see "NPR Stifles Evolution Debate by Canceling Guest Critical of Darwinian Evolution," Discovery Institute, June 1, 2004, http://www.discovery.org/scripts/viewDB/index.php?command=view&id=2059 (accessed July 25, 2005); John G. West and Robert Crowther, "Documentation of NPR Errors in Reporting on Teaching of Evolution," Discovery Institute, June 1, 2004, http://www.discovery.org/scripts/viewDB/index.php?command=view&id=2058 (accessed July 25, 2005); "NPR Criticized for Evolution Misinformation and 'AWOL' Ombudsman," Discovery Institute, June 2, 2004, http://www.discovery.org/scripts/viewDB/index.php?command=view&id=2062 (accessed July 25, 2005).

150. Charles Krauthammer, "Phony Theory, False Conflict: 'Intelligent Design' Foolishly Pits Evolution Against Faith," *Washington Post*, November 18, 2005, A23

151. "Rationale of the State Board for Adopting These Science Curriculum Standards." Krauthammer also inaccurately claimed that the Kansas board redefined science in order to include the supernatural, when in fact the board was reinstating the standard definition of science used in virtually all other states. See Jonathan Wells, "Analysis of Kansas Definition of Science Compared to All Other State Science Definitions," http://www.discovery.org/scripts/viewDB/index.php?command=view&id=2573 (accessed September 6, 2006).

152. John West, "CNN Sends Fabricator of Texas Textbook Story to Kansas; Nixes Debate between Meyer and Miller," Aug. 11, 2005, http://www.evolutionnews.org/2005/08/cnn_sends_fabricator_of_texas_textbook_s.html (accessed Feb. 27, 2007).

153. *Kitzmiller v. Dover Area Sch. Dist.*, Dec. 20, 2005, http://www.pamd.uscourts.gov/kitzmiller/kitzmiller_342.pdf (accessed March 13, 2007), 2.

154. See "Nightline: War In Dover," ABC television broadcast, Jan. 13, 2005; Transcript of Testimony of Sheila Harkins, Kitzmiller v. Dover, No. 4:04-CV-2688 (M.D. Pa., Nov. 2, 2005), 115–116. Even Judge Jones found that "most if not all of the Board members who voted in favor of the biology curriculum change conceded that they still do not know, nor have they ever known, precisely what ID is." *Kitzmiller v. Dover*, 131.

155. See Richard N. Ostling, Associated Press, "Darwin vs. Design in Ohio," *Wisconsin State Journal*, March 17, 2002, p. A3 ("Stephen Meyer of Seattle's Discovery Institute, the leading ID think tank, told the board that rather than making ID part of the curriculum it should merely encourage teachers to cover the disagreements about Darwinism."); and Discovery Institute, "Ohio Votes 13–5 to Approve Lesson Plan Critical of Evolution," March 9, 2004, http://www.discovery.org/scripts/viewDB/index.php?command=view&id=1898 ("Chapman added that the lesson plan is exactly the approach to teaching evolution that Discovery Institute has advocated all along, helping students learn both the scientific strengths and weaknesses of Darwin's theory . . . The lesson plan does not discuss religion or alternative scientific theories such as intelligent design.") (accessed March 6, 2007).

156. "Should public schools require the teaching of intelligent design? No. Instead of mandating intelligent design, Discovery Institute recommends that states and school districts focus on teaching students more about evolutionary theory, including telling them about some of the theory's problems that have been discussed in peer-reviewed science journals." Discovery Institute, "Top Questions," http://www.discovery.org/csc/topQuestions.php (accessed Oct. 30, 2006).

157. For an explanation of this view, see "Discovery Institute Opposes Proposed PA Bill on Intelligent Design," June 22, 2005, http://www.discovery.org/scripts/viewDB/index.php?command=view&id=2688 (accessed March 6, 2007).

158. Quoted in Andrew J. Petto, "Grantsburg Activists Budge School Board," http://www.ncseweb.org/resources/rncse_content/vol24/6105_grantsburg_activists_budge_sch_12_30_1899.asp (accessed March 6, 2007).

159. "The school board of Grantsburg, Wisconsin adopted a revised policy on the teaching of evolution at a special meeting on December 6, which states that "Students shall be able to explain the scientific strengths and weaknesses of evolutionary theory[.]" The new policy makes clear that the school board is not authorizing the teaching of either creationism or the scientific theory of intelligent design." Discovery Institute, "Wisconsin School Board Adopts Improved Policy Endorsing Fully Teaching Evolution, Not Creationism," Dec. 7, 2004, http://www.discovery.org/scripts/viewDB/index.php?command=view&id=2323 (accessed March 7, 2007).

160. "Amy DeLong, a pastor at Grantsburg's United Methodist Church, opposed the district's original policy and said she feels her concerns have been addressed. But she said it makes her nervous that the Discovery Institute, a national pro-intelligent design group, issued a news release Tuesday praising the new policy's critical analysis of evolution." Megan Boldt, "School clarifies policy on evolution: Religious theories won't be taught," *St. Paul Pioneer Press*, Dec. 8, 2004, http://www.twincities.com/mld/twincities/news/local/10362937.htm?1c [accessed Dec. 8, 2004].

161. Suzanne Quick, "Policy That Reopened Origins Debate Revised: Grantsburg Board Says Creationism Need Not Be Taught," Dec. 10, 2004, http://www.jsonline.com/story/index.aspx?id=283092 (accessed March 6, 2007).

162. Petto, "Grantsburg Activists."

163. In a 2007 letter to the editor of the student newspaper at Yale University, Zimmerman made the astonishing claim that "those of us who fought the [Granstburg] school board declared victory after their creationist policy was jettisoned and this new one was adopted [in December of 2004]. My only concern with the [revised] policy that was adopted was that it seemed to set an awfully ambitious agenda for middle and high school students . . . to claim that I opposed the policy whose passage I celebrated, and then to assert that I opposed 'analyzing Darwinism's strengths and weaknesses,' is laughable." (Zimmerman, "Writer Missed Point

of Evolution Sunday," http://www.yaledailynews.com/articles/view/19737 [accessed March 6, 2007].) What is truly laughable is Zimmerman's feeble attempt to rewrite history. In addition to his criticism of the revised policy documented in note 161, an article published nearly a year after the revised policy was enacted reported: "Zimmerman continues to organize letter-writing campaigns protesting Grantsburg's vote." (Lawrence Hardy, " 'Teaching the Controversy' in Wisconsin," *American School Board Journal*, November 2005, 29.) Finally, the website for the "Clergy Letter Project" that Zimmerman initiated contains the following claim from a letter signed by Zimmerman himself: "In a few weeks, nearly 200 clergy signed the statement [disputing the Grantsburg policy], which we sent to the Grantsburg school board on December 16, 2004." (Statement from Michael Zimmerman, http://www.butler.edu/clergyproject/clergy_project.htm [accessed March 6, 2007].) The Grantsburg board issued its revised policy on December 6. Thus, according to Zimmerman's own account, he was still trying to raise opposition to the policy some ten days after achieving his purported "victory."

164. The summary of Discovery Institute's role in the Dover controversy in this and following paragraphs draws on the author's personal knowledge as well as "Setting the Record Straight about Discovery Institute's Role in the Dover School District Case," Nov. 10, 2005, http://www.discovery.org/scripts/viewDB/index.php?command=view&id=3003 (accessed March 7, 2007); and "Statement by Seth L. Cooper Concerning Discovery Institute and the Decision in Kitzmiller v. Dover Area School Board Intelligent Design Case," Dec. 21, 2005, http://www.evolutionnews.org/2005/12/statement_by_seth_l_cooper_con.html (March 7, 2007).

165. "Discovery Calls Dover Evolution Policy Misguided, Calls For its Withdrawal," Dec. 14, 2004, http://www.discovery.org/scripts/viewDB/index.php?command=view&id=2341 (accessed March 6, 2007).

166. "AP Cites Discovery Institute's Opposition to Dover School Board Policy," Nov. 12, 2004, http://www.discovery.org/scripts/viewDB/index.php?command=view&id=2848 (accessed March 6, 2007).

167. Valerie Strauss, "Fresh Challenges in the Old Debate Over Evolution," Dec. 7, 2004, http://www.washingtonpost.com/wp-dyn/articles/A40839-2004Dec6.html (accessed Dec. 7, 2004).

168. E-mail to the author from Valerie Strauss, Dec. 7, 2004; also see John West, "Local PA Paper Gets DI Position Right, Washington Post and Major Media Don't," Dec. 21, 2004, http://www.evolutionnews.org/2004/12/local_pa_paper_gets_di_position_right_wa.html (accessed March 6, 2007).

169. The official citation of the case is *Kitzmiller v. Dover Area Sch. Dist.*, 400 F. Supp. 2d 707 (M.D. Pa. 2005), but the version used throughout this chapter is the publicly-accessible slip version of the opinion cited in note 155.

170. "Happy Birthday, Charles Darwin," Feb. 16, 2006, http://www.realclimate.org/index.php/archives/2006/02/happy-birthday-charles-darwin/ (accessed Nov. 25, 2006).

171. ". . . in which I defend the judiciary against barbaric assault," Aug. 14, 2006, http://timpanogos.wordpress.com/2006/08/14/in-which-i-defend-the-judiciary-against-barbaric-assault/ (accessed Nov. 25, 2006).

172. John Rennie, "Threw the Book at 'Em," Dec. 20, 2005, http://blog.sciam.com/index.php?title=threw_the_book_at_em&more=1&c=1&tb=1&pb=1 (accessed Nov. 25, 2006).

173. "I think the kitzmiller trial documents and decision are important as works of both scholarship and history." Comment by Andrew McClure, May 3, 2006, http://www.pandasthumb.org/archives/2006/05/judge_jones_mak.html - comment-99857 (ccessed Nov. 25, 2006).

174. "Jones came into the case with (apparently) only superficial knowledge of the issues at hand, and was able to absorb and assess them all very well. that makes him seem like a top-notch thinker to me." Comment by Mike Z, May 3, 2006, http://www.pandasthumb.org/archives/2006/05/judge_jones_mak.html - comment-99873 (accessed Nov. 25, 2006).

175. "When Judge Jones demonstrated clear thinking in abundance in his writing of a decision that seems to have put the ID movement on the run, much as Gen. Jackson's forces did to the British at the Battle of New Orleans, it qualifies him as an outstanding thinker." Comment by Ed Darrell, May 3, 2006, http://www.pandasthumb.org/archives/2006/05/judge_jones_mak.html - comment-99863 (accessed Nov. 25, 2006).

176. Comment by Andrew McClure.

177. Matt Ridley, "John Jones: The Judge Who Ruled for Darwin," May 6, 2006, http://www.time.com/time/magazine/article/0,9171,1187265,00.html (accessed Nov. 25, 2006).

178. For a detailed analysis of the legal and factual flaws of Judge Jones's decision, see DeWolf, West, Luskin, and Witt, *Traipsing into Evolution;* and David K. DeWolf, John G. West, and Casey Luskin, "Intelligent Design Will Survive *Kitzmiller v. Dover," Montana Law Review,* Winter 2007.

179. The discussion that follows incorporates material from John G. West and David K. DeWolf, "A Comparison of Judge Jones' Opinion in *Kitzmiller v. Dover* with Plaintiffs' Proposed 'Findings of Fact and Conclusions of Law,' " http://www.discovery.org/scripts/viewDB/filesDB-download.php?command=download&id=1186 (accessed Jan. 19, 2007).

180. See Robert Crowther "Judge's Copying of ACLU 'Highly Frowned Upon' by Courts," Dec. 13, 2006, http://www.evolutionnews.org/2006/12/judges_copying_of_aclu_highly.html (accessed March 13, 2007); Casey Luskin, "Darwinists Desperate to Defend Kitzmiller Copying," Dec. 18, 2006, http://www.evolutionnews.org/2006/12/defending_the_judge_jones_stud.html (accessed March 13, 2007); and Casey Luskin, "Analogical Legal Reasoning and Legal Policy Argumentation: A Response to Darwinist Defenders of Judge Jones' Copying from the ACLU," Jan. 26, 2007, http://www.discovery.org/scripts/viewDB/filesDB-download.php?command=download&id=1209 (accessed March 13, 2007).

181. *Kitzmiller v. Dover,* 78.

182. Michael Behe, Testimony Transcript, *Kitzmiller v. Dover,* afternoon session, Oct. 19, 2005, 19.

183. "Plaintiffs' Findings Fact and Conclusions of Law," *Kitzmiller v. Dover,* 35.

184. *Kitzmiller v. Dover,* 87. Compare to "Plaintiffs' Findings Fact and Conclusions of Law," *Kitzmiller v. Dover,* 51.

185. Scott Minnich, Testimony Transcript, *Kitzmiller v. Dover,* morning session, Nov. 4, 2005, 34; Stephen Meyer, "The Origin of Biological Information and the Higher Taxonomic Categories." For the full citation to the Meyer article see note 2.

186. Brief of Amicus Curiae Foundation for Thought and Ethics, at appendix D 8-18, at http://www.discovery.org/scripts/viewDB/filesDB-download.php?command=download&id=648 (accessed March 13, 2007).

187. *Kitzmiller v. Dover,* 30, 66, 67. Compare to language in "Plaintiffs' Findings Fact and Conclusions of Law," 21.

188. Ibid.

189. Scott Minnich, Testimony Transcript, *Kitzmiller v. Dover,* afternoon session, Nov. 3, 2005, 45–46. See also Michael Behe, Testimony Transcript, *Kitzmiller v. Dover,* morning session, Oct. 17, 2005, 86.

190. Comment by "Coin," Dec. 12, 2006, http://sandwalk.blogspot.com/2006/12/judge-jones-dover-trial.html#comment-8513107298111491492 (accessed March 13, 2007), emphasis in the original.

191. John West, "Judge Jones and the Shattering of Darwinist Illusions," Dec. 13, 2006, http://www.evolutionnews.org/2006/12/judge_jones_and_the_shattering.html (accessed March 13, 2007).

192. Albert Alschuler, "The Dover Intelligent Design Decision, Part I: Of Motive, Effect, and History," Dec. 21, 2005, http://uchicagolaw.typepad.com/faculty/2005/12/the_dover_intel.html (accessed March 13, 2007).

193. Amy Worden, "Bad Frog Beer to 'Intelligent Design': The Controversial Ex-Pa. Liquor Board Chief Is Now U.S. Judge in the Closely Watched Trial," *Philadelphia Inquirer,* October 16, 2005, at http://www.philly.com/mld/inquirer/news/local/states/pennsylvania/counties/montgomery_county/12912029.htm (accessed Nov. 6, 2005).

194. Alschuler, "The Dover Intelligent Design Decision."

195. In *Village of Euclid v. Amber Realty Co.,* the Supreme Court pointed out that it is the "traditional policy of this Court" to decide only the legal question most directly at issue, not all possible legal questions raised by a particular controversy: "In the realm of constitutional law, especially, this Court has perceived the embarrassment which is likely to result from an attempt to formulate rules or decide questions beyond the necessities of the immediate issue. It has preferred to follow the method of a gradual approach to the general by a systematically guarded application and extension of constitutional principles to particular cases as they arise, rather than by out of hand attempts to establish general rules to which future cases must be fitted." *Village of Euclid v. Ambler Realty Co.,* 272 U.S. 365 (1926), 397.

196. *Kitzmiller v. Dover,* 90–132.

197. See discussion of this point in DeWolf, West, and Luskin, "Intelligent Design Will Survive."

198. "Those who disagree with our holding will likely mark it as the product of an activist judge. If so, they will have erred as this is manifestly not an activist Court." *Kitzmiller v. Dover*, 137.

199. Jay D. Wexler, "Kitzmiller and the 'Is It Science?' Question," *First Amendment Law Review* 5 (Fall 2006), 93.

200. Casey Luskin, "Ohio State Board of Education Repeals Critical Analysis Policy; Sends to Subcommittee for Further Review and Recommendation," Feb. 15, 2006, http://www.evolutionnews.org/2006/02/ohio_board_rescinds_policy_ove.html (accessed March 13, 2007).

201. David Klinghoffer, "What's the Matter with Kansas? Dishonest Darwinists—Coming to a State Near You," *National Review Online*, Aug. 3, 2006, http://article.nationalreview.com/?q=OTA1Y2IzMTk1MDYzNzI5MGFhZmJjOWE3MWM0Y2ZkOGI (accessed March 13, 2007); John West, "Kansas Board of Education Adopts Dumbed-Down Curriculum Standards on Evolution and History of Science," Feb. 13, 2007, http://www.evolutionnews.org/2007/02/kansas_board_of_education_adop.htm (accessed March 13, 2007).

202. John West, "AP: 'ID Backer Knocks Tuskegee Deletion from Kansas Standards,'" Feb. 13, 2007, http://www.evolutionnews.org/2007/02/ap_id_backer_knocks_tuskegee_d.html (accessed March 13, 2007). See also "Kansas Board of Education Urged to Reject 'Shameful' Proposal to Delete Tuskegee Experiment and Other Science Abuses from State Curriculum," Feb. 12, 2007, http://www.discovery.org/scripts/viewDB/index.php?command=view&id=3903 (accessed March 13, 2007).

203. "South Carolina Praised for Requiring Students to Critically Analyze Evolutionary Theory," June 12, 2006, http://www.discovery.org/scripts/viewDB/index.php?command=view&id=3527 (March 13, 2007).

204. "Local Louisiana School Board Praised for Adopting Policy Protecting Teachers Who Teach Evolution Objectively," Dec. 1, 2006, http://www.discovery.org/scripts/viewDB/index.php?command=view&id=3823 (accessed March 13, 2007); John West, "California School District Adopts Policy Allowing Scientific Criticisms of Evolution," March 22, 2006, http://www.evolutionnews.org/2006/03/california_school_district_ado.html (accessed March 13, 2007); "'Academic Freedom Act' Moves to Senate," March 2, 2006, http://www.okhouse.gov/OkhouseMedia/pressroom.aspx?NewsID=519 (accessed March 13, 2007).

205. "Americans Overwhelmingly Support Teaching Scientific Challenges to Darwinian Evolution, Zogby Poll Shows," March 7, 2006, http://www.discovery.org/scripts/viewDB/index.php?command=view&id=3325 (accessed March 13, 2007). The official Zogby report can be downloaded at http://www.discovery.org/scripts/viewDB/filesDB-download.php?command=download&id=719 (accessed March 13, 2007).

206. Rebecca Wittman, "Zogby America Report [Survey on Evolution]" (New York: Zogby International, September 21, 2001), available online at http://www.reviewevolution.com/press/ZogbyFinalReport.pdf (accessed July 25, 2005). For the results of various state polls, see "Overwhelming Support in Ohio For Teaching Both Sides of Evolution, Zogby Poll Shows," Feb. 13, 2006, http://www.discovery.org/scripts/viewDB/index.php?command=view&id=3245 (accessed March 13, 2007); "Californians Say Teach Scientific Evidence Both For and Against Darwinian Evolution, Show New Polls," Discovery Institute, May 3, 2004, http://www.discovery.org/scripts/viewDB/index.php?command=view&id=2024 (accessed July 25, 2005); "Views of Texas Residents on Teaching Evolution," (Zogby International, September 8, 2003), available online at http://www.discovery.org/csc/texas/docs/TexasPollFinal.doc (accessed July 25, 2005); "Minnesota Becomes Third State to Require Critical Analysis of Evolution," Discovery Institute, June 8, 2004, http://www.discovery.org/scripts/viewDB/index.php?command=view&id=2043 (accessed July 25, 2005).

207. George Johnson, "A Free-for-All on Science and Religion," Nov. 21, 2006, http://www.nytimes.com/2006/11/21/science/21belief.html?ex=1321765200&en=1248e2f606e1e138&ei=5090 (accessed March 13, 2007).

208. For information about the film, see its official website, http://www.flockofdodos.com (accessed March 14, 2007). For extensive information documenting the film's inaccuracies and distortions, see http://www.hoaxofdodos.com (accessed March 14, 2007).

209. For a more detailed refutation of Olson's claims about Haeckel's embryo diagrams, see John G. West and Casey Luskin "Hoax of Dodos: Flock of Dodos Filmmaker Claims Haeckel's Embryo Drawings Weren't in Modern Textbooks—But He Knows Better, "http://www.discovery.org/scripts/viewDB/filesDB-download.php?command=download&id=1220 (accessed March 15, 2007).

210. Gould, "Abscheulich! Atrocious!"

211. James Glanz, "Biology Text Illustrations More Fiction Than Fact," *New York Times,* April 8, 2001, http://www.nytimes.com/2001/04/08/science/08EVOL.html (accessed March 14, 2007).

212. Quoted in *Icons of Evolution* (2002) documentary.

213. John West, "Filmmaker Randy Olson Backtracks on False Claim in Film, Admitting: 'apparently there are a few textbooks that have traces of Haeckel's embryos . . .'" Feb. 10, 2007, http://www.evolutionnews.org/2007/02/filmmaker_randy_olson_backtrac.html (accessed March 14, 2007). Also see Jonathan Wells, "Flock of Dodos, or Pack of Lies?," Feb. 9, 2007, http://www.evolutionnews.org/2007/02/flock_of_dodos_or_pack_of_fals_4.html (accessed March 14, 2007).

214. "AAAS Board Resolution on Intelligent Design Theory," Oct. 18, 2002, http://www.aaas.org/news/releases/2002/1106id2.shtml (accessed March 14, 2007).

215. In November 2002, the author e-mailed or faxed all of the board members of the AAAS, asking them the following question: "Prior to the adoption of the resolution, what books and/or articles did you read by those advocating intelligent design theory?" Four responses were received, including one from AAAS executive director, Alan Leshner, and a response from a staff member on behalf of the chairman of the board, biologist Peter Raven.

216. E-mail from Alan Leshner to the author, November 23, 2002.

217. Lenny Flank, "Tactics in Fighting Creationists and IDers," http://www.geocities.com/Cape-Canaveral/Hangar/2437/index.htm (accessed March 14, 2007).

218. Jason Cox, "Dover-trial film might evolve," Dec. 17, 2006, *York Daily Record/Sunday News,* http://www.ydr.com/doverbiology/ci_4855425 (accessed March 14, 2007).

Chapter 12: Junk Science in the Bedroom

1. "Sexual Politics: [Advice about losing virginity]," http://www.scarleteen.com/politics/adv_loseit.html (accessed June 8, 2004).

2. "About Scarleteen and Our Staff," http://www.scarleteen.com/about.html and http://www.scarleteen.com/ (accessed June 17, 2004).

3. For information about these products, see http://www.scarleteen.com/shop/index.html (accessed June 17, 2004).

4. "Families are Talking: For Young People," http://www.familiesaretalking.org/teen/teen0000.html (accessed June 16. 2004). This website is a project of SIECUS.

5. See "Phone Sex" (November 5, 1993) http://www.goaskalice.columbia.edu/0076 (accessed June 17, 2004); "S/M roleplaying" (July 6, 1998), http://www.goaskalice.columbia.edu/0646 (accessed June 16, 2004); "Menage a Trois? (July 5, 2002), http://www.goaskalice.columbia.edu/0673.html (accessed June 16, 2004); "Porn magazines by mail?" (August 11, 1998), http://www.goaskalice.columbia.edu/0374.html (accessed June 16, 2004); "Two couples + Four play = Okay?" (November 13, 1998), http://www.goaskalice.columbia.edu/1400.html (accessed June 16, 2004).

6. "Ask the Experts: My boyfriend wants to have anal sex. Will this just feel good for him and will it hurt me in any way?" http://wwwteenwire.com/ask/2003/as_20030425p552_anal.asp (accessed June 16, 2004); Jim Trzaska, "Be a Voice for Choice!" (September 13, 2002), http://www.teenwire.com/takingac/articles/ta_20020913p063_choice.asp (accessed June 16, 2004); John Platner, "The Global Gag Rule" (March 5, 2004), http://www.teenwire.com/views/articles/wv_20040305p073_Bush.asp (accessed June 16, 2004); Julie Sabatier, "Speaking Out About Sex Ed" (November 25, 2003), http://www.teenwire.com/infocus/2003/if_20031125p265_ed.asp (accessed June 16, 2004).

7. See Edward Westermarck, *The History of Human Marriage,* 3 vols. (London: Macmillan, 1921).

8. Following is a selection of passages in which Westermarck cites the importance of natural selection in the develop of human sexuality: "If we ask why in certain animal species male and female remain together not only during the pairing season but till after the birth of the offspring, I think that there can be no doubt as regards the true answer. They are induced to do so by an instinct which has been acquired through the process of natural selection beause it has

a tendency to preserve the next generation and thereby the species." (Westermark, *The History of Human Marriage*, 1:35); "Marriage . . . seems to be based upon a primeval habit. We have found reasons to believe that even in primitive times it was the habit for a man and a woman, or several women, to remain together till after the birth of the offspring, and that they were induced to do so by an instinct which had been acquired through natural selection because the offspring were in need of both maternal and paternal care." (Ibid., 3:365); Discussing the incest taboo, Westermarck made a general comment that "any satisfactory explanation of the normal characteristics of the sexual instinct, which is of such immense importance for the existence of the species, must be sought for in their specific usefulness. We may assume that in this, as in other cases, natural selection has operated, and by eliminating destructive tendencies and preserving useful variations has moulded the sexual instinct so as to meet the requirements of the species." (Edward Westermarck, *The Future of Marriage in Western Civilisation* [New York: Macmillan, 1936], 258–59, emphasis added.)

 More generally, for the influence of Darwin on Westermarck, see Edward Westermarck, *Memories of My Life*, trans. Anna Barwell, (New York: The Macaulay Co., 1929), 67–68, 77–79, 88–89, 91–92. Westermarck stated that "Darwin's book on the *Descent of Man* . . . proved of the greatest importance for my future work" (ibid., 67) and that he "found the theory of natural selection, especially in its application to instinct, to be of enormous importance for the solution of many of the problems with which I was occupied." (Ibid., 77) At the same time, he criticized Darwin's theory of sexual selection, and his Lamarckian explanation for the origin of new variations, preferring the mutation theory that later formed the basis of what became known as "neo-Darwinism." Darwin's greatest accomplishment according to Westermarck was the banishment of design from explanations of biology: "Darwin discovered one factor which must have influenced evolution in a greater or less degree; and his discovery was undeniably of such a nature that it surpassed all later biological discoveries in its effect upon our general view of life. He gave an explanation of the appearance of purpose in organic life without calling in the aid of the hypothesis of a providence that has created the world according to a certain plan drawn up after a human pattern." (Ibid., 78–79) For more on the relationship between Westermarck and Darwin's theory, see Timothy Stroup, *Westermarck's Ethics* (Åbo: Research Institute of the Åbo Akademi Foundation, 1982), 35–38, 66–68, 129–31; Timothy Stroup, "Introduction," in *Edward Westermarck: Essays on His Life and Works*, ed. Timothy Stroup (Helsinki: The Philosophical Society of Finland, 1982), xiii–xiv; Georg Henrik Von Wright, "The Origin and Development of Westermarck's Moral Philosophy," in Stroup, *Edward Westermarck: Essays*, 29–30, 39–41.

9. See discussion in Westermarck, *The Future of Marriage*, 151–71.

10. For Westermarck's development of moral relativism, see Edward Westermarck, *The Origin and Development of the Moral Ideas* (New York: Macmillan, 1906), vols. 1 and 2, and Edward Westermarck, *Ethical Relativity* (New York: Harcourt, Brace and Company, 1932). According to Westermarck, "the emotional origin of moral judgments consistently leads to a denial of the objective validity ascribed to them both by common sense and by normative theories of ethics." (Westermarck, *Ethical Relativity*, xvii.) Regarding homosexuality, Westermarck wrote: "Many physiologists are nowadays of opinion that each sex contains the latent characters of the other sex—in other words, is latently hermaphrodite. Among mammals the male possesses useless nipples, which occasionally even develop into breasts, and the female possesses a clitoris, which is merely a rudimentary penis, and may also develop. So, too, a homosexual tendency may be regarded as simply the psychical manifestation of special characters of the other sex, susceptible of being evolved under certain circumstances, such as may occur about the age of puberty . . . When facts of this kind become more commonly known, they can hardly fail to influence public opinion about homosexuality." (Westermarck, *The Future of Marriage*, 254–55.) Regarding bestiality, Westermarck contended that "the moral condemnation of bestiality" had "not a rational, but an entirely emotional foundation . . . it is considered immoral because it causes disgust. It is therefore an opinion which is nowadays gaining ground that it should not be punished at all . . . I was told by a group of mountaineers [in Morocco] that a person who has intercourse with another man's animal has to buy for it new shoes, a new pack-saddle, and new panniers, must feed it for a day, and if it becomes ill, will have to pay its price. When I then asked what would happen if the animal was his own, the answer was, amidst much laughter, 'Why should not a man be allowed to do with his animal whatever he likes?' " Westermarck, *The Future of Marriage*, 244–45.

11. Westermarck, *The Future of Marriage*, 201.
12. Thornhill and Palmer, *A Natural History of Rape*, 84.
13. Wright, *The Moral Animal*, 87.
14. David M. Buss, *Evolutionary Psychology: The New Science of the Mind* (Boston: Allyn and Bacon, 1999), 185.
15. See Wright, *The Moral Animal*, 37, 88. Wright on 37 says that "understanding the often unconscious nature of genetic control is the first step toward understanding that —in many realms, not just sex—we're all puppets, and our best hope for even partial liberation is to try to decipher the logic of the puppeteer." But if, as Wright also says, "free will is an illusion, brought to us by evolution" (350), how can we liberate ourselves from "the puppeteer"?
16. Wright, *The Moral Animal*, 350.
17. Malcolm Potts and Roger Short, *Ever Since Adams and Eve: The Evolution of Human Sexuality* (Cambridge: Cambridge University Press, 1999), 332.
18. Carol M. Ostrom, "Not Meant for Monogamy?" *Seattle Times*, February 3, 1998, D-1.
19. James Jones, *Alfred C. Kinsey: A Public/Private Life* (New York: W. W. Norton, 1997), 42.
20. Ibid., 32, 125.
21. Wardell Pomeroy, *Kinsey and the Institute for Sex Research* (New York: Harper and Row, 1972), 34.
22. For a discussion of Kinsey and eugenics, see Jones, *Alfred Kinsey*, 194–95. Despite Kinsey's support of eugenics, his high school biology textbook dealt with the topic far more discreetly than many other texts of the period. Eugenics books were included in the bibliography, and the discussion of heredity and environment raised the standard eugenic refrain that "[w]e only fool ourselves when we say that all men are born equal. While we may have equal rights legally, we greatly differ in our biologic equipments." But there were no appeals to forcably sterilize the unfit. The closest Kinsey came to directly promoting eugenics in the book was a study question that asked students: "Can you find (in some reference book on heredity, eugenics, or sociology) the story of the Jukes, the Kallikaks, the Darwin or the Edwards family, or others, and show how their heredities helped or hurt society?" Alfred C. Kinsey, *An Introduction to Biology* (Philadelphia: J. B. Lippincott Co., 1926), 174–75, 177, 539–40.
23. Jones, *Alfred Kinsey*, 423–38.
24. Alfred C. Kinsey, Wardell B. Pomeroy, Clyde E. Martin, *Sexual Behavior in the Human Male* (Philadelphia: W. B. Saunders Co., 1948).
25. Jones, *Alfred Kinsey*, 534–600.
26. Kinsey, *Sexual Behavior in the Human Male*, 585, 597, 598.
27. Ibid., 650.
28. Ibid., 651.
29. Ibid., 463. Even 4 percent of city boys supposedly participated in bestiality "between adolescence and 15." Ibid.
30. Ibid., 280.
31. Ibid., 660.
32. Alfred C. Kinsey, Wardell B. Pomeroy, Clyde E. Martin, Paul H. Gebhard, *Sexual Behavior in the Human Female* (Philadelphia: W. B. Saunders Co., 1953), 451.
33. Kinsey, *Sexual Behavior in the Human Male*, 678.
34. Ibid.
35. Ibid., 202.
36. Ibid., 667.
37. Ibid., 668, 670–74.
38. Ibid., 676.
39. "To those who believe, as children do, that conformance should be universal, any departure from the rule becomes an immorality. The immorality seems particularly gross to an individual who is unaware of the frequency with which exceptions to the supposed rule actually occur." Ibid., 667.
40. "The student of human folkways is inclined to see a considerable body of superstition in the origins of all such taboos, even though they may ultimately become religious and moral issues for whole nations and whole races of people." Ibid., 669.
41. Ibid., 677.
42. Ibid.

43. Ibid., 589.
44. Kinsey, *Sexual Behavior in the Human Female*, 448.
45. Ibid., 448; *Sexual Behavior in the Human Male*, 660.
46. Kinsey, *Sexual Behavior in the Human Male*, 222.
47. Kinsey, *Sexual Behavior in the Human Female*, 121.
48. Ibid.
49. Kinsey, *Sexual Behavior in the Human Male*, 592.
50. Ibid., 589.
51. Ibid., 211.
52. Ibid., 211, 209.
53. Ibid., 211.
54. Ibid., 203.
55. Ibid., 263. Kinsey offered this view as one of three possible interpretations of sex. Although he does not explicitly endorse this view on this page, it is clear from the rest of the book that this is Kinsey's own position. The other possible interpretations of sex given by Kinsey were the "hedonistic doctrine that sexual activity is justifiable for its immediate and pleasurable return" and the Judeo-Christian view of "sex primarily as the necessary means of procreation, to be enjoyed only in marriage, and then only if reproduction is the goal of the act." Ibid.
56. Ibid., 201.
57. Ibid., 263.
58. Kinsey, *Sexual Behavior in the Human Female*, 10.
59. Ibid., 10.
60. Ibid., 8. Kinsey compared such critics to people who when faced with an ocean in their path pretend it isn't there: "some who, finding the ocean an impediment to the pursuit of their designs, try to ignore its existence. If they are unable to ignore it because of its size, they try to legislate it out of existence, or try to dry it up with a sponge . . . There is no ocean of greater magnitude than the sexual function, and there are those who believe that we would do better if we ignored its existence, that we should not try to understand its material origins, and that if we sufficiently ignore it and mop at the flood of sexual activity with new laws, heavier penalties, more pronouncements, and greater intolerances, we may ultimately eliminate the reality." Kinsey, *Sexual Behavior in the Human Male*, 10.
61. Kinsey, *Sexual Behavior in the Human Male*, 224.
62. "For those who like the term, it is clear that there is a sexual drive which cannot be set aside for any large portion of the [male] population, by any sort of social convention. For those who prefer to think in simpler terms of action and reaction, it is a picture of an animal who, however civilized or cultured, continues to respond to the constantly present sexual stimuli, albeit with some social and physical restraints." Ibid., 269.
63. Kinsey, *Sexual Behavior in the Human Female*, 14.
64. Ibid., 21.
65. Kinsey, *Sexual Behavior in the Human Male*, 238.
66. Alan Gregg, "Preface," in Kinsey, *Sexual Behavior in the Human Male*, vii.
67. Quoted in Jones, *Alfred Kinsey*, 570.
68. Morris L. Ernst and David Loth, *American Sexual Behavior and The Kinsey Report* (New York: Thge Greystone Press, 1948), 11. For a comparison between Kinsey and Galileo, see comment of Dartmouth President John S. Dickey cited in Jones, *Alfred Kinsey*, 596.
69. Kinsey, *Sexual Behavior in the Human Female*, 9.
70. Judith A. Reisman, *Kinsey: Crimes and Consequences*, 2nd rev. ed. (Crestwood, KY: Institute for Media Education, 2000).
71. For a detailed description of these activities, see Jones, *Alfred Kinsey*, 601–15.
72. Quoted in Jones, *Alfred Kinsey*, 607. In addition to this person, Jones notes several others who had reservations about what Kinsey wanted them to do. Kinsey associate Clyde Martin and his wife refused to be filmed having sex. Martin says that he did finally agree to be filmed masturbating, although "the idea kind of offended me." William Dellenback, who was hired by Kinsey to film the sexual encounters, did not want to participate himself. But, again, he gave in to pressure and once participated in a group masturbation session "because Kinsey wanted it" even though he "didn't enjoy it." Ibid.

73. Jonathan Gathorne-Hardy, *Alfred C. Kinsey: Sex the Measure of All Things, A Biography* (London: Chatto & Windus, 1998), 228. Gathorne Hardy claims that today Nowlis believes Kinsey was correct. "Nowlis said that later he realised this was meant as a 'nutrturing experience', and indeed he now felt Kinsey was correct—this was the best way . . . to train interviewers to get at the truth in the difficult area of homosexuality." Also see account in Jones, *Alfred Kinsey,* 491–92.

74. Jones, *Alfred Kinsey,* 81–84, 609–10, 738–39.

75. Ibid., 610.

76. Quoted in ibid., 604.

77. Ibid., 514.

78. Gathorne-Hardy, *Alfred Kinsey,* 271.

79. Ibid., 31.

80. Pomeroy, *Kinsey,* 29.

81. Jones, *Alfred Kinsey,* 257.

82. Kinsey, *Sexual Behavior in the Human Male,* 6.

83. Reisman, *Kinsey: Crimes and Consequences,* 51.

84. "By 1946 he [Kinsey] Gebhard and I had interviewed about 1,400 convicted sex offenders in penal institutions scattered over a dozen states." Pomeroy, *Kinsey,* 208. In *Sexual Behavior in the Human Male,* Kinsey acknowledged having data for "more than 1200 persons who have been convicted of sex offenses." Kinsey, *Sexual Behavior in the Human Male,* 392. In *Sexual Behavior in the Human Female,* Kinsey reported that he had gathered data on 1300 convicted sex offenders, both male and female. Kinsey, *Sexual Behavior in the Human Female,* 19, 943.

85. Pomeroy, *Kinsey,* 211–12.

86. Paul H. Gebhard, et. al., *Sex Offenders: An Analysis of Types* (New York: Harper and Row, 1965); Reisman, *Kinsey: Crimes and Consequences,* 91. Gebhard claims that from 1940–45 Kinsey collected 37 percent of an eventual sample of "888 white males who had never been convicted for a sex offense, but who had been convicted for some other misdemeanor or felony (our prison group)." He said that "virtually all the prison group were interviewed while they were in an institution."

87. Reisman, *Kinsey: Crime and Consequences,* 91–92; Cornelia V. Christenson, *Kinsey: A Biography* (Bloomington, IN: Indiana University Press, 1971), 135; Pomeroy, *Kinsey,* 83–85.

88. Kinsey, *Sexual Behavior in the Human Male,* 78. Kinsey claimed to have interviewed members of "Underworld Communities" in Chicago, Peoria, Indianapolis, New York City, and Gary, Indiana. Ibid., 16.

89. Pomeroy, *Kinsey,* 75; Kinsey, *Sexual Behavior in the Male,* 16. The 450 figure comes from Kinsey's associate Pomeroy, who wrote that "by the end of 1940 he [Kinsey] had recorded more than 450 homosexual histories." [Pomeroy, *Kinsey,* 75] Because Kinsey and his associates continued to collect histories from homosexual communities, this number would have been considerably larger by the time Kinsey's study of male sexual behavior was published in 1948.

90. Reisman, *Kinsey: Crimes and Consequences,* 97.

91. Kinsey, *Sexual Behavior in the Male,* 38.

92. Quoted in Reisman, *Kinsey: Crimes and Consequences,* 56.

93. Kinsey, *Sexual Behavior in the Male,* 103–4.

94. Ibid., 93–102.

95. Ibid., 95.

96. Kinsey, *Sexual Behavior in the Human Female,* 29–30; Kinsey, *Sexual Behavior in the Human Male,* 95, 103–4.

97. Jones, *Alfred Kinsey,* 513–16.

98. Paul H. Gebhard and Alan B. Johnson, *The Kinsey Data: Marginal Tabulations of the 1938–1963 Interviews Conducted by the Institute for Sex Research* (Philadelphia: W. B. Saunders Co., 1979).

99. Ibid., 8.

100. Ibid., 9.

101. Kinsey, *Sexual Behavior in the Human Male,* 597.

102. Tom W. Smith, "American Sexual Behavior: Trends, Socio-Demographic Differences, and Risk Behavior," GSS Topical Report No. 25, Version 5 (National Opinion Research Center, University of Chicago, April 2003), 108.

103. Kinsey, *Sexual Behavior in the Human Male,* 585.

104. Smith, "American Sexual Behavior," 54. Regarding current rates of infidelity, Smith reports that "the best estimates are that about 3–4% of currently married people have a sexual partner besides their spouse in a given year and about 15–18% of ever-married people have had a sexual partner other than their spouse while married." Ibid., 6.
105. Kinsey, *Sexual Behavior in the Human Male*, 651.
106. Smith, "American Sexual Behavior," 7, 57.
107. Kinsey, *Sexual Behavior in the Human Male*, 650.
108. Smith, "American Sexual Behavior," 64.
109. Kinsey, *Sexual Behavior in the Human Male*, 53, emphasis in the original.
110. This can be seen in Kinsey's advice on how to extract further information from "feeble-minded individuals." His main concern was to encourage them to admit more sexual activities, even if that involved trickery on the part of the interviewer: "it is sometimes effective to pretend that one has misunderstood the negative replies and ask additional questions, just as though the original answers were affirmatives; whereupon the subject may then expose the truth by answering as though he had never given a negative reply, 'Yes, I know you have never done that, but how old were you the *first* time that you did it' is a question which, amazingly enough, may break down the cover-up of a feeble-minded individual." (Ibid., 55, emphasis in the original) Kinsey did add the caveat that "with such a technique . . . it is especially important to make sure that the subject's final admissions are not fictions which the interviewer has suggested to him." But Kinsey did not make clear how an interviewer was supposed to do this, and his overall instructions focus much more on pressuring interview subjects to admit activities they may be covering up than on figuring out if they are inventing experiences to please the interviewer.
111. Ibid.
112. Ibid.
113. Ibid., 56.
114. See discussion of Maslow's research on this point, cited by Reisman, *Kinsey: Crimes and Consequences*, 113.
115. For a good summary of contemporary criticisms of Kinsey's volume on male sexual behavior, see Erdman Palmore, "Published Reactions to the Kinsey Report," in *An Analysis of the Kinsey Reports on Sexual Behavior in the Human Male and Female*, ed. Donald Porter Geddes (New York: New American Library, 1954), 267–84. For a specific example of a critical review by a contemporary scholar, see G. Legman, "Minority Report on Kinsey," in Norman Lockridge, *The Sexual Conduct of Men and Women: A Minority Report* (New York: Hogarth House, 1948), 13–32.
116. See discussion in Jones, *Alfred Kinsey*, 638–65. Kinsey associate Paul Gebhard acknowledged years later that Kinsey "mistrusted random probability sampling," and he only pretended to make changes to his methods in order to meet the objections of statisticians. According to Gebhard, Kinsey "felt . . . it would be wise to seem to accommodate to them [his statistician critics] but continue as before." Gebhard and Johnson, *The Kinsey Data*, 25, 26.
117. For a detailed discussion of the Balluseck case, including excerpts from contemporaneous German news accounts (in translation), see Judith Reisman, *Kinsey: Crimes and Consequences*, 164–68, 328–33.
118. "Samthanschuhe für Kinderschander Balluseck [Velvet Gloves for Pedophile Balluseck]," *Der Morgenpost*, May 19, 1957, quoted in Reisman, *Kinsey: Crimes and Consequences*, 328.
119. " 'Balluseck-Report' an Kinsey ["Balluseck-Report" on Kinsey]," *Der Tagenspiegel*, May 16, 1957, quoted in Reisman, *Kinsey: Crimes and Consequences*, 333.
120. "Kinsey hatte Balluseck anzeigen sollen [Kinsey ought to have reported Balluseck to the authorities]," *Der Morgenpost*, May 16, 1957, quoted in Reisman, *Kinsey: Crimes and Consequences*, 330.
121. Reisman, *Kinsey: Crimes and Consequences*, 165.
122. "Kinsey hatte Balluseck anzeigen sollen," 330.
123. Kinsey, *Sexual Behavior in the Human Male*, 176.
124. Ibid., 178.
125. Ibid., 179.
126. Ibid., 180.
127. Ibid., 176–77.
128. Ibid., 177, emphasis added.

129. Privately, one member of Kinsey's staff objected to the use of data from a pedophile, as did an academic critic of Kinsey. See Jones, *Alfred Kinsey*, 513, 592.

130. Judith Reisman and Edward W. Eichel, *Kinsey, Sex and Fraud: The Indoctrination of a People* (Lafayette, LA: Lochinvar-Huntington House, 1990); Reisman, *Kinsey: Crimes and Consequences*. For Reisman's history of her role in raising questions about Kinsey's research, see Reisman, *Kinsey: Crimes and Consequences*, xx-xxiv, 262–63 [n. 72].

131. John Bancroft,"Kinsey Institute Response to Indiana State Rep. Woody Burton's Resolution," (1998), http://www.indiana.edu/~kinsey/about/cont-burton.html (accessed July 27, 2005).

132. Ibid.

133. In addition, there is evidence that Kinsey engaged in sperm collection studies involving at least some children. See Reisman, *Kinsey: Crimes and Consequences*, 63, 137.

134. "Allegations about Childhood Data in the 1948 book, *Sexual Behavior in the Human Male*," http://www.indiana.edu/~kinsey/about/controversy%202.htm (accessed July 27, 2005).

135. Reisman, *Kinsey: Crimes and Consequences*, 135–36, 161–64.

136. Kinsey, *Sexual Behavior in the Human Male*, 177.

137. Bancroft,"Kinsey Institute Response."

138. Ibid.

139. "Allegations about Childhood Data."

140. Paul Gebhard, quoted in Reisman, *Kinsey: Crimes and Consequences*, 165.

141. For Esther White's story, see Reisman, *Kinsey: Crime and Consequences*, 151–52.

142. According to the Kinsey Institute's website: "We have no reason to doubt that this woman [Esther White] was sexually abused. However, Kinsey did not ask people to fill out questionnaires. It is conceivable that this woman's father or grandfather wrote to Kinsey, as many people have done. Following that documentary, we checked through Kinsey's correspondence and could not find any that would match this story. We do know that there have been people who have used Kinsey's name to justify what they do sexually, even recently." ("Allegations about Childhood Data.")

143. John Bancroft, "Kinsey and Children," (1998), http://www.indiana.edu/~kinsey/about/cont-akchild.html (accessed July 27, 2005).

144. See discussion in Reisman, *Kinsey: Crimes and Consequences*, 203–68; Pomeroy, *Kinsey*, 210–11; Linda Jeffrey, "Restoring Legal Protections for Women and Children: A Historical Analysis of the States' Criminal Codes," The State Factor (American Legislative Exchange Council, April 2004).

145. Fred Rodell, "Our Unlovable Sex Laws," in *The Sexual Scene*, ed. John H. Gagnon and William Simon (Trans-Action Books, 1970.), 89. This article was originally published in May/June 1965 edition of *Trans-action* magazine.

146. Rodell, "Our Unlovable Sex Laws," 83.

147. Morris Ploscowe, "Sexual Patterns and the Law," in *Sex Habits of American Men: A Symposium on the Kinsey Report*, ed. Albert Deutsch (New York: Prentice-Hall, 1948), 126.

148. Ernst and Loth, *American Sexual Behavior*, 128.

149. See "Model Penal Code, Tentative Draft No. 4" (April 25, 1955) in The American Law Institute, Model Penal Code, Tentative Drafts Nos. 1, 2, 3 and 4 (Philadelphia: The Executive Office, The American Law Institute, 1956), 206, 207, 208, 252, 259, 276, 279, 281, 282, 283.

150. Gathorne-Hardy, *Alfred Kinsey*, 449.

151. See "Table 1. Legal Ages of Consent in the United States, 1885 and 1920," in Mary E. Odem, *Delinquent Daughters: Protecting and Policing Adolescent Female Sexuality in the United States, 1885–1920* (Chapel Hill, NC: University of North Carolina Press, 1995), 14–15. The age of consent had been raised in many states starting in the late nineteenth century as a result of reform efforts spearheaded by members of the Woman's Christian Temperance Union and other groups.

152. *Model Penal Code and Commentaries, Part II, Definition of Specific Crimes*, §§ 210.0 to 213.6 (Philadelphia: The American Law Institute, 1980), Art. 213.1 (Rape and Related Offenses), 274–75; Art. 213.3 (Corruption of Minors and Seduction), 376; Art. 213.4 (Sexual Assault), 397–98.

153. "Model Penal Code, Tentative Draft No. 4," Comments §207.6, 295.

154. The state of California funded research by Bowman into sexual deviations. According to Kinsey associate Wardell Pomeroy, Bowman's research drew on Kinsey's data and was part of a larger effort by Kinsey to study sex offenders. (See Pomeroy, *Kinsey*, 210–11.) Bowman's findings were published in the Final Report on California Sexual Deviation Research, State of

California Department of Mental Hygiene, March 1954 (Sacramento: Printed by The Assembly of the State of California, 1954). As for whether the "recidivism" of pedophiles is truly "low," the answer depends partly on what one regards as "low" and how one defines "recidivism." A massive federal study of nearly ten thousand male sex offenders released from prisons in fifteen states in 1994 found that 5.3 percent of the offenders were rearrested for a sex offense within three years of release, and 3.5 percent of sex offenders were reconvicted of a sex crime. Among child molesters only, 5.1 percent were rearrested for a new sex crime within three years and 3.5 percent were reconvicted of a sex crime. (See Patrick A. Langan, Erica L. Schmitt, Matthew R. Durose, *Recidivism of Sex Offenders Released from Prison in 1994* [Washington, DC: U.S. Department of Justice, Office of Justice Programs, Bureau of Justice Statistics, November 2003], 24].) A meta-review of sixty-one research studies on sex offenders found that "the average sex offense recidivism rate (as evidenced by rearrest or reconviction) was 18.9 percent for rapists and 12.7 percent for child molesters over a four to five year period." ("Recidivism of Sex Offenders" [Center for Sex Offender Management, May 2001], http://www.csom.org/pubs/recidsexof.pdf [accessed July 27, 2005], 10.) What about recidivism over the long term? A 1997 study of more than two hundred sex offenders found that 52 percent of the 115 child molesters in the study were rearrested for sex offenses during a twenty-five-year period. (Ibid., 6.) Since it is estimated that nearly 90 percent of rapes and 70 percent of sexual assaults are never reported to the police, these recidivism calculations likely tell only part of the story. ("Myths and Facts About Sex Offenders" [Center for Sex Offender Management, August 2000], http://www.csom.org/pubs/mythsfacts.html [accessed July 27, 2005].) A 1988 study that tried to use "both unofficial and official measures of recidivism (reconviction, new charge, or unofficial record)" found that 43 percent of "extra-familial child molesters . . . sexually reoffended within a four-year follow-up period." ("Recidivism of Sex Offenders," 9.) Regardless of whether those convicted of sex crimes against children are rearrested for new sex offenses, they are likely to be rearrested for some type of crime and end up back in jail. According to the federal tracking study of nearly ten thousand released sex offenders, 39.4 percent of child molesters and 49.9 percent of statutory rapists were rearrested for some crime within three years of release. Moreover, 38.2 percent of child molesters and 45.7 percent of statutory rapists were "[r]eturned to prison with or without a new sentence" within three years. (Langan, et. al., *Recidivism of Sex Offenders Released from Prison in 1994*, 15.) Non-sex offenders do appear to have lower overall rearrest and reconviction rates than sex offenders for all crimes (non-sex offenses as well as sex offenses). ("Myths and Facts.") But this data is not particularly relevant for weighing the relative risks of sex offenders and non-sex offenders to commit new sex offenses. According to the federal tracking study, released sex offenders in general are roughly four times more likely to be arrested for a sex crime than released non-sex offenders. Similarly, released child molesters are more than five times more likely to be arrested for a sex crime against a child than released non-sex offenders. (Langan, et. al., *Recidivism of Sex Offenders Released from Prison in 1994*, 11.) Thus, it is clear that sex offenders pose a much greater threat to the community of future sex crimes than other criminal offenders. Given this data, most people probably would not find the arguments on this point offered by drafters of the MPC as especially convincing.

155. See Carolyn E. Cocca, "The Politics of Statutory Rape Laws: Adoption andd Reinvention of Morality Policy in the States, 1971–1999," *Polity* 35, No. 1 (2002): 51ff. Retrieved through www.questia.com (accessed July 18, 2007).

156. Linda Brookover Bourque, Defining Rape (Durham, NC: Duke University Press, 1989), 116; also see Michelle Oberman, "Turning Girls into Women: Re-evaluating Modern Statutory Rape Law (Symposium: Gender Issues and the Criminal Law)," *Journal of Criminal Law and Criminology* 85, No. 1 (Summer 1994): 15–79.

157. For an overview and discussion of the case, see Oberman, "Turning Girls into Women."

158. Reisman, *Kinsey: Crimes and Consequences*, 205.

159. Ibid.

160. Dennis L. Carlson, "Ideological Conflict and Change in the Sexuality Curriculum," in *Sexuality and the Curriculum: The Politics and Practices of Sexuality Education*, ed. James T. Sears (New York: Teachers College Press, 1992), 51.

Chapter 13: Sex Miseducation

1. Kinsey, *Sexual Behavior in the Human Male*, 223.
2. Kinsey, *Sexual Behavior in the Human Female*, 16.
3. See Lester A. Kirkendall, "The Journey Toward SIECUS: 1964 A Personal Odyssey," *SIECUS Report* 17, no. 4 (March/April, 1989): 18–21; "Mary S(teichen) Calderone, 1904–1998," *Contemporary Authors Online* (Gale, 2005), reproduced in *Biography Resource Center* (Farmington Hills, MI: Thomson Gale, 2005), http://galenet.galegroup.com/servlet/BioRC (accessed July 26, 2005); "Lester A(llen) Kirkendall, 1903–1983," *Contemporary Authors Online* (Gale, 2005), reproduced in *Biography Resource Center* (accessed July 26, 2005).
4. See "Wardell B. Pomeroy, 1913–2001," *Contemporary Authors Online* (Gale, 2005), reproduced in *Biography Resource Center* (accessed July 26, 2005); "Harriet F(leischl) Pilpel, ?–1991," *Contemporary Authors Online* (Gale, 2005), reproduced in *Biography Resource Center* (accessed July 26, 2005).
5. "The SIECUS purpose," quoted in Mary Breasted, *Oh! Sex Education!* (New York: Praeger Publishers, 1970), 207.
6. Quoted in Howard A. Rusk, "Sex Education in U.S.," *New York Times*, October 5, 1969, 95.
7. Mary S. Calderone, "Sex and Social Responsibility," in *Sex Education: Issues and Directives*, ed. G. Pat Powers and Wade Baskin (New York: Philosophical Library, 1969), 104. This essay was originally published in 1965 in the *Journal of Home Economics*.
8. Mary S. Calderone, "The Development of Healthy Sexuality," in Powers and Baskin, *Sex Education*, 92. This essay was adapted from a speech Calderone originally gave in 1966.
9. Ibid., 93.
10. Breasted, *Oh! Sex Education!*, 234–239.
11. Quoted in John Leo, "Fight on Sex Education is Widening," *New York Times*, May 29, 1969, 49.
12. Lawrence Feinberg, "Md. Mother Charged Sex Lecture Brainwashed Her 9th Grade Son," *Washington Post*, July 30, 1969, C2.
13. Douglas Robinson, "Sex Education Battles Splitting Many Communities Across U.S.," *New York Times*, Sept. 14, 1969, 1.
14. Quoted in Richard Cohen, "Foes Fail to End Sex Education Spread," *Washington Post*, June 3, 1969, C3.
15. Ibid.
16. Robinson, "Sex Education Battles," 77.
17. Billy James Hargis, quoted in Fred M. Hechinger, "Storm Over the Teaching of Sex," *New York Times*, Sept. 7, 1969, E11.
18. Robinson, "Sex Education Battles," 77.
19. Cohen, "Foes Fail to End Sex Education," C1.
20. Wesley McCune, quoted in ibid.
21. Robinson, "Sex Education Battles," 77.
22. Quoted in Jim Shoop, " 'Tear Gas' Ends Hearing on Sex Education Bill," *Minneapolis Star*, April 24, 1969, 1B.
23. *Sexuality and Man: Compiled and Edited by Sex Information and Education Council of the United States* (New York: Charles Scribner's Sons, 1970).
24. Lester Kirkendall and Isadore Rubin, chap. 1, "Sexuality and the Life Cycle," in *Sexuality and Man*, 7–8.
25. Ibid., 11–12.
26. Ibid., 13–14.
27. Ibid., 16.
28. Isadore Rubin, chap. 6, "Homosexuality," 74.
29. Cornelia V. Christensen and Wardell B. Pomeroy, chap. 2, "Characteristics of Male and Female Sexual Responses," in *Sexuality and Man*, 34.
30. Ira L. Reiss, chap. 3, "Premarital Sexual Standards," in *Sexuality and Man*, 41.
31. John H. Gagnon and William Simon, chap. 7, "Sexual Encounters between Adults and Children," in *Sexuality and Man*, 85–86.
32. Ibid., 83.
33. Ibid., 95.

34. Ibid., 94–95, 97–98.
35. Ibid., 86.
36. Ibid., 90.
37. Ibid., 88–89.
38. Ibid., 88.
39. Harold T. Christensen, chap. 10, "Sex, Science, and Values," in *Sexuality and Man*, 139–40.
40. Lester Kirkendall, chap. 9, "Sex Education," in *Sexuality and Man*, 131.
41. Christenson, "Sex, Science, and Values," 150.
42. Isadore Rubin, chap. 11, "The Sex Educator and Moral Values," in *Sexuality and Man*, 171.
43. Quoted in ibid. Despite his advocacy of relativism on such issues as premarital sex, Isadore Rubin had a list of values that he thought should be promoted as morally correct. He thought that sex education should promote "universal" values such as "life itself, freedom from chronic pain or discomfort, freedom from severe fear or anxiety, freedom to make choices, and the condition of loving and being loved." Ibid., 161.
44. Reiss, "Premarital Sexual Standards," 45.
45. Christenson, "Sex, Science, and Values," 145.
46. Ibid., 141.
47. Ibid., 151.
48. Ibid., 145.
49. Kirkendall, "Sex Education," 125.
50. Lester A. Kirkendall and Robert N. Whitehurst, eds., *The New Sexual Revolution* (New York: Donald W. Brown, 1971).
51. Paul Kurtz, preface, "Humanism and the Sexual Revolution," in Kirkendall and Whitehurst, *The New Sexual Revolution*, ix.
52. Herb Seal, "Cross Cultural Sexual Practices," in Kirkendall and Whitehurst, *The New Sexual Revolution*, 19.
53. Ibid., 20.
54. Ibid., 21.
55. Ibid., 23.
56. Albert Ellis, "A Rational Sexual Morality," in Kirkendall and Whitehurst, *The New Sexual Revolution*, 51.
57. For biographical information about Ellis, see "A Brief Biography of Dr. Albert Ellis," http://www.rebt.ws/albertellisbiography.html (accessed July 26, 2005); "Albert Ellis," *Marquis Who's Who* (Marquis Who's Who, 2005), reproduced in *Biography Resource Center* (Farmington Hills, MI: Thomson Gale, 2005), http://galenet.galegroup.com/servlet/BioRC (accessed July 26, 2005).
58. Ellis, "A Rational Sexuality Morality," 53.
59. Ibid., 55.
60. Ibid.
61. Rustum and Della Roy, "Is Monogamy Outdated?" in Kirkendall and Whitehurst, *The New Sexual Revolution*, 131–48.
62. Ibid., 141–42.
63. Ibid., 142–43.
64. Ibid., 144.
65. Ibid., 144–45.
66. Lester A. Kirkendall, "Developing Human Sexuality," in Kirkendall and Whitehurst, *The New Sexual Revolution*, 151.
67. Ibid., 163.
68. For a brief biography, see Leah Cahan Schaefer, "Kirkendall, Lester A.," in *Human Sexuality: An Encyclopedia*, ed. Vern L. Bullough and Bonnie Bullough (New York: Garland Publishing, 1994), available online at http://www2.hu-berlin.de/sexology/GESUND/ARCHIV/SEN/CH15.HTM (accessed August 3, 2005).
69. Leonard Ramer, *Your Sexual Bill of Rights* (New York: Exposition Press, 1973), 78.
70. Ibid., 76.
71. Ibid., 80.
72. Ibid., 103–4.
73. Wardell B. Pomeroy, *Boys and Sex* (New York: Delacorte Press, 1968); Wardell B. Pomeroy, *Girls and Sex* (New York: Delacorte Press, 1969).

74. Ibid., 17.

75. Ibid., 36–47, 63.

76. Ibid., 108. The biggest problem with premarital sex, wrote Pomeroy, is the possibility of guilt. See ibid., 109–10. But much of his two books are devoted to getting young people to get over such feelings. Pomeroy does seem to have a slightly more cautious attitude about girls engaging in premarital sex than boys. In his book for girls, he notes that having intercourse is a major life event. "It is her crossing of the Rubicon, from which there is no turning back." Pomeroy, *Girls and Sex*, 108.

77. "Couples who have made a decision to have intercourse need to exercise their ingenuity in finding the best place to have it, where the chances of discovery will not be great . . . In the end, it seems, places like woods, a beach, or a secluded area one can walk to is likely to work out better than any of the others I have mentioned." Pomeroy, *Girls and Sex*, 107–8.

78. Pomeroy, *Boys and Sex*, 63.

79. Eleanor Hamilton, *Sex Before Marriage: Guidance for Young Adults—Ages 16 to 20* (New York: Meredith Press, 1969).

80. Ibid., v.

81. Ibid., 26–28.

82. Ibid., 19–20.

83. Ibid., 36.

84. Ibid., 41.

85. Ibid., 32, 38, 43–44.

86. Ibid., 38.

87. Helen Manley, "Sex Education: Where, When and How Should It Be Taught?" in Powers and Baskin, *Sex Education*, 73, 74, 75. Also see the "General Objectives of Family Life and Sex Education" listed in H. Frederick Kilander, *Sex Education in the Schools* (New York: Macmillan, 1970), 73. Objective number 8 is: "To emphasize the case for premarital chastity as the sexual standard approved by our society because chastity provides a positive goal for teenagers, linking human sexual behavior with love, marriage, parenthood, and family life and because of the individual, family and community problems associated with premarital or extramarital sexual relations."

88. Helen Manley, "Outline of a Sequential Program: Questions, Answers and Resources," in Powers and Baskin, *Sex Education*, 364.

89. S.P. Marland, "Placing Sex Education in the Curriculum," in Powers and Baskin, *Sex Education*, 400.

90. Florence Knepp Paterson, "How We Do It: Presenting a Unit on Human Reproduction," in Powers and Baskin, *Sex Education*, 422. This article was reprinted from a 1962 issue of *The American Biology Teacher*.

91. Arthur H. Steinhaus, "Teaching the Role of Sex in Life," in Powers and Baskin, *Sex Education*, 338. This article was originally printed in the *Journal of School Health*, October 1965.

92. Hilda Bryant, "Sex-Ed Film 'Inappropriate, But Not Porno,'" *Seattle Post-Intelligencer*, February 11, 1981, A1. Also see, Mike Layton, "Moral Majority to Sue Library," A3.

93. "Moral Majority Halts Lawsuit on Sex Education Film," *New York Times*, February 24, 1981, A14.

94. Gordon Jensen, *Youth and Sex: Pleasure and Responsibility* (Chicago: Nelson-Hall Company, 1973), 76; also see Patty Campbell, *Sex Guides: Books and Films About Sexuality for Young Adults* (New York: Garland Publishing, 1986), 161–62. Jensen also stated that "there are some strong arguments against it [premarital sex], too," but he proceeded to dismiss them (Jensen, *Youth and Sex*, 77). He argued that birth control can prevent unintended pregnancies and that claims that premarital sex will lead to weaker marriages are "beliefs . . . rather than substantiated facts." He further asserted that claims that premarital sex "will weaken the love relationship" were contradicted by the comments of "many unmarried girls" he interviewed (ibid., 78).

95. Heidi Handman and Peter Brennan, *The Sex Handbook: Information and Help for Minors* (1974), quoted in Campbell, *Sex Guides*, 163.

96. Eleanor Hamilton, *Sex, with Love: A Guide for Young People* (Boston: Beacon Press, 1978).

97. Alex Comfort and Jane Comfort, *The Facts of Love: Living, Loving, and Growing Up* (New York: Crown Publishers, 1979); Campbell, *Sex Guides*, 198. Assuming that young people take the proper precautions against disease and pregnancy, the only problem with recreational sex

according to the Comforts was if one partner did not really want it. The Comforts said it was wrong "to treat another person as a thing simply because you like their body and want to play with them, without making sure they want that too." Comfort and Comfort, *The Facts of Love*, 57. In other words, if both teenage partners wanted to have sex "simply because [they] like[d] their bod[ies] and want[ed] to play with them," that would be perfectly fine.

98. "Reading pornography won't hurt you . . . Some adults flip if they see children with this kind of reading matter. If the children have been taught the facts about sex, these adults really need not worry." Comfort and Comfort, *The Facts of Love*, 118.

99. Ibid., 85.

100. Ruth Bell, et. al., *Changing Bodies, Changing Lives: A Book for Teens on Sex and Relationships* (New York, Random House, 1980), 75–123.

101. Campbell, *Sex Guides*, 207.

102. Bell, *Changing Bodies*, 85.

103. "Quite a number of people, about 10 percent, are mainly attracted to people of their own sex." Ibid., 112.

104. Ibid., 54.

105. Ibid., 55.

106. Ibid., 56.

107. Judy Blume, *Forever* (Scarsdale, NY: Bradbury Press, 1975).

108. Campbell, *Sex Guides*, 217; see detailed discussion of Blume book by Campbell on 217–25.

109. Ibid., 188.

110. David R. Stronck, ed., *Discussing Sex in the Classroom: Readings for Teachers* (Washington, DC: National Science Teachers Association, 1982).

111. Adele D. Hofmann, "Adolescents, Sex, and Education," in Stronck, *Discussing Sex*, 15–16.

112. Ibid., 17.

113. Ibid., 18.

114. Peggy Brick, "Sex and Society: Teaching the Connection," in Stronck, *Discussing Sex*, 43.

115. Ibid., 44.

116. Peter Scales, "The New Opposition to Sex Education: A Powerful Threat to Democracy," in Stronck, *Discussing Sex*, 19.

117. Ibid.

118. Lorna Brown, ed., *Sex Education in the Eighties: The Challenge of Healthy Sexual Evolution* (New York: Plenum Press, 1981).

119. Ibid., v.

120. Mary S. Calderone, "From Then to Now—and Where Next?" in Brown, *Sex Education in the Eighties*, 249.

121. Floyd Martinson, "The Sex Education of Young Children," in Brown, *Sex Education in the Eighties*, 63–71.

122. Quoted in Martinson, "The Sex Education of Young Children," 67.

123. Ibid.

124. Ibid., 69.

125. Ibid., 70.

126. Ibid., 71.

127. "In the professional literature it is more and more apparent that sexuality is coming to be viewed as a competence, the capacity for which appears in infancy and the learning of which should proceed from birth and through childhood . . . Ethnological data from permissive societies is remarkably consistent as to the behavior of infants and children who are allowed to proceed at their own rate of exploration and development. From infancy on the children are free to observe nudity and adult sexual activity; their own activity begins with the earliest of genital fondling and eventually proceeds to include every kind of sexual behavior—masturbation, homosexual, heterosexual, and group play, and intercourse . . ." Ibid., 76.

128. Gary F. Kelly, "Parents as Sex Educators," in Brown, *Sex Education in the Eighties*, 111.

129. Harriet F. Pilpel and Laurie R. Rockett, "Sex Education and the Law," in Brown, *Sex Education in the Eighties*, 22–30.

130. Scales, "The New Opposition to Sex Education," 19.

131. For background on the AFLA see Rebekah Saul, "Whatever Happened to the Adolescent Family Life Act?" *The Guttmacher Report on Public Policy* 1, no. 2, April 1998, http://www.agi-usa.

org/pubs/journals/gr010205.html (accessed July 26, 2005); "Exclusive Purpose: Abstinence-Only Proponents Create Federal Entitlement in Welfare Reform," *Siecus Report* 24, no. 4 (Sexuality Information and Education Council of the United States), http://www.siecus.org/policy/SReport/srep0001.html (accessed July 26, 2005); *Toward a Sexually Healthy America: Roadblocks Imposed by the Federal Government's Abstinence-Only Until-Marriage Education Program* (Advocates for Youth and Sexuality Information Education Council of the United States, 2001), 5; *Bowen v. Kendrick*, 487 U.S. 589 (1988). The relevant provisions of the Adolescent Family Life Act can be found in 42 USC § 300z (2004).

132. From 1971–2005, the federal government spent nearly $6 billion on family planning programs under what is known as Title X, while from 1982–2004 it spent roughly $368 million on programs funded under the AFLA. The first year in which the AFLA was funded (1982), it received $11 million v. $124 million for Title X family planning programs. In 2004, the disparity was $30.7 million for AFLA programs v. $288 million for Title X family planning. See "Office of Family Planning, Funding History, FY 1971–2004," United States Department of Health and Human Services, Office of Public Health and Science, Office of Population affairs, http://opa.osophs.dhhs.gov/titlex/ofp-funding-history.html (accessed July 26, 2005); "Office of Adolescent Pregnancy Programs, Funding History, FY 1982–2005," United States Department of Health and Human Services, Office of Public Health and Science, Office of Population Affairs, http://opa.osophs.dhhs.gov/titlexx/oapp-funding-history.html (accessed July 26, 2005). Family planning programs are funded by the federal government under what is known as "Title X." See "Fact Sheet: America's Family Planning Program: Title X," Planned Parenthood Federation of America, http://www.plannedparenthood.org/library/FAMILYPLANNINGIS-SUES/TitleX_fact.html (accesed July 26, 2005). It should be noted that Title X is only one way the federal government supports family planning services; there are a variety of other federal programs which may be used to support such efforts. For an overview of all government spending on family planning, see Melissa G. Pardue, Robert E. Rector, and Shannan Martin, "Government Spends $12 on Safe Sex and Contraceptives for Every $1 Spent on Abstinence," The Heritage Foundation Backgrounder #1718 January 14, 2004, http://www.heritage.org/research/family/bg1718.cfm (accessed July 26, 2005).

133. William J. Bennett, "Sex and the Education of Our Children," in *School Based Clinics and Other Critical Issues in Public Education*, ed. Barrett L. Mosbacker (Westchester, IL: Crossway Books, 1987), 158.

134. John Leo, "Cradle-to-Grave Intimacy," *Time*, September 7, 1981, 69.

135. Title V, Sec. 510 42 U.S.C. 701; also see Ron Haskins and Carol Statuto Bevan, "Abstinence Education Under Welfare Reform," *Children and Youth Services Review* 19, no. 5/6 (1997): 465–84, posted at http://www.aei.org/publication17765 (accessed July 26, 2005); Pardue, et. al., "Government Spends $12 on Safe Sex and Contraceptives for Every $1 Spent on Abstinence."

136. Robert E. Rector, "Facts about Abstinence Education," *The Heritage Foundation Web Memo #461*, March 30, 2004, http://www.heritage.org/Research/Welfare/wm461.cfm (accessed July 26, 2005); also see Pardue, et. al., "Government Spends $12 on Safe Sex and Contraceptives for Every $1 Spent on Abstinence."

137. Barbara Devaney, Amy Johnson, Rebecca Maynard, Chris Trenholm, "The Evaluation of Abstinence Education Programs Funded Under Title V Section 510: Interim Report," (Mathematica Policy Research, Inc., 2002), http://aspe.hhs.gov/hsp/abstinence02/ (accessed July 26, 2005).

138. Ibid.

139. Robert E. Rector, Melissa G. Pardue, and Shannan Martin, "What Do Parents Want Taught In Sex Education Programs?" *The Heritage Foundation Backgrounder #1722*, January 28, 2004, http://www.heritage.org/Research/Welfare/bg1722.cfm (accessed July 26, 2005).

140. *Guidelines For Comprehensive Sexuality Education, Kindergarten–12th Grade*, 2nd ed. (Sexuality Information and Educational Council of the United States, 1996), 3.

141. Ibid., 14, 16, 20, 29, 30.

142. Ibid., 16.

143. Ibid.

144. Ibid., 30.

145. Ibid., 35.

146. Ibid., 30.

147. Ibid., 32.

148. Ibid., 16.
149. Ibid.
150. Ibid., 31.
151. Ibid., 32.
152. Ibid.
153. Ibid., 7.
154. Ibid., 18.
155. Ibid., 16–17.
156. Ibid., 35.
157. Ibid.
158. Although the guidelines do not call for learning about fetal development during the section on abortion, they do mention in the section on reproduction that students should learn that "[d]uring pregnancy, the fetus develops during a 40-week cycle that ends at birth." Ibid., 14. Conceivably, a teacher could use this standard to get into the specifics of fetal development, but there is no direction to include this as part of the discussion of abortion.
159. Ibid., 35.
160. Ibid., 42.
161. Ibid.
162. Although most Americans believe in legal equality for homosexuals, the vast majority of practicing religious adherents continue to believe that homosexual behavior is morally wrong. For example, according to Gallup surveys 58 percent of Catholics who attend weekly religious services consider homosexual behavior to be morally wrong, as do 80 percent of Protestants who attend weekly services. (See Jeffrey M. Jones, "Preaching to Another Church's Choir: Practicing Protestants' Moral Views More Consistent with Catholic Church," April 26, 2005, http://brain.gallup.com/content/default.aspx?ci=16027 [accessed July 26, 2005].) In fact, according to Gallup surveys taken in 2001, 2002, 2003, 2004, and 2005, between 52.3 and 55.2 percent of all Americans "personally believe that in general [homosexual behavior] . . . is . . . morally wrong." (Gallup Poll Social Series—Values and Beliefs, q39q_FormA, 5/2/2005-5/5/2005; qn39f, 5/2/2004-5/4/2004; qn35f, 5/5/2003-5/7/2003; qn56F, 5/6/2002-5/9/2002; qn49E_2, 5/10/2001-5/14/2001, http://brain.ballup.com [accessed July 26, 2005].) Given such data, it is false to create the impression that there is no clear consensus on this issue among religious believers. Moreover, the data from these public opinion surveys track with the official positions of America's largest religious groups. Catholic teaching considers homosexual activity to be objectively disordered. (See Congregation for the Doctrine of the Faith, "Letter to the Bishops of the Catholic Church on the Pastoral Care of Homosexual Persons," October 1, 1986, http://www.vatican.va/roman_curia/congregations/cfaith/documents/rc_con_cfaith_doc_19861001_homosexual-persons_en.html [accessed July 27, 2005].) The United Methodist Church's *Book of Discipline* classifies "the practice of homosexuality" as "incompatible with Christian teaching." (See "Human Sexuality," selection from the *United Methodist Book of Discipline* (2004), http://archives.umc.org/interior.asp?mid=1728 [accessed July 27, 2005].) The Presbyterian Church (U.S.A.) forbids the ordination of practicing homosexuals to church offices, requiring church officials "to live either in fidelity within the covenant of marriage between a man and a woman . . . or chastity in singleness." (Standard G-6.0106, The Constitution of the Presbyterian Church (U.S.A.), Part II: Book of Order, 2004–2005 [Office of the General Assembly, Presbyterian Church (U.S.A.)], 2004, http://www.pcusa.org/oga/publications/boo-04-05.pdf [accessed July 27, 2005].) The Southern Baptist Convention classifies homosexuality as a form of "sexual immorality." (*The Baptist Faith and Message* [2000], Section XV, The Christian and the Social Order, http://www.sbc.net/bfm/bfm2000.asp#xv [accessed July 27, 2005].) Only a tiny number of American religious denominations actually maintain that homosexual activity is morally acceptable, and some of those church bodies are in open conflict with members of their churches in the rest of the world. For example, the Episcopal Church in the United States has ordained practicing homosexuals in the face of overwhelming opposition from most of the rest of the Anglican communion. At the Lambeth Conference of 1998 representatives of the seventy million Anglican adherents worldwide adopted a resolution "rejecting homosexual practice as incompatible with Scripture." (Resolution I.10, Human Sexuality, http://www.anglicancommunion.org/lambeth/1/sect1rpt.html [accessed July 27, 2005].) The vote in favor of the resolution was a lopsided 570 to 70, with 45 abstentions. (*The Lambeth Daily,*

issue 14, August 6, 1998, archived at http://justus.anglican.org/resources/Lambeth1998/Lambeth-Daily/06/sextext.html [accessed July 27, 2005].)

163. For example: "Many boys and girls begin to masturbate for sexual pleasure during puberty." Ibid., 30. "Most people have masturbated at some time in their lives." Ibid. "Many people experience sexually erotic thoughts called fantasies." Ibid., 32. "Most women report no problems after having an abortion." Ibid., 35.

164. Ibid., 37.

165. Ibid., 31.

166. Ibid.

167. "Why Do You Encourage Sex without Love?" http://www.positive.org/Home/faq/love.html (accessed June 16, 2004).

168. "Just Say Yes Intro," http://www.positive.org/JustSayYes/index.html (accessed June 16, 2004).

169. "Sexual Politics: [Advice about losing virginity]."

170. "Is It Really Okay to Like Pornography?" http://www.scarleteen.com/boyfriend/porn_5.html (accessed June 17, 2004).

171. Ashley Rondini, "Abstinence: Who, What, When, Why, and How!" (April 30, 2003), http://www.teenwire.com/warehous/articles/wh_20030430p177_abstinence.asp (accessed June 16, 2004).

172. Minna Dubin, "Are You Experienced?" (June 8, 2004), http://www.teenwire.com/infocus/2004/if_20040608p296_experience.asp (accessed June 16, 2004).

173. Kristy Castora, "Slut . . . Just Another Four-Letter Word?" (January 22, 2002), http://www.teenwire.com/infocus/articles/if_20020122p144.asp (accessed June 16, 2004)

174. Sandy Olkowski, "You Got the Hookup" (January 15, 2003), http://www.teenwire.com/warehous/articles/wh_20030115p165_hookup.asp (accessed June 16, 2004); "On Again, Off Again" (March 2, 2004), http://www.teenwire.com/warehous/articles/wh_20040302p214_relationship.asp; "Playing the Field," http://www.teenwire.com/hothouse/issue_140_141/hh140_141_20020625p001.asp (accessed June 16, 2004).

175. "Ask The Experts: Is Having Sexual Curiosity about a Family Member Normal?" http://www.teenwire.com/ask/articles/as_20000421p041.asp (accessed June 16, 2004).

176. "Bestiality" (April 14, 2000), http://www.goaskalice.columbia.edu/0707.html (accessed June 16, 2004); "Sex with Animals and STDs?" (April 14, 2000), http://www.goaskalice.columbia.edu/1056.html (accessed June 17, 2004).

177. Elisa Klein, "The Truth about Virginity Pledges" (May 11, 2004), http://www.teenwire.com/infocus/2004/if_20040511p288_virginity.asp (accessed June 16, 2004).

178. "Making a private pledge to wait to have sexual intercourse until one is older significantly reduced an adolescents' risk of initiating vaginal intercourse over a 1-year period, even after controlling for important demographic and psychosocial variables. Adolescents who did not make a private pledge to wait until they were older to have sexual intercourse were nearly two and a half times more likely to initiate sexual intercourse than those who had made a promise." Melina M. Bersamin, Samantha Walker, Elizabeth D. Waiters, Deborah A. Fisher, and Joel W. Grube, "Promising to Wait: Virginity Pledges and Adolescent Sexual Behavior," *Journal of Adolescent Health* 36 (2005), 433. Virginity pledges are also associated with reduced risk of sexually transmitted diseases, contrary to the claims of one widely reported study. See Robert Rector and Kirk A. Johnson, "Adolescent Virginity Pledges, Condom Use, and Sexually Transmitted Diseases Among Young Adults," June 14, 2005, http://www.heritage.org/Research/Abstinence/whitepaper06142005-1.cfm (accessed Feb. 24, 2007).

179. Joyce, "Outercourse: Abstinence for Experts" (December 1, 1998), http://www.teenwire.com/warehous/articles/wh_19981201p053.asp (accessed June 16, 2004).

180. "Urging Teens to Stop Short of Vaginal Sex Could Cut Teen Pregnancy Rate" (May 9, 2004) http://www.teenwire.com/flash/fl_mainpage.asp (accessed June 8, 2004).

181. "Ask the Experts: Can My Girlfriend and I Masturbate Each Other?" (April 4, 2003), http://www.teenwire.com/ask/2003/as_20030404p542_masturbate.asp (accessed June 16, 2004).

182. For a good summary of these kinds of arguments, see *Toward a Sexually Healthy America*, 7–8, 16–21.

183. For the survey data cited in this paragraph, see Rector, et. al., "What Do Parents Want Taught In Sex Education Programs?" The poll was commissioned by Focus on the Family.

184. "Parents Prefer Abstinence Education 2 to 1," May 3, 2007, National Abstinence Education Association, http://www.abstinenceassociation.org/newsroom/pr_050307_parents_prefer_2to1.html (accessed July 19, 2007).

185. See "Federal Abstinence-Only-Until-Marriage Programs Not Proven Effective in Delaying Sexual Activity Among Young People," April 13, 2007, http://www.siecus.org/media/press/press0141.html (accessed July 19, 2007); Christopher Trenholm, Barbara Devaney, Ken Fortson, Lisa Quay, Justin Wheeler, Melissa Clark, "Impacts of Four Title V, Section 510 Abstinence Education Programs: Final Report," April 2007, Mathematica Policy Research, Inc., http://www.mathematica-mpr.com/publications/PDFs/impactabstinence.pdf (accessed July 18, 2007); Stan E. Weed, Irene H. Ericksen, Paul J. Birch, Joseph M. White, Matthew T. Evans, Nicole E. Anderson, "'Abstinence' or 'Comprehensive' Sex Education?—The Mathematica Study in Context," June 8, 2007, The Institute for Research and Evaluation, http://www.abstinenceassociation.org/docs/Abstinence_vs_Comprehensive_Sex_Ed.pdf (accessed July 19, 2007).

186. "Abstinence Works! Studies Validating the Efficacy of Abstinence Education," National Abstinence Education Association, April 12, 2007, http://www.abstinenceassociation.org/docs/NAEA_Abstinence_Works_041207.pdf (accessed July 19, 2007); Rector, "Facts about Abstinence Education"; also, Robert Rector, "The Effectiveness of Abstinence Education Programs in Reducing Sexual Activity Among Youth." The Heritage Foundation Backgrounder No. 1533, April 8, 2002, http://www.heritage.org/Research/Family/BG1533.cfm (accessed July 26, 2005); Melissa G. Pardue, "More Evidence of the Effectiveness of Abstinence Education Programs," May 5, 2005, http://www.heritage.org/Research/Abstinence/wm738.cfm (accessed Feb. 24, 2007).

187. See discussion in Breasted, *Oh! Sex Education!*, 328–30. Breasted makes the point that in the early years of the sex-education controversy both proponents and opponents of sex education made claims that outstripped the empirical evidence. Among the proponents who overstepped the evidence she identified Mary Calderone of SIECUS.

188. Weed, "'Abstinence' or 'Comprehensive' Sex Education?" 3.

189. Shannan Martin, Robert Rector, and Melissa G. Pardue, *Comprehensive Sex Education vs. Authentic Abstinence: A Study of Competing Curricula* (Washington, DC: Heritage Foundation, 2004), ix. This report is available online at http://www.heritage.org/Research/Welfare/abstinencereport.cfm (accessed July 26, 2005).

190. Ibid., vi.

191. Ibid., v, vi, viii.

192. For information about the campaign to require sex education curricula to be "medically accurate," see "Action across the Nation on Bills Requiring Medically Accurate Sexuality Education," *Policy Update*, February 2003, Sexuality Information and Education Council of the United States, http://www.siecus.org/policy/PUpdates/arch03/arch030049.html (accessed July 26, 2005); "Medical Accuracy Bill Introduced in U.S. House of Representatives," *Policy Update*, March 2003, Sexuality Information and Education Council of the United States, http://www.siecus.org/policy/PUpdates/arch03/arch030051.html (accessed July 26, 2005); "California Governor Signs Sexuality Education Bill," *Policy Update*, September 2003, Sexuality Information and Education Council of the United States, http://www.siecus.org/policy/PUpdates/arch03/arch030074.html (accessed July 26, 2005); "Family Life Education Act Introduced In the House of Representatives," *Policy Update*, April 2004, Sexuality Information and Education Council of the United States, http://www.siecus.org/policy/PUpdates/pdate0099.html (accessed July 26, 2005); "2005 State Sexual and Reproductive Health Legislative Reports," Sexuality Information and Education Council of the United States, http://www.siecus.org/policy/legislative/index.html (accessed July 26, 2005).

193. *Toward a Sexually Healthy America*, 15.

194. Ibid., 14.

195. For data on some of the consequences of childbearing out of wedlock, see Robert E. Rector, Patrick F. Fagan, and Kirk A. Johnson, "Marriage: Still the Safest Place For Women and Children," *The Heritage Foundation Backgrounder* #1732, March 9, 2004, http://www.heritage.org/Research/Family/bg1732.cfm (accessed July 26, 2005); and Patrick F. Fagan, Robert E. Rector, Kirk A. Johnson, and America Peterson, *The Positive Effects of Marriage: A Book of Charts* (Washington, DC: The Heritage Foundation, April 2002), http://www.heritage.org/research/features/marriage/index.cfm (accessed July 26, 2005).

196. *Toward a Sexually Healthy America*, 15.

197. This argument is highly questionable, and SIECUS presents no empirical data to corroborate it. Far from harming child victims of sex abuse, teaching that sexual relationships outside of

marriage are clearly inappropriate could reassure such children that what was done to them was wrong and that the blame rests squarely on the shoulders of the perpetrators.

198. See *Review of Comprehensive Sex Education Curriculum* (Washington, DC: The Administration for Children and Families, U.S. Department of Health and Human Services, May 2007), 7–8; *Straight from the Source: What So-Called "Comprehensive" Sex Education Teaches to America's Youth* (Washington, DC: National Abstinence Education Association, June 2007), 2–3.

199. "Youth Risk Behavior Surveillance—United States, 2005," *Morbidity and Mortality Weekly Report* (Centers for Disease Control and Prevention), June 9, 2006, 55, no. SS-5, 31.

Chapter 14: Redefining Life

1. Testimony of Dr. James Neel, May 20, 1981, in *The Human Life Bill: Hearings before the Subcommitte on Separation of Powers of the Committee on the Judiciary, United States Senate, Ninety-Seventh Congress, First Session, on S. 158, a Bill to Provide that Human Life Shall be Deemed to Exist from Conception, April 23, 24; May 20, 21; June 1, 10, 12 and 18.* Serial No. J-97-16 (Washington, D.C.: U.S. Government Printing Office, 1982), 77.

2. Wells, *Icons of Evolution*, 87–107; Stephen Jay Gould, *Ontogeny and Phylogeny* (Cambridge, MA: Belknap Press of Harvard University Press, 1977), especially 76–85.

3. Wells, *Icons of Evolution*, 106.

4. Lewis Wolpert, *The Triumph of the Embryo* (Oxford: Oxford University Press, 1991), 185.

5. See Wells, *Icons of Evolution*, 88–90, 101–9.

6. Gould, *Ontogeny and Phylogeny*, 211.

7. Ibid., 4. It should be noted that Gould deplored the redefinition of the term "recapitulation" as confused historical revisionism, which he thought attempted to synthesize "the two most contradictory views of developmental mechanics." Ibid.

8. See Wells, *Icons of Evolution*, 94–101; Michael K. Richardson, et. al., "There is no highly conserved embryonic stage in the vertebrates: implications for current theories of evolution and development," *Anatomy and Embryology* 196 (1997): 91–106.

9. Traditionally, a human "embryo" was regarded as existing from conception through the eighth week of pregnancy; a "fetus" referred to the unborn child after the eighth week of pregnancy until birth. However, sometimes the term "embryo" was applied to the developing child during the entire pregnancy. During the last several years, there has been an effort by some doctors and bioethicists to redefine the the term "embryo" as applied to human beings so that it starts after implantation, not conception. This effort to redefine the term is driven by politics, not science. See discussion in C. Ward Kischer and Dianne N. Irving, *The Human Development Hoax: Time to Tell the Truth*, 2nd ed. (Stafford, VA: American Life League, 1997), 26–27, 36–80.

10. Richard Weikart, *From Darwin to Hitler: Evolutionary Ethics, Eugenics, and Racism in Germany* (New York: Palgrave Macmillan, 2004), 147.

11. Cyril C. Means Jr., "Eugenic Abortion," letter to the editor, *New York Times*, April 16, 1965, 28.

12. Means, "Eugenic Abortion"; also see Cyril C. Means Jr., "A Fetus as Person," letter to the editor, *New York Times*, March 17, 1972, 40; and Linda Greenhouse, "Constitutional Question: Is There a Right to Abortion," *New York Times Magazine*, Jan. 25, 1970, 88.

13. Robert H. Williams, "Our Role in the Generation, Modification, and Termination of Life," *Archives of Internal Medicine* 124 (August 1969): 221.

14. Victor Eppstein, "When Destroying Life Is Morally Justified," letter to the editor, *New York Times*, October 9, 1980, A34.

15. John M. Oesterreicher, "Abortion, Evolution and an Untenable Biogenetic Law," letter to the editor, *New York Times*, October 24, 1980, A32.

16. *United States v. Vuitch*, 402 U.S. 62 (1971).

17. Letter from Milan M. Vuitch to Senator John East, April 22, 1981, in *The Human Life Bill Appendix: Hearings before the Subcommitte on Separation of Powers of the Committee on the Judiciary, United States Senate, Ninety-Seventh Congress, First Session, on S. 158, a Bill to Provide that Human Life Shall be Deemed to Exist from Conception, April 23, 24; May 20, 21; June 1, 10, 12 and 18.* Serial No. J-97-16 (Washington, D.C.: U.S. Government Printing Office, 1982), 105.

18. Letter from James G. Wilson to Sen. Max Baucus, April 23, 1981 in *The Human Life Bill Appendix*, 778.

19. Ibid., 778–79.

20. George Crile Jr., "When Does the Human Life Begin?" *Medical Tribune*, March 6, 1985, 22.

21. Rivers Singleton Jr., "Paradigms of Science/Society Interaction: The Abortion Controversy," *Perspectives in Biology and Medicine* 32, no. 2 (Winter 1989): 183.

22. Carl Sagan and Ann Druyan, "Is It Possible To Be Pro-Life And Pro-Choice?" *Parade Magazine* (April 22, 1990), 6.

23. Clayton Ward Kischer, "Quid Sit Veritas? The Odyssey of One Human Embryologist as a Modern Diogenes," in Kischer and Irving, *The Human Development Hoax*, 105–7.

24. Ibid., 112.

25. Ernest Van Den Haag, *National Review*, February 8, 1990, 6–8.

26. Prepared Statement of Dr. Jessica G. Davis, *The Human Life Bill*, 138.

27. Prepared Statement of Professor Clifford Grobstein, *The Human Life Bill*, 95; Singleton, "Paradigms of Science," 183.

28. Quoted in Greenhouse, "Constitutional Question," 89; Roy Lucas, "Federal Constitutional Limitations on the Enforcement and Administration of State Abortion Statutes," *North Carolina Law Review* 46, no. 16 (1968): 733.

29. Lucas, quoted in Greehnouse, "Constitutional Question," 88.

30. Michael Tooley, "Abortion and Infanticide," in *Applied Ethics*, ed. Peter Singer (New York: Oxford University Press, 1986), 76.

31. Peter Singer, *Practical Ethics* (New York: Cambridge University Press, 1979), 118.

32. Mary Anne Warren, "On the Moral and Legal Status of Abortion," in *Biomedical Ethics*, ed. Thomas A. Mappes and David DeGrazia,, 4th ed. (New York: McGraw-Hill, 1996), 437.

33. Lana Clarke Phelan, *Abortion Laws: The Cruel Fraud*. Speech presented at the California Conference on Abortion in Santa Barbara on February 10, 1968 (San Francisco: Society for Humane Abortion, Inc., 1968), 3.

34. Sagan and Druyan, "Is It Possible?," 6.

35. Pratt, *The Human Life Bill Appendix*, 657.

36. Prepared Statement of Sarah Weddington, *The Human Life Bill*, 955.

37. "Children From the Laboratory," interview with James D. Watson, *Prism* (May 1973), 34.

38. Phelan, "Abortion laws," 6.

39. Ibid., 4.

40. Roy Lucas, et. al, "Brief for Appellants," *Roe v. Wade*, Supreme Court of the United States, No. 70-18, 1971 Term, 19.

41. Christopher Hitchens, *God Is Not Great: How Religion Poisons Everything* (New York: Twelve, 2007), 221.

42. Cited in *Webster v. Reproductive Health Services*, 492 U.S. 490 (1989), 501.

43. Letter of Arno G. Motulsky to Sen. Max Baucus, April 24, 1981, *The Human Life Bill*, 71.

44. Letter of Robert H. Ebert to Sen. Max Baucus, April 22, 1981, *The Human Life Bill*, 69.

45. Letter of Frederick C. Robbins to Sen. Max Baucus, April 23, 1981, *The Human Life Bill*, 72.

46. Cited in the testimony of Dr. Lewis Thomas, *The Human Life Bill*, 74.

47. Letter of Arno Motulsky, 71.

48. Letter of Robert H. Ebert, 68, 69.

49. William Brennan, "How Should the Courts Interpret the Constitution? A Debate," in *The New Federalist Papers*, ed. J. Jackson Barlow, Dennis J. Mahoney, and John G. West Jr. (Lanham, MD: University Press of America, 1988), 237. Brennan's comment was originally directed at those who claim to discern and follow the original intentions of the Framers of the Constitution.

50. *Roe v. Wade*, 410 U.S. 113 (1973), 160.

51. "Texas urges that, apart from the Fourteenth Amendment, life begins at conception and is present throughout pregnancy, and that, therefore, the State has a compelling interest in protecting that life from and after conception." Ibid., 159.

52. Ibid., 163.

53. Ibid.

54. "With respect to the State's important and legitimate interest in potential life, the 'compelling' point is at viability. This is so because the fetus then presumably has the capability of meaningful life outside the mother's womb. State regulation protective of fetal life after viability

thus has both logical and biological justifications. If the State is interested in protecting fetal life after viability, it may go so far as to proscribe abortion during that period, except when it is necessary to preserve the life or health of the mother." *Roe v. Wade*, 163–64. For the purposes of this discussion I am accepting at face value the majority's claim in Roe that the state can protect the life of the fetus after viability "except when it is necessary to preserve the life or health of the mother." However, a strong case can be made that the Court in reality established birth as the only point at which the state may unequivocally protect the life of an infant. In the companion case of *Doe v. Bolton*, the Court's majority explained that the determination of whether an abortion is required for the "health of the mother" should include an investigation of all areas of a woman's physical and mental health. "We agree . . . that the medical judgment may be exercised in the light of all factors—physical, emotional, psychological, familial, and the woman's age—relevant to the wellbeing of the patient. All these factors may relate to health. This allows the attending physician the room he needs to make his best medical judgment. *Doe v. Bolton*, 410 U.S. 179 (1973), 192. Given the Court's expansive definition of a woman's health, it is difficult to see how a state could validly proscribe abortions even on viable fetuses. Surely every abortion could be justified as having at least some impact on a woman's "emotional, psychological" or "familial" well-being.

55. Quoted in Patricia Bartos, " 'Laboratory' Gets Bodies of Aborted, Nurse Claims," *Pittsburgh Catholic*, March 17, 1972, reprinted in *Fetal Research, 1974: Hearing before the Subcommittee on Health of the Committee on Labor and Public Welfare, United States Senate, Ninety-Third Congress, Second Session, July 19, 1974* (Washington, D.C.: U.S. Government Printing Office, 1974), 68. Also see *[Transcript of] Pennsylvania Abortion Law Commission Public Hearing, Pittsburgh, March 14, 1972, Chatham College Chapel Auditorium* (Pennsylvania Abortion Law Commission, 1972), 25–27.

56. "Post-Abortion Fetal Study Stirs Storm," *Medical World News*, June 8, 1973, 21.

57. Victor Cohn, "Scientists and Fetus Research," *The Washington Post*, April 15, 1973, A1.

58. Ibid., A6.

59. Bela A. Resch, et. al., "Comparison of Spontaneous Contraction Rates of In Situ and Isolated Fetal Hearts in Early Pregnancy," *American Journal of Obstetrics and Gynecology* 118 (January 1, 1974): 73–76.

60. Robert C. Goodlin, "Cutaneous Respiration in a Fetal Incubator," *American Journal of Obstetrics and Gynecology* 86 (July 1, 1963): 571–79.

61. Ibid., 574.

62. Ibid., 571.

63. "Students Picket Stanford Experiments on Live Aborted Fetuses; Doctor Admits Mothers of Some Children Were Not Informed," press release of Student Pro-Life Federation, San Francisco, California (May 19, 1972), reprinted in *Fetal Research, 1974*, 58.

64. Ibid., 60.

65. "California Bans Medical Testing on Living Fetuses," *Medical Tribune*, October 24, 1973, rprinted in *Fetal Research, 1974*, 67.

66. *Fetal Research, 1974*, 33–36; Antti Vaheri, et. al., "Isolation of Attenuated Rubella-Vaccine Virus from Human Products of Conception and Uterine Cervix," *New England Journal of Medicine* 286, no. 20 (May 18, 1972): 1071–74.

67. Quoted in William Brennan, *The Abortion Holocaust* (St. Louis: Gateway Press, 1983), 55.

68. See documentation in ibid., 55–56, 63–65.

69. See Harold M. Schmeck, Jr., "Conferees Agree to Ban Research on Live Fetuses," *New York Times*, June 8, 1974, 65; Morton Mintz, "Conferees Pass Pioneer Bill on Biomedicine," *Washington Post*, June 8, 1974, A4. For the actual statutory language of the moratorium, see National Research Act, Public Law 93-348 (July 12, 1974), 88 Stat 353.

70. Quoted in "Post-Abortion Fetal Study," 21.

71. Quoted in Victor Cohn, "Live-Fetus Research Debated," *Washington Post*, April 10, 1973, A9.

72. Quoted in Michael T. Malloy, "A 'No' to Research on Aborted, Live Fetuses," *The National Observer*, April 21, 1973, 4.

73. Joseph Fletcher, *Humanhood: Essays in Biomedical Ethics* (Buffalo, NY: Prometheus Books, 1979), 99.

74. Ibid., 104.

75. Quoted in Cohn, "Scientists and Fetus Research," A6.

76. Willard Gaylin and Marc Lappe, "Choosing Between Fetus and Child," *Washington Post*, April 27, 1975, C5.

77. Quoted in Cohn, "Live-Fetus Research Debated," A9.

78. Quoted in Maggie Scarf, "The Fetus as Guinea Pig," *New York Times*, October 19, 1975, 90. Also see Paul Ramsey, *The Ethics of Fetal Research* (New Haven, CT: Yale University Press, 1975).

79. Mary Anne Warren, "Can the Fetus be an Organ Farm?" *Hastings Center Report* (October 1978), 23–24.

80. Michael Lockwood, "The Warnock Report: A Philosophical Appraisal," in *Moral Dilemmas in Modern Medicine*, ed. Michael Lockwood, (New York: Oxford University Press, 1985): 168.

81. Testimony by Mary Tyler Moore, September 14, 2000, S. HRG. 106–413 STEM CELL RESEARCH, PART 3, Hearings before a Subcommittee of the Committee on Appropriations, United States Senate, 106th Congress, 2nd Session (Washington, D.C.: U.S. Government Printing Office, 2001), 120.

82. The recommendation for lifting the ban came from the National Commission for Protection of Human Subjects of Biomedical and Behavioral Research, a body established by congressional statute to recommend policies on ethical principles of human research, including research on fetuses. For the legislation establishing this commission, see National Research Act, 88 Stat 342–54. For the recommendations of the commission, see "Research on the Fetus," sections 8 and 9 from the National Commission for Protection of Human Subjects of Biomedical and Behavioral Research, *Report and Recommendations: Research on the Fetus (1975)*, in *Source Book in Bioethics*, ed. Albert R. Jonsen, Robert M. Veatch, LeRoy Walters (Washington, D.C.: Georgetown University Press, 1998): 29–39. Especially relevant are recommendations 4–6 and 12. Only one commissioner, David W. Louisell, dissented from the provisions on nontherapeutic research on living fetuses. Louisell complained that the commission's recommendations "would permit nontherapeutic research on the fetus in anticipation of abortion and during the abortion procedure, and on a living infant afer abortion when the infant is considered nonviable, even though such research is precluded by recognized norms governing human research in general. Although the Commission uses adroit language to minimize the appearance of violating standard norms, no facile verbal formula can avoid the reality that under these Recommendations the fetus and nonviable infant will be subjected to nontherapeutic research from which other humans are protected" (ibid., 35). Secretary of Health, Education, and Welfare Caspar Weinberger ultimately adopted the commission's recommendations without major changes. See Victor Cohn, "Fetal Research Ban Lifted," *Washington Post*, July 30, 1975, A1; "Weinberger Moves to End Ban on Research Involving Fetuses," *New York Times*, July 30, 1975, 11. For a sympathetic treatment of the history of the National Commission's work on fetal research by one of the commissioners, see Albert R. Jonsen, *The Birth of Bioethics* (New York: Oxford University Press, 1998), 94–101.

83. "Research on the Fetus," recommendation 6, 34. The report also allowed nontherapeutic research on "the possibly viable infant" as long as "no additional risk to the well-being of the infant will be imposed by the research." Ibid., recommendation 7, 34–35.

84. Ibid., recommendations 4–6, 34.

85. Ibid., recommendation 12, 35.

86. The relevant regulatory language, eventually codified as 45 CFR 46.209, stated: "No nonviable fetus may be involved as a subject in an activity covered by this subpart unless: (1) Vital functions of the fetus will not be artificially maintained, (2) Experimental activities which of themselves would terminate the heartbeat or respiration of the fetus will not be employed, and (3) The purpose of the activity is the development of important biomedical knowledge which cannot be obtained by other means" ("Federal Regulations Regarding the Protection of Human Subjects of Research," in *Source Book in Bioethics*, 71). This language was superceded by a new version of the regulations adopted in 2001. (See n. 87 for citation.)

87. 42 USC § 289g (2004), originally adopted as part of P.L. 99–158 (November 20, 1985).

88. The revised regulations can be found in 45 CFR.205 (2004). They were issued in November, 2001. See 66 FR 56775, 56779 (November 13, 2001).

89. Singer, *Practical Ethics*, 122–23.

90. Fletcher, *Humanhood*, 144.

91. Williams, "Our Role in the Generation," 221.

92. Letter from James Wilson, *The Human Life Bill Appendix*, 778.

93. "Children From the Laboratory," 13.

94. Tooley, "Abortion and Infanticide," 83.

95. Singer, *Practical Ethics*, 123. On the other hand, other evolutionists have made the argument that infanticide also has an evolutionary basis in both human beings and animals. See, for example, Steven Pinker, "Why They Kill Their Newborns," *New York Times Magazine*, November 2, 1997; Sarah Blaffer Hrdy, *Mother Nature: A History of Mothers, Infants, and Natural Selection* (New York: Pantheon Books, 1999), 88–90; Martin Daly and Margo Wilson, *The Truth about Cinderella: A Darwinian View of Parental Love* (New Haven, CT: Yale University Press, 1998), 8–17. Peter Singer has also noted the widespread practice of infanticide in other human societies. Although he concedes that this fact does not make infanticide morally right in and of itself, he surely thinks it is in some way relevant. According to Singer, "it is worth knowing that from a cross-cultural perspective it is our tradition, not that of the Kung or the Japanese, that is unusual in its official morality about infanticide. Recognizing that fact puts the modern medical practice of infanticide in a broader perspective." Peter Singer, *Rethinking Life and Death: The Collapse of our Traditional Ethics* (New York: St. Martin's Press, 1994), 129–30. Singer's caveat notwithstanding, it seems that he is appealing to the widespread practice of infanticide in other cultures as a way of eroding opposition to infanticide in our society.

96. Peter Singer, "Sanctity of Life or Quality of Life?" *Pediatrics* 72, no. 1 (July 1983): 129.

97. Quoted in Johann Hari, "Peter Singer—An Interview," originally run in *The Independent*, Jan. 7, 2004, now available online at http://www.johannhari.com/archive/article.php?id=410 (accessed July 27, 2005).

98. See Roger E. Stevenson, "Anencephaly," in *The Gale Encyclopedia of Genetic Disorders*, ed. Stacey L. Blachford (Farmington Hills, MI: Gale Group, 2001).

99. See Joyce L. Peabody, Janet R. Emery, Stephen Ashwal, "Experience with Anencephalic Infants as Prospective Organ Donors," *The New England Journal of Medicine* 321, no. 6 (August 10, 1989): 344–50; Louis Sahagun, "Harvesting of Baby Organs Not Feasible; Loma Linda Tells Why It Dropped Anencephalic Infants Program," *Los Angeles Times*, August 24, 1988, I-25. For the standard accepted view of "brain death," see "Defining Death: Medical, Legal, and Ethical Issues in the Definition of Death," in *Source Book in Bioethics*, 118–42.

100. For information about this experiment and the resulting controversy, see Peabody, "Experience with Anencephalic Infants"; Sahagun, "Harvesting of Baby Organs Not Feasible"; "New Policy Adopted on the Use of Babies As Donors of Organs," Associated Press report, *New York Times*, December 21, 1987, A19; Robert Steinbrook, "Frank Admissions End Infant Organ Harvesting," *Los Angeles Times*, August 19, 1988, I-3. For criticism of using anencephalic infants as organ donors, see D. Alan Shewmon, "The Use of Anencephalic Infants as Organ Sources: A Critique," *Journal of the American Medical Association* 261, no. 12 (March 24–31, 1989): 1773–81.

101. Quoted in Sandra Blakeslee, "Law Thwarts Effort to Donate Infants' Organs," *New York Times*, September 9, 1986, C-9; also see, Allan Parachini, "Science, Ethics Clash Over Infant Organ Donations Bill," *Los Angeles Times*, December 2, 1986, V-1.

102. Mike Clary, "Baby Theresa's Gift: Debate Over Organ-Harvesting Laws," *Los Angeles Times*, April 16, 1992, A-5.

103. Diane Gianelli, "Calling Anencephalic Donors Dead," *American Medical News* 35, no. 25 (June 29, 1992): 11.

104. For a sampling of the debate over the proposed AMA policy, see Katherine Dowling, "Slaughtering Babies, Ethically," *Los Angeles Times*, January 10, 1996, B-9; Charles Krauthammer, ". . . And a Troubling Development," *Washington Post*, June 9, 1995, A27; "An Exception to the Rule; Parents Should Be Able to Donate Organs of Infants Born without Brains," *Los Angeles Times*, June 8, 1995, B-6; Don Colburn, "AMA Reverses Its Stand on Organs from Infants," *Washington Post*, June 6, 1995, Z5; Cindy Schreuder, "AMA Takes Walk in an Ethical Minefield," *Chicago Tribune*, May 28, 1995, Chicagoland section, 1; "Good Intentions Spawn a Bad Idea," *Chicago Tribune*, May 27, 1995, Editorial section, 20.

Chapter 15: Redefining Death

1. Unless otherwise noted, details about the Schiavo case provided here were taken from "A Comprehensive Timeline and Archive of the Terri Schindler-Schiavo Case," and "Most Common Misconceptions about Terri's Situation," http://www.terrisfight.org/ (accessed May 27, 2004); International Task Force on Euthanasia and Assisted Suicide Update 17, no. 3 (2003), http://wwwinternationaltaskforce.org/iua29.htm#16 (accessed May 24, 2004). For additional background information, also see Rich McKay and Maya Bell, "Friends Call Michael Schiavo a 'Gentle Giant,' Devoted to Wife," (KRT), October 25, 2003, http://www.centredaily.com/mld/centredaily/news/7104921.htm; Sean Mussenden, "Schiavo's Condition Tested, Then Fortified, Family's Faith," (KRT), October 25, 2003, http://www.centredaily.com/mld/centredaily/news/7104920.htm (accessed October 27, 2003); Rev. Robert Johansen, "Killing Terri Schiavo," *Crisis* (January 8, 2004), http://www.crisismagazine.com/january2004/johansen.htm (accessed May 24, 2004).

2. Bill No. HB 35-E (2003), available online at http://news.findlaw.com/cnn/docs/schiavo/flsb35e102103.pdf (accessed August 1, 2005).

3. The Multi-Society Task Force on PVS, "Medical Aspects of the Persistent Vegetative State," *New England Journal of Medicine* 330, no. 21 (May 26, 1994): 1500, 1501. For problems with the definition of PVS, see Alan Shewmon, "The ABC of PVS: Problems of Definition," in Machado and Shewmon, eds., *Brain Death and Disorders of Consciousness* (New York: Kluwer Academic/Plenum Publishers, 2004), 215–28.

4. The print media repeatedly reported as fact the disputed claim that Terri Schiavo was in a persistent vegetative state. Associated Press reporters were particularly responsible for misreporting this issue and never providing the other side. Following is a sampling of some of the inaccurate and one-sided claims made in news stories: "Terri Schiavo has been in what doctors call a persistent vegetative state for 14 years." ["Law Keeping Brain-Damaged Woman Alive Struck Down," *CNN.com*, May 6, 2004, http://www.cnn.com/2004/LAW/05/06/schiavo.case (accessed May 31, 2004)]; "Doctors say Terri Schiavo has been in a 'persistent vegetative state' since 1990 . . ." [Megan O'Matz and Diane Lade, "ACLU Joins Husband in Battle to Stop Feeding of Brain-Damaged Woman," *Sun-Sentinel.com*, October 24, 2003, http://www.sun-sentinel.com/ (accessed October 27, 2003)]; "Terri Schiavo has been in what doctors call a 'persistent vegetative state' and on a feeding tube since 1990 . . ." ["Brain-Damaged Woman Terri Schiavo Again Being Fed through Tube," (AP), October 24, 2003, http://www.brunei-online.com/bb/fri/oct24w22.htm (accessed October 27, 2003)]; "A heart attack in 1990 left Terri Schiavo, now 39, in a persistent vegetative state and court-appointed doctors believe there is no hope of recovery." [Vickie Chachere, "Schiavo Judge Vilified, Praised," (AP) *The Tampa Tribune*, October 26, 2003, http://www.tampatrib.com/ (accessed October 27, 2003)]; "Terri Schiavo, 39, has been in what doctors call a 'persistent vegetative state' since 1990 . . . Her eyes are open, but doctors say she has no consciousness." [Mitch Stacy, "Attorney for Husband of Brain-Damaged Woman Outraged at Florida Governor," (AP), *Daily Herald*, October 23, 2003, http://www.harktheherald.com (accessed October 27, 2003)]; "Court-Appointed Doctors Have Described Schiavo as Being in a Vegetative State Since Her Heart Stopped in 1990 . . ." [Jackie Hallifax, "Florida House passes law to restore comatose woman's feeding tube," (AP), *Daily Herald*, October 21, 2003, http://www.harktheherald.com (accessed October 27, 2003)]. On the rare occasion when reporters mentioned a dispute over Terri Schiavo's condition, it was often framed as a dispute between doctors who agreed that Terri was in a PVS and her parents or other family members who insisted otherwise. Because no mention was made in such stories of the doctors who disagreed with the PVS diagnosis, the views of family members were clearly depicted as less than credible. [For examples of this type of report see, Vickie Chachere, "Schiavo's Husband Vows Continued Fight," Talahassee.com, October 24, 2003, http://www.tallahassee.com/mld/democrat/news/local/7089767.htm (accessed October 27, 2003); Gloria Campisi, "Terry Schiavo's Family Insists She's Not Comatose," *Philadelphia Daily News*, October 24, 2003, http://www.philly.com/mld/dailynews/news/local/7090958.htm (accessed October 27, 2003)] For a rare mainstream news story that mentions some of the medical experts who challenge the diagnosis of PVS for Terri, see Bob LaMendola, "Schiavo Case Rekindles Debate on Reviving People with Severe Brain Damage," *Sun-*

Sentinel.com, October 27, 2003, http://www.sun-sentinel.com/ (accessed October 27, 2003). For a good summary of inaccuracies and biased news coverage about the Terri Schiavo case, see Diana Lynne, "Media 'Getting It Wrong' on Terri Schiavo Story," *WorldNetDaily.com*, October 25, 2003, http://www.wnd.com/news/article.asp?ARTICLE_ID=35270 (accessed October 27, 2003) and Wesley J. Smith, "Life, Death, and Silence: Why the Media Elites Won't Tell the Full Story on Terri's Prognosis and Michael Schiavo," *The Daily Standard*, October 31, 2003, http://www.weeklystandard.com/Content/Public/Articles/000/000/003/329sghqk.asp (accessed October 31, 2003). Also see "A comprehensive Timeline and Archive of the Terri Schindler-Schiavo Case," and "Most Common Misconceptions about Terri's Situation." Copies of the medical affidavits filed that disputed the diagnosis that Terri Schiavo was in a PVS can be found at http://www.terrisfight.org/mainlinks.php?tablesingle=main_terri_story&id=7 (accessed Feb. 24, 2007).

5. "Media 'Getting It Wrong' on Terri Schiavo Story."
6. Flordia Statutes 765.101 §12a, available online at http://www.flsenate.gov/Statutes/ (accessed May 31, 2004).
7. "Most common misconceptions about Terri's situation"; Chachere, "Schiavo's husband vows continued fight"; Johansen, "Killing Terri Schiavo."
8. Clips from these videos are available online at http://www.terrisfight.org/ (accessed May 27, 2004).
9. Quoted in Rob Stein, "Woman's Sustaining Care Resumed, But Experts Say No Hope Exists For Her Recovery," *Washington Post*, October 23, 2003, A3.
10. For detailed criticisms of these claims, see "New Schiavo Document Gives Cause For Independent Autopsy Review," Nov. 25, 2006, http://www.northcountrygazette.org/articles/112506SchiavoReview.html (accessed Feb. 24, 2007); Sherry Eros, MD and Steven Eros, "Terri Schiavo's Autopsy: The Blind Spot," June 21, 2005, http://eroscoloredglasses.blogspot.com/2005/06/terri-schiavos-autopsy-blind-spot.html (accessed Feb. 24, 2007). A copy of the Medical Examiner's Report of Autopsy can be downloaded from http://www.blogsforterri.com/archives/5050439_autopsy%20report%20and%20supporting%20documents.pdf (accessed Feb. 24, 2007).
11. Ronald E. Cranford, "The Case of Mr. Stevens," *Issues in Law and Medicine* 7 no. 2 (Fall 1991): 201.
12. Judge Stein, unpublished opinion in In re Jobes, No. C-4971-85E, slip op. at 8-9 (N.J. Super. Ct. Ch. Div. Apr. 23, 1986), quoted in ibid.
13. *Cruzan v. Director, Missouri Department of Health*, 497 U.S. 261 (1990).
14. See testimony of Dr. Saad M. Al-Shatir, Transcript in the case of Nancy Beth Cruzan et. al. vs. Robert Harmon and Donald Lampkins, Case No. CV384-9P, Circuit Court of Jasper County, Missouri (March 9-11, 1988), 281-86; Testimony of Dr. Hish S. Majzoub, Transcript in the Case of Nancy Beth Cruzan, 111.
15. See testimony of Nurses Lisa Perrin, Alice Johnson, and Janie Bowker, Transcript in the Case of Nancy Beth Cruzan, 374, 595-97, 644.
16. Testimony of Dr. James Dexter, Transcript in the Case of Nancy Beth Cruzan, 764.
17. Ibid., 751.
18. Ibid., 755. Dexter also stated that Nancy's ability to orient herself toward sounds involved the higher brain. Ibid., 758.
19. Ibid., 758-59.
20. Testimony of Dr. Ronald Cranford, Transcript in the Case of Nancy Beth Cruzan, 126-27, emphasis added.
21. Ibid., 137.
22. Ibid., 141-42.
23. Ibid., 142.
24. Ibid., 140.
25. Testimony of Dr. James Dexter, Transcript in the Case of Nancy Beth Cruzan, 758.
26. Testimony of Dr. Ronald Cranford, Transcript in the Case of Nancy Beth Cruzan, 145.
27. Ibid. 145-46, emphasis added.
28. Ibid., 146.
29. Ibid., 146-47.
30. Ibid., 162.
31. Testimony of Dr. James Dexter, Transcript in the Case of Nancy Beth Cruzan, 762.

32. Testimony of Dr. Ronald Cranford, Transcript in the Case of Nancy Beth Cruzan, 157.

33. "Tragedy Compounded: Joe Cruzan Commits Suicide," *International Anti-Euthanasia Task Force Update* vol. 10, no. 4 (1996), available online at http://www.internationaltaskforce.org/iua4.htm#8 (accessed August 1, 2005).

34. Ibid.

35. Nat Hentoff, "Not a 'Hopeless Case' After All," *Washington Post*, April 29, 1989, A25.

36. See A. Dahleen Glanton, "Judge Modifies Right-to-Die Order; Woman May Be Moved," *San Diego Union*, April 16, 1987, A-1; Pat Flynn, "Right-to-Die Lawyer Wants Doctor Silenced," *San Diego Union*, April 18, 1987; Jim Schachter, "Right-to-Die Move Shifted From Doctor to Daughter," *Los Angeles Times*, April 16, 1987, II-6; Leslie Bond, "Doctor, Nursing Home Refuse To Starve Anna Hirth," *National Right to Life News* 14, no.7,(April 16, 1987): 1; Leslie Bond, "Anna Hirth Dies nine Days After Physician Fired and Anna Moved to New Locale," *National Right to Life News* 14, no. 10 (May 28, 1987): 1.

37. "Michael Martin Case," International Task Force on Euthanasia and Assisted Suicide website, http://www.internationaltaskforce.org/mm.htm#12 (accessed August 1, 2005).

38. For details of this case see, ibid., 73–78; Letter from Janie Hickok Siess to *Good Morning America*, January 23, 2001, available at http://www.internationaltaskforce.org/wend1.htm (accessed August 1, 2005); Transcript of "A Life in Limbo: Should Robert Wendland be Allowed to Die?" Burden of Proof, aired January 15, 2001 on CNN, available at http://edition.cnn.com/TRANSCRIPTS/0101/15/bp.00.html (accessed May 28, 2004); Harriet Chiang, "New Twist in Cases over Right to Die, Patient Is in Twilight State, Not in a Coma," *San Francisco Chronicle*, May 30, 2001, http://sfgate.com (accessed June 1, 2004); Harriet Chiang, "Justices Struggle with New Right-to-Die Case; State High Court Hearing on Barely Conscious Man," *San Francisco Chronicle*, May 31, 2001, http://sfgate.com (accessed June 1, 2004); Chuck Squatriglia, "Man at Center of Right-to-Die Fight Is Dead; Controversial Case Was Pending in California Supreme Court," *San Francisco Chronicle*, July 18, 2001, http://sfgate.com (accessed June 1, 2004); Harriet Chiang, "6-0 Court Tightens Right-to-Die Standard; Patient Has to Leave Clear Direction on Ending Care," *San Francisco Chronicle*, August 9, 2001, http://sfgate.com (accessed June 1, 2004); Maura Dolan, "Injured Man in Life Support Case Dies," *Los Angeles Times*, July 18, 2001, B-1.

39. Letter from Janie Hickok Siess to *Good Morning America*.

40. Florence Wendland, comments on "A Life in Limbo."

41. Dolan, "Injured Man"; Squatriglia, "Man at center of right-to-die fight is dead."

42. Testimony of Dr. Ronald Cranford, Transcript in the Case of Nancy Beth Cruzan, 163.

43. Ibid., 230.

44. Ibid., 229.

45. Ibid., 230, emphasis added.

46. Ibid., 168. Cranford did not believe that spoon and syringe feeding should always be considered medical treatment. For example, he thought that such feedings for someone who had suffered "a small stroke who can still eat" would be a part of basic care, not medical treatment. Ibid.

47. Quoted in Johansen, "Killing Terri Schiavo."

48. Quoted in Sarah Foster, "Terri Schiavo wants to live," *WorldNetDaily.com*, October 14, 2003, http://www.wnd.com/news/article.asp?ARTICLE_ID=35077 (October 27, 2003).

49. "Michael Martin Case."

50. William F. Stone, quoted in Smith, *Culture of Death*, 71. The details about the case of Marjorie Nighbert are taken from ibid., 70–71.

51. Robert Barry, *Protecting the Medically Dependent: Social Challenge and Ethical Imperative* (Stafford, VA: Castello Institute, 1988), 20.

52. E-mail from Tia Willin, April 1, 2005.

53. Ibid., 21.

54. For a criticism of the term "vegetative state" along these lines, see William P. Whitenouse, "Vegetative State: Term Should Be Changed," July 16, 2003, bmj.com, http://bmjjournals.com/cgi/eletters/327/7406/67#34499 (accessed May 31, 2004); Keith Andrews, "Managing the Persistent Vegetative State," *British Medical Journal* 305 (August 29, 1992): 486; Albert E. Gunn, "Risk-Benefit Ratio: The Soft Underbelly of Patient Autonomy," *Issues in Law and Medicine* 7, no.2 (Fall 1991): 146.

55. According to the Uniform Determination of Death Act (UDDA), "An individual who has sustained either (1) irreversible cessation of circulatory and respiratory functions, or (2) irrevers-

ible cessation of all functions of the entire brain, including the brain stem, is dead. A determination of death must be made in accordance with accepted medical standards." Quoted in "Defining Death," 136. The UDDA is a model statute developed and/or endorsed by a variety of groups, including the American Medical Association, the American Bar Association, the American Academy of Neurology, and the President's Commission for the Study of Ethical Problems in Medicine and Biomedical and Behavioral Research. For a thoughtful critical analysis of brain-based diagnoses of death, see Alan D. Shewmon, " 'Brainstem Death,' 'Brain Death' and Death: A Critical Re-evaluation of the Purported Equivalence," *Issues in Law and Medicine* 14, no. 2 (Fall 1998): 125–45.

56. Fletcher, *Humanhood*, 135.

57. Ibid., 163.

58. Ibid., 164.

59. See, for example, Robert M. Veatch, "The Impending Collapse of the Whole-Brain Definitions of Death," in *Biomedical Ethics*, 319–27. Veatch has been critical of the "whole-brain" definition of brain death since the 1970s. Although he advocates expanding the definition of brain death, he also favors a "conscience clause" which would allow people to choose their own definition of death for themselves and their children.

60. R. Hoffenbert, et. al., "Should Organs from Patients in Permanent Vegetative State Be Used for Transplantation?" *The Lancet* 350, no. 9087 (November 1, 1997): 1321.

61. Payne et. al., "Physicians' Attitudes about the Care of Patients in the Persistent Vegetative State: A National Survey," *Annals of Internal Medicine* 125, no. 2 (July 15, 1996): 105.

62. Ibid., 107.

63. Transcript, "A Life in Limbo."

64. Madelaine Lawrence, *In a World of Their Own: Experiencing Unconsciousness* (Westport, CT: Bergin and Garvey, 1997), 6.

65. Ibid., 70.

66. Affidavit of Kate Adamson, December 11, 2003, 2, available at http://www.terrisfight.org/ (accessed May 27, 2004).

67. Ibid., 1.

68. Rus Cooper-Dowda, "All I Ever Wanted to Do . . ." International Task Force on Euthanasia and Assisted Suicide Update 17, no. 3 (2003), online edition at http://www.internationaltaskforce.org/iua29.htm#16 (accessed August 1, 2005).

69. Keith Andrews, et. al., "Misdiagnosis of the Vegetative State: Retrospective Study in a Rehabilitation Unit," *British Medical Journal* 313, no. 7048 (July 6, 1996): 14.

70. Ibid., 15.

71. Quoted in Terence Monmaney, "Many So-Called Vegetative Patients Aware, Study Finds," *Los Angeles Times*, July 6, 1996, A1.

72. "Dubious Diagnosis of PVS," *American Journal of Nursing* 94, no. 2 (February 1994): 11. For another study about the problem of PVS misdiagnosis, see Lucy H. A. Strens, Gordan Mazibrada, John S. Duncan, and Richard Greenwood, "Misdiagnosing the Vegetative State after Severe Brain Injury: The Influence of Medication," *Brain Injury* 18, no. 2 (February 2004): 213–18.

73. *Defining Death: A Report on Medical, Legal and Ethical Issues in the Determination of Death* (Washington, D.C.: President's Commission for the Study of Ethical Problems in Medical and Biomedical and Behavioral Research, July, 1981), 40; also see 17. Despite the commission's skepticism about the higher-brain formulation of brain death, it accepted the claim that PVS patients do not have any "awareness of self or environment." Ibid., 18.

74. Shewmon, "The Use of Anencephalic Infants," references omitted, 1776.

75. Keith Andrews, "Patients in the Persistent Vegetative State: Problems in Their Long Term Management," *British Journal of Medicine* 306 (June 12, 1993): 1601.

76. Chris Borthwick, "The Proof of the Vegetable: A Commentary on Medical Futility," *Journal of Medical Ethics* 21, no. 4: 205–8.

77. Ronald E. Cranford and David Randolph Smith, "Consciousness: The Most Critical Moral (Constitutional) Standard for Human Personhood," *American Journal of Law and Medicine* 13, no. 2–3 (1987): 240.

78. Ibid., 241.

79. For example, a recent study found that current diagnostic practices may underestimate the amount of auditory processing in patients with extreme brain injuries, including those in a

PVS. "In extreme conditions such as coma or vegetative state, this may lead to mislabeling a patient as 'non-responsive' while in fact his (her) auditory differentiation ability is not lost." [B. Kotchoubey, et. al., "Stimulus Complexity Enhances Auditory Discrimination in Patients with Extremely Severe Brain Injuries," *Neuroscience Letters* 352 (2003): 131.] According to another study, the brains of PVS patients show evidence of some types of pain-processing, although whether they are conscious of the pain remains unclear: "regional activity found at the cortical level indicates that a residual pain-related cerebral network remains active in long-term PVS patients." [Jan Kassubek, et. al., "Activation of a Residual Cortical Network during Painful Stimulation in Long-Term Postanoxic Vegetative State," *Journal of the Neurological Sciences* 212 (2003): 85.] A third article discusses the evidence that seems to corroborate complete unawareness on the part of PVS patients, but then concludes that conceptually "we cannot be sure that patients with severely damaged brains, supporting some residual cerebral activity, are unaware. The absence of evidence of awareness does not amount to proof of the absence of awareness." [Adam Zeman, "Theories of Visual Awareness," *Progress in Brain Research* 144 (2004): 326.]

80. Barbara A. Wilson, Fergus Gracey and Kate Bainbridge, "Cognitive Recovery from 'Persistent Vegetative State': Psychological and Personal Perspectives," *Brain Injury* 15, no. 12 (2001): 1085.

81. Ronald Cranford, "Misdiagnosing the Persistent Vegetative State," *British Medical Journal* 313, no. 7048 (July 6, 1996): 5–6.

82. Ronald Cranford, quoted in Smith, *Culture of Death*, 77.

83. Cranford and Smith, "Consciousness," 246.

84. Ian Dowbiggin, *A Merciful End: The Euthanasia Movement in Modern America* (New York: Oxford University Press, 2003), 25, 53–56.

85. Ibid., 54.

86. Ibid., 55.

87. Mitchell, quoted in ibid.

88. Ibid.

89. See, for example, Karl Binding and Alfred Hoche, "Permitting the Struction of Unworthy Life: Its Extent and Form," *Issues in Law and Medicine* 8, no.2 (Fall 1992): 231–65. This essay was originally published in Germany in 1920. For the influence of Darwinism on the euthanasia movement in Germany, see Richard Weikart, *From Darwin to Hitler* (New York: Pallgrave Macmillan, 2004). Also see discussion in Gunn, "Risk-Benefit Ratio," 146.

90. George Gallup with William Proctor, *Adventures in Immortality* (New York: McGraw-Hill, 1982), 143–44, 207, 209, 183. Leading scientists and doctors were randomly sampled from those listed in Marquis' *Who's Who in America*. According to surveys during the 1990s, only 7.9 percent of scientists surveyed from the National Academy of Sciences believe in personal immortality, while 38 percent of scientists surveyed from American Men and Women of Science do. See Witham, *Where Darwin Meets the Bible*, 272.

91. Bryan Zepp Jamieson, "Nobody's Home. Why do fundies insist on tormenting the dead?" October 28, 2003, http://www.topplebush.com/oped73.shtml (accessed November 3, 2003).

Conclusion: The Abolition of Man

1. Edward Bellamy, *Looking Backward* (1887), http://eserver.org/fiction/bellamy/contents.html (accessed July 26, 2005), 37. [Pagination given is from the 1917 Vanguard edition.]

2. Ibid., 199–203.

3. Ibid., 268.

4. Ibid., 286–87.

5. Ibid., 292.

6. Nathaniel Hawthorne, "Chiefly About War Matters, By a Peaceable Man" (1862) http://www.eldritchpress.org/nh/cawm.html#top (accessed July 26, 2005).

7. "Eugenics in Politics," *New York Times*, October 9, 1921, 93.

8. C. S. Lewis, "Is Progress Possible? Willing Slaves of the Welfare State," in *God in the Dock: Essays on Theology and Ethics* (Grand Rapids, MI: Eerdmans, 1970), 315.

9. See, for example, Chris Mooney, *The Republican War on Science* (New York: Basic Books, 2005). The effort to demonize normal democratic dissent in the area of science and public policy has been fomented by Fenton Communications, the public relations firm for such groups as MoveOn.org, Planned Parenthood, the American Trial Lawyers Association, Greenpeace, and the National Abortion Rights Action League (NARAL; http://www.fenton.com/pages/3_ourwork/1_clients/clients.htm [accessed September 9, 2006]). With funding from the Tides Center, Fenton has set up a group bearing the Orwellian name of the "Campaign to Defend the Constitution" ("DefCon"; see http://www.defconamerica.org/ [accessed September 9, 2006]; "Introducing the Campaign to Defend the Constitution: Because the Religious Right is Wrong," *Fenton Communique*, Spring 2006, p. 3, available at http://www.fenton.com/pages/5_resources/pdf/Spring2006_newsletter.pdf. For more information about Fenton Communications, see Chuck Colson, "Bankrolling Hostility," *Breakpoint* commentary, September 8, 2006, http://www.breakpoint.org/listingarticle.asp?ID=2432 [accessed September 9, 2006]). According to DefCon, good science is coextensive with the political agenda of the Left, and anyone who says otherwise is a theocrat who opposes "scientific progress." However, even some conservatives seem tempted by the idea that scientists rather than ordinary people should rule. *National Review's* John Derbyshire, for example, has claimed that "if only PH.D.s in the sciences had the vote, we should have Libertarian govt."—which to him would be a very good thing, indeed. Although Derbyshire later retracted this claim after being challenged, he has elsewhere demonstrated an apparently boundless trust in scientific expertise that is breathtaking. (John Derbyshire, "The Darwinian Mandarinate," http://www.nationalreview.com/thecorner/05_02_06_corner-archive.asp - 055833 [accessed September 11, 2006]; John Derbyshire, "Scientists Libertarian," http://www.nationalreview.com/thecorner/05_02_06_corner-archive.asp - 055846 [accessed September 11, 2006].)
10. James Wilson, "Presidential Address," 56.
11. Herbert Walter, "Human Conservation", in Newman, *Evolution, Genetics and Eugenics*, 531. The essay was reprinted from a book published by Walter in 1913.
12. Laughlin, *Legal Status of Eugenical Sterilization*, 79.
13. Eliot, "The Fruits, Prospects and Lessons of Recent Biological Science," 928.
14. James Madison, Federalist Paper #51 in Alexander Hamilton, James Madison, John Jay, *The Federalist Papers* (New York: New American Library, 1961), 322.
15. George Washington, "[Proposed Address to Congress]," in *The Writings of George Washington from the Original Manuscript Sources, 1745–1799*, ed. John C. Fitzpatrick (Washington, D.C.: United States George Washington Bicentennial Commission), 30:301–302.
16. Nathaniel Hawthorne, "Earth's Holocaust," http://www.eldritchpress.org/nh/holo.html (accessed July 26, 2005).
17. Cantor, *Crime, Criminals and Criminal Justice*, 265.
18. East, *Heredity and Human Affairs*, 29.
19. Sanger in Perry, *Pivot of Civilization*, 214.
20. Richard Weikart, *From Darwin to Hitler*.
21. Wilson, *The New Freedom*, 41–42.
22. Democratic Debate between Al Gore and Bill Bradley, March 1, 2000.
23. Timothy White, "Letter to the University of Idaho Faculty, Staff and Students," Oct. 4, 2005, http://www.president.uidaho.edu/default.aspx?pid=85947 (accessed Feb. 28, 2007).
24. Robert Crowther, "University of Idaho Admits Targetting Professor by Banning Differing Viewpoints on Evolution," Oct. 13, 2005, http://www.evolutionnews.org/2005/10/university_of_idaho_admits_targetting_pr.html (accessed Feb. 28, 2007).
25. This comment from June 14, 2005 can be found at http://www.pandasthumb.org/archives/2005/06/a_new_recruit.html - c35130 (accessed September 9, 2006).
26. This comment from August 4, 2005 can be found at http://pharyngula.org/index/weblog/comments/perspective/ (accessed September 9, 2006).
27. Weisz, *The Science of Biology* (1959), 21.
28. Stephen C. Meyer, "Word Games: DNA, Design, and Intelligence," in Dembski and Kushiner, *Signs of Intelligence*, 104. For a more comprehensive treatment of the same topic, see Stephen C. Meyer, "DNA and the Origin of Life: Information, Specification, and Explanation," in Meyer and Campbell, *Darwinism, Design, and Public Education*, 223–85.

29. Meyer, "Word Games," 105–6; Wells, *Icons of Evolution*, 9–27.

30. Meyer, "Word Games," 108.

31. Stephen Meyer, quoted in the documentary, *Unlocking the Mystery of Life* (Illustra Media, 2002).

32. Transcript of ColdWater Media interview with Stephen Meyer, October 12, 2001.

33. Meyer, quoted in *Unlocking the Mystery of Life*.

34. Henry Stapp, "Preface to The Mindful Universe," (August 5, 2004), http://www-physics.lbl. gov/~stapp/MUI.pdf (accessed July 26, 2005), 1, 6.

35. Ibid., 5.

36. Ibid., 6.

37. Ibid., 2.

38. David J. Chalmers, "Consciousness and its Place in Nature," http://jamaica.u.arizona.edu/ ~chalmers/papers/nature.html (accessed July 26, 2005).

39. See Jeffrey M. Schwartz, Henry P. Stapp, Mario Beauregard, "Quantum Physics in Neuroscience and Psychology: A Neurophysical Model of Mind/Brain Interaction," (June 1, 2004), http:// www-physics.lbl.gov/~stapp/PTB8.doc (accessed July 26, 2005). For an overview of the non-materialist findings of recent neuroscience, see Mario Beauregard and Denyse O'Leary, *The Spiritual Brain: How Neuroscience Is Revealing the Existence of God* (New York: HarperOne, 2007).

40. William Dembski, "Challenging Materialism's 'Chokehold,' on Neuroscience," http://www. designinference.com/documents/2003.02.Schwartz_Review.htm (accessed July 26, 2005).

41. Edward O. Wilson, *Consilience: The Unity of Knowledge* (New York: Vintage Books, 1998), 12.

42. Ibid., 279.

43. Lewis, *The Abolition of Man*, 90.

Afterword to the Paperback Edition: Scientism in the Age of Obama—and Beyond

1. "President Barack Obama's Inaugural Address," White House Blog, January 21, 2009, www. whitehouse.gov/blog/inaugural-address.

2. "Remarks by the President at the National Academy of Sciences Annual Meeting," April 27, 2009, www.whitehouse.gov/the_press_office/Remarks-by-the-President-at-the-National-Academy-of-Sciences-Annual-Meeting.

3. Charles Hurt, "White House's Botched 'Op,'" *New York Post*, April 6, 2009, nypost. com/2009/10/06/white-houses-botched-op/.

4. Jonathan Last quoted in Ed Driscoll, "Interview: Jonathan Last on America's Looming Demographic Crisis," *PJ Media*, February 18, 2013, pjmedia.com/eddriscoll/2013/02/18/interview-jonathan-last/?singlepage=true.

5. Paul R. Ehrlich, *The Population Bomb* (New York: Ballantine Books, 1968), xi.

6. See Jonathan Last, *What to Expect When No One's Expecting: America's Coming Demographic Disaster* (New York: Encounter Books, 2014), especially 7–8; see also Dan Gardner, *Future Babble: Why Expert Predictions Are Next to Worthless, and You Can Do Better* (New York: Dutton, 2011), especially 7–8.

7. John P. Holdren and Paul R. Ehrlich, *Global Ecology: Readings toward a Rational Strategy for Man* (New York: Harcourt Brace Jovanovich, 1971), 77.

8. The words are Paul Ehrlich's, but he is summarizing what he says is John Holdren's view. Paul Ehrlich, *The Machinery of Nature* (New York: Simon & Schuster, 1986), 274.

9. The "scientifically solid" quote comes from Kelly Sims Gallagher and John Holdren, "Sea Change in the Politics of Climate," *Newsweek/Washington Post*, April 19, 2007, archived at www. hks.harvard.edu/news-events/news/news-archive/sea-change-in-the-politics-of-climate. Information on Holdren's help in the creation of *An Inconvenient Truth* comes from John Lauerman, "Holdren's Letterman Warning Is No Laughing Matter (Update1)," Bloomberg, February 10, 2009, www.bloomberg.com/apps/news?pid=newsarchive&sid=aT9Qr0JjWycs. For a thorough critique of the factual errors of *An Inconvenient Truth*, see Marlo Lewis Jr., *Al Gore's Science Fiction: A Skeptic's Guide to "An Inconvenient Truth,"* CEI Congressional Briefing Paper

(Washington, DC: Competitive Enterprise Institute, 2007), available for download at cei.org/pdf/5820.pdf.

10. Paul R. Ehrlich, Anne H. Ehrlich, and John P. Holdren, *Ecoscience: Population, Resources, Environment* (New York: W. H. Freeman, 1977), 783.

11. Ibid., 773.

12. Ibid., 784–86, 838.

13. Ibid., 943, 917.

14. "The Polar Vortex Explained in 2 Minutes," YouTube, posted January 8, 2014, www.youtube.com/watch?v=5eDTzV6a9F4.

15. John M. Wallace, Isaac M. Held, David W. J. Thompson, Kevin E. Trenberth, and John E. Walsh, "Global Warming and Winter Weather," *Science* 343 (February 14, 2014): 729–30, available online at www.sciencemagazinedigital.org/sciencemagazine/14_february_2014?pg=29#pg29.

16. Marc Morano, "Obama Science Adviser John Holdren: 'Weather Practically Everywhere Is Being Caused by Climate Change'—Prof. Roger Pielke Jr. Slams Holdren for Promoting 'Zombie Science,'" *Climate Depot*, February 14, 2014, www.climatedepot.com/2014/02/14/obama-science-adviser-john-holdren-weather-practically-everywhere-is-being-caused-by-climate-change-prof-roger-pielke-jr-slams-holdren-for-promoting-zombie-science/.

17. Charles C. W. Cooke, "Burn the Witch! Roger Pielke Jr. Out at FiveThirtyEight," *National Review Online*, July 29, 2014, www.nationalreview.com/corner/384057/burn-witch-roger-pielke-jr-out-fivethirtyeight-charles-c-w-cooke.

18. Roger Pielke Jr., "An Obama Advisor Is Attacking Me for Testifying that Climate Change Hasn't Increased Extreme Weather," *New Republic*, March 5, 2014, www.newrepublic.com/article/116887/does-climate-change-cause-extreme-weather-i-said-no-and-was-attacked.

19. "It Only Takes Three Minutes to See Why We Must Act on Climate Change," WhiteHouse.gov, August 5, 2014, www.whitehouse.gov/share/it-only-takes-three-minutes-see-why-we-must-act-climate-change.

20. Tom Banse, "Is Wildfire Severity Really Getting Worse?" *NW News Network*, August 8, 2014, nwnewsnetwork.org/post/wildfire-severity-really-getting-worse.

21. "Year-to-Date Statistics" and "Annual Average Prior 10 years," National Interagency Fire Center, November 7, 2014, www.nifc.gov/fireInfo/nfn.htm; Valerie Richardson, "Lukewarm Wildfire Season Throws Damper on Climate-Change Predictions," *Washington Times*, September 7, 2014, www.washingtontimes.com/news/2014/sep/7/lukewarm-wildfire-season-throws-damper-on-climate-/.

22. Kyle Olson, "More Schools Mull Dropping Michelle O's Lunch Rules," EAGnews.org, September 25, 2014, eagnews.org/more-schools-mull-dropping-michelle-os-lunch-rules/; Kyle Olson, "Ohio School Drops Michelle O's Lunch Program: 'We're Keeping Our Standards Very High,'" EAGnews.org, October 2, 2014, eagnews.org/ohio-school-drops-michelle-os-lunch-program-were-keeping-our-standards-very-high/; Kyle Olson, "Parents Slam Paltry 'Rabbit Food' School Lunches," EAGnews.org, October 3, 2014, eagnews.org/parents-slam-paltry-rabbit-food-school-lunches/; Mia Summerson, "New Lunch Rules Cause Concern," TonawandaNews.com, October 4, 2014, www.tonawanda-news.com/news/nt/article_91104be0-b920-59bd-a921-504e89c9f0ee.html; "'I'll Never Forgive Michelle Obama for This School Lunch': Students Sick of Smaller, Healthier Portions Take to Social Media to Rail against First Lady's Pet Project," *Daily Mail*, April 6, 2014, www.dailymail.co.uk/news/article-2598525/Ill-never-forgive-Michelle-Obama-school-lunch-Students-sick-smaller-healthier-portions-social-media-rail-against-ladys-pet-project.html.

23. "We Are Hungry," YouTube, September 17, 2012, www.youtube.com/watch?v=2IB7NDUSBOo.

24. Michelle Obama, "The Campaign for Junk Food," *New York Times*, May 28, 2014, www.nytimes.com/2014/05/29/opinion/michelle-obama-on-attempts-to-roll-back-healthy-reforms.html.

25. Craig Bannister, "Potato Council Debunks First Lady's 'Science' Claims," CNSNews.com, May 29, 2014, cnsnews.com/mrctv-blog/craig-bannister/potato-council-debunks-first-ladys-science-claims.

26. See Rebecca Schleicher, "Family's Anger over School Lunch Reveals More Widespread Issues," Fox25, KOKH-TV, Oklahoma City, October 15, 2014, www.okcfox.com/story/26799348/familys-anger-over-school-lunch-reveals-more-widespread-issues; "We Are Hungry."

27. U.S. Government Accountability Office, *School Lunch: Implementing Nutrition Changes Was Challenging and Clarification of Oversight Requirements Is Needed* (Washington, DC: U.S. Government

Accountability Office, January 2014), 15, n. 31, available at www.gao.gov/assets/670/660427.
pdf.

28. See "Committee Investigation into EPA's Secret Science," U.S. House of Representatives
Committee on Science, Space, and Technology website, science.house.gov/issue/committee-
investigation-epa-secret-science; Letter from the Honorable Lamar Smith, Chairman, House
Committee on Science, Space, and Technology to the Honorable Eddie Bernice Johnson, Ranking Member, Committee on Science, Space, and Technology, August 8, 2013, science.house.
gov/sites/republicans.science.house.gov/files/documents/08–08–2013%20Chairman%20
Smith%20Response%20to%20Ranking%20Member%20Johnson.pdf; Letter from the Honorable
Lamar Smith, Chairman, House Committee on Science, Space, and Technology to the Honorable Gina McCarthy, Administrator, U.S. Environmental Protection Agency, September 3,
2013, science.house.gov/sites/republicans.science.house.gov/files/documents/09–03–2013%20
Chairman%20Smith%20to%20McCarthy.pdf.

29. "Administrator Gina McCarthy, Remarks at the National Academy of Sciences, as Prepared,"
United States Environmental Protection Agency, April 28, 2014, yosemite.epa.gov/opa/
admpress.nsf/8d49f7ad4bbcf4ef852573590040b7f6/2c0a15a30105f16185257cc8004be075!Open
Document.

30. "The Committee recognized the privacy issues potentially implicated by the documents, and
accordingly granted you the option to reply with de-identified data." Letter from Lamar Smith
to Gina McCarthy, September 3, 2013, 3.

31. "Administrator Gina McCarthy, Remarks at the National Academy of Sciences."

32. Letter from EPA Administrator Gina McCarthy to the Honorable Lamar Smith, Chairman,
Committee on Science, Space, and Technology, U.S. House of Representatives, March 7, 2014,
2, available at www.cnsnews.com/sites/default/files/documents/EPA%20letter%20to%20
Smith%20March%207%202014%20%281%29.pdf. See also Barbara Hollingsworth, "EPA Concedes: We Can't Produce All the Data Justifying Clean Air Rules," CNSNews.com, April 11, 2014,
www.cnsnews.com/news/article/barbara-hollingsworth/epa-concedes-we-can-t-produce-all-
data-justifying-clean-air-rules.

33. "Obama Ends Stem Cell Research Ban," CBSNews.com, March 9, 2009, www.cbsnews.com/
news/obama-ends-stem-cell-research-ban/.

34. See "On Embryonic Stem Cell Research: A Statement of the United States Conference of
Catholic Bishops," June 13, 2008, stemcellresearch.org/wp-content/uploads/2013/11/
usccb_2008–06–13.pdf; J. C. Willke, "I'm Pro-Life and Oppose Embryonic Stem Cell Research,"
LifeIssues.org, www.lifeissues.org/cloningstemcell/article.html; Ryan C. Enke, "Opposition to Human Embryonic Stem Cell Research," PM&R Resident, May 2009, www.aapmr.org/
members/residents/newsletter/Pages/archives/Opposition-to-Human-Embryonic-Stem-
Cell-Research.aspx; "The Case Against Embryonic Stem Cell Research: An Interview with
Yuval Levin," Pew Research Religion & Public Life Project, July 17, 2008, www.pewforum.
org/2008/07/17/the-case-against-embryonic-stem-cell-research-an-interview-with-yuval-
levin/.

35. "FAQs: Becket Fund's Lawsuits Against HHS," Becket Fund for Religious Liberty, www.
becketfund.org/faq/; Edward Whelan, "The HHS Contraception Mandate vs. the Religious
Freedom Restoration Act," Ethics and Public Policy Center, eppc.org/publications/the-hhs-con-
traception-mandate-vs-the-religious-freedom-restoration-act/; "Institutional Religious Freedom
Alliance Letter to HHS Secretary Sebelius," June 11, 2012, www.becketfund.org/wp-content/
uploads/2012/01/Letter-to-HHS-secretary-Signed-6–11–2012.pdf.

36. Robert Pear, "Obama Reaffirms Insurers Must Cover Contraception," New York Times, January
20, 2012, www.nytimes.com/2012/01/21/health/policy/administration-rules-insurers-must-
cover-contraceptives.html.

37. Sharyl Attkisson, "Full Disclosure: Did Government's Experiment on Preemies Hide Risks?"
Daily Signal, June 3, 2014, dailysignal.com/2014/06/03/uninformed-consent-nih-sacrifice-
preemies-sake-research/; Sharyl Attkisson, "Full Disclosure: 'Input' Stalls Agency's Ethics Probe in Baby Oxygen Trials," Daily Signal, June 4, 2014, dailysignal.com/2014/06/04/
baby-part-2/; Letter from Public Citizen to U.S. Secretary of Health and Human Services
regarding Unethical Trial Involving Premature Infants, January 27, 2014, www.citizen.org/
documents/2177.pdf; Letter from Public Citizen to Secretary of Health and Human Services
regarding Unethical Trial Involving Premature Infants, April 10, 2013, www.citizen.org/

documents/2111.pdf; Michael Carome, Sidney Wolfe, and Ruth Macklin, *Analysis of the Complete Protocol and Consent Form for the SUPPORT Study: Lack of Informed Consent and a Failure to Ensure that Risks Were Minimized* (Washington, DC: Public Citizen, May 8, 2013), www.citizen. org/documents/2124.pdf. For a detailed listing of documents and information related to the controversy, see the SUPPORT Study page at www.citizen.org/support-study.

38. Carome, quoted in Attkisson, "Full Disclosure: Did Government's Experiment on Preemies Hide Risks?"

39. Letter from Public Citizen to the HHS Inspector General Calling for an Investigation into Inappropriate Conduct by Senior HHS Officials Related to the SUPPORT Study, May 20, 2014, 7, available at www.citizen.org/documents/2201.pdf.

40. Ibid., 12.

41. Ibid.

42. For example, Collins castigated Christians who questioned Darwinian evolution for peddling "lies" and promoting "anti-scientific thinking" (Francis Collins, "Foreword" to Karl Giberson, *Saving Darwin* [New York: HarperOne, 2008], v, vii). He warned that the idea of intelligent design "is potentially threatening . . . to America's future" (Francis Collins, back cover blurb for Kenneth Miller, *Only a Theory: Evolution and the Battle for America's Soul* [New York: Viking, 2008]). He even declined to condemn eugenic abortions of infants with Down syndrome, telling Beliefnet.com that "in our current society, people are in a circumstance of being able to take advantage of those technologies [i.e., abortions]. And we have decided as a society that that choice needs to be defended" (Francis Collins quoted in "God Is Not Threatened by Our Scientific Adventures," interview with Francis Collins, BeliefNet.com, August 2008, www.beliefnet. com/News/Science-Religion/2006/08/God-Is-Not-Threatened-By-Our-Scientific-Adventures. aspx?p=3). Collins also championed the unrestricted funding of embryonic stem-cell research ("Statement of Francis Collins," *The Promise of Human Embryonic Stem Cell Research, Hearing before a Subcommittee of the Committee on Appropriations, United States Senate, 111th Congress, Second Session*, September 16, 2010 (Washington, DC: U.S. Government Printing Office, 2011), 6–32; Cathy Lynn Grossman, "Embryonic Stem Cell Research Is a Moral Choice: Francis Collins," *USA Today*, July 27, 2011, content.usatoday.com/communities/Religion/post/2011/07/embryonic-stem-cell-research-is-a-moral-choice-francis-collins/1#.VG7sloe71fQ). For more details about Collins's views on life issues, see David Klinghoffer, "Francis Collins on Abortion: Obama's Pick for NIH and His 'Devout' Views on Terminating Down Syndrome Children," BeliefNet.com, July 9, 2009, www.beliefnet.com/columnists/kingdomofpriests/2009/07/francis-collins-on-abortion.html.

43. Letter from Public Citizen to the HHS Inspector General Calling for an Investigation into Inappropriate Conduct by Senior HHS Officials Related to the SUPPORT Study.

44. Ibid.

45. Kathy L. Hudson, Alan E. Guttmacher, and Francis S. Collins, "In Support of SUPPORT—A View from the NIH," *New England Journal of Medicine* 368 (June 20, 2013): 2349–51, available at www.nejm.org/doi/full/10.1056/NEJMp1306986.

46. Letter from Public Citizen to U.S. Secretary of Health and Human Services regarding Unethical Trial Involving Premature Infants, January 27, 2014, 12, available at www.citizen.org/documents/2177.pdf.

47. Andrew Kirell, "Neil deGrasse Tyson: We Didn't Ask Obama to Intro Cosmos," *Mediaite*, June 10, 2014, www.mediaite.com/tv/neil-degrasse-tyson-we-didnt-ask-obama-to-intro-cosmos/.

48. For a critique of the new *Cosmos*, see David Klinghoffer, editor, *The Unofficial Guide to "Cosmos": Fact and Fiction in Neil deGrasse Tyson's Landmark Science Series* (Seattle: Discovery Institute Press, 2014). For a critique of the original *Cosmos* with Carl Sagan, see Mark G. McKim, "The Cosmos according to Carl Sagan (1934–1996): Review and Critique," *Perspectives on Science and the Christian Faith* 45 (March 1993): 18–25, available at www.asa3.0rg/ASA/PSCF/1993/PSCF3-93McKim.html.

49. Tyson quoted in "God of the Gaps & Frontier of Knowledge—Neil deGrasse Tyson," YouTube, posted March 1, 2011, www.youtube.com/watch?v=HooeZrC76s0.

50. Brandon Braga, "Star Trek as Atheist Mythology," YouTube, posted March 13, 2012, www.youtube.com/watch?v=iJm6vCs6aBA.

51. Seth McFarlane quoted in Scott Brown, "The 'Family' Guy Commences to Harvard," *Entertainment Weekly*, June 13, 2006, popwatch.ew.com/2006/06/13/the_family_guy_/.

52. For documentation of these and other errors, see Klinghoffer, *The Unofficial Guide*.

53. Kenneth P. Green and Hiwa Alaghebandian, "Science Turns Authoritarian," *The American*, July 27, 2010, www.american.com/archive/2010/july/science-turns-authoritarian.

54. Eric R. Pianka, "The Vanishing Book of Life on Earth," 17, available at www.zo.utexas.edu/courses/bi0373/Vanishing.Book.pdf; Jamie Mobley, "Doomsday: UT Prof Says Death Is Imminent," *Seguin Gazette-Enterprise*, February 27, 2010.

55. Wesley J. Smith, *The War on Humans* (Seattle: Discovery Institute Press, 2014).

56. Pianka, "Vanishing Book," 21.

57. Ibid., 10.

58. Ibid., 19.

59. Christopher Manes, *Green Rage: Radical Environmentalism and the Unmaking of Civilization* (Boston: Little, Brown and Company, 1990), 142.

60. James Lee, "The Discovery Channel MUST Broadcast to the World Their Commitment to Save the Planet and to Do the Following IMMEDIATELY," archived at web.archive.org/web/20101220024835/savetheplanetprotest.com/.

61. Peter Singer, *A Darwinian Left: Politics, Evolution, and Cooperation* (New Haven: Yale University Press, 2000).

62. Gregg Re, "*National Review* Fires Writer John Derbyshire for Saying Blacks Are Dangerous, Unintelligent," *Daily Caller*, April 8, 2012, dailycaller.com/2012/04/08/national-review-fires-writer-john-derbyshire-author-of-outlandish-article-on-blacks-racism/. The original article written by Derbyshire was "The Talk: Nonblack Version," *Taki's Magazine*, April 5, 2012, takimag.com/article/the_talk_nonblack_version_john_derbyshire/page_2#axzz1rPInwjt2.

63. John Derbyshire, "Equality: The Elusive Ideal," JohnDerbyshire.com, November 26, 2008, www.johnderbyshire.com/Opinions/HumanSciences/equality.html; John Derbyshire, "What's So Scary About Evolution?—For Both Right and Left, a Lot," JohnDerbyshire.com, May 19, 2008, www.johnderbyshire.com/Opinions/HumanSciences/darwin.html.

64. Derbyshire, "What's So Scary About Evolution?," emphasis in the original.

65. Quoted in Kyle Drennen, "NBC: It's 'Pro-Science' to Abort Children with Genetic Defects," *LifeNews.com*, June 12, 2012, www.lifenews.com/2012/06/12/nbc-its-pro-science-to-abort-children-with-genetic-defects/.

66. Jon Entine, "Let's (Cautiously) Celebrate the 'New Eugenics,'" *Huffington Post*, October 30, 2014, www.huffingtonpost.com/jon-entine/lets-cautiously-celebrate_b_6070462.html.

67. Nick Bostrom, "Transhumanist Values," www.nickbostrom.com/ethics/values.html.

68. "Letter from the Attorney General [Eric Holder] to Congress on Litigation Involving the Defense of Marriage Act," February 23, 2011, www.justice.gov/opa/pr/letter-attorney-general-congress-litigation-involving-defense-marriage-act.

69. See, for example, Stanton L. Jones, "Same-Sex Science," *First Things*, February 2012, www.firstthings.com/article/2012/02/same-sex-science.

70. For specific examples of coercion against same-sex marriage dissenters, see Kirsten Anderson, "Catholic Couple Fined $13,000 for Refusing to Host Same-Sex 'Wedding' at Their Farm," *LifeSiteNews.com*, August 20, 2014, www.lifesitenews.com/news/catholic-couple-fined-13000-for-refusing-to-host-same-sex-wedding-at-their; Kelsey Harkness, "They Lost Their Bakery, Now Face Bankruptcy: Government's 'Discrimination' Fine Brings Baker to Tears," *Daily Signal*, September 29, 2014, dailysignal.com/2014/09/29/lost-bakery-now-face-bankruptcy-governments-discrimination-fine-brings-oregon-baker-tears/; Ryan Anderson, "The Government Should Stop Waging War on Those against Same-Sex Marriage," *Daily Signal*, November 2, 2014, dailysignal.com/2014/11/02/government-penalizing-simply-believe-marriage/; Samuel Smith, "Four More North Carolina Judges Resign over Refusal to Conduct Same-Sex Weddings," *Christian Post*, October 30, 2014, www.christianpost.com/news/four-more-north-carolina-judges-resign-over-refusal-to-conduct-same-sex-weddings-128890/; "Ingersoll v. Arlene's Flowers, State of Washington v. Arlene's Flowers, and Arlene's Flowers v. Ferguson," Alliance Defending Freedom, October 28, 2013, www.adfmedia.org/News/PRDetail/8608. For broader discussions of the impact on religious liberty of same-sex marriage, see Matthew J. Franck, "Same-Sex Marriage and Religious Freedom, Fundamentally at Odds," *Public Discourse*, June 18, 2013, www.thepublicdiscourse.com/2013/06/10393/; Thomas M. Messner, "Same-Sex Marriage and the Threat to Religious Liberty," *Heritage Foundation Backgrounder* #2201, October 30, 2008, www.heritage.org/research/reports/2008/10/same-sex-

marriage-and-the-threat-to-religious-liberty. For a discussion of the constitutional and legal issues involved, see "Brief Amicus Curiae of the Becket Fund for Religious Liberty in Support of Hollingsworth and the Bipartisan Legal Advisory Group Addressing the Merits," January 2013, available at www.americanbar.org/content/dam/aba/publications/supreme_court_pre-view/briefs-v2/12–144–12–307_becket_fund.authcheckdam.pdf. For cases of similar coercion taking place in Great Britain, see Andrew Pierce, "The Real Victims of Bigotry: A Family of Bakers Dragged to Court and How Opponents of Gay Marriage Are Being Persecuted—Even though the Tories Vowed to Protect Them," *Daily Mail*, November 6, 2014, www.daily-mail.co.uk/news/article-2824510/The-real-victims-bigotry-family-bakers-dragged-court-opponents-gay-marriage-persecuted-Tories-vowed-protect-them.html.

71. For examples, see Joel Gehrke, "Mozilla CEO Brendan Eich Forced to Resign for Supporting Traditional Marriage Laws," *Washington Examiner*, April 3, 2014, www.washingtonexaminer.com/mozilla-ceo-brendan-eich-forced-to-resign-for-supporting-traditional-marriage-laws/article/2546770; Thomas M. Messner, "The Price of Prop 8," *Heritage Foundation Backgrounder* #2328, October 22, 2009, www.heritage.org/research/reports/2009/10/the-price-of-prop-8; Lisa Respers France, "Benham Brothers Lose HGTV Show after 'Anti-Gay' Remarks," CNN.com, May 9, 2014, www.cnn.com/2014/05/08/showbiz/tv/benham-brothers-hgtv/.

72. The text of the California law can be found at "SB-1172 Sexual orientation change efforts," approved September 30, 2012, leginfo.legislature.ca.gov/faces/billNavClient.xhtml?bill_id=201120120SB1172. The text of the New Jersey law can be found at "New Jersey Assembly Bill 3371," approved August 19, 2013, legiscan.com/NJ/text/A3371/2012.

73. Michael Brown, "Did a Gay Activist Lie to the New Jersey Senate?" *Townhall.com*, March 25, 2013, townhall.com/columnists/michaelbrown/2013/03/25/did-a-gay-activist-lie-to-the-new-jersey-senate-n1548304/page/full.

74. "Do NARTH Institute Therapists Advocate the Use of Aversion or Shock Therapy?," NARTH Institute, www.narth.com/#!faq/cirw.

75. "SB-1172 Sexual Orientation Change Efforts"; "New Jersey Assembly Bill 3371."

76. *Complaint for Declaratory and Injunctive Relief in the Case of John Doe v. Christopher Christie* (U.S. District Court for the District of New Jersey, Camden Division), 7–19.

77. Nicholas A. Cummings, "Sexual Reorientation Therapy Not Unethical," *USA Today*, July 30, 2013, www.usatoday.com/story/opinion/2013/07/30/sexual-reorientation-therapy-not-unethical-column/2601159/.

78. Ritch C. Savin-Williams and Kara Joyner, "The Dubious Assessment of Gay, Lesbian, and Bisexual Adolescents of Add Health," *Archives of Sexual Behavior* 43, no. 3 (April 2014): 416.

79. Ibid. For a critique, see Gu Li, Sabra Katz-Wise, and Jerel Calzo, "The Unjustified Doubt of Add Health Studies on the Health Disparities of Non-Heterosexual Adolescents: Comment on Savin-Williams and Joyner (2014)," *Archives of Sexual Behavior* 43, no. 6 (August 1, 2014): 1023–26.

80. See discussion in "Can Sexual Orientation Change?," (chapter 12 in Neil Whitehead and Briar Whitehead, *My Genes Made Me Do It! Homosexuality and the Scientific Evidence* (3rd ed., October 2013), www.mygenes.co.nz/PDFs/Ch12.pdf.

81. "Lisa Diamond on Sexual Fluidity of Men and Women," YouTube, posted on December 6, 2013, youtu.be/m2rTHDOuUBw. The comment quoted is at 43:15.

82. Quoted in Andrew Gilligan, " 'Paedophilia Is Natural and Normal for Males,' " *The Telegraph*, July 5, 2014, www.telegraph.co.uk/comment/10948796/Paedophilia-is-natural-and-normal-for-males.html.

83. Ibid. See also Warren Throckmorton, "Does the APA Consider Hebephilia to Be Normal?," *Patheos.com*, May 16, 2013, www.patheos.com/blogs/warrenthrockmorton/2013/05/16/does-the-apa-consider-hebephilia-to-be-normal/; Ray Blanchard, "A Dissenting Opinion on DSM-5 Pedophilic Disorder," *Archives of Sexual Behavior* 42, no. 5: 675–78, also available at wp.patheos.com.s3.amazonaws.com/blogs/warrenthrockmorton/files/2013/05/Blanchard-DissentingDSM5.pdf; Hunter Stuart, "Not All Pedophiles Have Mental Disorder, American Psychiatric Association Says in New DSM," *Huffington Post*, November 1, 2013, www.huffing-tonpost.com/2013/11/01/dsm-pedophilia-mental-disorder-paraphilia_n_4184878.html.

84. David Barash, "God, Darwin, and My Biology Class," *New York Times*, September 27, 2014, www.nytimes.com/2014/09/28/opinion/sunday/god-darwin-and-my-college-biology-class.html.

85. For documentation about this course, see Letter to Ball State University President about

Academic Freedom, September 10, 2013, www.discovery.org/scripts/viewDB/filesDB-download.php?command=download&id=9701, 6–8.

86. "Faith in Flux: Changes in Religious Affiliation in the U.S.," Pew Forum on Religion and Public Life, April 2009, www.pewforum.org/files/2009/04/fullreport.pdf, 16.

87. Quoted in "More Young People Are Moving Away from Religion, but Why?" NPR, January 15, 2013, www.npr.org/2013/01/15/169342349/more-young-people-are-moving-away-from-religion-but-why.

88. "Scientific Achievements Less Prominent Than a Decade Ago; Public Praises Science; Scientists Fault Public, Media," July 9, 2009, Pew Research Center for the People and the Press survey in collaboration with the American Association for the Advancement of Science, www.people-press.org/files/legacy-pdf/528.pdf, 18.

89. "Six Reasons Young Christians Leave Church," Barna Group, September 28, 2011, www.barna.org/teens-next-gen-articles/528-six-reasons-young-christians-leave-church.

90. 2004 College Students' Beliefs and Values (CSBV) Survey, conducted by the Higher Education Research Institute, University of California, Los Angeles. A description of the survey can be found at spirituality.ucla.edu/background/methodology/longitudinal-study.php; a summary of the results can be downloaded at www.ats.msstate.edu/pdf/reports/2004CSBV_Report.pdf.

91. Deborah Kelemen, "Are Children 'Intuitive Theists'?," *Psychological Science* 15, no. 5 (2004): 299.

92. Deborah Kelemen, Natalie A. Emmons, Rebecca Seston Schillaci, and Patricia A. Ganea, "Young Children Can Be Taught Basic Natural Selection Using a Picture-Storybook Intervention," *Psychological Science* 25, no. 4 (2014): 895, 902.

93. Testimony of John Holdren, *An Overview of the Administration's Federal Research and Development Budget, Hearing before the Committee on Science, Space, and Technology, House of Representatives, 112th Congress, First Session, February 17, 2011,* Serial No. 112–2 (Washington, DC: U.S. Government Printing Office, 2011), 39.

94. Paul Thornton, "On Letters from Climate-Change Deniers," *Los Angeles Times*, October 8, 2013, www.latimes.com/opinion/opinion-la/la-ol-climate-change-letters-20131008-story.html.

95. Lawrence Torcello, "Is Misinformation about the Climate Criminally Negligent?," *The Conversation*, March 13, 2014, theconversation.com/is-misinformation-about-the-climate-criminally-negligent-23111.

96. Adam Weinstein, "Arrest Climate-Change Deniers," *Kinja*, March 28, 2014, gawker.com/arrest-climate-change-deniers-1553719888.

97. For information about the litigation, see "No. 14-CV-126, In the District of Columbia Court of Appeals . . . Brief of Appellant National Review, Inc.," August 4, 2014, www.nationalreview.com/sites/default/files/NR_Opening-Brief.pdf; Jonathan Adler, "Media and Rights Organizations Defend *National Review*, et al. against Michael Mann," *Washington Post*, August 13, 2014, www.washingtonpost.com/news/volokh-conspiracy/wp/2014/08/13/media-and-rights-organizations-defend-national-review-et-al-against-michael-mann/; Steven Hayward, "Climate High-Sticking: The Unmanly Man," *Powerline*, August 22, 2012, www.powerlineblog.com/archives/2012/08/climate-high-sticking-the-unmanly-mann.php.

98. E-mail from University of Kentucky physicist Thomas Troland, quoted in Casey Luskin, "E-Mails in Gaskell Case Show that Darwin Skeptics Need Not Apply to the University of Kentucky," *Evolution News and Views*, February 10, 2011, www.evolutionnews.org/2011/02/e-mails_in_gaskell_case_show_t043711.html.

99. Casey Luskin, "Evidence of Discrimination against Martin Gaskell Due to His Views on Evolution," *Evolution News and Views*, December 15, 2010, www.evolutionnews.org/2010/12/evidence_of_discrimination_aga041621.html.

100. "Refereed Publications [of Eric Hedin]," Ball State University, cms.bsu.edu/-/media/WWW/DepartmentalContent/Physics/PDFs/Hedin/PublicationsHedin%20%283%29.pdf.

101. John G. West, "Misrepresenting the Facts about Eric Hedin's 'Reading List,'" *Evolution News and Views*, July 11, 2013, www.evolutionnews.org/2013/07/misrepresenting074291.html.

102. Joshua Youngkin, "What Does Eric Hedin Really Teach? Self-Professed Agnostic Speaks Out about 'Boundaries of Science' Seminar," *Evolution News and Views*, August 2, 2013, www.evolutionnews.org/2013/08/what_does_eric075081.html; Joshua Youngkin, "Hedin Witness #3: 'This Course Made Me a Better Learner,'" *Evolution News and Views*, August 9, 2013, www.evolutionnews.org/2013/08/hedin_witness_3075271.html; Joshua Youngkin, "Dr. Hedin's Stu-

dent Could Teach Ball State University a Thing or Two," *Evolution News and Views*, July 16, 2013, www.evolutionnews.org/2013/07/what_happened_i074521.html.

103. David Klinghoffer, "At Ball State University, Intimidation Campaign against Physicist Gets Troubling Results," *Evolution News and Views*, May 22, 2013, www.evolutionnews.org/2013/05/at_ball_state_u072381.html.

104. John G. West, "Questions Raised about Impartiality of Panel Reviewing Ball State University Professor's Course," *Evolution News and Views*, June 25, 2013, www.evolutionnews.org/2013/06/review_panel_or073761.html; Joshua Youngkin, "Indiana Professors Question Ball State University's Disregard for Rules on Academic Freedom," *Evolution News and Views*, August 25, 2013, www.evolutionnews.org/2013/08/indiana_profess075781.html; John G. West, "Clarifying the Issues at Ball State: Some Questions and Answers," *Evolution News and Views*, September 13, 2013, www.evolutionnews.org/2013/09/clarifying_the_076591.html.

105. "Atheist Rift!!," BSU Freethought Alliance: The Official Blog of Ball State University Freethought Alliance, October 23, 2009, freethoughtbsu.blogspot.com/2009/10/atheist-rift.html.

106. John G. West, "Ball State President's Orwellian Attack on Academic Freedom," *Evolution News and Views*, August 1, 2013, www.evolutionnews.org/2013/08/ball_state_pres075041.html.

107. For information and documentation about the Coppedge case, see Robert Crowther, "Trial to Begin in Intelligent Design Discrimination Lawsuit against NASA's Jet Propulsion Lab," *Evolution News and Views*, March 5, 2012, www.evolutionnews.org/2012/03/trial_to_begin_057061.html; "Facts of the Coppedge Lawsuit Contradict the Spin from Jet Propulsion Lab and National Center for Science Education," *Evolution News and Views*, March 12, 2012, www.evolutionnews.org/2012/03/facts_of_the_c0057321.html; Joshua Youngkin, "Why Did NASA's JPL Discriminate against David Coppedge and Why Does It Matter?" *Evolution News and Views*, November 22, 2011, www.evolutionnews.org/2011/11/what_happened_t053331.html.

108. "California Science Center Pays $110,000 to Settle Intelligent Design Discrimination Lawsuit," *Evolution News and Views*, August 29, 2011, www.evolutionnews.org/2011/08/california_science_center_pays050081.html; Robert Crowther, "*Los Angeles Times* Reporting on Lawsuit against California Science Center for Cancelling Intelligent Design Film," *Evolution News and Views*, December 29, 2009, www.evolutionnews.org/2009/12/los_angeles_times_reporting_on030391.html; John G. West, "Why the California Science Center's Censorship of Pro–Intelligent Design Film Is a Big Deal," *Evolution News and Views*, December 30, 2009, www.evolutionnews.org/2009/12/why_the_california_science_cen030421.html; John G. West, "California Science Center Engaged in Illegal Cover-Up to Hide the Truth about Its Censorship of Pro-Intelligent Design Film," *Evolution News and Views*, January 4, 2010, www.evolutionnews.org/2010/01/california_science_center_enga030441.html.

109. Anika Smith, "Darwinists Launch Cyber Attack against Intelligent Design Website," *Evolution News and Views*, October 28, 2009, www.evolutionnews.org/2009/10/darwinists_launch_cyber_attack027431.html. For a later cyber attack, see John G. West, "Cyber Attacks Attempt to Shut Down Discovery Institute's Websites on Day of Event Challenging Darwinism," *Evolution News and Views*, September 23, 2013, www.evolutionnews.org/2010/09/cyber_attacks_attempt_to_shut_038631.html.

110. Casey Luskin, "Bullies-R-Us: How 'Freethought Oasis' Threatened 'Disruption' and Pressured a College into Cancelling Intelligent Design Course," *Evolution News and Views*, December 10, 2013, www.evolutionnews.org/2013/12/freethought-oasis-bullies079991.html.

111. Thomas S. Nagel, *Mind and Cosmos: Why the Materialist Neo-Darwinian Conception of Nature Is Almost Certainly False* (New York: Oxford University Press, 2012).

112. "Thomas Nagel: Biography," New York University School of Law, its.law.nyu.edu/facultyprofiles/profile.cfm?section=bio&personID=20156.

113. Brian Leiter and Michael Weisberg, "Do You Only Have a Brain? On Thomas Nagel," *The Nation*, October 3, 2012, www.thenation.com/article/170334/do-you-only-have-brain-thomas-nagel; Victor Stenger, "The Argument from Ignorance," *Huffington Post*, October 23, 2012, www.huffingtonpost.com/victor-stenger/the-argument-from-ignoran_b_2004743.html.

114. Steven Pinker, "What Has Gotten into Thomas Nagel?," Twitter, October 16, 2012, twitter.com/sapinker/status/258350644979695616; Jerry Coyne, "Tom Nagel's Antievolution Book Gets Thrice Pummeled," *Why Evolution Is True*, January 23, 2013, whyevolutionistrue.wordpress.com/2013/01/23/tom-nagels-antievolution-book-gets-thrice-pummeled/.

115. Mark Vernon, "The Most Despised Science Book of 2012 Is . . . Worth Reading," *The Guardian*,

January 4, 2013, www.theguardian.com/commentisfree/belief/2013/jan/04/most-despised-science-book-2012.

116. Leon Wieseltier, "A Darwinist Mob Goes after a Serious Philosopher," *New Republic*, March 8, 2013, www.newrepublic.com/article/112481/darwinist-mob-goes-after-serious-philosopher.

117. Jennifer Schuessler, "An Author Attracts Unlikely Allies," *New York Times*, February 6, 2013, www.nytimes.com/2013/02/07/books/thomas-nagel-is-praised-by-creationists.html?pagewanted=all.

118. Andrew Ferguson, "The Heretic," *Weekly Standard*, March 25, 2013, www.weeklystandard.com/articles/heretic_707692.html.

119. Nagel, *Mind and Cosmos*, 10.

120. Schuessler, "An Author Attracts Unlikely Allies."

121. "Thomas Nagel and Stephen C. Meyer's Signature in the Cell," *Times Literary Supplement*, August 22, 2011, www.the-tls.co.uk/tls/public/article706905.ece; Stephen C. Meyer, *Signature in the Cell: DNA and the Evidence for Intelligent Design* (New York: HarperOne, 2009).

122. "Thomas Nagel and Stephen C. Meyer's *Signature in the Cell*."

123. Nagel, *Mind and Cosmos*, 12.

124. Ibid., 27.

125. See Arthur J. Balfour, *Theism and Humanism*, edited by Michael W. Perry (Seattle: Inkling Books, 2000); C. S. Lewis, *Miracles: A Preliminary Study*, 1960 ed. (New York: Macmillan, 1978); Alvin Plantinga, *Where the Conflict Really Lies: Science, Religion, and Naturalism* (New York: Oxford University Press, 2011). For a further discussion of C.S. Lewis's use of this argument, see John G. West, ed., *The Magician's Twin: C. S. Lewis on Science, Scientism, and Society* (Seattle: Discovery Institute Press, 2012), 129–31, 179–232.

126. See "An Integrated Encyclopedia of DNA Elements in the Human Genome," *Nature* 489 (September 6, 2012): 57–74; Casey Luskin, "Junk No More: ENCODE Project Nature Paper Finds 'Biochemical Functions for 80% of the Genome,'" *Evolution News and Views*, September 5, 2012, www.evolutionnews.org/2012/09/junk_no_more_en_1064001.html; "Decoding ENCODE: A Q&A with Jonathan Wells," *Evolution News and Views*, September 24, 2012, www.evolutionnews.org/2012/09/decoding_encode064611.html; James A. Shapiro, "Bob Dylan, ENCODE, and Evolutionary Theory: The Times They Are A-Changin'," *Huffington Post*, November 12, 2012, www.huffingtonpost.com/james-a-shapiro/bob-dylan-encode-and-evol_b_1873935.html.

127. For a history of the idea of "junk DNA" in Darwinian biology and how it is now being overturned, see Jonathan Wells, *The Myth of Junk DNA* (Seattle: Discovery Institute Press, 2010).

128. For an early prediction of the functionality of so-called junk DNA by an intelligent design proponent, see William Dembski, "Science and Design," *First Things*, October 1998, www.firstthings.com/article/1998/10/001-science-and-design. Biologist Richard Sternberg made similar predictions. See J. A. Shapiro and R. Sternberg, "Why Repetitive DNA Is Essential to Genome Function," *Biological Reviews* 80, no. 2 (May 2005): 227–50.

129. George Church endorsement from the *Darwin's Doubt* back cover, DarwinsDoubt.com, www.darwinsdoubt.com/blurbs/.

130. Mark McMenamin endorsement from the *Darwin's Doubt* back cover, DarwinsDoubt.com, www.darwinsdoubt.com/blurbs/.

131. Stephen Meyer, quoted in Casey Luskin, "A Listener's Guide to the Meyer-Marshall Radio Debate: Focus on the Origin of Information Question," *Evolution News and Views*, December 4, 2013, www.evolutionnews.org/2013/12/a_listeners_gui079811.html.

132. C.S. Lewis, *The Discarded Image* (Cambridge: Cambridge University Press, 1964), 223.

133. G. M. A. Grube, trans., *Plato's Republic* (Indianapolis: Hackett Publishing Co., 1974), 394d, 65.

Index